T0183167

# Lecture Notes in Computer Science 9984

Commenced Publication in 1973
Founding and Former Series Editors:
Gerhard Goos, Juris Hartmanis, and Jan van Leeuwen

## Editorial Board

More information about this series at http://www.springer.com/series/7408

Martin Fränzle · Deepak Kapur
Naijun Zhan (Eds.)

# Dependable
# Software Engineering

## Theories, Tools, and Applications

Second International Symposium, SETTA 2016
Beijing, China, November 9–11, 2016
Proceedings

 Springer

*Editors*
Martin Fränzle
Carl von Ossietzky Universität
Oldenburg
Germany

Naijun Zhan
Chinese Academy of Sciences
Beijing
China

Deepak Kapur
University of New Mexico
Albuquerque, NM
USA

ISSN 0302-9743                    ISSN 1611-3349  (electronic)
Lecture Notes in Computer Science
ISBN 978-3-319-47676-6            ISBN 978-3-319-47677-3  (eBook)
DOI 10.1007/978-3-319-47677-3

Library of Congress Control Number: 2016953662

LNCS Sublibrary: SL2 – Programming and Software Engineering

Printed on acid-free paper

This Springer imprint is published by Springer Nature
The registered company is Springer International Publishing AG
The registered company address is: Gewerbestrasse 11, 6330 Cham, Switzerland

# Preface

This volume contains the papers presented at the second in the SETTA (the Symposium on Dependable Software Engineering: Theories, Tools and Applications) series of conferences – held during November 9–11, 2016, in Beijing, China. The symposium series was inaugurated in 2015 to build a forum for computer scientists and software engineers from Chinese and international communities to exchange and inform each other of research ideas and activities, building new collaborations and strengthening existing collaborations among formal methods researchers inside and outside China. A key goal of SETTA is to especially encourage and nourish young researchers working in the use of formal methods in building software and cyber-physical systems.

SETTA 2016 received over 58 submissions of abstracts, with 45 of them as full-paper submissions by the submission deadline. These submissions were coauthored by researchers from 22 countries. Each submission was reviewed by at least three Program Committee (PC) members with help from reviewers outside the PC. After two weeks of online discussions, the committee decided to accept 20 papers for presentation at the conference (with the acceptance rate of 44 %); this includes 17 full papers and three short papers. It was decided to include short papers to provide a forum for participants to present research in progress. To alleviate presentational issues on some papers of good technical quality, shepherding by PC members was employed while preparing revisions. Such submissions were accepted after an additional round of reviewing leading to one submission being rejected.

We would like to express our gratitude to all the researchers who submitted their work to the symposium. We arc particularly thankful to all colleagues who served on the PC, as well as the external reviewers, whose hard work in the review process helped us prepare a high-quality conference program. The international diversity of the PC as well as external reviewers is noteworthy as well: PC members and external reviewers have affiliations with institutes in 17 countries. Special thanks go to the invited speakers, Prof. Lee from the University of California, Berkeley, Prof. Sankaranarayanan of the University of Colorado, and Prof. Ying of the University of Sydney and Tsinghua University, for agreeing to present their research. The abstracts of the invited talks are included in this volume.

Like SETTA 2015, SETTA 2016 also had a young SETTA Researchers Workshop, which was held on November 12, 2016. Another inaugural event – the first National Conference on Formal Methods and Applications in China – was held during November 12–13, 2016.

A number of colleagues worked very hard to make this conference a success. We wish to express our gratitude especially to Prof. Shuling Wang for taking care of numerous activities related to the publicity, conference proceedings, and other aspects of the conference. We thank the conference chair, Huimin Lin, publicity chairs, Nils Muellner and Lijun Zhang, and the local Organizing Committee of Andrea Turini, Shuling Wang, Peng Wu, and Zhilin Wu. Finally, we enjoyed great institutional and financial support from the Institute of Software, the Chinese Academy of Sciences (ISCAS), without which an international conference like SETTA and the colocated events could not have been be successfully organized. We also thank the Chinese Computer Federation (CCF) and the Natural Science Foundation of China (NSFC) for financial support.

August 2016                                                    Martin Fränzle
                                                                      Deepak Kapur
                                                                      Naijun Zhan

# Organization

## Program Committee

| | |
|---|---|
| Erika Abraham | RWTH Aachen University, Germany |
| Farhad Arbab | CWI and Leiden University, The Netherlands |
| Sanjoy Baruah | University of North Carolina, Chapel Hill, USA |
| Michael Butler | University of Southampton, UK |
| Deepak D'Souza | Indian Institute of Science, Bangalore, India |
| Yuxin Deng | East China Normal University, China |
| Xinyu Feng | University of Science and Technology of China, Suzhou, China |
| Goran Frehse | University of Grenoble Alpes - Laboratoire Verimag, Grenoble, France |
| Martin Fränzle | University of Oldenburg, Germany |
| Lindsay Groves | Victoria University of Wellington, New Zealand |
| Dimitar Guelev | Bulgarian Academy of Sciences, Bulgaria |
| Fei He | Tsinghua University, China |
| Holger Hermanns | Saarland University, Germany |
| Deepak Kapur | University of New Mexico, USA |
| Axel Legay | IRISA/Inria, Rennes, France |
| Xuandong Li | Nanjing University, China |
| Shaoying Liu | Hosei University, Tokyo, Japan |
| Zhiming Liu | Southwest University, Chongqing, China |
| Xiaoguang Mao | National University of Defense Technology, Changsha, China |
| Markus Müller-Olm | Westfälische Wilhelms-Universität Münster, Germany |
| Raja Natarajan | Tata Institute of Fundamental Research, Bombay, India |
| Jun Pang | University of Luxembourg, Luxenmbourg |
| Shengchao Qin | Teesside University, UK |
| Sriram Rajamani | Microsoft Research India, Bangalore, India |
| Jean-François Raskin | Université Libre de Bruxelles, Belgium |
| Stefan Ratschan | Czech Academy of Sciences, Prague, Czech |
| Martin Steffen | University of Oslo, Norway |
| Zhendong Su | UC Davis, USA |
| Cong Tian | Xidian University, China |
| Tarmo Uustalu | Tallinn University of Technology, Estonia |
| Chao Wang | University of Southern California, Los Angeles, USA |
| Farn Wang | National Taiwan University, ROC |
| Heike Wehrheim | University of Paderborn, Germany |
| Wang Yi | Uppsala University, Sweden |

| Naijun Zhan | Institute of Software, Chinese Academy of Sciences, Beijing, China |
| Lijun Zhang | Institute of Software, Chinese Academy of Sciences, Beijing, China |

## Additional Reviewers

Bu, Lei
Dan, Li
Fahrenberg, Uli
Fu, Ming
Geeraerts, Gilles
Guo, Shengjian
Hahn, Ernst Moritz
Hoang, Thai Son
Huang, Yanhong
Kekatos, Nikolaos
Kim, Jin Hyun
Krämer, Julia
Kucera, Antonin
Le Bouder, Hélène
Lei, Suhua
Liu, Bo
Moszkowski, Ben
Mukherjee, Suvam
Nordhoff, Benedikt
Norman, Gethin

Randour, Mickael
Ray, Rajarshi
Salehi Fathabadi, Asieh
Singh, Abhishek Kr
Sun, Chengnian
Sung, Chungha
Tacchella, Armando
Tinchev, Tinko
Travkin, Oleg
Turrini, Andrea
van Breugel, Franck
Velez, Martin
Veltri, Niccolò
Wu, Meng
Xu, Zhiwu
Yu, Hengbiao
Zhang, Miaomiao
Zhang, Qirun
Zhao, Jianhua

# Keynote Abstracts

# Dependable Cyber-physical Systems

Edward A. Lee

Electrical Engineering and Computer Sciences Department,
University of California, Berkeley, USA

**Abstract.** Cyber-physical systems are integrations of computation, communication networks, and physical dynamics. Applications include manufacturing, transportation, energy production and distribution, biomedical, smart buildings, and military systems, to name a few. Increasingly, today, such systems leverage Internet technology, despite a significant mismatch in technical objectives. A major challenge today is to make this technology reliable, predictable, and controllable enough for "important" things, such as safety-critical and mission-critical systems. In this talk, I will analyze how emerging technologies can translate into better models and better engineering methods for this evolving Internet of Important things.

# From Finitely Many Simulations to Flowpipes

Sriram Sankaranarayanan

Computer Science Department, University of Colorado Boulder, Boulder, USA

**Abstract.** Flowpipe construction techniques generalize symbolic execution for continuous-time models by computing future trajectories for sets of inputs and initial states. In doing so, they capture infinitely many behaviors of the underlying system, thus promising exhaustive verification. We examine the progress in this area starting from techniques for linear systems to recent progress in reasoning about nonlinear dynamical systems. We demonstrate how this area of research transforms fundamental results from dynamical systems theory into useful computational techniques for reasoning about cyber-physical systems. This progress has led to increasingly popular tools for verifying cyber-physical systems with applications to important verification problems for medical devices and automotive software. We demonstrate how recent approaches have exploited commonly encountered properties of the underlying continuous models such as monotonicity, incremental stability and structural dependencies to verify properties for larger and more complex systems. Despite this progress, many challenges remain. We present some of the key theoretical and practical challenges that need to be met before flowpipe construction can be a true "technology" for verifying industrialscale systems.

# Toward Automatic Verification
## of Quantum Programs
## (Extended Abstract)

Mingsheng Ying[1,2]

[1] Centre for Quantum Computation and Intelligent Systems,
University of Technology Sydney, Ultimo, Australia
Mingsheng.Ying@uts.edu.au
[2] Department of Computer Science and Technology,
Tsinghua University, Beijing, China
yingmsh@tsinghua.edu.cn

**Keywords:** Quantum programming · Hoare logic · Invariant generation · Algorithmic analysis of termination · Synthesis of ranking functions

Programming is error-prone. Programming a quantum computer and designing quantum communication protocols are even worse due to the weird nature of quantum systems [11]. Therefore, verification techniques for quantum programs and quantum protocols will be indispensable whence commercial quantum computers and quantum communication systems are available. In the last 10 years, various verification techniques for classical programs including program logics and model-checking have been extended to deal with quantum programs. This talk summaries several results obtained by the author and his collaborators in this line of research.

## 1 Quantum Hoare Logic

In quantum programming, the state space of a program variable is a Hilbert space. A quantum predicate in a Hilbert space was defined by D'Hondt and Panangaden in [4] as a Hermtian operator, i.e. an observable, between the zero and identity operators. A proof system for partial and total correctness of the Floyd-Hoare style was developed and its (relative) completeness was proved in [10] for the following quantum extension of **while**-language:

$$P ::= \textbf{skip} \mid P_1; P_2 \mid q := |0\rangle \mid \bar{q} := U[\bar{q}] \mid \textbf{if } (\text{W}_m \, M[\bar{q}] = m \rightarrow P_m) \textbf{ fi}$$
$$\mid \textbf{while } M[\bar{q}] = 1 \textbf{ do } P \textbf{ od}$$

The command "$q := |0\rangle$" is an initialisation that sets quantum variable $q$ to a basis state $|0\rangle$. The statement "$\bar{q} := U[\bar{q}]$" means that unitary transformation $U$ is performed on quantum register $\bar{q}$, leaving the states of the variables not in $\bar{q}$ unchanged. The construct "**if** $\cdots$ **fi**" is a quantum generalisation of case or switch statement. In executing it, measurement $M = \{M_m\}$ is performed on $\bar{q}$, and then a subprogram $P_m$ is selected to be executed next according to the outcomes $m$ of measurement. The statement "**while** $\cdots$ **od**" is a quantum generalisation of **while**-loop. The measurement in it has only two possible outcomes 0, 1. If the outcome 0 is observed, then the program terminates, and if the outcome 1 occurs, the program executes the loop body $P$ and continues the loop. It is interesting to carefully compare the Hoare rule for loops:

$$\frac{\{\varphi \wedge b\}P\{\varphi\}}{\{\varphi\} \text{ while } b \text{ do } P \text{ od}\{\varphi \wedge \neg b\}}$$

with the rule for quantum loops given in [10]:

$$\frac{\{B\}P\left\{M_0^\dagger AM_0 + M_1^\dagger BM_1\right\}}{\{M_0^\dagger AM_0 + M_1^\dagger BM_1\} \text{ while } M[\bar{q}] = 1 \text{ do } P \text{ od } \{A\}}$$

A theorem prover was built by Liu, Li, Wang et al. in [8] for quantum Hoare logic based on Isabelle/HOL.

## 2   Invariants of Quantum Programs

A super-operator in a Hilbert space is a completely positive mapping from (linear) operators to themselves. The control flow of a quantum program can be represented by a super-operator-valued transition system (SVTS):

**Definition 1.** *An SVTS is a 5-tuple $\mathcal{S} = \langle \mathcal{H}, L, l_0, T, \Theta \rangle$, where: (1) $\mathcal{H}$ is a Hilbert space; (2) $L$ is a finite set of locations; (3) $l_0 \in L$ is the initial location; (4) $\Theta$ is a quantum predicate in $\mathcal{H}$ denoting the initial condition; and (5) $T$ is a set of transitions. Each transition $\tau \in T$ is written as $\tau = l \xrightarrow{\mathcal{E}} l'$ with $l, l' \in L$ and $\mathcal{E}$ being a super-operator in $\mathcal{H}$. For each $l \in L$, it is required that $\mathcal{E}_l = \sum \{|\mathcal{E} : l \xrightarrow{\mathcal{E}} l' \in T|\}$ is trace-preserving, i.e. $tr(\mathcal{E}_l(\rho)) = tr(\rho)$ for all $\rho$.*

The notion of invariant for quantum programs was recently introduced in [14]. A set $\Pi$ of paths is said to be prime if for each $\pi = l_1 \xrightarrow{\mathcal{E}_1} \ldots \xrightarrow{\mathcal{E}_{n-1}} l_n \in \Pi$, its proper initial segments $l_1 \xrightarrow{\mathcal{E}_1} \ldots \xrightarrow{\mathcal{E}_{k-1}} l_k \notin \Pi$ for all $k < n$. We write $\mathcal{E}_\pi$ for the composition of $\mathcal{E}_1, \ldots, \mathcal{E}_{n-1}$ and $\mathcal{E}_\Pi = \sum \{|\mathcal{E}_\pi : \pi \in \Pi|\}$.

**Definition 2.** *Let $\mathcal{S} = \langle \mathcal{H}, L, l_0, T, \Theta \rangle$ be an SVTS and $l \in L$. An invariant at location $l \in L$ is a quantum predicate $O$ in $\mathcal{H}$ satisfying the condition: for any density operator $\rho$ and prime set $\Pi$ of paths from $l_0$ to $l$, we have:*

$$tr(\Theta\rho) \leq 1 - tr(\mathcal{E}_\Pi(\rho)) + tr(O\mathcal{E}_\Pi(\rho)).$$

In [14], it was shown that invariants can be used to establish partial correctness of quantum programs, and by generalising the constraint-based technique of Colón et al. [3, 9], invariant generation for quantum programs is reduced to an SDP (Semidefinite Programming) problem.

## 3 Terminations of Quantum Programs

Algorithmic analysis of termination for quantum programs was first considered in [13] where the Jordan decomposition of complex matrices was employed as the main tool. It was further studied by the author and his collaborators in a series of papers [7, 15–17] by introducing quantum Markov chains as a semantic model of quantum programs and using matrix representation of super-operators.

The notion of ranking function was defined in [10] for proving total correctness of quantum programs. The synthesis problem of ranking functions for quantum programs was recently investigated in [12] where the fundamental Gleason theorem [6] in quantum foundations was used to determine the template of ranking functions. In the last few years, (super)martingales have been employed as a powerful mathematical tools for termination analysis of probabilistic programs [1, 2, 5]. It seems that the ideas of this line of research can be generalised to deal with quantum programs, but we need to systematically develop a mathematical theory of quantum (super)martingales first.

**Acknowledgment.** This work was partly supported by the Australian Research Council (Grant No: DP160101652) and the Overseas Team Program of Academy of Mathematics and Systems Science, Chinese Academy of Sciences.

## References

1. Chakarov, A., Sankaranarayanan, S.: Probabilistic program analysis with martingales. In: CAV 2013. LNCS, vol. 8044, pp. 511–526. Springer, Berlin (2013)
2. Chatterjee, K., Fu, H.F., Novotný, P., Hasheminezhad, R.: Algorithmic analysis of qualitative and quantitative termination problems for affine probabilistic programs. In: Proceedings of the 43rd Annual ACM Symposium on Principles of Programming Languages (POPL), pp. 327–342 (2016)
3. Colón, M.A., Sankaranarayanan, S., Sipma, H.B.: Linear invariant generation using non-linear constraint solving. In: CAV 2003. LNCS, vol. 2725, pp. 420–433. Springer, Berlin (2003)
4. D'Hondt, E., Panangaden, P.: Quantum weakest preconditions. Math. Struct. Comput. Sci. **16,** 429–451 (2006)
5. Fioriti, L.M.F., Hermanns, H.: Probabilistic termination: soundness, completeness, and compositionality. In: Proceedings of the 42nd Annual ACM Symposium on Principles of Programming Languages (POPL), pp. 489–501 (2015)
6. Gleason, A.M.: Measures on the closed subspaces of a Hilbert space. J. Math. Mech. **6,** 885–893 (1957)

7. Li, Y.J., Yu, N.K., Ying, M.S.: Termination of nondeterministic quantum programs. Acta Informatica **51**, 1–24 (2014)
8. Liu, T., Li, Y.J., Wang, S.L. et al.: A theorem prover for quantum Hoare logic and its applications. arXiv:1601.03835
9. Sankaranarayanan, S., Sipma, H.B., Manna, Z.: Non-linear loop invariant generation using Gröbner bases. In: Proceedings of the 31st ACM Symposium on Principles of Programming Languages (POPL), pp. 318–329 (2004)
10. Ying, M.S.: Floyd-Hoare logic for quantum programs. ACM Trans. Program. Lang. Syst. **39**(19) (2011)
11. Ying, M.S.: Foundations of Quantum Programs. Morgan-Kaufmann (2016)
12. Ying, M.S.: Ranking function synthesis for quantum programs, Draft
13. Ying, M.S., Feng, Y.: Quantum loop programs. Acta Informatica **47**, 221–250 (2010)
14. Ying, M.S., Ying, S.G., Wu, X.D.: Invariants of quantum programs: characterisations and generation, Draft.
15. Ying, M.S., Yu, N.K., Feng, Y., Duan, R.Y.: Verification of quantum programs. Sci. Comput. Program. **78**, 1679–1700 (2013)
16. Ying, S.G., Feng, Y., Yu, N.K., Ying, M.S.: Reachability analysis of quantum Markov chains. In: Proceedings of the 24th International Conference on Concurrency Theory (CONCUR), pp. 334–348 (2013)
17. Yu, N.K., Ying, M.S.: Reachability and termination analysis of concurrent quantum programs. In: Proceedings of the 23th International Conference on Concurrency Theory (CONCUR), pp. 69–83 (2012)

# Contents

# Place Bisimulation and Liveness for Open Petri Nets

Xiaoju Dong[1(✉)], Yuxi Fu[1], and Daniele Varacca[2]

[1] BASICS, Department of Computer Science, Shanghai Jiao Tong University,
Shanghai, China
{xjdong,yxfu}@sjtu.edu.cn
[2] PPS - CNRS and Université Paris Diderot, Paris, France
varacca@pps.jussieu.fr

**Abstract.** Petri nets are a kind of concurrent models for distributed and asynchronous systems. However they can only model closed systems, but not open ones. We extend Petri nets to model open systems. In Open Petri Nets, the way of interaction is achieved by composing nets. Some places with labels, called open or external, are considered as an interface with environment. Every external places are both input and output ones. Two such open Petri nets can be composed by joining the external places with the same label. In addition, we focus on the operational semantics of open nets and study observational properties, especially bisimulation properties. We define place bisimulations on nets with external places. It turns out that the largest bisimulation, i.e. the bisimilarity, is a congruence. A further result is that liveness is preserved by bisimilarity.

**Keywords:** Open Petri net · Interaction · Bisimulation · Liveness

## 1 Introduction

Petri nets [30] are a kind of concurrent models for distributed and asynchronous systems. However they can only model closed systems, but not open ones. A closed system is a system in the state of being isolated from its surrounding environment. It is always a theoretical assumption. In practice no system can be completely isolated. A system is bound to interact with the environment in one way or another. Now open systems are everywhere and continuously interact with their environment. In this paper we extend Petri nets to model open systems [19].

Interactions always occur at the interfaces. For instance, channels are the interfaces in process calculi. There are two kinds of objects which are places and transitions in a net. If a net could interact with its environment, one has three choices for interfaces: places, transitions or both.

Interactions between systems usually are in different ways [1,8–10,12–18,28, 33]. Interactions in process calculi are communications. For nets interfaces help to define composition. If places are interfaces, one could compose the shared places of nets to get a "sharing" net. If transitions are interfaces, one could compose or

M. Fränzle et al. (Eds.): SETTA 2016, LNCS 9984, pp. 1–17, 2016.
DOI: 10.1007/978-3-319-47677-3_1

synchronize the same transitions of nets to get a "synchronized" net. Another method is if two nets share the same subnets, then one can compose the two nets, or the two nets are glued together by the shared subnets.

Open Petri nets are a variant of Petri nets first introduced by Baldan et al. [2–6] that are quite different from the variant proposed by [24]. In such nets, some places with labels, called *open* or *external*, are considered as an interface with environment. For these external places, some are input ones, others are output ones. An external place can be both an input and an output one at the same time. This could be told by the dangling arcs attached to the places. Besides external places, there could be some synchronization transitions. When two nets are composed, the connections of transitions to their pre-set and post-set should be preserved. New connections cannot be added. In the larger net, a new arc may be attached to a place only if the corresponding place of the subnet has a dangling arc in the same direction. Dangling arcs may be removed, but cannot be added in the larger net. Baldan et al. study the denotational semantics of such open nets.

Our approach is to label some places as external ones which is the interface of a net. However, we don't distinguish input and output ones. In other words, every external places are both input and output ones. Therefore, they do not have dangling arcs. Two such open Petri nets can be composed by joining the external places with the same label. In addition, we focus on the operational semantics of open nets and study observational properties [7, 21, 22, 26, 29, 31, 32], especially bisimulations [11, 20, 23, 25, 27].

Bisimulations on "labeled" nets have been studied widely. In such nets, transitions are labeled. This is quite different from the nets we discuss here. We regard transitions as internal computation and external places as observable objects. One can classify the definitions of bisimulations into two categories. One is bisimulations between markings, the other is bisimulations between some places.

We define place bisimulations on nets with external places. It turns out that the largest bisimulation, i.e. the bisimilarity, is a congruence. A further result is that liveness is preserved by bisimilarity.

Section 2 introduces the structure of open Petri nets and gives the way of composition of two nets. Section 3 defines bisimulations and proves that the largest bisimulation is a congruence. Since the concept of liveness changes for an open net, Sect. 4 studies liveness in detail. We give some conclusion in Sect. 5 and a long proof in Appendix.

## 2    Open Petri Nets

### 2.1    Definitions

The definition we give here is essentially the one presented in [4,6]. The main difference is that we label places, not transitions.

**Definition 1.** *An* open P/T-net *is a 6-tuple* $N = \langle P, T, F, W, L, M \rangle$ *where*

- *$P$ and $T$ are the finite set of* places *and* transitions, *respectively. $P \cap T = \emptyset$;*
- *$F \subseteq P \times T \cup T \times P$ is the* flow relation;
- *$W : F \to \mathbb{N} \setminus \{0\}$ is the* weight function;
- *$L : P \rightharpoonup \mathcal{N}$ is the* labeling function, *a partial function injective on its domain of definition;*
- *$M : P \to \mathbb{N}$ is the* initial marking.

Since every $P/T$-net can be transformed to a net with unlimited capacities without affecting its behavior, we will assume open nets $N$ with unlimited capacities for all places $s \in P_N$. A place of an open P/T-net is *external* if it is labeled by a name; it is *internal* if it is unlabeled. We shall write $P^e$ and $P^i$ for the set of external places respectively the set of internal places.

If $N$ is an open P/T-net, we write $P_N$, $T_N$, $F_N$, $W_N$, $L_N$, $M_N$ for the components.

The following definitions are the same as for standard place/transition nets.

**Definition 2.** *Let $N$ be an open P/T-net.*

(i) *A function $M : P_N \to \mathbb{N}$ is a* marking *of $N$.*
(ii) *For $x \in P_N \cup T_N$, $^\bullet x = \{y \mid yFx\}$ is called the* preset *of $x$; $x^\bullet = \{y \mid xFy\}$ is called the* postset *of $x$.*
(iii) *A transition $t \in T_N$ is $M$-enabled, notation $M[t\rangle$, if $\forall s \in {}^\bullet t.M(s) \geq W_N(s,t)$.*
(iv) *An $M$-enabled transition $t$ may produce a* follower marking *$M'$ of $M$ defined by the following*

$$
M(s) \stackrel{\text{def}}{=}
\begin{cases}
M(s) - W_N(s,t), & \text{if } s \in {}^\bullet t \setminus t^\bullet \\
M(s) + W_N(t,s), & \text{if } s \in t^\bullet \setminus {}^\bullet t \\
M(s) - W_N(s,t) + W_N(t,s), & \text{if } s \in {}^\bullet t \cap t^\bullet \\
M(s), & \text{otherwise}
\end{cases}
\tag{1}
$$

*We write $M[t\rangle M'$ to indicate that $M$ evolves into $M'$ by firing $t$.*

Figure 1 is an example of open net.

## 2.2 Composition of Open P/T Nets

An open net could interact with its environment via its external places. Two nets can interact by sharing some external places. Nets are thus composed by merging all places having the same label.

We shall always assume that two open P/T-nets $N_1, N_2$ have disjoint sets of places ($P_{N_1} \cap P_{N_2} = \emptyset$) and disjoint sets of transitions ($T_{N_1} \cap T_{N_2} = \emptyset$). The following notations are used.

$$
P_{N_1 \setminus N_2} = \{s \mid s \in P_{N_1} \wedge L_{N_1}(s) \notin L_{N_2}[P_{N_2}]\}
\tag{2}
$$
$$
P_{N_1 \cap N_2} = \{\langle s_1, s_2 \rangle \mid s_1 \in (P_{N_1})^e \wedge s_2 \in (P_{N_2})^e \wedge L_{N_1}(s_1) = L_{N_2}(s_2)\}
\tag{3}
$$

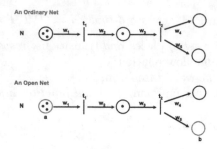

**Fig. 1.** An ordinary net and an open one

In words, $P_{N_1 \setminus N_2}$ is the set of all internal places of $N_1$ plus all external places whose label is not in $N_2$, while $P_{N_1 \cap N_2}$ is the set of pairs of external places that share the same labels. The idea is that such pairs of places will represent the "merged" places in the composition. The following definition formalises this intuition.

**Definition 3.** *The composition of two open P/T-nets $N_1, N_2$, denoted by $N_1 \mid N_2$, is the open P/T-net $N$ defined as follows:*

*– $P_N = P_N^i \cup P_N^e$ where*

$$P_N^i = P_{N_1}^i \cup P_{N_2}^i \tag{4}$$

$$P_N^e = P_{N_1 \setminus N_2}^e \cup P_{N_2 \setminus N_1}^e \cup P_{N_1 \cap N_2} \tag{5}$$

*– $T_N = T_{N_1} \cup T_{N_2}$;*
*– $F_N$ is the following relation*

$$F_{N_1} \upharpoonright P_{N_1 \setminus N_2} \cup F_{N_2} \upharpoonright P_{N_2 \setminus N_1}$$
$$\cup \{ \langle \langle s_1, s_2 \rangle, t \rangle \mid \langle s_1, t \rangle \in F_{N_1}, \langle s_1, s_2 \rangle \in P_{N_1 \cap N_2} \}$$
$$\cup \{ \langle t, \langle s_1, s_2 \rangle \rangle \mid \langle t, s_1 \rangle \in F_{N_1}, \langle s_1, s_2 \rangle \in P_{N_1 \cap N_2} \}$$
$$\cup \{ \langle \langle s_1, s_2 \rangle, t \rangle \mid \langle s_2, t \rangle \in F_{N_2}, \langle s_1, s_2 \rangle \in P_{N_1 \cap N_2} \}$$
$$\cup \{ \langle t, \langle s_1, s_2 \rangle \rangle \mid \langle t, s_2 \rangle \in F_{N_2}, \langle s_1, s_2 \rangle \in P_{N_1 \cap N_2} \}$$

*– $W_N$ is the function defined as follows:*

$$W_N(\langle s, t \rangle) = \begin{cases} W_{N_1}(\langle s, t \rangle), & \text{if } s \in P_{N_1 \setminus N_2} \\ W_{N_2}(\langle s, t \rangle), & \text{if } s \in P_{N_2 \setminus N_1} \\ W_{N_1}(\langle s_1, t \rangle) & \text{if } s = \langle s_1, s_2 \rangle \ \& \ t \in T_{N_1} \\ W_{N_2}(\langle s_2, t \rangle) & \text{if } s = \langle s_1, s_2 \rangle \ \& \ t \in T_{N_2} \end{cases}$$

*and dually for $W_N(\langle t, s \rangle)$.*
*– $L_N$ is the function defined as follows:*

$$L_N(s) = \begin{cases} L_{N_1}(s), & \text{if } s \in P_{N_1 \setminus N_2} \\ L_{N_2}(s), & \text{if } s \in P_{N_2 \setminus N_1} \\ L_{N_1}(s_1), & \text{if } s = \langle s_1, s_2 \rangle \end{cases}$$

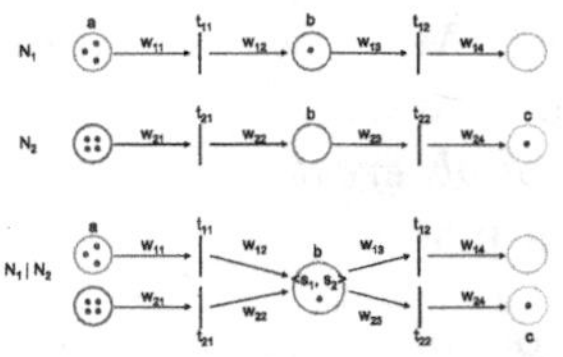

**Fig. 2.** Composition of two open nets

– $M_N$ is the function defined as follows:

$$M_N(s) = \begin{cases} M_{N_1}(s), & \text{if } s \in P_{N_1 \setminus N_2} \\ M_{N_2}(s), & \text{if } s \in P_{N_2 \setminus N_1} \\ M_{N_1}(s_1) + M_{N_2}(s_2), & \text{if } s = \langle s_1, s_2 \rangle \end{cases}$$

Figure 2 is a simple example of the composition of two nets.

We also define a restriction operator, to close some open places. This amounts to a restriction on the labelling function.

**Definition 4.** *The restriction of an open P/T-net N at a name a, denoted by (a)N, is the open P/T-net obtained from N by modifying the labeling function as follows:*

$$L_{(a)N}(s) \stackrel{\text{def}}{=} \begin{cases} L_N(s), \text{ if } L_N(s) \neq a \\ \uparrow, \qquad \text{otherwise} \end{cases} \tag{6}$$

## 3   Bisimulation for Open P/T Nets

The notion of bisimulation we propose differs from the one by Baldan et al. in that it does not observe the identity of the transitions - in our work, transitions are not labelled. What can be observed is whether a transition changes the external marking of the net or not.

As a consequence of our approach, we do not model the interaction with the environment by means of special transition that change the external marking - we simply state that external places can at any moment receive or lose tokens. Formally this is done using the notion of substitution.

### 3.1   Observations on Open Nets

Traditionally, behavioural properties on nets are based on the principle that every part of the net can be observed. Whether each transition has its own identity or whether transitions are labelled, everything that happens can be seen from the outside.

In open nets, there is a formal notion of "outside". The open places are the communication interface. It is thus reasonable, in this framework, to consider open places as the only observable part of a net. The definitions that follow are a consequence of this approach. What we indeed observe are the changes that transitions impose on external marking.

We first define the notion of *observation* of a marked net: it is the restriction of the marking to the external places.

**Definition 5.** *The* observation *of an open P/T-net $N$ is the function $M^e$ : $P_N^e \to \mathbb{N}$, the restriction of $M$ to $P_N^e$.*

We say that two observations $M_1^e$ and $M_2^e$ are *equivalent*, notation $M_1^e \asymp M_2^e$, if there is a bijection $\iota : P_{N_1}^e \to P_{N_2}^e$ satisfying the following conditions:

- $M_1^e = M_2^e \circ \iota$;
- $L_1^e = L_2^e \circ \iota$.

The environment can add or remove tokens from external places. Formally this is represented by the notion of assignment.

**Definition 6.** *An* assignment $\sigma : \mathcal{N} \rightharpoonup \mathbb{N}$ *is a partial function such that $dom(\sigma)$ is finite.*

Suppose $\sigma$ is an assignment and $M$ is a marking of $N$. The marking $M\sigma$ is obtained from $M$ as follows:

$$M\sigma(s) \stackrel{\text{def}}{=} \begin{cases} \sigma(a), & \text{if } s \in P_N^e \text{ and } L_N(s) = a \\ M(s), & \text{otherwise} \end{cases} \tag{7}$$

The open P/T-net $N\sigma$ is obtained from $N$ by replacing $M_N$ by $M_N\sigma$.

An assignment can arbitrarily change the observation of an open Petri nets. We take this into account by defining an extended notion of reachability.

**Definition 7.** *Let $[M\rangle^e$ be the set inductively defined as follows: (i) $M \in [M\rangle^e$; (ii) if $M' \in [M\rangle^e$ and $M'[t\rangle M''$ for some $t \in T_N$ then $M'' \in [M\rangle^e$; and (iii) if $M' \in [M\rangle^e$ then $M'\sigma \in [M\rangle^e$ for any assignment $\sigma$.*

Then it is easy to get the following conclusion.

**Theorem 1.** *The reachability of an P/T open net is decidable.*

*Proof.* Suppose $N = \langle P_N, T_N, F_N, W_N, L_N, M_N \rangle$ is an open P/T net. Accordingly, we could construct a normal P/T net $N'$ as follows:

- $P_{N'} = P$;
- $T_{N'} = T \bigcup \{t_s, t_s' \mid \text{for all external places } s \in P_N\}$;
- $F_{N'} = F_N \bigcup \{(t_s, s), (s, t_s') \mid \text{for all external places } s \in P_N\}$;
- $W_{N'} = W_N$;
- $M_{N'} = M_N$.

That is, for all external place $s \in P_N$, we add two transitions $t_s$ and $t_s'$ into the set of transitions. Meanwhile, the flow relation $F$ is extended by two pairs $(t_s, s), (s, t_s')$. It is obvious that the reachability sets of $N$ and $N'$ are the same. Since the reachability of $N'$ is decidable, so does $N$.

## 3.2   Bisimulation

The notion of equivalence we are going to define must be closed under interactions with the environment.

**Definition 8.** *A binary relation $\mathcal{R}$ on nets is* equipotent *if $M_{N_1}^e \asymp M_{N_2}^e$ whenever $N_1 \mathcal{R} N_2$. An equipotent relation $\mathcal{R}$ is* fully equipotent *if $N_1 \mathcal{R} N_2$ implies that $N_1 \sigma \mathcal{R} N_2 \sigma$ for all $\sigma$.*

Bisimulation relations are usually defined on labelled transition systems. The notion defined by Baldan et al. takes into account the labels of the transitions of the nets. In our framework, transitions are unlabelled. The only observation we make on transition is whether they are *external* or *internal*, that is whether or not they change the external marking.

**Definition 9.** *Let $M_1, M_2$ be two markings of the open P/T-net $N$ and suppose $M_1[t\rangle M_2$. If $M_1^e = M_2^e$ we write $N_1 \longrightarrow N_2$ (the transition is* internal. *Otherwise we write $N_1 \overset{\tau}{\longrightarrow} N_2$ (the transition is* external*) We will write $\Longrightarrow$ for the reflexive and transitive closure of $\longrightarrow$, $\overset{\tau}{\Longrightarrow}$ for the composition $\Longrightarrow \overset{\tau}{\longrightarrow} \Longrightarrow$, $\overset{*}{\longrightarrow}$ for either $\longrightarrow$ or $\overset{\tau}{\longrightarrow}$ and $\overset{*}{\Longrightarrow}$ for either $\Longrightarrow$ or $\overset{\tau}{\Longrightarrow}$.*

We have the ingredients to define bisimulation:

**Definition 10.** *A fully equipotent symmetric binary relation $\mathcal{R}$ on nets is a* bisimulation *if the following properties hold whenever $N_1 \mathcal{R} N_2$:*
*(i) if $N_1 \longrightarrow N_1'$ then $N_2 \Longrightarrow N_2' \mathcal{R} N_1'$ for some $N_2'$;*
*(ii) if $N_1 \overset{\tau}{\longrightarrow} N_1'$ then $N_2 \overset{\tau}{\Longrightarrow} N_2' \mathcal{R} N_1'$ for some $N_2'$.*
*The bisimilarity $\approx$ is the largest bisimulation.*

The main result of this section is that bisimilarity is congruence with respect to the composition of Open nets.

**Theorem 2.** *The equivalence $\approx$ is a congruence.*

The proof is in the appendix.

# 4   Liveness

We are interested in studying which properties are invariant under bisimulation. Traditionally, one of the main property considered in the study of Petri nets is *liveness*. We will provide in the following notion of liveness for open Petri net, and show that this property is invariant under bisimulation.

As we know, the liveness of a net depends on the liveness of all the transitions. Let $N$ be an ordinary P/T net and $t \in T_N$. Then,

1. a transition $t$ is called live iff $\forall M \in [M_N\rangle$ $exists M' \in [M\rangle$ s.t. $t$ is $M'$-enabled.
2. The net $N$ is called live iff $\forall t \in T_N : t$ is live.

According to the definition of livenss, the following net in Fig. 3 is not live.

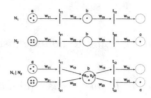

**Fig. 3.** Liveness

However, when another net $N'$ communicates with the above net $N$ and puts one token in the place $a$, $t$ could be fired. Hence, the definition of liveness for an open net should be different from it for an ordinary one since interactions would have effect on the liveness. Moreover, liveness should be observed by the environment. Accordingly, we refine liveness for open nets.

Firstly, in an open net, the transitions could be classified into two categories by the effect on the markings. If the trigger of a transition modifies the number of tokens in the external places, we call them observable. Otherwise, it is unobservable.

**Definition 11.** *A transition $t \in T_N$ is observable in $N$, if there exist $M, M' \in [M_N\rangle^e$ such that $M[t\rangle M'$ and $M^e \neq M'^e$.*

Therefore, from the observational view, we would focus on the liveness of observable transitions. The following property of observable transitions are obvious.

**Corollary 1.** *If a transition $t \in T_N$ is observable in $N$, then $O = ({}^\bullet t \cup t^\bullet) \cap P^e \neq \emptyset$ and $\exists s \in O, w(s,t) \neq w(t,s)$.*

For convenience, we have $({}^\bullet t)^e = {}^\bullet t \cap P^e$ and $(t^\bullet)^e = t^\bullet \cap P^e$.

Since some observable transitions have the same effect on the marking, we could construct an equivalent relation on observable transitions. The following definition is based on the idea that if two transitions have the same influence on the external places in the whole net at a moment, then they are equivalent.

**Definition 12.** *Transitions $t_1, t_2 \in T_N$ are instanteous globally equivalent, noted as $t_1 \backsim_{OBSig} t_2$, if $\forall M_1, M_1' \in [M_N\rangle^e.M_1[t_1\rangle M_1'$, $\exists M_2, M_2' \in [M_N\rangle^e.M_2[t_2\rangle M_2'$ such that $M_1^e = M_2^e$ and $M_1'^e = M_2'^e$, and vice versa.*

**Lemma 1.** *Suppose $M_1[t_1\rangle M_1'$ and $M_2[t_2\rangle M_2'$, where $M_1^e = M_2^e$ and $M_1'^e = M_2'^e$, for some $t_1, t_2 \in T_N$ and $M_1, M_1', M_2, M_2' \in [M_N\rangle^e$. Then $t_1$ and $t_2$ are instanteous globally equivalent.*

*Proof.* Since $M_1[t_1\rangle M_1'$, $M_2[t_2\rangle M_2'$, $M_1^e = M_2^e$ and $M_1^e = M_2^e$, one has $({}^\bullet t_1)^e = ({}^\bullet t_2)^e$, $(t_1^\bullet)^e = (t_2^\bullet)^e$ and $\forall s \in ({}^\bullet t_1)^e.W(s,t_1) = W(s,t_2)$, $\forall s \in (t_1^\bullet)^e.W(t_1,s) = W(t_2,s)$. Then $\forall M, M' \in [M_N\rangle^e$ which satisfy $M[t_1\rangle M'$, there exist assignments $\sigma, \sigma'$ such that $M^e = M_1^e \sigma$ and $M'^e = M_1'^e \sigma'$, i.e.$M_1\sigma[t_1\rangle M_1\sigma'$ . Hence, we also have $M_2\sigma[t_2\rangle M_2\sigma'$. Therefore, $t_1 \backsim_{OBSig} t_2$.

Then we have to define the liveness of a transition.

**Definition 13.** *A transition $t \in T_N$ is initially live in $N$, if there exists $M \in [M_N\rangle^e$ such that $M[t\rangle$.*

Now the liveness of an open net could be defined as follows:

**Definition 14.** *$(N, M_N)$ is live if for all observable and initially live transition $t \in T_N$, $\forall M \in [M_N\rangle^e$, $\exists M' \in [M\rangle^e$ and $\exists t' \frown_{OBSig} t$ such that $M'[t'\rangle$.*

The rest of the section focuses on the properties of liveness.

**Lemma 2.** *If $N_1 \approx N_2$ and $t_1 \in T_{N_1}$ is observable and initially live, i.e. $\exists M_1, M_1' \in [M_{N_1}\rangle^e$ such that $M_1[t_1\rangle M_1'$ and $M_1^e \neq M_1'^e$, then $\exists M_2, M_2' \in [M_{N_2}\rangle^e$ and $t_2 \in T_{N_2}$ such that $M_2[t_2\rangle M_2'$, $M_2^e \asymp M_1^e$ and $M_2'^e \asymp M_1'^e$.*

*Proof.* Since $t_1 \in T_{N_1}$ is observable and initially live, there should be a sequence of assignments $\sigma^0, \sigma^1, ..., \sigma^{n-1}$ such that $N_1\sigma^0 \xrightarrow{*} N_1^1$, $N_1^1\sigma^1 \xrightarrow{*} N_1^2$, ..., $N_1^{n-1}\sigma^{n-1} \xrightarrow{\tau} N_1^n$ where $M_1$ is the marking of $N_1^{n-1}\sigma^{n-1}$ and $M_1'$ is the marking of $N_1^n$. For $N_1 \approx N_2$, we have $N_2\sigma^0 \xRightarrow{*} N_2^1 \approx N_1^1$, $N_2^1\sigma^1 \xRightarrow{*} N_2^2 \approx N_1^2$, ..., $N_2^{n-1}\sigma^{n-1} \xRightarrow{*} N_2^n \approx N_1^n$ for some $t_2 \in T_{N_2}$. Suppose $M_2$ is the marking of $N_2^{n-1}\sigma^{n-1}$ and $M_2'$ is the marking of $N_2^n$, then one has $M_2[t_2\rangle M_2'$, $M_2^e \asymp M_1^e$ and $M_2'^e \asymp M_1'^e$.

Now we come to the most important result of the section.

**Definition 15.** *$E$ is the equivalence relation on $[M\rangle^e$ if $\forall M_1, M_2 \in [M\rangle^e.\langle M_1, M_2\rangle \in E$ iff $M_1^e = M_2^e$.*

**Definition 16.** *An equivalence class of $M \in [M_N\rangle^e$ under $E$, denoted $[M]$, is the subset of $[M_N\rangle^e$ for which every element $M'$, $\langle M, M'\rangle \in E$. The quotient set on $[M_N\rangle^e$ by $E$ is the set of all equivalence classes of $[M_N\rangle^e$ by $E$, denoted $[M_N\rangle^e/E = \{[M] \mid M \in [M_N\rangle^e\}$.*

**Lemma 3.** *If $N_1 \approx N_2$, then $[M_{N_1}\rangle^e/E_1$ and $[M_{N_2}\rangle^e/E_2$ are isomorphic.*

*Proof.* Suppose $M_{N_1}$ and $M_{N_2}$ are initial markings of $N_1$ and $N_2$, respectively. Because $N_1 \approx N_2$, $M_{N_1}^e \asymp M_{N_2}^e$ and $M_{N_1}^e\sigma \asymp M_{N_2}^e\sigma$. For every element $[M_1] \in [M_{N_1}^e\rangle/E_1$, we take $M_1'$ as a representive element of $[M_1]$. Then there should be a sequence of assignments $\sigma^0, \sigma^1, ..., \sigma^{n-1}$ such that $N_1\sigma^0 \xrightarrow{*} N_1^1$, $N_1^1\sigma^1 \xrightarrow{*} N_1^2$, ..., $N_1^{n-1}\sigma^{n-1} \xrightarrow{*} N_1^n$ and $M_1$ is the marking of $N_1^n$. For $N_1 \approx N_2$, we have $N_2\sigma^0 \xRightarrow{*} N_2^1 \approx N_1^1$, $N_2^1\sigma^1 \xRightarrow{*} N_2^2 \approx N_1^2$, ..., $N_2^{n-1}\sigma^{n-1} \xRightarrow{*} N_2^n \approx N_1^n$. Assume $M_2 \in [M_{N_2}\rangle^e$ is the marking of $N_2^n$. Hence, $M_1^e \asymp M_2^e$. Therefore $[M_2]$ is corresponding to $[M_1]$. We can also prove that for every element $[M_2] \in [M_{N_2}^e\rangle/E_2$, there is an element $[M_1] \in [M_{N_1}^e\rangle/E_1$ is corresponding to it.

**Proposition 1.** *If $N_1 \approx N_2$ and $N_2$ is live, then $N_1$ is live.*

*Proof.* Assume $N_1$ was not live. Then for some observable and initially live transition $t_1 \in T_{N_1}$, $\exists M_1 \in [M_{N_1}\rangle^e$ such that $\forall M_1' \in [M_1\rangle^e.\forall t \frown_{OBSig} t_1$, $M_1'$ could not fire $t_1$. According to Lemma 2 there exists an observable and initially live transition $t_2 \in T_{N_2}$ such that if $M_{f_1}[t_1\rangle M'_{f_1}$ for some $M_{f_1}, M'_{f_1} \in [M_{N_1}\rangle^e$ then $M_{f_2}[t_2\rangle M'_{f_2}$ for some $M_{f_2}, M'_{f_2} \in [M_{N_2}\rangle^e$, where $M^e_{f_2} \asymp M^e_{f_1}$ and $M'^e_{f_2} \asymp M'^e_{f_1}$.

By the proof of Lemma 3, there is a marking $M_2 \in [M_{N_2}\rangle^e$ satisfying $M^e_1 = M^e_2$. Since $N_2$ is live, we have $(\exists M'_2, M''_2 \in [M_2\rangle^e)(\exists t'_2 \frown t_2).M'_2[t'_2\rangle M''_2$. From Lemma 2, $\exists t'_1 \in T_{N_1}, \exists M'_p, M''_p \in [M_1\rangle^e$ such that $M'_p[t'_1\rangle M''_p$ and $M'^e_p \asymp M'^e_2$, $M''^e_p \asymp M''^e_2$. By Lemma 1, $t_1 \frown_{OBSig} t'_1$. It is inconsistent to the assumption. It is done.

## 5   Conclusion

We devide the places in a Petri net into two categories: external and internal. The external places are used to communicate with other nets. Accordingly, it results in the composition between nets. We define a bisimulation on nets to model communication between nets and indicate their behaviors. The laregest bisimulation is proved to be a congruence. Petri nets have many properties. However the definitions and characteristics of these properties are quite different for an open nets. It shows the reachability of an open P/T is decidable. The liveness has been redefined. We have proved that liveness is preserved by composition. The other properties will be revisited in our future research.

**Acknowledgments.** The work is supported by the National Nature Science Foundation of China(61472239, 61100053). The authors would like to thank the unknown reviewers for the comments.

## Appendix

**Theorem 3.** *The equivalence $\approx$ is a congruence.*

*Proof.* Suppose $N_1 \approx N_2$. Then $N_1 \mid N_0 \approx N_2 \mid N_0$ and $(x)N_1 \approx (x)N_2$.

1. Suppose $\mathcal{R} = \{(N_1 \mid N_0, N_2 \mid N_0) \mid N_1 \approx N_2\}$. Then $\mathcal{R}$ is a bisimulation. Let $N_{10}$ and $N_{20}$ denote $N_1 \mid N_0$ and $N_2 \mid N_0$ respectively.
$N_{10} = \langle P_{N_{10}}, T_{N_{10}}, F_{N_{10}}, K_{N_{10}}, W_{N_{10}}, L_{N_{10}}, M_{N_{10}} \rangle$, where
   - $P_{N_{10}} = P^i_{N_{10}} \cup P^e_{N_{10}}$ where

$$P^i_{N_{10}} = P^i_{N_1} \cup P^i_{N_0} \tag{8}$$
$$P^e_{N_{10}} = P^e_{N_1 \setminus N_0} \cup P^e_{N_0 \setminus N_1} \cup P_{N_1 \cap N_0} \tag{9}$$

   - $T_{N_{10}} = T_{N_1} \cup T_{N_0}$;
   - $F_{N_{10}}$ is the following relation

$$F_{N_1} \upharpoonright P_{N_1 \setminus N_0} \cup F_{N_0} \upharpoonright P_{N_0 \setminus N_1}$$
$$\cup \{\langle \langle s_1, s_0 \rangle, t \rangle \mid \langle s_1, t \rangle \in F_{N_1}, \langle s_1, s_0 \rangle \in P_{N_1 \cap N_0}\}$$
$$\cup \{\langle t, \langle s_1, s_0 \rangle \rangle \mid \langle t, s_1 \rangle \in F_{N_1}, \langle s_1, s_0 \rangle \in P_{N_1 \cap N_0}\}$$
$$\cup \{\langle \langle s_1, s_0 \rangle, t \rangle \mid \langle s_0, t \rangle \in F_{N_0}, \langle s_1, s_0 \rangle \in P_{N_1 \cap N_0}\}$$
$$\cup \{\langle t, \langle s_1, s_0 \rangle \rangle \mid \langle t, s_0 \rangle \in F_{N_0}, \langle s_1, s_0 \rangle \in P_{N_1 \cap N_0}\}$$

– $W_{N_{10}}$ is the function defined as follows:

$$W_{N_{10}}(\langle s,t\rangle) = \begin{cases} W_{N_1}(\langle s,t\rangle), & \text{if } s \in P_{N_1 \setminus N_0} \\ W_{N_0}(\langle s,t\rangle), & \text{if } s \in P_{N_0 \setminus N_1} \\ \begin{array}{c} W_{N_1}(\langle s_1,t\rangle) \\ + \\ W_{N_0}(\langle s_0,t\rangle) \end{array} & , \text{if } s = \langle s_1,s_0\rangle \end{cases}$$

and

$$W_{N_{10}}(\langle t,s\rangle) = \begin{cases} W_{N_1}(\langle t,s\rangle), & \text{if } s \in P_{N_1 \setminus N_0} \\ W_{N_0}(\langle t,s\rangle), & \text{if } s \in P_{N_0 \setminus N_1} \\ \begin{array}{c} W_{N_1}(\langle t,s_1\rangle) \\ + \\ W_{N_0}(\langle t,s_0\rangle) \end{array} & , \text{if } s = \langle s_1,s_0\rangle \end{cases}$$

– $L_{N_{10}}$ is the function defined as follows:

$$L_{N_{10}}(s) = \begin{cases} L_{N_1}(s), & \text{if } s \in P_{N_1 \setminus N_0} \\ L_{N_0}(s), & \text{if } s \in P_{N_0 \setminus N_1} \\ L_{N_1}(s_1), & \text{if } s = \langle s_1,s_0\rangle \end{cases}$$

– $M_{N_{10}}$ is the function defined as follows:

$$M_{N_{10}}(s) = \begin{cases} M_{N_1}(s), & \text{if } s \in P_{N_1 \setminus N_0} \\ M_{N_0}(s), & \text{if } s \in P_{N_0 \setminus N_1} \\ M_{N_1}(s_1) + M_{N_0}(s_0), & \text{if } s = \langle s_1,s_0\rangle \end{cases}$$

$N_{20} = \langle P_{N_{20}}, T_{N_{20}}, F_{N_{20}}, K_{N_{20}}, W_{N_{20}}, L_{N_{20}}, M_{N_{20}}\rangle$, where

– $P_{N_{20}} = P_{N_{20}}^i \cup P_{N_{20}}^e$ where

$$P_{N_{20}}^i = P_{N_2}^i \cup P_{N_0}^i \tag{10}$$
$$P_{N_{20}}^e = P_{N_2 \setminus N_0}^e \cup P_{N_0 \setminus N_2}^e \cup P_{N_2 \cap N_0} \tag{11}$$

– $T_{N_{20}} = T_{N_2} \cup T_{N_0}$;
– $F_{N_{20}}$ is the following relation

$$F_{N_2} \upharpoonright P_{N_2 \setminus N_0} \cup F_{N_0} \upharpoonright P_{N_0 \setminus N_2}$$
$$\cup \{\langle\langle s_2,s_0\rangle,t\rangle \mid \langle s_2,t\rangle \in F_{N_2}, \langle s_2,s_0\rangle \in P_{N_2 \cap N_0}\}$$
$$\cup \{\langle t,\langle s_2,s_0\rangle\rangle \mid \langle t,s_2\rangle \in F_{N_2}, \langle s_2,s_0\rangle \in P_{N_2 \cap N_0}\}$$
$$\cup \{\langle\langle s_2,s_0\rangle,t\rangle \mid \langle s_0,t\rangle \in F_{N_0}, \langle s_2,s_0\rangle \in P_{N_2 \cap N_0}\}$$
$$\cup \{\langle t,\langle s_2,s_0\rangle\rangle \mid \langle t,s_0\rangle \in F_{N_0}, \langle s_2,s_0\rangle \in P_{N_2 \cap N_0}\}$$

– $W_{N_{20}}$ is the function defined as follows:

$$W_{N_{20}}(\langle s,t\rangle) = \begin{cases} W_{N_2}(\langle s,t\rangle), & \text{if } s \in P_{N_2 \setminus N_0} \\ W_{N_0}(\langle s,t\rangle), & \text{if } s \in P_{N_0 \setminus N_2} \\ \begin{array}{c} W_{N_2}(\langle s_2,t\rangle) \\ + \\ W_{N_0}(\langle s_0,t\rangle) \end{array} & , \text{if } s = \langle s_2,s_0\rangle \end{cases}$$

and

$$W_{N_{20}}(\langle t,s \rangle) = \begin{cases} W_{N_2}(\langle t,s \rangle), & \text{if } s \in P_{N_2 \setminus N_0} \\ W_{N_0}(\langle t,s \rangle), & \text{if } s \in P_{N_0 \setminus N_2} \\ W_{N_2}(\langle t,s_2 \rangle) \\ \quad + & \text{, if } s = \langle s_2,s_0 \rangle \\ W_{N_0}(\langle t,s_0 \rangle) \end{cases}$$

– $L_{N_{20}}$ is the function defined as follows:

$$L_{N_{20}}(s) = \begin{cases} L_{N_2}(s), & \text{if } s \in P_{N_2 \setminus N_0} \\ L_{N_0}(s), & \text{if } s \in P_{N_0 \setminus N_2} \\ L_{N_2}(s_2), & \text{if } s = \langle s_2,s_0 \rangle \end{cases}$$

– $M_{N_{20}}$ is the function defined as follows:

$$M_{N_{20}}(s) = \begin{cases} M_{N_2}(s), & \text{if } s \in P_{N_2 \setminus N_0} \\ M_{N_0}(s), & \text{if } s \in P_{N_0 \setminus N_2} \\ M_{N_2}(s_2) + M_{N_0}(s_0), & \text{if } s = \langle s_2,s_0 \rangle \end{cases}$$

Since $N_1 \approx N_2$, $N_1\sigma \approx N_2\sigma$ and $M_{N_1}^e \asymp M_{N_2}^e$. Assume the bijection $\iota : P_{N_1}^e \to P_{N_2}^e$ satisfies the following conditions:

– $M_{N_1}^e = M_{N_2}^e \circ \iota$;
– $L_{N_1}^e \upharpoonright P_{N_1}^e = L_{N_2}^e \circ \iota$;
– $K_{N_1}^e \upharpoonright P_{N_1}^e = K_{N_2}^e \circ \iota$.

(1) A bijection $\iota' : P_{N_{10}}^e \to P_{N_{20}}^e$ can be defined as follows:

$$\iota'(s) \stackrel{\text{def}}{=} \begin{cases} \iota(s), & \text{if } s \in P_{N_1 \setminus N_0}^e \\ s, & \text{if } s \in P_{N_0 \setminus N_1}^e \\ \langle \iota(s_1),s_0 \rangle, & \text{if } s = \langle s_1,s_0 \rangle \end{cases} \tag{12}$$

Then $M_{N_{10}}^e \asymp M_{N_{20}}^e$. Therefore, $\mathcal{R}$ is equipotent. If $N_{10}\mathcal{R}N_{20}$, then $(N_{10}\sigma)\mathcal{R}(N_{20}\sigma)$ for $N_{10}\sigma \equiv N_1\sigma \,|\, N_0\sigma'$ and $N_{20}\sigma \equiv N_2\sigma \,|\, N_0\sigma'$, where

$$\sigma'(a) \stackrel{\text{def}}{=} \begin{cases} \sigma(a), \text{ if } a = L(s_0), \\ \quad \text{where } s_0 \in P_{N_0 \setminus N_1}^e \\ 0, \quad \text{if } a = L(s_0), \\ \quad \text{where } \langle s_1,s_0 \rangle \in P_{N_1 \cap N_0} \end{cases} \tag{13}$$

Hence, $\mathcal{R}$ is fully equipotent.

(2) Suppose $N_{10} \longrightarrow N_{10}'$. Then there exists a marking $M_{N_{10}}'$ of $N_{10}$ such that $M_{N_{10}}[t\rangle M_{N_{10}}'$ and $M_{N_{10}}^e = M_{N_{10}}'^e$.

– If $t \in T_{N_1}$, then there are two cases.
   • If there is a marking $M_{N_1}'$ such that $M_{N_1}[t\rangle M_{N_1}'$ and $M_{N_1}^e = M_{N_1}'^e$, i.e. $N_1 \longrightarrow N_1'$, then $N_{10}' \equiv N_1' \,|\, N_0$. Since $N_1 \approx N_2$, $N_2 \Longrightarrow N_2'$ such that $N_1' \approx N_2'$. Then $N_{20} \Longrightarrow N_2' \,|\, N_0$ and $(N_1' \,|\, N_0)\mathcal{R}(N_2' \,|\, N_0)$. Otherwise,

- $O = (^\bullet t \cup t^\bullet) \cap P_{N_1 \cap N_0} \neq \emptyset$ and $\forall s = \langle s_1, s_0 \rangle \in O$, $s \in^\bullet t \cap t^\bullet$ and $W(s,t) = W(t,s)$. We define

$$\sigma(a) \stackrel{def}{=} \begin{cases} M_1(s_1) + M_0(s_0), \\ \quad \text{if } a = L(s_1), \text{where } s = \langle s_1, s_0 \rangle \in O \\ M_1(s_1), \\ \quad \text{if } a = L(s_1), \text{where } s_1 \in P^e_{N_1} \setminus O \end{cases} \tag{14}$$

$$\sigma_0(a) \stackrel{def}{=} \begin{cases} 0, \\ \quad \text{if } a = L(s_0), \text{where } s = \langle s_1, s_0 \rangle \in O \\ M_0(s_0), \\ \quad \text{if } a = L(s_0), \text{where } s_0 \in P^e_{N_0} \setminus O \end{cases} \tag{15}$$

Then $N_{10} \equiv N_1\sigma \,|\, N_0\sigma_0$. Obviously, $N_1\sigma \longrightarrow N_1'$ such that $N_{10}' \equiv N_1' \,|\, N_0\sigma_0$. Since $N_1\sigma \approx N_2\sigma, N_2\sigma \Longrightarrow N_2'$ and $N_1' \approx N_2'$. Then $N_{20} \equiv N_2\sigma \,|\, N_0\sigma_0 \Longrightarrow N_2' \,|\, N_0\sigma_0$ and $(N_1' \,|\, N_0\sigma_0)\mathcal{R}(N_2' \,|\, N_0\sigma_0)$.
- If $t \in T_{N_0}$, then there are also two cases.
  - If there is a marking $M_{N_0}'$ such that $M_{N_0}[t\rangle M_{N_0}'$ and $M_{N_0}^e = M_{N_0}'^e$, i.e. $N_0 \longrightarrow N_0'$, then $N_{10}' \equiv N_1 \,|\, N_0'$ and $N_{20} \longrightarrow N_2 \,|\, N_0'$. Since $N_1 \approx N_2$, $(N_1 \,|\, N_0')\mathcal{R}(N_2 \,|\, N_0')$. Otherwise,
  - $O = (^\bullet t \cup t^\bullet) \cap P_{N_1 \cap N_0} \neq \emptyset$ and $\forall s = \langle s_1, s_0 \rangle \in O$, $s \in^\bullet t \cup t^\bullet$ and $W(s,t) = W(t,s)$. It can be defined as

$$\sigma(a) \stackrel{def}{=} \begin{cases} M_0(s_0) + M_1(s_1), \\ \quad \text{if } a = L(s_0), \text{where } s = \langle s_1, s_0 \rangle \in O \\ M_0(s_0), \\ \quad \text{if } a = L(s_0), \text{where } s_0 \in P^e_{N_0} \setminus O \end{cases} \tag{16}$$

$$\sigma_1(a) \stackrel{def}{=} \begin{cases} 0, \\ \quad \text{if } a = L(s_1), \text{where } s = \langle s_1, s_0 \rangle \in O \\ M_1(s_1), \\ \quad \text{if } a = L(s_1), \text{where } s_1 \in P^e_{N_1} \setminus O \end{cases} \tag{17}$$

Then $N_{10} \equiv N_1\sigma_1 \,|\, N_0\sigma$. Obviously, $N_0\sigma \longrightarrow N_0'$, $N_{10}' \equiv N_1\sigma_1 \,|\, N_0'$ and $N_{20} \equiv N_2\sigma_1 \,|\, N_0\sigma \longrightarrow N_2\sigma_1 \,|\, N_0'$. Since $N_1 \approx N_2$, $(N_1\sigma_1 \,|\, N_0')\mathcal{R}(N_2\sigma_1 \,|\, N_0')$.

Suppose $N_{10} \stackrel{\tau}{\longrightarrow} N_{10}'$. Then there exists a marking $M_{N_{10}}'$ of $N_{10}$ such that $M_{N_{10}}[t\rangle M_{N_{10}}'$ and $M_{N_{10}}^e \neq M_{N_{10}}'^e$.

- If $t \in T_{N_1}$, then there are two cases.
  - If there is a marking $M_{N_1}'$ such that $M_{N_1}[t\rangle M_{N_1}'$ and $M_{N_1}^e \neq M_{N_1}'^e$, i.e. $N_1 \stackrel{\tau}{\longrightarrow} N_1'$, then $N_{10}' \equiv N_1' \,|\, N_0$. Since $N_1 \approx N_2$, $N_2 \stackrel{\tau}{\Longrightarrow} N_2'$ such that $N_1' \approx N_2'$. Then $N_{20} \stackrel{\tau}{\Longrightarrow} N_2' \,|\, N_0$ and $(N_1' \,|\, N_0)\mathcal{R}(N_2' \,|\, N_0)$. Otherwise,

- $O = ({}^\bullet t \cup t^\bullet) \cap P_{N_1 \cap N_0} \neq \emptyset$. We define

$$
\sigma(a) \stackrel{\text{def}}{=}
\begin{cases}
M_1(s_1) + M_0(s_0), \\
\quad \text{if} a = L(s_1), \text{where } s = \langle s_1, s_0 \rangle \in O \\
M_1(s_1), \\
\quad \text{if } a = L(s_1), \text{where } s_1 \in P^e_{N_1} \setminus O
\end{cases}
\tag{18}
$$

$$
\sigma_0(a) \stackrel{\text{def}}{=}
\begin{cases}
0, \\
\quad \text{if} a = L(s_0), \text{where } s = \langle s_1, s_0 \rangle \in O \\
M_0(s_0), \\
\quad \text{if } a = L(s_0), \text{where } s_0 \in P^e_{N_0} \setminus O
\end{cases}
\tag{19}
$$

Then $N_{10} \equiv N_1\sigma \,|\, N_0\sigma_0$. Obviously, $N_1\sigma \xrightarrow{\tau} N_1'$ such that $N_{10}' \equiv N_1' \,|\, N_0\sigma_0$. Since $N_1\sigma \approx N_2\sigma$, $N_2\sigma \stackrel{\tau}{\Longrightarrow} N_2'$ and $N_1' \approx N_2'$. Then $N_{20} \equiv N_2\sigma \,|\, N_0\sigma_0 \stackrel{\tau}{\Longrightarrow} N_2' \,|\, N_0\sigma_0$ and $(N_1' \,|\, N_0\sigma_0)\mathcal{R}(N_2' \,|\, N_0\sigma_0)$.

- If $t \in T_{N_0}$, then there are also two cases.
  - If there is a marking $M_{N_0}'$ such that $M_{N_0}[t\rangle M_{N_0}'$ and $M_{N_0}^e \neq M_{N_0}'^e$, i.e. $N_0 \xrightarrow{\tau} N_0'$, then $N_{10}' \equiv N_1 \,|\, N_0'$ and $N_{20} \xrightarrow{\tau} N_2 \,|\, N_0'$. Since $N_1 \approx N_2$, $(N_1 \,|\, N_0')\mathcal{R}(N_2 \,|\, N_0')$. Otherwise,
  - We have $O = ({}^\bullet t \cup t^\bullet) \cap P_{N_1 \cap N_0} \neq \emptyset$. We also define

$$
\sigma(a) \stackrel{\text{def}}{=}
\begin{cases}
M_0(s_0) + M_1(s_1), \\
\quad \text{if} a = L(s_0), \text{where } s = \langle s_1, s_0 \rangle \in O \\
M_0(s_0), \\
\quad \text{if } a = L(s_0), \text{where } s_0 \in P^e_{N_0} \setminus O
\end{cases}
\tag{20}
$$

$$
\sigma_1(a) \stackrel{\text{def}}{=}
\begin{cases}
0, \\
\quad \text{if} a = L(s_1), \text{where } s = \langle s_1, s_0 \rangle \in O \\
M_1(s_1), \\
\quad \text{if } a = L(s_1), \text{where } s_1 \in P^e_{N_1} \setminus O
\end{cases}
\tag{21}
$$

Then $N_{10} \equiv N_1\sigma_1 \,|\, N_0\sigma$. We get $N_0\sigma \xrightarrow{\tau} N_0'$, $N_{10}' \equiv N_1\sigma_1 \,|\, N_0'$ and $N_{20} \equiv N_2\sigma_1 \,|\, N_0\sigma \xrightarrow{\tau} N_2\sigma_1 \,|\, N_0'$. Since $N_1 \approx N_2$, $(N_1\sigma_1 \,|\, N_0')\mathcal{R}(N_2\sigma_1 \,|\, N_0')$.

2. Suppose $\mathcal{R} = \{((x)N_1, (x)N_2) \mid N_1 \approx N_2\}$. Then $\mathcal{R}$ is a bisimulation.
   - Suppose $(x)N_1 \longrightarrow N_{1\setminus x}'$ and $M_{1\setminus x}$ is the initial marking of $(x)N_1$. Then there is a marking $M_{1\setminus x}'$ of $(x)N_1$ such that $M_{1\setminus x}[t\rangle M_{1\setminus x}'$ for some $t$ and $M_{1\setminus x}^e = M_{1\setminus x}'^e$.
     - If $\forall s \in P^e_{N_1}.x \neq L(s)$, then $(x)N_1 \equiv N_1$ and $(x)N_2 \equiv N_2$. Then $N_1 \longrightarrow N_1'$ and $N_{1\setminus x}' \equiv N_1' \equiv (x)N_1'$. Since $N_1 \approx N_2$, $N_2 \Longrightarrow N_2'$ such that $N_1' \approx N_2'$. Hence $(x)N_2 \equiv N_2 \Longrightarrow N_2' \equiv (x)N_2'$ and $(x)N_1'\mathcal{R}(x)N_2'$.
     - If $\exists s \in P^e_{N_1}.x = L(s)$, then there are two cases:
       * If $\forall s \in {}^\bullet t \cup t^\bullet.x \neq L(s)$, then $N_1 \longrightarrow N_1'$ and $N_{1\setminus x}' \equiv (x)N_1'$. Since $N_1 \approx N_2$, $N_2 \Longrightarrow N_2'$ such that $N_1' \approx N_2'$. Hence $(x)N_2 \Longrightarrow (x)N_2'$ and $(x)N_1'\mathcal{R}(x)N_2'$.

* If $\exists s \in^\bullet t \cup t^\bullet.x = L(s)$, then $O = (^\bullet t \cup t^\bullet) \cap P^e_{N_1} \neq \emptyset$. We can conclude that $\forall s \in O \wedge L(s) \neq x.W(s,t) = W(t,s)$. If $(\exists s \in^\bullet t \cap t^\bullet.x = L(s))$ and $W(s,t) = W(t,s)$, then $N_1 \longrightarrow N'_1$ and $N'_{1\backslash x} \equiv (x)N'_1$. Since $N_1 \approx N_2$, $N_2 \Longrightarrow N'_2$ such that $N'_1 \approx N'_2$. Hence $(x)N_2 \Longrightarrow (x)N'_2$ and $(x)N'_1\mathcal{R}(x)N'_2$. Otherwise, if $(\exists s \in^\bullet t \cap t^\bullet.x = L(s))$ and $W(s,t) \neq W(t,s)$ or $(\forall s \in^\bullet t \cap t^\bullet.x \neq L(s))$, then $N_1 \xrightarrow{\tau} N'_1$ and $N'_{1\backslash x} \equiv (x)N'_1$. It is obvious that the external place $s_1$ of $N_1$, which satisfies $x = L(s_1)$, is the unique external one with $M_1(s_1) \neq M'_1(s_1)$. Since $N_1 \approx N_2$, $N_2 \Longrightarrow N_{21} \xrightarrow{\tau} N_{22} \Longrightarrow N'_2$ such that $N'_1 \approx N'_2$. Then there must be an external place $s_2$ of $N_2$, which satisfies $x = L(s_2)$, is the unique external one with $M_2(s_2) \neq M'_2(s_2)$. The change is induced by the transition $N_{21} \xrightarrow{\tau} N_{22}$. Hence $(x)N_2 \Longrightarrow (x)N_{21} \longrightarrow (x)N_{22} \Longrightarrow N'_2 \equiv (x)N'_2$ and $(x)N'_1\mathcal{R}(x)N'_2$.

– Suppose $(x)N_1 \xrightarrow{\tau} N'_{1\backslash x}$ and $M_{1\backslash x}$ is the initial marking of $(x)N_1$. Then there is a marking $M'_{1\backslash x}$ of $(x)N_1$ such that $M_{1\backslash x}[t\rangle M'_{1\backslash x}$ for some $t$ and $M^e_{1\backslash x} \neq M'^e_{1\backslash x}$.

  • If $\forall s \in P^e_{N_1}.x \neq L(s)$, then $(x)N_1 \equiv N_1$ and $(x)N_2 \equiv N_2$. Then $N_1 \xrightarrow{\tau} N'_1$ and $N'_{1\backslash x} \equiv N'_1 \equiv (x)N'_1$. Since $N_1 \approx N_2$, $N_2 \Longrightarrow N'_2$ such that $N'_1 \approx N'_2$. Hence $(x)N_2 \equiv N_2 \xrightarrow{\tau} N'_2 \equiv (x)N'_2$ and $(x)N'_1\mathcal{R}(x)N'_2$.

  • If $\exists s \in P^e_{N_1}.x = L(s)$, then there are two cases:

    * If $\forall s \in^\bullet t \cup t^\bullet.x \neq L(s)$, then $N_1 \xrightarrow{\tau} N'_1$ and $N'_{1\backslash x} \equiv (x)N'_1$. Since $N_1 \approx N_2$, $N_2 \xrightarrow{\tau} N'_2$ such that $N'_1 \approx N'_2$. Hence $(x)N_2 \Longrightarrow (x)N'_2$ and $(x)N'_1\mathcal{R}(x)N'_2$.

    * If $\exists s \in^\bullet t \cup t^\bullet.x = L(s)$, then $O = (^\bullet t \cup t^\bullet) \cap P^e_{N_1} \neq \emptyset$. We can conclude that $\exists s \in O \wedge L(s) \neq x.M_{1\backslash x}(s) \neq M'_{1\backslash x}(s)$. Hence $N_1 \xrightarrow{\tau} N'_1$ and $N'_{1\backslash x} \equiv (x)N'_1$. Since $N_1 \approx N_2$, $N_2 \Longrightarrow N_{21} \xrightarrow{\tau} N_{22} \Longrightarrow N'_2$ and $N'_1 \approx N'_2$. Hence $(x)N_2 \Longrightarrow (x)N_{21} \longrightarrow (x)N_{22} \Longrightarrow N'_2 \equiv (x)N'_2$ and $(x)N'_1\mathcal{R}(x)N'_2$.

# References

1. van der Aalst, W.M.P.: Pi calculus versus petri nets: let us eat humble pie rather than further inflate the pi hype (2003)
2. Baldan, P., Bonchi, F., Gadducci, F.: Encoding asynchronous interactions using open petri nets. In: Bravetti, M., Zavattaro, G. (eds.) CONCUR 2009. LNCS, vol. 5710, pp. 99–114. Springer, Heidelberg (2009). doi:10.1007/978-3-642-04081-8_8
3. Baldan, P., Corradini, A., Ehrig, H., Heckel, R.: Compositional modeling of reactive systems using open petri nets. In: Larsen, K.G., Nielsen, M. (eds.) CONCUR 2001. LNCS, vol. 2154, pp. 502–518. Springer, Heidelberg (2001). doi:10.1007/3-540-44685-0_34
4. Baldan, P., Corradini, A., Ehrig, H., Heckel, R.: Compositional semantics for open petri nets based on deterministic processes. Math. Struct. Comput. Sci. **15**(1), 1–35 (2005)

5. Baldan, P., Corradini, A., Ehrig, H., Heckel, R., König, B.: Bisimilarity and behaviour-preserving reconfigurations of open petri nets. In: Mossakowski, T., Montanari, U., Haveraaen, M. (eds.) CALCO 2007. LNCS, vol. 4624, pp. 126–142. Springer, Heidelberg (2007). doi:10.1007/978-3-540-73859-6_9
6. Baldan, P., Corradini, A., Ehrig, H., König, B.: Open petri nets: non-deterministic processes and compositionality. In: Ehrig, H., Heckel, R., Rozenberg, G., Taentzer, G. (eds.) ICGT 2008. LNCS, vol. 5214, pp. 257–273. Springer, Heidelberg (2008). doi:10.1007/978-3-540-87405-8_18
7. Best, E., Devillers, R., Hall, J.G.: The box calculus: a new causal algebra with multi-label communication. In: Rozenberg, G. (ed.) Advances in Petri Nets 1992. LNCS, vol. 609, pp. 21–69. Springer, Heidelberg (1992). doi:10.1007/3-540-55610-9_167
8. Best, E., Devillers, R., Koutny, M.: A unified model for nets and process algebras. In: Bergstra, J., Ponse, A., Smolka, S. (eds.) Handbook of Process Algebra, pp. 875–944. Elsevier Science, Amsterdam (2001)
9. Busi, N., Gorrieri, R.: A Petri net semantics for pi-calculus. In: Lee, I., Smolka, S.A. (eds.) CONCUR 1995. LNCS, vol. 962, pp. 145–159. Springer, Heidelberg (1995). doi:10.1007/3-540-60218-6_11
10. Busi, N., Gorrieri, R.: Distributed semantics for the -calculus based on Petri nets with inhibitor arcs. J. Logic Algebraic Programm. **78**(3), 138–162 (2009)
11. Bergstra, J.A., Klop, J.W.: The algebra of recursively defined processes and the algebra of regular processes. In: Paredaens, J. (ed.) ICALP 1984. LNCS, vol. 172, pp. 82–94. Springer, Heidelberg (1984). doi:10.1007/3-540-13345-3_7
12. Cao, M., Wu, Z., Yang, G.: Pi net - a new modular higher petri net. J. Shanghai Jiaotong Univ. **38**(1), 52–58 (2004). in Chinese
13. Devillers, R., Klaudel, H., Koutny, M.: Petri net semantics of the finite pi-calculus. In: Frutos-Escrig, D., Núñez, M. (eds.) FORTE 2004. LNCS, vol. 3235, pp. 309–325. Springer, Heidelberg (2004). doi:10.1007/978-3-540-30232-2_20
14. Degano, P., Nicola, R.D., Montanari, U.: A distributed operational semantics for CCS based on C/E systems. Acta Informatica **26**(1–2), 59–91 (1988)
15. Fu, Y., Lv, H.: On the expressiveness of interaction. Theoret. Comput. Sci. **411**, 1387–1451 (2010)
16. Fu, Y.: Theory of interaction. Theoret. Comput. Sci. **611**, 1–49 (2016)
17. Guo, X., Hao, K., Hou, H., Ding, J.: The representation of petri nets with prohibition arcs by Pi+ calculus. J. Syst. Simul. **S2**, 9–12 (2002)
18. Goltz, U.: CCS and petri nets. In: Guessarian, I. (ed.) LITP 1990. LNCS, vol. 469, pp. 334–357. Springer, Heidelberg (1990). doi:10.1007/3-540-53479-2_14
19. Hao, K.: Open nets - a model for interative concurrent systems. J. Northwest Univ. (Nat. Sci. Ed.) **27**(6), 461–466 (1997)
20. Hoare, C.A.R.: Communicating Sequential Processes. Commun. ACM **21**(8), 666–677 (1978)
21. Koutny, M., Best, E.: Operational and denotational semantics for the box algebra. Theoret. Comput. Sci. **211**(1–2), 1–83 (1999)
22. Koutny, M., Esparza, J., Best, E.: Operational semantics for the petri box calculus. In: Jonsson, B., Parrow, J. (eds.) CONCUR 1994. LNCS, vol. 836, pp. 210–225. Springer, Heidelberg (1994). doi:10.1007/978-3-540-48654-1_19
23. Kindler, E.: A compositional partial order semantics for Petri net components. In: Azéma, P., Balbo, G. (eds.) ICATPN 1997. LNCS, vol. 1248, pp. 235–252. Springer, Heidelberg (1997). doi:10.1007/3-540-63139-9_39
24. Liu, G., Jiang, C., Zhou, M.: Interactive petri nets. IEEE Trans. Syst. Man Cybern. Syst. **43**(2), 291–302 (2013)

25. Milner, R.: A Calculus of Communicating systems. LNCS. Springer, Berlin (1980)
26. Meseguer, J., Montanari, U.: Petri nets are monoids. Inf. Comput. **88**, 105–155 (1990)
27. Milner, R., Parrow, J., Walker, D.: A calculus of mobile processes. Inf. Comput. 100, 1–40(Part I), 41–77 (Part II) (1992)
28. Nielsen, M.: CCS — and its relationship to net theory. In: Brauer, W., Reisig, W., Rozenberg, G. (eds.) ACPN 1986. LNCS, vol. 255, pp. 393–415. Springer, Heidelberg (1987). doi:10.1007/3-540-17906-2_32
29. Nielsen, M., Priese, L., Sassone, V.: Characterizing behavioural congruences for Petri nets. In: Lee, I., Smolka, S.A. (eds.) CONCUR 1995. LNCS, vol. 962, pp. 175–189. Springer, Heidelberg (1995). doi:10.1007/3-540-60218-6_13
30. Peterson, J.L.: Petri Net Theory and the Modelling of Systems. Prentice Hall, Englewood Cliffs (1981)
31. Priese, L., Wimmel, H.: A uniform approach to true-concurrency and interleaving semantics for Petri nets. Theoret. Comput. Sci. **206**(1–2), 219–256 (1998)
32. Winskel, G.: Event structures. In: Brauer, W., Reisig, W., Rozenberg, G. (eds.) ACPN 1986. LNCS, vol. 255, pp. 325–392. Springer, Heidelberg (1987). doi:10.1007/3-540-17906-2_31
33. Yu, Z., Cai, Y., Xu, H.: Petri net semantics for Pi Calculus. Control Decis. **22**(8), 864–868 (2007). in Chinese

# Divergence Detection for CCSL Specification via Clock Causality Chain

Qingguo Xu[1,2(✉)], Robert de Simone[3], and Julien DeAntoni[3]

[1] School of Computer Engineering and Science, Shanghai University,
No. 99, Shangda Road, Shanghai 200444, China
qgxu@mail.shu.edu.cn
[2] Shanghai Key Laboratory of Computer Software Testing and Evaluating,
Shanghai 201114, China
[3] Inria Sophia Antipolis Méditerranée AOSTE,
University of Nice Sophia Antipolis, I3S, 06902 Sophia Antipolis Cedex, France
Robert.de_Simone@inria.fr,
Julien.Deantoni@polytech.unice.fr

**Abstract.** The Clock Constraint Specification Language (CCSL), first introduced as a companion language for Modeling and Analysis of Real-Time and Embedded systems (MARTE), has now evolved beyond the time specification of MARTE, and has become a full-fledged domain specific modeling language widely used in many domains. A CCSL specification is a set of constraints, which symbolically represents a set of valid clock schedules, where a schedule represents the order of the actions in a system. This paper proposes an algorithm to detect the divergence behavior in the schedules that satisfy a given CCSL specification (*i.e.* it proposes to detect the presence of infinite but non periodic schedules in a CCSL specification). We investigate the divergence by constructing causality chains among the clocks resulting from the constraints of the specification. Depending on cycles in the causality chains, a bounded clock set built by our proposed algorithm can be used to decide whether the given specification is divergence-freedom or not. The approach is illustrated on one example for architecture-driven analysis.

**Keywords:** CCSL · Divergence · Clock causality chain · Bounded Clock Set · PVS

## 1 Introduction

The Unified Modeling Language (UML) Profile for Modeling and Analysis of Real-Time and Embedded systems (MARTE) [1], adopted in November 2009, has introduced a *Time Model* [2] that extends the informal Simple Time of UML2. This time model is general enough to support different forms of time (discrete or dense,

This work is supported by the Natural Science Foundation of China (Grant No. 61572306, 61502294).

M. Fränzle et al. (Eds.): SETTA 2016, LNCS 9984, pp. 18–37, 2016.
DOI: 10.1007/978-3-319-47677-3_2

chronometric or logical). Its so-called clocks allow enforcing as well as observing the occurrences of events and the behavior of annotated UML elements. The *Time Model* comes with a companion language named the Clock Constraint Specification Language (CCSL) [3] defined in the annex of the MARTE specification. Initially devised as a language for expressing constraints between clocks of a MARTE model, CCSL has evolved and has been developed independently of the UML. CCSL is now equipped with a formal semantics [3] and is supported by a software environment (TimeSquare [4]) that allows for the specification, solving, and visualization of clock constraints.

MARTE promises a general modeling framework to design and analyze systems. Lots of works have been published on the modeling capabilities offered by MARTE, much less on verification techniques supported. Inspired by the works about state-based semantics interpretation for the kernel CCSL operators [5], this paper focuses on the **divergence** (see Sect. 3.1) detection of some CCSL specifications. This issue was addressed by [6, 7] but their propositions were applying parallel composition of individual CCSL constraints, making the propositions unsuitable for industrial size systems. In this work, we significantly reduce the complexity of the divergence detection by constructing and analyzing clock causality chains. Additionally our algorithm can point out what constraint can be added to make the specification divergence-free. In order to acquire clock causality chains, we first highlighted some interesting properties about causal relation between clocks. Furthermore, we propose an algorithm to decide if a given CCSL is divergence-free or not, by constructing the proposed a "Bounded Clock Set" (BCS) based on clock delay expression as well as the causality relation between clocks.

Section 2 introduces a state-transition based semantics for CCSL. Section 3 shows how to detect the divergence and make sure the specification is divergence-freedom based on the notion of divergence of CCSL specifications. Also, an algorithm, which is used to build the bounded clock set for determining the convergence, is depicted in this section. Section 4 illustrates, by using an example from architecture-driven analysis, the use of our algorithm on a CCSL specification. It shows how to improve a divergent specification such that it becomes a convergent one by adding clock constraints hinted by the algorithm. Finally, Sect. 5 makes a comparison with related works and Sect. 6 concludes the contribution and outlines some future works.

## 2 Preliminaries

This section briefly introduces the logical time model [2] of MARTE and the Clock Constraint Specification Language (CCSL). A technical report [3] and it latest update [8] describes the syntax and the semantics of a kernel set of CCSL constraints. We describe in this paper only the constraints that are used for our discussion.

The notion of multiform logical time has first been used in the theory of synchronous languages [9] and its polychronous extensions. CCSL is a formal declarative language to specify polychronous clock specification. It provides a concrete syntax to make the clocks and clocks constraints first-class citizens of UML-like models. Clocks in CCSL are used to measure the number of occurrences of events in a system. Logical clocks replace physical times by a logical sequencing. A CCSL specification do not need for clocks to be relative to a global physical time. They are by default independent of each other and what matter is the partial ordering of their ticks induces by the constraints between them.

A clock belongs to a clock set $\mathcal{C}$. During the execution of a system, an execution step is defined and at a given step, every clock in $\mathcal{C}$ can tick or not according to the constraints defined in the specification. A schedule captures what happens during one particular execution.

**Definition 1 (Schedule):** A *schedule* is defined as a function *Sched*: $\mathbb{N}_{>0} \to 2^{\mathcal{C}}$. ■

Given an execution step $i \in \mathbb{N}_{>0}$, and a schedule $\sigma \in$ *Sched*, $\sigma(i)$ denotes the set of clocks that tick at step $i$.

For a given schedule, the configurations of the clocks tell us the advance of the clocks, relative to the others.

**Definition 2 (Clock Configuration):** For a given schedule $\sigma$, clock $c \in \mathcal{C}$ and a natural number $n \in \mathbb{N}$, the configuration $\chi_\sigma$: $\mathcal{C} \times \mathbb{N} \to \mathbb{N}$ is defined recursively as:

$$\chi_\sigma(c, n) = \begin{cases} 0, & \text{if } n = 0 \\ \chi_\sigma(c, n-1), & \text{if } c \notin \sigma(n) \\ \chi_\sigma(c, n-1) + 1, & \text{if } c \in \sigma(n) \end{cases} \tag{F.1}$$

■

For a clock $c \in$, and a step $n \in \mathbb{N}$, $\chi_\sigma(c, n)$ counts the number of times the clock $c$ has ticked at step $n$ for the given schedule $\sigma$.

CCSL is used to specify a set of valid schedules. There is usually more than one valid schedule that satisfies a given specification. If there is no satisfying schedule, then we say that the specification is ill-formed. The detail properties about the schedules against the given specification is investigated in [10]. Clocks can be finite of infinite. Since divergence problem only makes sense on infinite clocks, we do not care about constraints that make clock terminating (see [8] for details). Consequently, every clock in $\mathcal{C}$ are infinite (will never terminate), i.e., $\forall c \in \mathcal{C}$, $\chi_\sigma(c, n)$ is boundless with $n$ increasing.

**Definition 3 (CCSL Specification):** A CCSL specification $\mathcal{SPEC}$ is a pair <$\mathcal{C}$, *CConstr*> , where $\mathcal{C}$ is a set of clocks, *CConstr* is a set of formulae (see Definition 5) used to specify the relations among the clocks in the set $\mathcal{C}$. ■

**Definition 4 (Clock Set)** An element in the clock set $\mathcal{C}$ can be given by the specification writer explicitly (**explicit clock**), or by one of the following clock expressions (**implicit clock**):

$Clock := a + b \mid a * b \mid a \backslash b \mid sup(a, b) \mid inf(a, b) \mid a\$n \mid SampledOn(a, b) \mid FilteredBy(a, u, v)$

$$\tag{F.2}$$

where $a$, $b \in C$ are clocks, $u$, $v \in (0 + 1)^*$ are finite binary words, and $n \in \mathbb{N}$ is a natural number.    ■

Once we write one clock expression in the form of (F.2), a new clock is created and added into the clock set $C$. For example, if we give the explicit clock set $\{a, b, c\}$, and clock expression set $\{d = a + b, e = c \, \$1\}$, then the considered clock set is $C = \{a, b, c, d, e\}$. Note that there may not be any given name for the implicit clock of the clock expression (*e.g.* if it occurs in one clock relation) (see Definition 5). It should be noted that the new clock will be scheduled depending on the clock(s) that occur in that expression.

It is convenient to define a clock as an Abstract Data Type (ADT) [11] in Prototype Verification System (PVS) [12, 13] if we treat the clock expression as an element in a clock set instead of assigning them a new clock name.

The CCSL constraint defined over the clock set includes both the **explicit** clocks and **implicit** ones.

**Definition 5 (CCSL Relation):** For a given clock set $C$, the corresponding clock relation set $\Phi(C)$ over $C$ is defined recursively as:

$$\psi := a \prec b \mid a \preccurlyeq b \mid a \subset b \mid a \,\#\, b \mid \psi \wedge \psi$$
where $a, b \in C$.    ■

Every clock relation in the set $\Phi(C)$, is a primitive formula that relates a clock pair or their conjunction.

**Definition 6 (CCSL Specification Satisfaction).** For a given CCSL specification $\mathcal{SPEC} = \,<C, CConstr>$, A schedule $\sigma$ over $C$ satisfies $\mathcal{SPEC}$, denoted $\sigma \vDash \mathcal{SPEC}$, if and only if for every formulae $r$ in $CConstr$, $\sigma$ evaluates to true according to the following definition postulated by the CCSL semantics:

$\sigma \vDash r$ *if and only if cases $r$'s form of*

| | | |
|---|---|---|
| $a \prec b : \forall n \in \mathbb{N},$ | $\chi_\sigma(a, n) = \chi_\sigma(b, n) \Rightarrow b \notin \sigma(n + 1)$ | (**Precedence**) |
| $a \preccurlyeq b : \forall n \in \mathbb{N},$ | $\chi_\sigma(a, n) \geq \chi_\sigma(b, n)$ | (**Causality**) |
| $a \subset b : \forall n \in \mathbb{N}_{>0},$ | $a \in \sigma(n) \Rightarrow b \in \sigma(n)$ | (**Subclock**) |
| $a \,\#\, b : \forall n \in \mathbb{N}_{>0},$ | $a \notin \sigma(n) \vee b \notin \sigma(n)$ | (**Exclusion**) |
| $\psi_1 \wedge \psi_2 : \sigma \vDash \psi_1 \wedge \sigma \vDash \psi_2$ | | (**Conjunction**) |

$$(\text{F.3})$$

where $a, b \in C$.    ■

It's straightforward to prove that both *Causality* and *Subclock* are pre-orders on $C$, i.e., they are reflexive and transitive. For simplicity, we can write $a \preccurlyeq b \preccurlyeq c$ for $a \preccurlyeq b \wedge b \preccurlyeq c$, and so do other transitive clock relation. The transitivity property of the *Causality* relation is of importance in determining the boundedness of a specification (see Sect. 3.2). It is also straightforward to prove that *Exclusion* is neither reflexive nor transitive. The transitive property of *Precedence* is very tedious to prove from this form of definition, although it is obvious in other semantics model [8].

The *implicit clocks* defined using clock expressions (F.2), are constrained according to the parameters of the clock expression. In other words, a clock expression is a clock generator where the output clock ticks or not according to the input clock(s) state and other arguments, if any.

**Definition 7 (Clock Expression Satisfaction).** Whether an implicit clock (denoted $c$ in the following) can tick or not in a schedule $\sigma$, is determined by the behaviors of the input clock(s) in $\sigma$,

$\sigma \vDash c$ *if and only if cases $c$ is defined by*

$$a + b : \forall n \in \mathbb{N}_{>0}, c \in \sigma(n) iff a \in \sigma(n) \vee b \in \sigma(n) \quad \textbf{(Union)}$$
$$a * b : \forall n \in \mathbb{N}_{>0}, c \in \sigma(n) iff a \in \sigma(n) \wedge b \in \sigma(n) \quad \textbf{(Intersection)}$$
$$a \backslash b : \forall n \in \mathbb{N}_{>0}, c \in \sigma(n) iff a \in \sigma(n) \wedge b \notin \sigma(n) \quad \textbf{(Minus)}$$
$$sup(a, b) : \forall n \in \mathbb{N}, \chi_\sigma(c,n) = \min(\chi_\sigma(a,n), \chi_\sigma(b,n)) \quad \textbf{(Supremum)}$$
$$inf(a, b) : \forall n \in \mathbb{N}, \chi_\sigma(c,n) = \max(\chi_\sigma(a,n), \chi_\sigma(b,n)) \quad \textbf{(Infimum)}$$
$$a\$d : \forall n \in \mathbb{N}, \chi_\sigma(c,n) = \max(\chi_\sigma(a,n) - d, 0) \quad \textbf{(Delay)}$$

(F.4)

*SampledOn(a, b)*: $\forall n \in \mathbb{N}_{>0}, c \in \sigma(n) iff$

$$(b \in \sigma(n) \wedge (\exists j \in [1..n], a \in \sigma(j) \wedge \forall m : [j..n - 1], b \notin \sigma(m))) \quad \textbf{(SampledOn)}$$

*FilteredBy (a, u, v)*: $\forall n \in \mathbb{N}_{>0}, c \in \sigma(n) iff$

$$(a \in \sigma(n) \wedge (if k \leq |u| \text{then } u[k] = 1 \; else \; v[(k - |u|) mod |v|] = 1), where \, k = \mathcal{X}_\sigma(a, n))$$

**(FilteredBy)**

where $a, b \in C$, $u, v \in (0 + 1)^*$, and $d \in \mathbb{N}_{>0}$. $|u|$ (resp. $|v|$) represents the length of binary word $u$ (resp. $v$), $k = \chi_\sigma(a, n)$ is the number of tick of clock $a$ from step 1 to $n$. ∎

By composing the relations and the expressions provided in Definitions 4 and 5, some user-defined clock relations can be further defined, as below:

$$a \sim b := a \prec b \prec a \$ 1 \quad \textbf{(Alternation)} \tag{F.5}$$

$$a \prec_n b := a \prec b \prec a \$ n \quad \textbf{(Bounded precedence)} \tag{F.6}$$

$$a \equiv b := a \preccurlyeq b \wedge b \preccurlyeq a \quad \textbf{(Coincidence)} \tag{F.7}$$

Obviously, *Alternation* (F.5) is a special case of *bounded precedence* (F.6). *Alternation* (which is frequently used in the CCSL specifications), is a kind of bounded relation that is discussed in detail in Sect. 3.

From the definitions above, some proved propositions can be listed below. The reader who is interested the proof can get the details in the report [5].

**Proposition 1 (Precedence Implies Causality).** The Precedence is a stronger form of causality:

$$\sigma \vDash a \prec b \Longrightarrow \sigma \vDash a \preccurlyeq b \; \blacksquare$$

**Proposition 2 (Subclock Implies Causality).** When $a$ is a *Subclock* of $b$, then $b$ is faster than $a$:

$$\sigma \vDash a \subset b \Longrightarrow \sigma \vDash b \preccurlyeq a \blacksquare$$

**Proposition 3 (Union and Causality).** The union of two clocks is faster than both clocks:

$$\sigma \vDash u \coloneqq a + b \Longrightarrow \sigma \vDash u \preccurlyeq a \wedge \sigma \vDash u \preccurlyeq b \blacksquare$$

**Proposition 4 (Intersection and Causality).** The intersection of two clocks is slower than both clocks:

$$\sigma \vDash i \coloneqq a * b \Longrightarrow \sigma \vDash a \preccurlyeq i \wedge \sigma \vDash b \preccurlyeq i \blacksquare$$

**Proposition 5 (Infimum and Causality).** The infimum of two clocks is always faster than both clocks:

$$\sigma \vDash f \coloneqq inf(a, b) \Longrightarrow \sigma \vDash f \preccurlyeq a \wedge \sigma \vDash f \preccurlyeq b \blacksquare$$

**Proposition 6 (Supremum and Causality).** The supremum of two clocks is always slower than both clocks:

$$\sigma \vDash s \coloneqq sup(a, b) \Longrightarrow \sigma \vDash a \preccurlyeq s \wedge \sigma \vDash b \preccurlyeq s \blacksquare$$

**Proposition 7 (Delay and Causality).** The delayed clock is always slower than the base clock:

$$\sigma \vDash c \coloneqq a \, \$ \, d \Longrightarrow \sigma \vDash a \preccurlyeq c \blacksquare$$

Propositions 1 to **7** defines new implicit *Causality* relations from all the other relations (*Precedence* and *Subclock*) and expressions (*Union, Intersection, Infimum, Supremum* and *Delay*). The Definitions 1 to **7** have been formalized in PVS, and the Propositions 1 to **7** have been proved (in most cases by induction).

Based on the definitions of *SampledOn* and *FilteredBy* in (F.3), $c \coloneqq SampledOn(a, b)$ implies $a \preccurlyeq c$ and $c \subset a$, $c \coloneqq$ FilteredBy$(a, u, v)$ implies $c \subset a$. Hence, some *Causality* relations between the new implicit clock and their input clock can indirectly be deduced from the clock expression *SampledOn* and *FilteredBy*. Therefore, we can also get *Causality* relations using *SampledOn* and *FilteredBy*, besides using Propositions 1 to **7**.

# 3  Divergence Detection

## 3.1  Divergence and Bounded Relation

Let's consider a very simple CCSL specification $\mathcal{SPEC}_1 = <\{sending, ack\}, \{sending \preccurlyeq ack\}>$. It is **divergent** as we don't know what will happen at a certain execution step $i$ in a schedule $\sigma$ provided that $\sigma$ satisfies $\mathcal{SPEC}_1$ from step 1 to step $i - 1$. That is to say, there are many possibilities of clock ticking (tick *sending*, tick *ack*, or both) to ensure the satisfaction of $\mathcal{SPEC}_1$ at step $i$. Furtherly, whether the clock *ack* ticks or not is uncontrollable and unpredicatable since there is nothing to trigger its firing. Therefore, some expected properties may eventually not be guaranteed, if it is implied by the implementation of *ack*'s tick.

**Definition 8 (Divergent Specification).** We say a CCSL specification $SPEC = <C,$ $CConstr>$ is divergent, if an expected clock in $C$ has the possibility never ticks, formally,

$\exists \sigma, (\sigma \vDash SPEC \wedge \exists g, r \in, \forall i: \{k: \mathbb{N}_{>0} | r \in \sigma(k)\}, \exists j > i, g \notin \sigma(j))$.    ∎

Here $g$ (means *goal*) is the expected action that may be an acknowledge signal/operation of other clock $r$ (means *request*). That is to say, after performing some necessary operation(s) by ticking the some clocks in $C\backslash\{g\}$ containing clock $r$, there must be an expected final result (ticking clock $g$) appears in the admissible future, but unfortunately only God knows $g$ will tick or not in the divergent specification $SPEC$. Therefore, the divergence of $SPEC_\infty$ is asserted by choosing $g = ack, r = sending$ and the given schedule $\sigma$ such that $\forall i \in \mathbb{N}_{>0}, \sigma(i) = \{sending\}$.

On the contrary, another example $SPEC_\in = <\{a, b\}, \{a \sim b\}>$ is not divergent because the constraint $a \sim b$ postulates that the schedule that satisfies $SPEC_\in$ must execute by alternating $a$ with $b$ forever. In this case, we say the behavior and the corresponding CCSL specification are **convergent** if it is also free of some unexpected properties, such as deadlock and livelock and so on.

The divergent specification behavior is not predicable in the sufficient future. It is possible for some expected action(s) to be delayed infinitely in the future. This *bad* behavior is unexpected since some actions can never happen after a certain simulation step. Therefore, a specification that contains divergence is unsafe. How to detect **divergence** existence in a given CCSL specification is the main subject in this section. Additionally, we provide some suggestions to modify the CCSL specification so that it becomes divergence-free.

For a given CCSL specification $<C, CConstr>$, if the difference between the speeds of two clocks $a, b \in C$ is limited in an allowed boundary, we say the clock pair $(a, b)$ has a **bounded relation**.

**Definition 9 (Bounded Relation).** For a given clock set $C$, two clocks $a, b \in C$, and a schedule $\sigma$ over $C$, $a$ and $b$ has the bounded relation with a given boundary $m \in \mathbb{N}$, denotes $|a, b| \leq m$:

$\sigma \vDash |a, b| \leq m$ *iff* $\forall i \in \mathbb{N}_{>0}, \exists j \in \mathbb{N}_{>0}, | \mathcal{X}_\sigma(a, i) - \mathcal{X}_\sigma(b, j) | \leq m$

$m$ (resp. $- m$) can be called *upper* (resp. *lower) bound*.    ∎

When such an unbounded clock pair is found, we say that there is **divergence** in the specification. We say the specification free of divergence is a bounded specification.

**Definition 10 (Bounded Specification).** For a given CCSL specification $SPEC = <C,$ $CConstr>$, $\forall a, b \in C$, $SPEC$ is **bounded** if and only if any clock pair has the **bounded relation**:

$\forall \sigma, \sigma \vDash SPEC \implies \exists m \in \mathbb{N}, \sigma \vDash |a, b| \leq m$    ∎

A bounded specification is divergence-free because there are no possible boundless drifts between any clock pair (*i.e.* between the actions).

If we check every clock pairs among all clocks in $C$ to decide whether a specification is divergence-free or not, there are $\binom{|C|}{2} = |C| \times (|C| - 1)/2$ pairs need to be checked. The number of checks then totals to $(|C|^2)$. But in practice many checks can be safely neglected when the bounded relation is implied by the already checked one.

The **Bounded Relation** can directly be derived from the formula for most of the constraints. Let us show how to get the **Bounded Relation** between two clocks implied by the existing clock relations and expressions.

**Bounded Relation** is an equivalence relation over $C$, i.e., it is reflexive, symmetric and transitive. Note that they do not necessarily have the same boundaries for different bounded relations.

**Proposition 8 (Bounded Extension).** The **Bounded Relation** is transitive and the transitive resulting boundary is the sum of the original ones.

$$|a, b| \leq m_1 \wedge |b, c| \leq m_2 \implies |b, c| \leq m_1 + m_2 \qquad \blacksquare$$

Proposition 8 can be proved by using classical properties of inequalities addition. The proof is obvious and omitted here.

**Proposition 9 (Bounded Restriction).** Bounded relation can be restricted w.r.t. *Causality*, and the resulting boundary is not greater than the original one.

$$|a, b| \leq m \wedge \forall c \in C, (a \preccurlyeq c \preccurlyeq b \vee b \preccurlyeq c \preccurlyeq a) \implies |a, c| \leq m \wedge |b, c| \leq m \qquad \blacksquare$$

*Proof of **Proposition 9**:*
*Let $maxDrift\_ab(n) = \chi_\sigma(a, n) - \chi_\sigma(b, n)$,*
*$maxDrift\_ac(n) = \chi_\sigma(a, n) - \chi_\sigma(c, n)$,*
*$maxDrift\_bc(n) = \chi_\sigma(b, n) - \chi_\sigma(c, n)$, for all $n \in \mathbb{N}$,*
*from the* Definition 9, *we have*
*$|a, b| \leq m \implies |maxdrift\_ab(n)| \leq m \implies -m \leq maxdrift\_ab(n) \leq m$, for all $n \in \mathbb{N}$,*
*Assume $a \preccurlyeq c \preccurlyeq b$, from the definition 6, we have $\forall n \in \mathbb{N}$, $\chi_\sigma(a, n) \geq \chi_\sigma(c, n) \geq \chi_\sigma(b, n)$, then $maxDrift\_ac(n) \geq 0 \geq -m$.*

$$from\ maxdrift\_ab(n) \leq m \qquad (1)$$

$$maxdrift\_bc(n) \leq 0 \qquad (2)$$

*(1) + (2) gets*
*$maxDrift\_ac(n) = maxdrift\_ab(n) + maxdrift\_bc(n) \leq 0 + m = m$*
*therefore, $|maxDrift\_ac(n)| \leq m$.*
*similarly, from (2) we have $\forall n \in \mathbb{N}$, $maxDrift\_bc(n) \leq 0 \leq m$,*

*— $maxDrift\_bc(n) \leq maxdrift\_ab(n) \leq m \implies maxDrift\_bc(n) \geq -m$*

There exists the similar proof under the cases $b \preccurlyeq c \preccurlyeq a$. $\qquad \blacksquare$
From Definition 7, we can see *Delay* implies the *Bounded Relation*:

**Proposition 10 (Delay Implies Bounded).** The delayed clock $a$ and the corresponding base clock $b$ has the **Bounded Relation** with a lower bound $0$ and a upper bound $d$:
$$\sigma \models a := b \$ d \implies \sigma \models |a, b| \leq d \qquad \blacksquare$$

Proof of **Proposition 10**: This is direct conclusion from Definition 7.

The primitive clock relations and the other clock expressions except *Delay* and *FilteredBy* don't imply bounded relation. But their composition may deduce.

**Proposition 11 (Alternate Implies Bounded).** The clock pair involves in *Alternation relation* has the **bounded relation** with the boundary 1:
$$\sigma \vDash a \sim b \Longrightarrow \sigma \vDash |a, b| \leq 1 \qquad \blacksquare$$

Proof of **Proposition 11**:

$a \sim b \Leftrightarrow$ by **Alternation** definition (F.5)

$a \preccurlyeq b \preccurlyeq a \$ 1 \Longrightarrow$ *by Proposition 1 (Precedence Implies Causality)*

$a \preccurlyeq b \preccurlyeq a \$ 1 \Longrightarrow$ *by Proposition 10 (Delay Implies Bounded)*

$|a, a \$ 1| \leq 1 \Longrightarrow$ *by Proposition 9 (Bounded Restriction)*

$|a, b| \leq 1 \wedge |a \$ 1, b| \leq 1 \Longrightarrow$ *by proposition calculus*

$|a, b| \leq 1$ $\qquad \blacksquare$

Similarly, one can prove *Bounded Precedence* (F.6) is also a **Bounded Relation**. *FilteredBy* expression $c := a \blacktriangledown(u.v^{\omega})$, which defines a clock new clock $c$ as a **Subclock** of $a$ according to two binary words $u$ and $v$, implies a *Bounded Relation* between clock pair $(c, a)$ if there exists at least one 1-bit in the periodical pattern $v$.

**Proposition 12 (FilteredBy may Imply Bounded).** The clock pair $(a, c)$ involves in the expression $c := a \blacktriangledown(u.v^{\omega})$ has the **Bounded Relation** with the boundary $|u| + p \times (|v| - 1)$ if and only if $\exists i \in [1..|v|]$ such that $v[i] = 1$. Where $p$ is the number of periodical patterns that have passed from the initial configuration. $\qquad \blacksquare$

Proof of **Proposition 12**:

*Suppose a schedule $\sigma$ against by $c := a \blacktriangledown(u.v^{\omega})$, For some schedule step m, suppose clock a has ticked $|u| + |v|$ times from start, then*

$$\chi_{\sigma}(a, m) = |u| + |v|$$

*By **FilteredBy** definition (F.4), during the process, clock a ticks at step j if and only if both a ticks and $u.(v)^{\omega}[j] = 1$. Because $\exists i \in [1..|v|]$, $v[i] = 1$, we get:*

$$1 \leq \chi_{\sigma}(c, m) \leq |u| + |v|$$

*With the passage of schedule $\sigma$, $\chi_{\sigma}(c, n)$ will increase at least one while $\chi_{\sigma}(a, n)$ increase every $|u| + |v|$ after passing a periodical pattern. Therefore, when the schedule $\sigma$ reaches step n by the time p periodical patterns has passed:*

$$\chi_{\sigma}(a, n) = |u| + p \times |v| \text{ and } p \leq \chi_{\sigma}(c, n) \leq |u| + p \times |v| \Longrightarrow |\chi_{\sigma}(a, n) - \chi_{\sigma}(c, n)| \leq |u| + p \times (|v| - 1)$$ $\qquad \blacksquare$

### 3.2   Clock Causality Chain and Bounded Clock Set

In order to determine whether a CCSL specification is divergent or not, according to Definition 9, one must show *every* clock pair has *Bounded Relation*. Due to most clock constraints without the bound information, we try to capture the *Bounded Relation* from the *Causality* relation.

**Definition 11 (Clock Causality Chain).** For a given clock set $C$, some clocks in $C$ may form a Clock Causality Chain (CCC), which is a finite sequence $c_1, c_2, ..., c_n$ such that $\forall i \in [1..n-1], c_i \preccurlyeq c_{i+1}$ and $n \geq 2$. It is called Bounded Clock Causality Chain (BCCC) if $\exists m \in \mathbb{N}, |c_1, c_n| \leq m$, and $m$ is called the chain's boundary. It is an Unbounded CCC (UCCC) otherwise. ∎

**Proposition 13 (BCCC Boundedness).** ABCCC $\rho = c_1, c_2, ..., c_n$ with a boundary $m$ implies that any two clocks in $\rho$ has the **Bounded Relation**: $\forall i, j \in [1..n], |c_i, c_j| \leq m$. ∎

Proposition 13 is easily proved by using Proposition 9 **(Bounded Restriction)** as well as the *Causality* transition.

Due to Proposition 13, Checking $\binom{n}{2}$ clock pairs bounded relation is replaced by checking only one clock pair and additionally sorting $n$ clocks with respect to *Causality* relation. Moreover, the *Causality* relation is much easier to get than Bounded Relation as the former is implied by some clock constraints (see Propositions 1–7).

Obviously, a specification is divergent if we can get a UCCC.

**Theorem 1 (UCCC Implies Divergence).** A CCSL specification $\mathcal{SPEC} = <\mathcal{C},$ $CConstr>$ is divergent, if there exists a UCCC $\rho = c_1, c_2, ..., c_n$ ($\forall i \in [1..n], c_i \in C$) induced by the *CConstr*. ∎

**Proof of Theorem 1:**

$\rho$ *is unbounded* $\Longrightarrow \forall i \in [1..n-1], c_i \preccurlyeq c_{i+1} \Longrightarrow$ *That $c_n$ doesn't tick forever in some schedule $\sigma$ has nothing about asserting the true value of $\sigma \models \mathcal{SPEC}$.*

*We can find such a schedule $\sigma$ against $\mathcal{SPEC}$ that $c_1$ tick at least once. $\sigma$ is indeed the witness ($g = c_n$ and $r = c_1$) for uncovering the divergence of $\mathcal{SPEC}$ according to Definition 8.* ∎

*Let us illustrate* Theorem 1 *on a toy example,* $\mathcal{SPEC}_s = <\mathcal{C}, CConstr>$, *where* $\mathcal{C} = \{a, b, sup(a, b), c\}$ and $CConstr = \{a \sim c, sup(a, b) \preccurlyeq c\}$. Let $sup = sup(a, b)$, we get

$|a, c| \leq 1$, by Proposition 11

$a \prec b$ and $a \preccurlyeq b$, by **Alternation** definition and Proposition 1.

$a \preccurlyeq sup \preccurlyeq c$ and $b \preccurlyeq sup$, by Proposition 6.

The Causality Clock Graph (CCG) [5] in Fig. 1 shows the *Causality* relations among clocks. The *Causality* line from clock $a$ to $c$ can be safely removed because of causality transition.

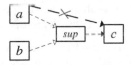

**Fig. 1.** CCG for $\mathcal{SPEC}_s$

The CCC $a$, $sup$, $c$ is a BCCC, while the CCC $\rho = b$, $sup$ is a UCCC. The existence of $\rho$ draws the conclusion $\mathcal{SPEC}_s$ is a divergent specification.

Via the analysis of $\mathcal{SPEC}_s$, Theorem 1 tell us that the existence of a UCCC in a CCSL specification witnesses its divergence. Unfortunately, this is a sufficient, but not necessary, condition for deciding specification's divergence. We say nothing about the convergence even if we get one or more BCCCs from a CCSL specification. The **Bounded Clock Set** can be used to determine a specification's convergence.

**Definition 12 (Bounded Clock Set).** For a given clock set $\mathcal{C}$, a clock set $\mathfrak{B} = \{c_\perp, c_1, ..., c_m, c^\top, c_{m+1}..., c_{m+n}\}$, subset of $\mathcal{C}$, which contains at least 2 elements, is called a Bounded Clock Set (BCS), if all the following conditions hold:

(i)  **(Lower-upper Bound)** $\exists d \in \mathbb{N}, |c_\perp, c^\top| \leq d$,
(ii) **(Causality Bound)** $\forall i \in [1..m], c_\perp \preccurlyeq c_i \preccurlyeq c^\top$,
(iii) **(Absence Unbounded Clock)** $\forall i \in [m+1..m+n], \exists c \in \mathfrak{B}_R$ *(see below)*, $d \in \mathbb{N}, |c, c_i| \leq d$.

The clock $c_\perp$ (resp. $c^\top$) is called the bottom (resp. top) clock. The subset $\mathfrak{B}_R = \{c_\perp, c_1, ..., c_m, c^\top\}$ of $\mathfrak{B}$ is called Causal Bounded Clock Set (CBCS). ∎

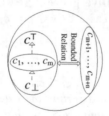

**Fig. 2.** The BCS structure

A BCS $\mathfrak{B} = \{c_\perp, c_1, ..., c_m, c^\top, c_{m+1}..., c_{m+n}\}$, as shown in Fig. 2, includes two disjoint subsets:

- $\mathfrak{B}_R = \{c_\perp, c_1, ..., c_m, c^\top\}$ contains the fastest clock $c_\perp$ as the bottom, and the slowest clock $c^\top$ as the top, while all other ones' speed lies between $c_\perp$ and $c^\top$. Note that $m = 0$ in some cases.
- $\mathfrak{B} \backslash \mathfrak{B}_R = \{c_{m+1}..., c_{m+n}\}$ We don't care about the speed of these clocks in relation to the one in $\mathfrak{B}_R$. But a common fact is that all of them must be bounded by speed of one of the clocks in $\mathfrak{B}_R$. Note that $n = 0$ in some cases, i.e., this set is an empty set.

The bottom (resp. top) clock is probably not a "proper" bottom (resp. top) clock because maybe there exists a clock in $\mathfrak{B} \backslash \mathfrak{B}_R$ which is faster (resp. slower) than it.

**Theorem 2 (BCS Implies Divergence-Freedom).** A CCSL specification $\mathcal{SPEC} = <\mathcal{C}, CConstr>$ is free of divergence, if there exists a BCS $\mathfrak{B} \supseteq \mathcal{C}$ implied by the $CConstr$. ∎

**Proof of Theorem 2:**

Let $\mathfrak{B} = \{c_\perp, c_1, \ldots, c_m, c^\top, c_{m+1}\ldots, c_{m+n}\}$, and $\mathfrak{B}_R = \{c_\perp, c_1, \ldots, c_m, c^\top\}$

*By applying* Proposition 9 *(**Bounded restriction**) to condition (i) and (ii) in* Definition 12, *we have*

$$\forall i \in [1,\ldots,m], \; clock\,pairs\,(c_\perp,c_i)\; and \;(c_i,c^\top)\; are\,bounded. \tag{B\_1}$$

*By* Proposition 8 *(**Bounded extension**), via the $c_\perp$ or $c^\top$ as the middle clock $b$ occurs, we have*

$$\forall i,j \in [1,\ldots,m], \; clock\,pair\,(c_i,c_j)\; is\,bounded. \tag{B\_2}$$

*Summarizing* (B\_1) *and* (B\_2),

$$\forall c_i,c_j \in \mathfrak{B}_R, \; clock\,pair\,(c_i,c_j)\; is\,bounded. \tag{B\_3}$$

*From condition (iii) in* Definition 12, $\forall i \in [m + 1..m + n]$, $\exists c \in \mathfrak{B}_R$, *clock pair $(c, c_i)$ is bounded,*

*By* Proposition 8 *again via the $c$ occurs in the last line as the middle clock $b$, we have*

$$\forall c_i \in \mathfrak{B}_R, c_j \in \mathfrak{B}\backslash\mathfrak{B}_R clock\,pair\,(c_i,c_j)\; is\,bounded. \tag{B\_4}$$

$$\forall c_i,c_j \in \mathfrak{B}\backslash\mathfrak{B}_R, \; clock\,pair\,(c_i,c_j)\; is\,bounded. \tag{B\_5}$$

*Summarizing* (B\_3), (B\_4) *and* (B\_5), *we conclude*
$\forall c_i, c_j \in \mathfrak{B}$, *clock pair $(c_i, c_j)$ is bounded.*
*Because $\mathfrak{B} \supseteq C$, $\forall c_i, c_j \in C$, clock pair $(c_i, c_j)$ is bounded.*

*By* Definition 10 *(**Bounded Specification**), $SPEC$ is divergence-freedom specification.* ∎

If we cannot derive a bounded relation from $C$ obviously, introducing some external clock(s) that can form BCCC is allowed. Therefore, we use $\mathfrak{B} \supseteq C$ rather than $\mathfrak{B} = C$ in Theorem 2.

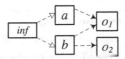

**Fig. 3.** Causality Clock Graph for $SPEC_i$

Let us illustrate Theorem 2 on a simple example, $SPEC_i = \;<C, CConstr>$ , where $C = \{a, b, inf(a, b), o_1, o_2\}$ and $CConstr = \{\; a \preccurlyeq o_1,\; inf(a, b) \sim o_1, b \sim o_2, b \preccurlyeq o_1\}$.

Let $inf = inf(a, b)$, via some given propositions above, we can get the explicit and implicit *Causality* relation list from *CConstr* (Fig. 3):

$inf \preccurlyeq a \preccurlyeq o_1, b \preccurlyeq o_1, b \preccurlyeq o_2$

By Proposition 11, we know clock pairs $(inf, o_1)$ and $(b, o_2)$ are bounded. Then we can construct a BCS $\mathfrak{B}_{speci} = \{c_\perp = inf, c_1 = a, c_2 = b, c^\top = o_1, c_3 = o_2\}$ shown in Fig. 4a with $m = 2$ and $n = 1$ in Definition 12. Because $\mathfrak{B}_{speci} \supseteq (SPEC_i)$, we assert $SPEC_i$ is free of divergence by Theorem 2.

Note that maybe there several other possibilities to assign the bottom or top clock. For example, $\mathfrak{B}_{speci\_alt} = \{c_\perp = inf, c^\top = b, c_1 = a, c_2 = o_1, c_3 = o_2\}$ shown in Fig. 4b is another same member set but assigned with different top clock. Here $m = 0$ and $n = 3$.

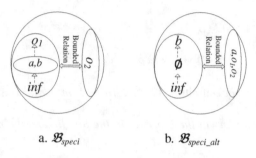

a. $\mathfrak{B}_{speci}$      b. $\mathfrak{B}_{speci\_alt}$

**Fig. 4.** Two represents for BCS of $SPEC_i$

$\mathfrak{B}_{speci}$ is a "much better" BCS than $\mathfrak{B}_{speci\_alt}$ in the sense of determining the boundedness among clocks. We can easily assert $\mathfrak{B}_{speci}$ is a BCS satisfied by conditions (i), (ii) and (iii) in Definition 12. On the contrary, As to $\mathfrak{B}_{speci\_alt}$, both conditions (i) and (iii) are not straightforwardly asserted using the *CConstr*( $SPEC_i$) as well as the associated propositions list above. In fact, $\mathfrak{B}_{speci\_alt}$ is constructed dedicatedly to show its inconvenience when I know the fact convergence. In purpose of the efficiency for convergence assertion, Subsect. 3.3 will design an algorithm for constructing BCS.

All the proofs about these propositions and theorems in Sects. 3.1 and 3.2 are completed with the help of PVS. Hence, we can try to solve some problem without any doubt about its correctness. The following theorem, written in PVS specification, is a special case[1] of Theorem 2, can be used to assert divergence-free of CCSL specifications.

---

> *bounded_set_boundSPEC:* **theorem exists** *(m: nat):*
>
>   **forall** *sgm: ((bot, top: Aclock): (sgm |= (bot <= top)) &*
>
>     *maxDrift(sgm)(bot, top)(m) & (forall (c: clock): sgm |= (bot <= c & c <= top)))*
>
>   => **forall** *(a, b: clock): maxDrift(sgm)(a, b)(m)*

---

Note:

---
[1] condition (iii) in Definition 12 is not necessarily be considered.

(1) *Clock* is type to represent *explicit* clock set.
(2) *AClock* is a defined ADT includes both *Clock* and *implicit* clocks defined in Definition 4.
(3) "⊨" is interpreted via Definitions 6 and 7.
(4) *maxDrift* is *Bounded relation* in Definition 9.

### 3.3   Detection Algorithm

For a given specification $\mathcal{SPEC}$ = <$\mathcal{C}$, *CConstr*>, there are three simply **rules** to prevent $\mathcal{SPEC}$ from divergence:

---

**Rule list for avoiding divergence**

- **Rule 1 (Redundant Clock Nonexistence)** There exists no a clock $c \in \mathcal{C}$ does not occur in any clock constraints in *CConstr*,   clock $c$ can tick arbitrarily otherwise.

- **Rule 2 (Bounds Existence)** At least one expression in the form of $a \$ n$ or *FilteredBy* can be found in *CConstr*. None of clock pair is bounded otherwise. Note that the delay expression may be in the implicit form. For instance, it can be found only by expanding the alternation in $\mathcal{SPEC}_i$.

- **Rule 3 (Absence of disordered clock)** A disorder clock $c \in \mathcal{C}$ is a clock which occurs only in the form of $c \# b$, cannot be found in other clock constraints. A disorder clock can tick arbitrarily except the requirement its exclusion with a certain clock $b$.

---

One violation of **Rules 1–3** causes the specification's divergence. The following parts will consider only the specification that follows **Rules 1–3**.

If we have no enough faith to assert the convergence CCSL specification, we can try to witness its divergence via discovering a UCCC by Theorem 1 because it is much easier to find an unbounded clock pair than to determine all the clock pairs are bounded.

When there is no obvious UCCC be found from a CCSL specification, we need to design an algorithm to try to construct a BCS, for the purpose of using Theorem 2 to guarantee specification's convergence.

For a given $\mathcal{SPEC} = <\mathcal{C}, CConstr>$, we first sort the clock based on the *Causality*-related clock constraints (includes *Causality* relation and those imply it) in *CConstr* with regard to *Causality*.

---

**Algorithm 1**   Divergence analysis

INPUT: SPEC=$<\mathcal{C}$ , $CConstr>$,

OUTPUT: YES if SPEC is Divergence-freedom, NO otherwise.

$\mathfrak{B} = \emptyset$.

I ) Get a clock pair $(a, b)$ in $\mathcal{C}$ s.t. $|a\text{-}b| \leq m$ for some $m \in \mathbb{N}$ and $a \leqslant b$, let $c_\perp = a$, $c^\top = b$

$\mathfrak{B} = \mathfrak{B} \cup \{c_\perp, c^\top\}$, $\mathcal{C} = \mathcal{C}\backslash\{c_\perp, c^\top\}$

If $\mathcal{C} = \emptyset$ return YES

If $\exists c \in \mathcal{C}$, $c^\top \leqslant c$ and $(c^\top, c)$ is bounded

$\mathfrak{B} = \mathfrak{B} \cup \{c\}$ , $\mathcal{C} = \mathcal{C}\backslash\{c\}$, $c^\top = c$

If $\mathcal{C} = \emptyset$ return YES

If $\exists c \in \mathcal{C}$, $c \leqslant c_\perp$ and $(c_\perp, c)$ is bounded

$\mathfrak{B} = \mathfrak{B} \cup \{c\}$ , $\mathcal{C} = \mathcal{C}\backslash\{c\}$, $c_\perp = c$

If $\mathcal{C} = \emptyset$ return YES

II ) For each clock $c \in \mathcal{C}$, such that $c_\perp \leqslant c \leqslant c^\top$,

$\mathfrak{B} = \mathfrak{B} \cup \{c\}$ , $\mathcal{C} = \mathcal{C}\backslash\{c\}$

If $\mathcal{C} = \emptyset$ return YES

III) For each clock $c \in \mathcal{C}$, such that $\exists b \in \mathfrak{B}$, pair $(b, c)$ is bounded

$\mathfrak{B} = \mathfrak{B} \cup \{c\}$ , $\mathcal{C} = \mathcal{C}\backslash\{c\}$

If $\mathcal{C} = \emptyset$ return YES

Return NO

---

Algorithm 1 includes three parts:

(I)   Get (and update if required) the bottom and top clock.
(II)  Get the clock set in which it is not faster than the bottom and not slower than the top.
(III) Analyze the boundedness of left clocks.

We can assert that $\mathfrak{B}$ constructed in Algorithm 1 must be a BCS by Propositions 8 and 9. So this algorithm's correctness is ensured by Theorem 2 since $\mathcal{C} = \emptyset \implies \mathfrak{B} \supseteq \mathcal{C}$.

Algorithm 1 must terminate because of the finiteness of clock set. The complexity of Algorithm 1 is $(|\mathcal{C}|)$. If most clocks can be dealt with in part I and II, the efficiency of algorithm is very high. Therefore, via Algorithm 1, determining boundedness among all clock pairs is converted into checking boundedness on much fewer clock pairs and additionally sorting clocks w.r.t. *Causality* relation.

## 4   Case Study

To illustrate the approach, we take an example inspired by [14], that was used for flow latency analysis on Architecture Analysis and Design Language (AADL) specifications. However, with CCSL we are conducting different kinds of analyses.

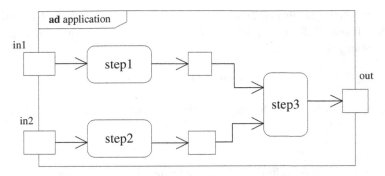

**Fig. 5.** Simple application

Figure 5 considers a simple application described as a UML activity. This application captures two inputs *in1* and *in2*, performs some calculations (*Step1*, *Step2* and *Step3*) and then produces a result *out*. This application has the possibility to compute *Step1* and *Step2* concurrently depending on the chosen execution platform. This application runs in a streaming-like fashion by continuously capturing new inputs and producing outputs.

To abstract this application as a CCSL specification, we assign one clock to each action. The clock has the exact same name as the associated action (e.g., *step1*). We also associate one clock with each input, this represents the capturing time of the inputs, and one clock with the production of the output (*out*). The successive instants of the clocks represent successive executions of the actions or input sensing time or output release time. The basic CCSL specification is $\mathcal{SPEC}_{simp} = <\mathcal{C}, \text{CConstr}>$ , where $\mathcal{C} = \{in1, in2, step1, step2, step3, out\}$, *CConstr* includes the following clock constraints:

$$in1 \preccurlyeq step1 \bigwedge step1 \prec step3 \tag{F.8}$$

$$in2 \preccurlyeq step2 \bigwedge step2 \prec step3 \tag{F.9}$$

$$step3 \preccurlyeq out \tag{F.10}$$

(F.8) specifies that *step1* may begin as soon as an input *in1* is available. Executing *step3* also requires *step1* to have produced its output. (F.9) is similar for *in2* and *step2*. (F.10) states that an output can be produced as soon as *step3* has executed. Note that CCSL precedence is well adapted to capture infinite FIFOs denoted on the figure as object nodes. Such a specification is clearly not convergent because of its violation of *Rule 2* (**Bounds Existence**) in **Rule list for avoiding divergence**. After the sorting the clocks w.r.t. *Causality* relation, we get two CCCs:

$$\rho_1 : in \preccurlyeq step1 \preccurlyeq step3 \preccurlyeq out$$

$$\rho_2 : in \preccurlyeq step1 \preccurlyeq step3 \preccurlyeq out$$

It is also stated again that $\mathcal{SPEC}_{\text{simp}}$ is divergent by Theorem 1 as both $\rho_1$ and $\rho_2$ are unbounded. If one CCSL constraint like (F.11) is added into $CConstr(\mathcal{SPEC}_{\text{simp}})$ as a test like that in [5].

$$sup(in1, in2) \sim out \qquad\qquad (F.11)$$

By Proposition 6, the following are two new UCCCs are acquired since the addition of (F.11):

$in1 \preccurlyeq sup(in1, in2)$

$in2 \preccurlyeq sup(in1, in2)$

*Now we try to use* Algorithm 1 *to check whether* $\mathcal{SPEC}_{\text{simp}}$ *is free of divergence or not. Let* $sup_{in12}/sup(in1, in2)$, by Expanding the Alternation definition, (F.11) becomes

$sup_{in12} \prec out \wedge out \prec (sup_{in12} \$ 1)$

Now $(\mathcal{SPEC}_{\text{simp}}) = \{in1, in2, step1, step2, step3, out, sup_{in12}, sup_{in12} \$ 1\}$ correspondingly, by part I of Algorithm 1, we get $\mathfrak{B} = \{c_\perp = sup_{in12}, c^\top = sup \$1\}$, then using part II of Algorithm 1, the clock *out* is added into $\mathfrak{B}$. Furtherly, because of $\rho_1, \rho_2$ and the fact $sup_{in12}$ is the fastest clock that is slower than *in1* and *in2* by Definition 7, we can deduce $\forall c \in \{step1, step2, step3\}$, $c_\perp \preccurlyeq c \preccurlyeq c^\top$ by Propositions 1 to 7. Therefore, the clocks *step1, step2 and step3* can also be added into $\mathfrak{B}$ via part II of Algorithm 1. Up to now, $\mathfrak{B} = \{c_\perp = sup_{in12}, step1, step2, step3, out, c^\top = sup_{in12} 1\}$, and none of other clock can be added into $\mathfrak{B}$ further. Therefore, $\mathcal{SPEC}_{\text{simp}}$ *is divergent because* $\mathfrak{B} \not\supseteq C$ witnessed by $in1, in2 \in C$ but $in1, in2 \notin \mathfrak{B}$. This is caused by that bounds on **Supremum** do not imply bounds on *in1(or in2)* and *out*, not to mention to the bound on *in1* and *in2*.

To become a bounded(or safe) system $\mathcal{SPEC}_{\text{simp\_safe}}$, we can for instance replace (F.11) by (F.12).

$$inf(in1, in2) \sim out \qquad\qquad (F.12)$$

Let $in f_{in12}/inf(in1, in2)$, in fact, (F.12) equals

$in f_{in12} \prec out \wedge out (inf_{in12} \$ 1)$

Because of introducing new clock constraint (F.12), some new clocks (implicit clocks) are introduced as well, now $C(\mathcal{SPEC}_{\text{simp\_safe}}) = C(\mathcal{SPEC}_{\text{simp}}) \cup \{inf_{in12}, inf_{in12} \$ 1\}$.

Let's check its divergence-freedom again by Algorithm 1. By part I of Algorithm 1, we get $\mathfrak{B} = \{c_\perp = inf_{in12}, c^\top = inf_{in12} \$ 1\}$, then using part II of Algorithm 1, all the clocks in $C$ can be added into $\mathfrak{B}$ because their speed are constrained between the slowest clock $inf_{in12}$ and the fastest clock $inf_{in12} \$ 1$ *as revealed by the following CCCs deduced by the* Propositions 1, 5 and 7 as well as $\rho1$ and $\rho2$ above.

$inf_{in12} \preccurlyeq in1 \preccurlyeq step1 \preccurlyeq step3 \preccurlyeq out \preccurlyeq (inf_{in12} \$ 1)$

$inf_{in12} \preccurlyeq in2 \preccurlyeq step2 \preccurlyeq step3 \preccurlyeq out \preccurlyeq (inf_{in12} \$ 1)$

The resulting $\mathfrak{B} = \{c_\perp = inf_{in12}, in1, step1, in2, step2, step3, out, c^\top = inf_{in12} \$ 1\}$, as shown in Fig. 6, is obviously a superset of $C(\mathcal{SPEC}_{\text{simp\_safe}})$. Hence, the CCSL specification $\mathcal{SPEC}_{\text{simp\_safe}}$ is free of divergence. Note that $\mathfrak{B}$ is also a CBCS because all of clock $c \in \mathfrak{B}$, $c_\perp \preccurlyeq c \preccurlyeq c^\top$.

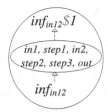

**Fig. 6.** BCS of simple application

With the help of *bounded_set_boundSPEC* in Sect. 3.2, we can complete the proof of theorem *simp_safe* for asserting the boundedness of $\mathcal{SPEC}_{\text{simp\_safe}}$ in PVS by the lemma *simp_bcs* as follows. In fact, $\mathcal{SPEC}_{\text{simp\_safe}}$ is a specification free of divergence.

---

*simp_bcs:* **lemma** **let** *bot = in1 ∧ in2 , top =delay(bot,1)* **in**

    *maxDrift(sgm)(bot,top)(1)* **&** *(forall (c: clock): sgm |= (bot <= c & c <= top)))*

*simp_safe:* **theorem** **forall** *(a,b:Clock) : maxDrift(sgm)(bot,top)(1)*

---

# 5 Related Work

Some techniques were provided as an effort to analyze CCSL specifications. [15] implemented the automatic analysis by translating CCSL into signal, for the purpose of generating executable specifications through discrete controller synthesis. However, this work did not consider the Infimum and Supremum operators that introduce unbounded counters and did not address the problem of deciding whether the specification is divergence-freedom or not. Exhaustive analysis of CCSL through a transformation into labeled transition systems has already been attempted in [16]. However, in those attempts, the CCSL operators were bounded because the underlying model-checkers cannot deal with infinite labeled transition systems.

In [6], the authors showed that even though the primitive constraints were unbounded, the composition of these primitive constraints could lead to a system where only a finite number of states were accessible. [7] defines a notion of safety for CCSL and establish a condition to decide whether a specification is safe on the transformed marked graph from CCSL.

All the above works share one common point: the specification analysis were done by some transformation and performed on the transformed target. The results were dependent on the correctness and efficiency of the mechanized transformation.

Our contribution is straightforward based on the clock *Causality* relation used to sort and the clock expression used to determine the clock pair's boundary. It is not necessary for the reader to have the other mathematic theory preliminaries except the basic set-theory.

## 6  Conclusion and Future Works

Based on the state-based semantics of a kernel subset of CCSL, We have presented a sufficient condition to discover the CCSL divergence existence, and an easily constructed Bounded Clock Set (BCS) for deciding the convergence of CCSL. An algorithm is proposed to actually build BCS of a given specification with the help of sorted the bounded clock chain with respect to causality relation and the clock delay expression used to decide clock pair's boundary. Therefore, determining boundedness among all clock pairs is converted into checking boundedness on much fewer clock pairs and additionally sorting clocks with respect to causality relation. Finally, a simple application's convergence is investigated. We first discover its divergence by the existence of unbounded clock causality chain. Consequently, a BCS is built to ensure the new specification is convergent by adding suitable constraint.

As a future work, we plan to extend and prove the extensive application of clock causality chain further. For instance, the potential causality conflict between clocks may be found if the specification is not deadlock-free or ill-formed, the periodicity in a schedule may be revealed in the chain in some style, etc.

## References

1. OMG, UML Profile for MARTE v1.0. Object Management Group (2009). formal/02-11-2009
2. André, C., Mallet, F., de Simone, R.: Modeling time(s). In: Engels, G., Opdyke, B., Schmidt, D.C., Weil, F. (eds.) MODELS 2007. LNCS, vol. 4735, pp. 559–573. Springer, Heidelberg (2007)
3. André, C.: Syntax and Semantics of the Clock Constraint Specification Language (CCSL), p. 37. Inria I3S Sophia Antipolis (2009)
4. DeAntoni, J., Mallet, F.: TimeSquare: treat your models with logical time. In: Furia, C.A., Nanz, S. (eds.) TOOLS 2012. LNCS, vol. 7304, pp. 34–41. Springer, Heidelberg (2012)
5. Mallet, F., Millo, J.-V., Romenska, Y.: State-based representation of CCSL operators (2013)
6. Mallet, F., Millo, J.-V.: Boundness issues in CCSL specifications. In: Groves, L., Sun, J. (eds.) ICFEM 2013. LNCS, vol. 8144, pp. 20–35. Springer, Heidelberg (2013)
7. Mallet, F., Millo, J.-V., De Simone, R.: Safe CCSL specifications and marked graphs. In: MEMOCODE - 11th IEEE/ACM International Conference on Formal Methods and Models for Codesign. IEEE CS (2013)
8. Deantoni, J., André, C., Gascon, R.: CCSL denotational semantics, p. 29. Inria I3S Sophia Antipolis (2014)
9. Benveniste, A., et al.: The synchronous languages 12 years later. Proc. IEEE **91**(1), 64–83 (2003)
10. Ling, Y., et al.: Schedulability analysis with CCSL specifications. In: 2013 20th Asia-Pacific Software Engineering Conference (APSEC) (2013)
11. Owre, S., Shankar, N.: Abstract Datatypes in PVS. Computer Science Laboratory, SRI International, Menlo Park (1993)
12. Owre, S., Rushby, J.M., Shankar, N.: PVS: A prototype verification system. In: Kapur, D. (ed.) CADE 2011. LNCS (LNAI), vol. 607, pp. 748–752. Springer, Heidelberg (1992)

13. Computer Science Laboratory, SRI International. PVS: Specification and Verification System. http://pvs.csl.sri.com/
14. Feiler, P.H., Hansson, J.: Flow latency analysis with the architecture analysis and design language (AADL), CMU (2007)
15. Mallet, F., Andre, C.: UML/MARTE CCSL, Signal and Petri nets (2008)
16. Gascon, R., Mallet, F., DeAntoni, J.: Logical time and temporal logics: comparing UML MARTE/CCSL and PSL. In: 2011 Eighteenth International Symposium on Temporal Representation and Reasoning (TIME) (2011)

# Performance Evaluation of Concurrent Data Structures

Hao Wu[1(✉)], Xiaoxiao Yang[2], and Joost-Pieter Katoen[1]

[1] Software Modelling and Verification Group, RWTH Aachen University,
Aachen, Germany
`hao.wu@cs.rwth-aachen.de`
[2] State Key Laboratory of Computer Science, Institute of Software,
Chinese Academy of Sciences, Beijing, China

**Abstract.** The speed-ups acquired by concurrent programming heavily rely on exploiting highly concurrent data structures. This has led to a variety of coarse-grained and fine-grained locking to lock-free data structures. The performance of such data structures is typically analysed by simulation or implementation. We advocate a *model-based* approach using *probabilistic model checking*. The main benefit is that our models can also be used to check the correctness of the data structures. The paper details the approach, and reports on experimental results on several concurrent stacks, queues, and lists. Our analysis yields worst- and best-case bounds on performance metrics such as expected time and probabilities to finish a certain number of operations within a deadline.

## 1 Introduction

*Background and Motivation.* Multi-core computers are ubiquitous. However shared concurrent data structures [11] are an important obstacle. The downside of lock-based data structures is that they are a sequential bottleneck. Lock-free data structures are resilient to failures, are more complex, and require special synchronisation primitives. Modern multi-core architectures support compare-and-swap operations to allow threads to read, modify and write atomically. Correctness of concurrent data structures is a key issue and typically addressed by a semi-formal pencil proof; performance is typically assessed by simulation or implementation [2,4]. We propose to carry out both correctness and performance analysis using a single, *model-based*, approach. This has the advantage that both analyses use the same model, and that results are coherent. In addition, it allows for using *a single technique*—model checking—*for both types of analysis.*

*Modelling approach.* The starting point for our approach is to model the concurrent data structure at hand, together with the threads that perform operations on it. We use the LOTOS NT language (LNT[1], for short) which is a

---

Supported by the CDZ project CAP (GZ 1023) and the A. von Humboldt-Foundation.

[1] The LNT language is a formal description technique standardized by ISO OSI (1989), please refer to: http://www.iso.org/iso/catalogue_detail.htm?csnumber=16258.

© Springer International Publishing AG 2016
M. Fränzle et al. (Eds.): SETTA 2016, LNCS 9984, pp. 38–49, 2016.
DOI: 10.1007/978-3-319-47677-3_3

compositional modelling language with process algebraic roots, that supports abstract data specifications. The structure of the entire system has the form $(T_1 \;|||\; \cdots \;|||\; T_n) \;||_G\; D$ where the $n$ threads $T_1$ through $T_n$ which are independent (hence indicated by $|||$, a shorthand for $||_\emptyset$) and communicate with the concurrent data structure $D$ via the communication gates $G$. To keep the state space finite, we bound the number of operations by applying a monitor process $M$ that keeps track of the number of read and write operations. Using the CADP toolbox [7], the underlying state space can be generated and be analysed for checking functional correctness. We focus here on performance evaluation.

*Performance modelling and evaluation.* To enable performance evaluation, we assume that delays between reads and writes are random in nature and are governed by negative exponential distributions. We insert these random delays into the model by renaming such read and write actions to Markovian delay with different rates in CADP. As a result, CADP yields (after a mild post-processing) a *Markov automaton* [3,6]. These automata have random delay transitions, and allow for non-determinism, a feature that we exploit to model the concurrency among the threads. Prior to the performance evaluation, the state space is minimised using branching bisimulation minimisation in CADP, which preserves the important properties in the performance evaluation [15]. The analysis of Markov automata is enabled using recently developed

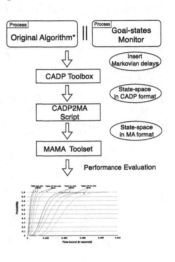

**Fig. 1.** Our approach

algorithms by Guck *et al.* [8,9], which allow for determining the expected time until a certain state is reached—like "what is the expected time until each thread has completed 10 reads and writes?"—and the likelihood to reach a state within a given deadline—like "what is the probability that all reads and writes finish within 10 min?". Due to the inherent non-determinism (due to concurrency), the analysis does not obtain *the* expected time, or *the* probability, but rather obtains *bounds*. These bounds represent the best- and worst-case scenarios. The quantitative analysis of MA is supported by the MAMA tool-set[2]. The entire approach is given in Fig. 1.

*Experimentation.* We have applied our approach to the modelling and performance evaluation of several concurrent stacks, queues, and lists. We treat the Treiber stack [13] and its variant with hazard pointers, see e.g., [16]. In addition, we cover the Michael-Scott two-lock (MS 2L) queue [14], its lock-free variant (MS LF) [14], and an improvement on this lock-free variant [5] (known as DGLM). Finally, we consider the coarse-grained synchronisation list [11, Chap. 9.4], a fine-grained synchronization list [11, Chap. 9.5], the lazy list [10] and an optimistic

---

[2] http://www.home.cs.utwente.nl/~timmer/mama/.

list [11, Chap. 9.6]. Our performance evaluation treats models of up to 378 million states. The experiments show that—as expected—lock-based data structures have a rather deterministic performance, whereas lock-free and fine-grained lock-based show more variance in their performance. In addition, fine-grained lists may yield a lower throughput as under intense race conditions many unsuccessful operations are carried out. To the best of our knowledge this is the first work that formally models concurrent data structures and uses this for a performance evaluation. Related works are the probabilistic model checking of low-level OS kernels including spin-locks [1] and the modelling and performance evaluation of mutual exclusion algorithms [12].

```
1  type Qnode is                              2  type Memory is
      Qnode (next: Nat)                             array [0 .. 10] of Qnode
      with "get", "set"                             with "get", "set"
   end type                                      end type

3  process Thread[M: Queue_Ops,              6  process Queue[M: Queue_Ops] is
   HL, TL: Lock_Ops, complete,                  var m: Memory, head, tail, size, hd,
   T: Thread_Ops](pid: Pid) is                  pos, pos_next: Nat, pid: Pid in
   loop                                         size := 10;  head := 0; tail := 0;
     select                                     m := Memory(Qnode (0 of Nat));
       Enq[M, TL, T](pid); complete             loop select
     []                                           M(read_head, head, ?any Pid)
       Deq[M, HL, T](pid); complete              []
     end select                                   M(read_next, ?pos, ?pos_next, ?pid)
   end loop                                         where (m[pos].next == pos_next)
   end process                                     []
                                                 M(set_tail_next, ?pid);
4  process Enq [M: Queue_Ops, TL: Lock_Ops,      m[tail] := m[tail].{next => (tail + 1)}
             T: Thread_Ops] (pid: Pid) is        []
   var locked : Bool in                          M(set_tail, ?pid); tail := m[tail].next
     TL(lock_tail, pid);                          []
     loop G in                                   M(set_head, ?hd, ?pid);
       TL(test_and_set, ?locked, pid);           head := m[head].next
       if (locked == false) then break G       end select end loop
       else T(thr_delay, pid)                  end var end process
       end if
     end loop;                               7  process MAIN [M: Queue_Ops, HL, TL: Lock_Ops,
     M(set_tail_next, pid);                     finish, complete, T: Thread_Ops] is
     M(set_tail, pid);                          par M, HL, TL, complete, T in
     TL(unlock_tail, pid)                       par M in
   end var end process                          par HL, TL in
                                                 par
5  process H_Lock[HL: Lock_Ops] is               Thread[M, HL, TL, complete, T] (1 of Pid)
   var locked : Bool, pid: Pid in                ||
     locked := false;                            Thread[M, HL, TL, complete, T] (2 of Pid)
     loop select                                 end par
       HL(lock_head, ?pid)                       ||
       []                                        par
       HL(test_and_set, false, ?pid)             H_Lock[HL]
         where (locked == false);                ||
       locked := true                            T_Lock[TL]
       []                                        end par
       HL(test_and_set, true, ?pid)              end par
         where (locked == true)                  ||
       []                                        Queue [M]
       HL(unlock_head, ?pid);                    end par
       locked := false                           ||
     end select end loop                         Monitor [M, HL, TL, finish, complete, T]
   end var end process                         end par end process
```

**Fig. 2.** The (partial) LNT code of the MS 2L queue

## 2   Modeling Concurrent Data Structures in LNT

For space sake we only show the LNT model[3] of the MS 2L queue [14] here in
Fig. 2. Furthermore, the references of LNT language and various tools (e.g. state
space minimisation) provided by CDAP can be found under http://cadp.inria.fr.
   Figure 2 - ☐1☐ ☐2☐ define the data structure of the MS 2L queue. It consists
of a bounded array of Qnodes, since mutable dynamic data structure is not
supported by LNT. Moreover, since to model a pointer is also not possible here,
the next field of the Qnode stores the index (a natural number) of its next node
in the array. A thread (Fig. 2 -☐3☐) identified by pid needs to synchronize with
the lock (process) for head (i.e., H_Lock), the lock for tail (i.e., T_Lock) and the
queue (process) to perform enqueue and dequeue operations, hence the gates (M,
HL, TL) required for synchronization with these processes are declared. Further,
the gates (complete, T) indicate the process's own operations. The behavior
of a thread is to repeatedly perform (enclosed by loop) either an enqueue or
a dequeue operation and emits a complete signal (synchronized with monitor
process for counting) if the operation is finished. Figure 2 -☐4☐ defines the enqueue
operation (for dequeue similarly) of a thread. First it tries to acquire the tail
lock (via gate TL) before performing operations on the queue. The two locks
are assumed to be simple *test-and-set* locks with *fixed* back-off delay. The delay
(T(thr_delay, pid)) are inserted after each unsuccessful try of test-and-set,
then the test-and-set is restarted. If it is succeed, the loop is exited and the
operations to enqueue (e.g., set_tail_next (set tail's next to new node) and
set_tail) are performed via the synchronization with gate M. Finally, the lock
is released (unlock_tail). Figure 2 -☐5☐ defines the head lock, which consists of a
boolean variable locked and operations to lock with the test_and_set operation
(it returns previous value of locked and set it to true atomically) and unlock
via the gate HL. Note that complex locks can be modelled similarly. Figure 2 -☐6☐
is the queue process which consists of the queue (array) and the head and tail
(as indexes in the array) with auxiliary variables for synchronization. Note that
enqueue a new node is simply represented by setting the tail's next to current
tail's index + 1 in our model. Figure 2 -☐7☐ defines the whole MS 2L queue:
the threads, the two locks, the queue and the monitor process are composed in
parallel with corresponding gates. Note that the threads are independent hence
they do not synchronize.

## 3   Towards Performance Evaluation

In performance evaluation, we bound number of operations on the data structures to keep the state space finite. The monitor process synchronises with the

---

[3] The complete LNT models and their corresponding scripts of all aforementioned concurrent data structures follow the same principle described here and can be found
under  the  link:  https://moves.rwth-aachen.de/wp-content/uploads/LNT-models.
zip.

`complete` actions from threads and counts the successfully completed operations. Goal states indicate when the number of operations has reached a certain bound. The performance of the concurrent data structure is then evaluated based on reaching a goal state. Since we assume that performing the elementary operations (e.g., read/write, test_and_set, compare_and_swap) cause a random delays governed by negative exponential distribution. We use the renaming rules to replace such operations with Markovian transitions with rates in the state space.

*Experimental setup.* To conduct experiments, we used the workflow as depicted in Fig. 1. The CADP tool [7] is used to generate the state space of the parallel composition of the LNT models of the data structures, the threads, the monitor. CADP also supports branching bisimulation minimisation that we exploit to reduce the models prior to performance analysis. We developed a CADP2MA script that transforms the state space as generated by CADP into a Markov automaton (MA) [3,6]. MA are state-transition systems that cater for non-determinism and support random delays. The performance evaluation of MA is done using the recent MAMA tool-set [8]. This tool supports the numerical computation of several quantitative objectives on MA. Our experiments focus on two measures: (1) the expected time until the system completes a certain number of operations requested by the overall threads and (2) the probability of finishing all these operations within a given deadline. As a thread repeatedly performs enqueue (push, add) and dequeue (pop, remove) operations in a *non-deterministic* manner[4], our analysis does not yield a single number but yields two *bounds*. A lower bound on the expected time gives the minimal time that is needed on average to complete all operations; an upper bound gives the maximal time. The former can be understood as the best achievable scenario; the latter as the worst one.

*Assumptions and parameter settings.* As concurrent programs with shared data structures (and possibly pointers) in principle have an unbounded state space, we make the following assumptions and modelling choices so as to enable a performance evaluation on a finite state space: (1) The number of threads is fixed. These threads invoke pre-defined operations (like pop and push) of the concurrent data structure. The invocation of such operations is modelled by means of synchronisation. (2) We bound the number of performed operations. (3) As LNT does not natively support pointers, fixed-size arrays with pre-defined elements are used, an index (a natural number) is used to locate the node, and a pointer is treated as an index record. The delays that are used in our experiments are adopted from [12] where the performance of mutual exclusion algorithms was analysed. The analysis is not based on a specific processor architecture, it is assumed that local caches are absent, and all operations are carried out on global memory. The rates of the exponential distributions are: a read operation from global memory (rate 3000, i.e., on average $1/3000$ time units), write to a global memory (2000), and complex operations (1200) on variables in global memory

---

[4] Thus, the thread behavior is not biased to certain scenarios.

e.g., compare_and_swap or test_and_set actions. The experiments are conducted on a computer with $4 \times 12$-core AMD CPU @ 2.1 GHz and 192 GB memory under 64-bit Debian 7.6.

# 4  Experimental Results

This section reports on applying our approach to the modelling and performance evaluation of several concurrent stacks, queues, and lists. For each data structure, we modelled and compared several variants from the literature. We report on the state spaces, the performance analysis results, and discuss them.

**Concurrent queues.** We cover the MS 2L queue [14], its lock-free variant (MS LF) [14], and an improvement on this lock-free variant (known as DGLM) [5].

*State space size and analysis times.* Table 1 shows the state spaces of different configurations of the three queues, the reduced state space by probabilistic branching bisimulation minimisation [7], the reduction factor, and the times (in seconds) for generating + reducing the state space and analysing expected time objectives. For four threads, three operations are considered (due to state space explosion). As expected, the state space grows exponentially in the number of threads and number of operations. State spaces up to about 378 million states have been generated (MS LF, 3 threads and 15 operations). The bisimulation reduction times for large state spaces are about 50 % of the generation times. Since the actions in the resulting system cannot be delayed by any other actions, all actions (except delay transitions) are turned into $\tau$-transitions. This gives rise to state space reductions of up to 99.9 %. The MS 2L queue has a relatively small state space, due to its nature of low concurrency.

*Expected time results.* Figure 3 shows the analysis results of the expected time to finish a number of operations on the queues. Observe that the MS 2L queue is rather deterministic as the minimal and maximal values are quite close. This does not hold for the lock-free queues. The performance of the lock-based queue thus

**Table 1.** State space and the analysis time of expected time for concurrent queues

| | MS 2L queue | | | | MS LF queue | | | | DGLM queue | | | |
|---|---|---|---|---|---|---|---|---|---|---|---|---|
| #Thr.-<br>#Oper. | state<br>space | red.<br>st. space | red.<br>rate | gen.+red./<br>comp.<br>time (in s) | state<br>space | red.<br>st. space | red.<br>rate | gen.+red./<br>comp.<br>time (in s) | state<br>space | red.<br>st. space | red.<br>rate | gen.+red./<br>comp.<br>time (in s) |
| 2-3 | 1623 | 273 | 0.8318 | 25/0.03 | 6206 | 502 | 0.9191 | 26/0.07 | 6079 | 483 | 0.9205 | 21/0.06 |
| 2-5 | 4862 | 561 | 0.8846 | 26/0.1 | 27023 | 1376 | 0.9491 | 30/0.47 | 25552 | 1350 | 0.9472 | 23/0.38 |
| 2-10 | 25942 | 1895 | 0.9270 | 27/0.97 | 292k | 5296 | 0.9819 | 39/9.1 | 017500 | 5061 | 0.0700 | 27/0.3 |
| 2-15 | 74712 | 3956 | 0.9471 | 29/4.7 | 1.39M | 11471 | 0.9917 | 52/28 | 1.06M | 11440 | 0.9893 | 47/29 |
| 2-20 | 163k | 6765 | 0.9584 | 32/11.7 | 4.49M | 20066 | 0.9955 | 497/124 | 3.13M | 19991 | 0.9936 | 104/95 |
| 2-30 | 502k | 14585 | 0.9709 | 35/63 | 25.3M | 44186 | 0.9983 | 964/580 | 15.2M | 44031 | 0.9971 | 656/600 |
| 3-3 | 22832 | 753 | 0.9670 | 27/0.17 | 350k | 3905 | 0.9888 | 29/3.5 | 326k | 3796 | 0.9883 | 29/2.8 |
| 3-5 | 65540 | 1857 | 0.9717 | 29/1.1 | 2.21M | 11649 | 0.9947 | 70/44 | 1.87M | 11674 | 0.9938 | 65/36 |
| 3-10 | 332k | 6346 | 0.9809 | 31/14.7 | 47.9M | 46993 | 0.9990 | 2018/1183 | 31.7M | 46452 | 0.9985 | 1398/1624 |
| 3-15 | 934k | 13267 | 0.9858 | 46/83 | 378.3M | 103k | 0.9997 | 21135/10088 | 200M | 101k | 0.9995 | 9826/10130 |
| 4-3 | 286k | 1814 | 0.9936 | 29/1.0 | 21.78M | 23042 | 0.9989 | 853/204 | 19.3M | 23631 | 0.9988 | 780/317 |

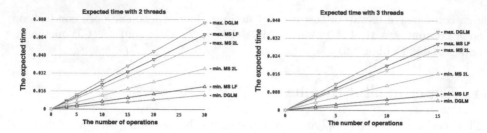

**Fig. 3.** Min./max. expected time versus # operations for concurrent queues (Color figure online)

provides a more stable service than its lock-free variants. Comparing Fig. 3 (left) and (right), we see that the expected time of finishing 3 operations with 2 threads is improved by 11 %/7 % in best/worst case of MS 2L queue, 26 %/10 % of MS LF queue, 27 %/9 % of DGLM queue with 3 threads (due to more concurrency), receptively. In best case the expected time for lock-free queues is much better than for the MS 2L queue comparing to the difference between lock-free queues and the MS 2L queue in worst case. The lock-free queues have—as expected—a much higher overhead than the lock-based queue. Thus, lock-free queues have a lower throughput in worst case than the lock-based one. The DGLM queue outperforms the MS LF queue in best case. Finally, we observe that expected times grow linearly in the number of operations.

*Probability of timely completion.* To get more insight into the performance of queues, we analyse the probability of finishing a certain number of operations within a (varying) deadline, see Fig. 4. Note that the faster the curve goes to one, the better the queue's performance. Since the number of opera-

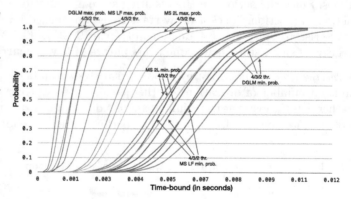

**Fig. 4.** How likely do concurrent queues complete three operations on time? (Color figure online)

tions only causes a linear increase in the expected delay, we consider three operations. We vary the number of threads from two to four. The three left-most groups of curves indicate the maximal probabilities of the three queues with 4/3/2 threads, respectively. We can observe that the algorithms have a very strong impact on these results: lock-free queues perform much better than the lock-based one in best case. However in worst case, the performance of algorithms is not that distinguishable (the rightmost curves), since they are close to each other. The number of threads in all queues influence these curves quite consistently, the more threads the much quicker the queue will finish the operations. As for the expected times, the lock-based algorithm behaves rather deterministic and its performance is quite stable when varying the number of threads.

*Evaluation.* In the best-case scenario, the two lock-free queues behave much better than the lock-based queue, in worst case however the lock-based queue outperforms the lock-free queues. Lock-based queues have a stable performance and are less vulnerable to the number of threads. The DGLM queue has better expected times than the MS LF in best case, but does not process operations within a deadline more likely.

**Concurrent stacks.** We consider two variants of the (lock-free) Treiber stack: one with hazard pointers (HPs) and one without. Hazard pointers [13] prevents the well-known ABA problem. HPs is used to keep certain locations as hazard and prohibit other threads to deallocate them. During garbage collection, only locations not pointed to by HPs can be freed. Our model of the Treiber-HP stack is based on [16] and includes a memory allocation thread. Our analyses focus on evaluating the performance influence of hazard pointers.

*State space size and analysis times.* Table 2 shows the state spaces for different parameter settings. The Treiber-HP stack causes a state space explosion due to the frequent scanning of HPs to find free locations after each pop()-operation. The branching bisimulation reduces the state space significantly (up to 0.999 in case of 4 threads). In cases for which the state space

**Table 2.** The state spaces and analysis times of expected time of Treiber stacks

| #Thr.-#Oper. | Treiber Stack | | | | Treiber Stack + Hazard Pointers | | | |
|---|---|---|---|---|---|---|---|---|
| | state space | red. st. space | red. rate | comp. time (in s) | state space | red. st. space | red. rate | comp. time (in s) |
| 2-3 | 1081 | 57 | 0.947 | 16/0.005 | 7303 | 320 | 0.956 | 26/0.01 |
| 2-5 | 2876 | 165 | 0.943 | 17/0.02 | 70763 | 2445 | 0.905 | 38/1.5 |
| 2-10 | 13597 | 960 | 0.929 | 19/0.3 | 8.04M | 258k | 0.968 | 841/23591 |
| 2-15 | 37411 | 2978 | 0.920 | 21/2 | M.O | M.O | - | - |
| 2-20 | 79582 | 6744 | 0.915 | 25/9.7 | M.O | M.O | - | - |
| 2-30 | 240k | 21941 | 0.909 | 22/112 | M.O | M.O | - | - |
| 3-3 | 16989 | 145 | 0.991 | 17/0.01 | 712k | 2154 | 0.997 | 61/0.97 |
| 3-5 | 53320 | 527 | 0.990 | 19/0.1 | 18.35M | 83972 | 0.995 | 1998/3580 |
| 3-10 | 358k | 5146 | 0.986 | 21/7.3 | M.O | M.O | - | - |
| 3-15 | 1.29M | 21011 | 0.984 | 39/113 | M.O | M.O | - | - |
| 4-3 | 265k | 313 | 0.999 | 18/0.03 | M.O | M.O | - | - |

could be generated, the analysis time for Treiber-HP is prohibitive (taking hours).

*Expected time results.* Figure 5 shows that the HPs cause significant overhead both in best and worst case. This is due to the additional operations of the HPs including setting/comparing with the HPs in pop(). It is interesting to observe that minimal expected times vary less than maximal ones. This is possibly due to the fact that HPs are not effective for push()-operations.

*Probability of timely completion.* As for the concurrent queues, we compute the time-bounded reachability of finishing three operations with 2/3/4 threads. We notice that increasing the number of threads in the system affect the minimal probability more than the maximal probability for both stacks. Moreover, the

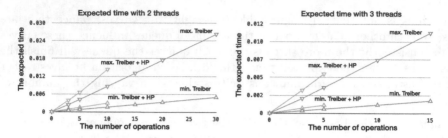

**Fig. 5.** The expected execution time versus # operations for Treiber stacks (Color figure online)

difference of the maximal time-bounded reachability between both stacks is quite small compared with the minimal values (Fig. 6).

*Evaluation.* We obverse that increasing the number of threads in the system has a significant impact on the minimal probability of finishing three operations for both stacks. Adding the HPs to Treiber stack is expensive in the worst case scenario, since the curves of minimal reachability probabilities of Treiber stacks with HPs (blue lines) are quite away from the corresponding curves (yellow lines) representing the minimal reachability probabilities of Treiber stacks without HPs. However, increasing the number of threads may alleviate this problem. Note that we did not have any result of the minimal probability of Treiber stack + HP with 4 threads due to memory out.

**Concurrent lists.** We consider four lock-based lists: a coarse-grained synchronization list (cgs) [11], a fine-grained synchronization list (fgs) [11], the Heller *et al.* lazy list (lazy) [10] and the optimistic list (opt.) [11]. All these concurrent lists are list-based implementations of a concurrent set object, where we can add (or remove) a value[5] to (or from) the list (set). Adding an element which is already

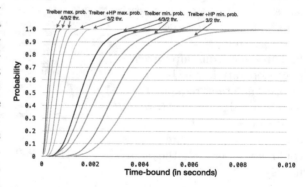

**Fig. 6.** The prob. of finishing 3 oper. within a given time-bound of Treiber stacks (Color figure online)

in the list is unsuccessful. The lists are *data-dependent*, and their performance strongly depends on the data to be added or removed. Thus, we model an additional process to generate pseudo random numbers to be added or removed to/from the list. To test extreme situations, we set the generated random number to be only 1 or 2 in our experiment. This will cause a large number of unsuc-

---

[5] In our experiments, we consider lists of natural numbers.

cessful operations. Since these data structures employ different granularities of locks, we discuss the effect of such granularities of locks on the performance under this setting. For space reasons, we focus only on the expected time of finishing different number of operations with several threads.

**Table 3.** The state spaces and comp. time of max/min expected time of lists

| #Thr.-#Oper. | Coarser grained sync. list (cgs) | | | | Fine grained sync. list (fgs) | | | | Lazy sync. list (lazy) | | | | Optimistic sync. list (opt.) | | | |
|---|---|---|---|---|---|---|---|---|---|---|---|---|---|---|---|---|
| | state space | red. st. space | red. rate | gen.+red./comp. time (in s) | state space | red. st. space | red. rate | gen.+red./comp. time (in s) | state space | red. st. space | red. rate | gen.+red./comp. time (in s) | state space | red. st. space | red. rate | gen.+red./comp. time (in s) |
| 2-3 | 5871 | 541 | 0.908 | 26/0.09 | 6205 | 906 | 0.854 | 22/0.2 | 41147 | 3171 | 0.923 | 31/1.6 | 46859 | 3587 | 0.923 | 29/1.8 |
| 2-5 | 17343 | 1546 | 0.911 | 26/0.5 | 17713 | 2883 | 0.837 | 23/1.5 | 168k | 11031 | 0.934 | 36/21 | 193k | 13202 | 0.932 | 34/24 |
| 2-10 | 52305 | 4135 | 0.921 | 26/4 | 52063 | 8231 | 0.842 | 24/14 | 633k | 42863 | 0.932 | 51/565 | 736k | 43457 | 0.941 | 54/576 |
| 2-15 | 87393 | 6916 | 0.921 | 28/12 | 86983 | 13773 | 0.842 | 26/50 | 1.11M | 78279 | 0.929 | 53/3485 | 1.29M | 111k | 0.914 | 62/11643 |
| 2-20 | 122k | 9505 | 0.922 | 30/21 | 121k | 19121 | 0.842 | 29/108 | 1.58M | 112k | 0.929 | 64/9158 | 1.84M | 155k | 0.916 | 90/15251 |
| 2-30 | 192k | 14875 | 0.923 | 33/71 | 191k | 30011 | 0.843 | 32/338 | 2.53M | 181k | 0.928 | 103/24475 | 2.95M | 249k | 0.916 | 128/58909 |
| 3-3 | 156k | 2269 | 0.985 | 30/1.2 | 148k | 3831 | 0.974 | 28/3.3 | 5.93M | 57926 | 0.990 | 244/1375 | 6.82M | 42497 | 0.994 | 294/422 |
| 3-5 | 445k | 7624 | 0.983 | 33/14 | 409k | 16033 | 0.961 | 32/89 | 33.7M | 276k | 0.992 | 49604/98989 | 40.2M | 442k | 0.989 | T.O. |
| 3-10 | 1.35M | 23729 | 0.982 | 62/465 | 1.21M | 57722 | 0.952 | 3524/266 | M.O. | - | - | - | M.O. | - | - | - |
| 3-15 | 2.27M | 39212 | 0.983 | 94/1160 | 2.0M | 98207 | 0.951 | 1980/7243 | M.O. | - | - | - | M.O. | - | - | - |

*State space size and analysis times.* Table 3 shows the generated state spaces and analysis times of the lists. State spaces in some scenarios can not be generated due to either memory out at state space generation or time out of computing the expected time. The generated state spaces correctly reflect the granularities of locks[6]: cgs > (is coarser than) fgs > lazy > opt. The bisimulation reduction times for the large state spaces are about 60 % of the generation times.

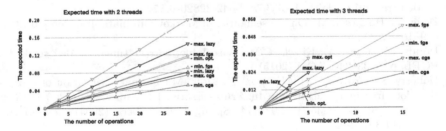

**Fig. 7.** The expected execution time versus # read/write oper. of concurrent lists (Color figure online)

*Expected time results.* One would expect that the fine-grained concurrency (=finer granularity of lock) allows more operations to be performed per unit time than coarse-grained concurrency. Hence one expects for the throughput: opt > (the expected time is smaller) > lazy > fgs > cgs. However, in our experiment this is not the case. Reversely, we can easily observe that cgs finishes 3 operations sooner than the others in both best and worst scenarios (Fig. 7).

---

[6] If for a given scenario, the number of states of a data structure is higher than for another data structure, it allows for more concurrency and has finer lock granularity.

*Evaluation.* The results of expected times above indicate that the finer-grained concurrency will not always achieve a higher throughput when the data race is (extremely) intensive. The more unsuccessful operations will lower the overall throughput and in such scenario the fine-grained implementation could be worse than the coarse-grained implementation.

## 5   Conclusion

This paper presented the modelling and performance analysis of various concurrent data structures using a combination of the CADP and MAMA tools. We emphasise that probabilistic model checkers such as PRISM/MRMC are not appropriate as they do not support non-deterministic continuous-time stochastic models. Future work consists of validating our model-based results against concurrent data structure implementations (e.g., in Java).

**Acknowledgments.** We thank Wendelin Serwe for his support in CADP.

## References

1. Baier, C., Daum, M., Engel, B., Härtig, H., et al.: Locks: picking key methods for a scalable quantitative analysis. J. Comput. Syst. Sci. **81**(1), 258–287 (2015)
2. Cederman, D., Chatterjee, B., Tsigas, P.: Understanding the performance of concurrent data structures on graphics processors. In: Kaklamanis, C., Papatheodorou, T., Spirakis, P.G. (eds.) Euro-Par 2012. LNCS, vol. 7484, pp. 883–894. Springer, Heidelberg (2012). doi:10.1007/978-3-642-32820-6_87
3. Deng, Y., Hennessy, M.: On the semantics of Markov automata. Inf. Comput. **222**, 139–168 (2013)
4. Dodds, M., Haas, A., Kirsch, C.M.: A scalable, correct time-stamped stack. In: POPL, pp. 233–246. ACM (2015)
5. Doherty, S., Groves, L., Luchangco, V., Moir, M.: Formal verification of a practical lock-free queue algorithm. In: Frutos-Escrig, D., Núñez, M. (eds.) FORTE 2004. LNCS, vol. 3235, pp. 97–114. Springer, Heidelberg (2004). doi:10.1007/978-3-540-30232-2_7
6. Eisentraut, C., Hermanns, H., Zhang, L.: On probabilistic automata in continuous time. In: LICS, pp. 342–351 (2010)
7. Garavel, H., Lang, F., Mateescu, R., Serwe, W.: CADP 2011: a toolbox for the construction and analysis of distributed processes. STTT **15**(2), 89–107 (2013)
8. Guck, D., Hatefi, H., Hermanns, H., Katoen, J., Timmer, M.: Analysis of timed and long-run objectives for Markov automata. Log. Methods Comput. Sci. **10**(3) (2014)
9. Guck, D., Timmer, M., Hatefi, H., Ruijters, E., Stoelinga, M.: Modelling and analysis of Markov reward automata. In: Cassez, F., Raskin, J.-F. (eds.) ATVA 2014. LNCS, vol. 8837, pp. 168–184. Springer, Heidelberg (2014). doi:10.1007/978-3-319-11936-6_13
10. Heller, S., Herlihy, M., Luchangco, V., Moir, M., Scherer III, W.N., Shavit, N.: A lazy concurrent list-based set algorithm. Parallel Process. Lett. **17**(4), 411–424 (2007)

11. Herlihy, M., Shavit, N.: The Art of Multiprocessor Programming. Morgan Kaufmann (2008)
12. Mateescu, R., Serwe, W.: Model checking and performance evaluation with CADP illustrated on shared-memory mutual exclusion protocols. Sci. Comput. Program. **78**(7), 843–861 (2013)
13. Michael, M.M.: Hazard pointers: safe memory reclamation for lock-free objects. IEEE Trans. Parallel Distrib. Syst. **15**(6), 491–504 (2004)
14. Michael, M.M., Scott, M.L.: Simple, fast, and practical non-blocking and blocking concurrent queue algorithms. In: PODC, pp. 267–275. ACM (1996)
15. Neuhäußer, M.R., Katoen, J.-P.: Bisimulation and logical preservation for continuous-time Markov decision processes. In: Caires, L., Vasconcelos, V.T. (eds.) CONCUR 2007. LNCS, vol. 4703, pp. 412–427. Springer, Heidelberg (2007). doi:10. 1007/978-3-540-74407-8_28
16. Tofan, B., Schellhorn, G., Reif, W.: Formal verification of a lock-free stack with hazard pointers. In: Cerone, A., Pihlajasaari, P. (eds.) ICTAC 2011. LNCS, vol. 6916, pp. 239–255. Springer, Heidelberg (2011). doi:10.1007/978-3-642-23283-1_16

# GPU-Accelerated Steady-State Computation of Large Probabilistic Boolean Networks

Andrzej Mizera[1], Jun Pang[1,2(✉)], and Qixia Yuan[1]

[1] Faculty of Science, Technology and Communication, University of Luxembourg, Luxembourg, Luxembourg
{andrzej.mizera,jun.pang,qixia.yuan}@uni.lu
[2] Interdisciplinary Centre for Security, Reliability and Trust, University of Luxembourg, Luxembourg, Luxembourg

**Abstract.** Computation of steady-state probabilities is an important aspect of analysing biological systems modelled as probabilistic Boolean networks (PBNs). For small PBNs, efficient numerical methods can be successfully applied to perform the computation with the use of Markov chain state transition matrix underlying the studied networks. However, for large PBNs, numerical methods suffer from the state-space explosion problem since the state-space size is exponential in the number of nodes in a PBN. In fact, the use of statistical methods and Monte Carlo methods remain the only feasible approach to address the problem for large PBNs. Such methods usually rely on long simulations of a PBN. Since slow simulation can impede the analysis, the efficiency of the simulation procedure becomes critical. Intuitively, parallelising the simulation process can be an ideal way to accelerate the computation. Recent developments of general purpose graphics processing units (GPUs) provide possibilities to massively parallelise the simulation process. In this work, we propose a *trajectory-level parallelisation framework* to accelerate the computation of steady-state probabilities in large PBNs with the use of GPUs. To maximise the computation efficiency on a GPU, we develop a dynamical data arrangement mechanism for handling different size PBNs with a GPU, and a specific way of storing predictor functions of a PBN and the state of the PBN in the GPU memory. Experimental results show that our GPU-based parallelisation gains a speedup of approximately 400 times for a real-life PBN.

## 1 Introduction

Systems biology aims to model and analyse biological systems using mathematical and computational methods from a holistic perspective in order to provide a comprehensive, system-level understanding of cellular behaviour. Recent developments in systems biology have greatly promoted the discovery of unknown biological information, leading to the revealing of more and more large biological systems. This brings a significant challenge to computational modelling in

Q. Yuan—Supported by the National Research Fund, Luxembourg (grant 7814267).

M. Fränzle et al. (Eds.): SETTA 2016, LNCS 9984, pp. 50–66, 2016.
DOI: 10.1007/978-3-319-47677-3_4

terms of the state-space size of the system under study. Developed in 2002 by Shmulevich et al. [1,2], probabilistic Boolean networks (PBNs) is a well-suited framework for modelling large-size biological systems. Originally, PBNs is introduced as a probabilistic generalisation of the standard Boolean networks (BNs) to model gene regulatory networks (GRNs). The framework of PBNs not only takes the advantage of BNs to incorporate rule-based dependencies between genes and allow the systematic study of global network dynamics, but also is capable of dealing with uncertainty, which naturally occurs at different levels in the study of biological systems.

One of the key aspects of analysing biological systems, especially for those modelled as PBNs, is the comprehensive understanding of their long-run (steady-state) behaviour. This is vital in many contexts, e.g., attractors of a GRN were considered to characterise cellular phenotype [3]. There have been a lot of studies in analysing the steady-state behaviours of biological systems modelled as PBNs. As the dynamics of a PBN can be viewed as a discrete-time Markov chain (DTMC), it can be studied with the use of the rich theory of DTMCs. Relying on this, many numerical methods exist to compute steady-state probabilities for small-size PBNs [4,5]. In the case of large-size PBNs, however, numerical methods face the state-space explosion problem. The use of statistical methods and Monte Carlo methods are then proposed to estimate the steady-state probabilities. These methods require simulating the PBN under study for a certain length and the simulation speed is an important factor in the performance of these approaches. For large PBNs and long trajectories, a slow simulation speed could render these methods infeasible as well. A natural way to address this problem is to parallelise the simulation process.

Recent improvements in the computing power and the general purpose graphics processing units (GPUs) enable the possibilities to massively parallelise this process. In this work, we propose a *trajectory-level parallelisation framework* to accelerate the computation of steady-state probabilities in large PBNs with the use of GPUs. The architecture of a GPU is very different from that of a central processing unit (CPU), and the performance of a GPU-based program is highly related to how the synchronisation between cores is processed and how memory accessing is managed. Our framework reduces the time-consuming synchronisation cost by allowing each core to simulate one trajectory. Regarding to the memory management, we contributes in two aspects. We first develop a dynamical data arrangement mechanism for handling different size PBNs with a GPU to maximise the computation efficiency on a GPU for relatively small-size PBNs. We then propose a specific way of storing predictor functions of a PBN and the state of the PBN in the GPU memory to reduce the memory consumption and to improve the accessing speed. We show with experiments that our GPU-based parallelisation gains a speedup of more than two orders of magnitudes.

**Structure of the Paper.** We present preliminaries on PBNs and the architecture of GPUs in Sect. 2. The difficulties of parallelising the simulation of a PBN and how to overcome them are discussed in Sect. 3. We evaluate our GPU implementation in Sect. 4 and conclude our paper with some discussions in Sect. 5.

# 2    Preliminaries

## 2.1    Probabilistic Boolean Networks (PBNs)

A PBN $G(X, F)$ consists of a set of binary-valued nodes (also known as genes) $X = \{x_1, x_2, \ldots, x_n\}$ and a list of sets $\mathcal{F} = (F_1, F_2, \ldots, F_n)$. For each $i \in \{1, 2, \ldots, n\}$, the set $F_i = \{f_1^{(i)}, f_2^{(i)}, \ldots, f_{\ell(i)}^{(i)}\}$ is a collection of Boolean functions, known as *predictor functions*, for node $x_i$, where $\ell(i)$ is the number of predictor functions for node $x_i$. Each $f_j^{(i)}$ is a Boolean function defined using a subset of the nodes, referred to as *parent nodes* of $x_i$. At each time point $t$, the value of each node $x_i$ is updated with one of its predictor functions. The predictor function is selected in accordance with a probability distribution $C_i = (c_1^{(i)}, c_2^{(i)}, \ldots, c_{\ell(i)}^{(i)})$, where the individual probabilities are the *selection probabilities* for the respective elements of $F_i$ and they sum to 1. Several variants of PBNs exist due to the different way of selecting predictor functions and the synchronisation of nodes update. In this paper, we focus on the *independent synchronous* PBNs, i.e., the choice of predictor functions for each node is made independently and the values of all the nodes are updated synchronously. We use $x_i(t)$ to denote the value of node $x_i$ at time point $t$, and $s(t) = (x_1(t), x_2(t), \ldots, x_n(t))$ to denote the state of the PBN at time point $t$. The state space of the PBN is $S = \{0, 1\}^n$ and it is of size $2^n$. The transition from state $s(t)$ to state $s(t+1)$ is performed by randomly selecting a predictor function for each node $x_i$ from $F_i$ and by applying those selected predictor functions to update the values of all the nodes synchronously. Let $f(t)$ be the combination of all the selected predictor functions at time point $t$. The transition of state $s(t)$ to $s(t+1)$ can then be denoted as

$$s(t+1) = f(t)(s(t)). \tag{1}$$

A PBN can therefore be viewed as a discrete-time Markov chain (DTMC) with state space $S = \{0, 1\}^n$ and transition relation defined by Eq. 1.

In a PBN with *perturbations*, a perturbation rate $p \in (0, 1)$ is introduced and the dynamics of a PBN is guided with both perturbations and predictor functions: at each time point $t$, the value of each node $x_i$ is flipped with probability $p$; and if no flip happens, the value of each node $x_i$ is updated with selected predictor functions synchronously. Let $\gamma(t) = (\gamma_1(t), \gamma_2(t), \ldots, \gamma_n(t))$ be a perturbation vector, where each element is a Bernoulli distributed random variable with parameter $p$, i.e., $\gamma_i(t) \in \{0, 1\}$ and $\mathbb{P}(\gamma_i(t) = 1) = p$ for all $t$ and $i \in \{1, 2, \ldots, n\}$. Extending Eq. 1, the transition from $s(t)$ to $s(t+1)$ in PBNs with perturbations is given as

$$s(t+1) = \begin{cases} s(t) \oplus \gamma(t) & \text{if } \gamma(t) \neq 0 \\ f(t)(s(t)) & \text{otherwise,} \end{cases} \tag{2}$$

where $\oplus$ is the element-wise exclusive or operator for vectors. According to Eq. (2), from any state, the system can move to any other state with one transition due to perturbations. Therefore, the underlying Markov chain is in fact irreducible and aperiodic. Thus, the dynamics of a PBN with perturbations can be viewed as an ergodic DTMC [1]. Based on the ergodic theory, the long-run dynamics of a PBN with perturbations is governed by a unique limiting distribution, convergence to which is independent of the choice of the initial state.

The density of a PBN is measured with the number of predictor functions and the number of parent nodes for each predictor function. For a PBN $G$, its density is defined as $\mathcal{D}(G) = \frac{1}{n} \sum_{i=1}^{M} \phi(i)$, where $n$ is the number of nodes in $G$, $M$ is the total number of predictor functions in $G$, and $\phi(i)$ is the number of parent nodes for the $i$th predictor function.

## 2.2 GPU Architecture

We review the basics of GPU architecture and its programming approach, i.e., common unified device architecture (CUDA) released by NVIDIA.

At the physical hardware level, an NVIDIA GPU usually contains tens of streaming multiprocessors (SMs, also abbreviated as MPs), each containing a fixed number of streaming processors (SPs), fixed size of *registers*, fast *shared memory* (as shown in Fig. 1, with $N$ being the number of MPs).

Accessing registers and shared memory is fast, but the size of these two types of memory is very limited. In addition, a large size *global memory*, a small size *texture memory* and *constant memory* are available outside the MPs. Global memory has a high bandwidth (128 bytes in our GPU), but also a high latency. Accessing global memory is usually orders of magnitude slower than accessing

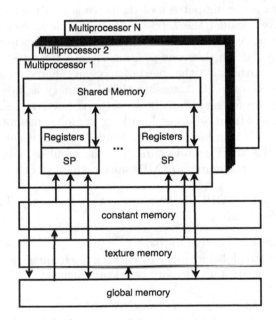

**Fig. 1.** Architecture of a GPU.

registers or shared memory. Constant memory and texture memory are memories of special type which can only store read-only data. Accessing constant memory is most efficient if all threads are accessing exactly the same data; and texture memory is better for dealing with random access. We refer to registers and shared memory as *fast memory*; global memory as *slow memory*; and constant memory and texture memory as *special memory*.

At the programming level, the programming interface CUDA is in fact an extension of C/C++. A segment of code to be run in a GPU is put into a function called a *kernel*. The kernels are then executed as a grid of blocks of threads. A thread is the finest granularity in a GPU and each thread can be viewed as a copy of the kernel. A block is a group of threads executed together in a batch. Each thread is executed in an SP and threads in a block can only be executed in one MP. One MP, however, can launch several blocks in parallel. Communications between threads in the same block are possible via shared memory. NVIDIA GPUs use a processor architecture called single instruction multiple thread (SIMT), i.e., a single instruction stream is executed via a group of 32 threads, called a *warp*. Threads within a warp are bounded together, i.e., they always execute the same instruction. Therefore, branch divergence can occur within a warp: if one thread within a warp moves to the 'if' branch of an 'if-then-else' sentence and the others choose the 'else' branch, then actually all the 32 threads will "execute" both branches, i.e., the thread moving to the 'if' branch will wait for other threads when they execute the 'else' branch and vice versa. If both branches are long, then the performance penalty is huge. Therefore, branches should be avoided as much as possible in terms of performance. Moreover, the data accessing pattern of the threads in a warp should be taken care of as well. We consider the access pattern of shared memory and global memory in this work. Accessing shared memory is most efficient if all threads in a warp are fetching data in the same position or each thread is fetching data in a different position. Otherwise, the speed of accessing shared memory is reduced by the so-called *bank conflict*. Accessing global memory is most efficient if all threads in a warp are fetching data in a *coalesced* pattern, i.e., all threads in a warp are reading data in adjacent locations in global memory. In principle, the number of threads in a block should always be an integral multiple of the warp size due to the SIMT architecture; and the number of blocks should be an integral multiple of the number of MPs since each block can only be executed in one MP.

An important task for GPU programmer is to hide latency. This can be done via the following four ways:

1. increase the number of active warps;
2. reduce the access to global memory by caching the frequently accessed data in fast memory, or in constant memory or texture memory, if the access pattern is suitable;
3. reduce bank conflict of shared memory access;
4. coalesce accesses to the global memory to use the bandwidth more efficiently.

However, the above four methods often compete with one another due to the restrictions of the hardware resources. For example, using more shared memory would restrict the number of active blocks and hence the number of active warps is limited. Therefore, a trade-off between the use of fast memory and the number of threads has to be considered carefully. We discuss this problem and provide our solution to it in Sect. 3.2.

# 3  PBN Simulation in GPU

In this section, we present how simulation of a PBN is performed in a GPU, while addressing the problems identified at the end of Sect. 2.

## 3.1  Trajectory-Level Parallelisation

In general, there are two ways for parallelising the PBN simulation process. One way is to update all nodes synchronously, i.e., each GPU thread only updates one node of a PBN; the other way is to simulate multiple trajectories simultaneously. The first way requires synchronisation among the threads, which is time-consuming in the current GPU architecture. Therefore, in our implementation, we take the second way to simulate multiple trajectories concurrently. Samples from multiple trajectories can be merged together to compute steady-state probabilities of a PBN using a combination of the two-state Markov chain approach [6] and the Gelman and Rubin method [7]. A detailed description for this combination can be found in [8]. Note that merging is performed in a CPU and no synchronization is required. We show in Fig. 2 the workflow for computing steady-state probabilities based on trajectory-level parallelisation.

Each blue box represents a kernel to be parallelised in a GPU. The first and second blue boxes perform the same task except that trajectories in the first

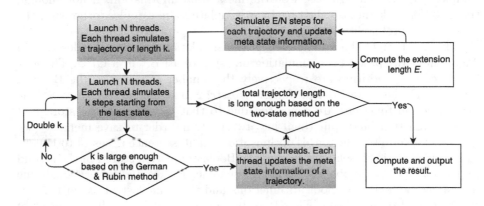

**Fig. 2.** Workflow of steady-state analysis using trajectory-level parallelisation. (Color figure online)

---

**Algorithm 1.** Simulate one step of a PBN in a GPU

```
 1: procedure SIMULATEONESTEP(n, F, extraF, p, S)
 2:     perturbed := false;
 3:     for (i := 0; i < n; i++) do
 4:         if rand() < p then perturbed := true; S[i/32] := S[i/32] ⊕ (1 ≪ (i%32));
 5:         end if                                  //the result of i/32 is ⟨⟨
 6:     end for
 7:     if perturbed then return S;
 8:     else
 9:         set array nextS to 0;
10:         for (i := 0; i < n; i++) do
11:             index := nextIndex(i); //sample the Boolean function index for node i
12:             compute the entry of the Boolean function based on index and S;
13:             v := F[index];
14:             if entry > 31 then                             //entry starts with 0
15:                 get index of the Boolean function in extraF;      //see Sect. 3.3
16:                 v := extraF[index]; entry := entry%32;
17:             end if
18:             v := v ≫ entry; nextS[i/32] := nextS[i/32] | ((v&1) ≪ (i%32));
19:         end for
20:     end if
21:     S := nextS; return S.
22: end procedure
```

---

blue box are abandoned while those in the second blue box are stored in global memory. This is due to the requirement of the Gelman and Rubin method [7] that only the second half samples are used for computing steady-state probabilities. Based on the last $k$ samples simulated in the second blue box, the third blue box computes the *meta state* information required by the two-state Markov chain approach [6]. The two-state Markov chain approach determines whether the samples are large enough based on the meta state information. If not enough, the last (forth) kernel is called again to simulate more samples; otherwise, the steady-state probability is computed.

The key part of the four kernels is the simulation process. We describe in Algorithm 1 the process for simulating one step of a PBN in a GPU. The four inputs of this algorithm are respectively the number of nodes $n$, the Boolean functions $F$, the extra Boolean functions $extraF$ and the current state $S$. The extra Boolean functions are generated due to that we optimise the storage of Boolean functions and split them into two parts in order to save memory (see Sect. 3.3 for details). Due to this optimisation, an 'if' sentence (lines 14 to 17) has to be added. This 'if' sentence fetches the Boolean function stored in the second part ($extraF$). The probability that this sentence is executed is very small due to the way we split the Boolean functions and the time cost of executing this sentence is also very small. Therefore, by paying a small penalty in terms of computational time, we are able to store Boolean functions in fast memory and gain much more speedups with the use of fast memory.

## 3.2 Data Arrangement

As mentioned in Sect. 2.2, suitable strategy for hiding latency should be carefully considered for a GPU program. Since the simulation process requires accessing the PBN information (in a random way) in each simulation step and the latency cost for frequently accessing data in slow memory is really huge, caching those information in fast and special memory results in a more efficient computation comparing to allowing more active warps. Therefore, we first try to arrange all frequently accessed data in fast and special memory as much as possible; then based on the remaining resources we calculate the optimal number of threads and blocks to launch. Since the size of fast memory is limited and the memory required to store a PBN varies from PBN to PBN, a suitable data arrangement policy is necessary. In this section, we discuss how we dynamically arrange the data in different GPU memories for different PBNs.

In principle, frequently accessed data should be put in fast memory as much as possible. We list all the frequently used data and how we arrange them in GPU memories in Table 1. As the size of the fast memory is limited and has different advantages for different data accessing modes, we save different data in different memories. Namely, those read-only data that are always or most likely to be accessed simultaneously by all threads in a warp, are put in constant memory; other read-only data are put in shared memory if possible; and the rest of the data are put in registers if possible. Since the memory required to store the frequently used data varies a lot from PBN to PBN, we propose to use a dynamic decision process to determine how to arrange some of the frequently accessed data, i.e., the data shown in the last four rows of Table 1. The dynamic process calculates the memory required to store all the data for a given PBN,

**Table 1.** Frequently accessed data arrangement.

| Data | Data type | Stored in |
|---|---|---|
| random number generator | CUDA built in | registers |
| node number | integer | constant memory |
| perturbation rate | float | constant memory |
| cumulative number of functions | short array | constant memory |
| selection probabilities of functions | float array | constant memory |
| indices of positive nodes | integer array | constant memory |
| indices of negative nodes | integer array | constant memory |
| cumulative number of parent nodes | short array | shared memory |
| Boolean functions | integer array | shared memory |
| indices for extra Boolean functions | short array | shared memory |
| parent nodes indices for each function | short array | shared/texture memory |
| current state | integer array | registers/global memory |
| next state | integer array | registers/global memory |

and determines where to put them based on their memory size. If the shared memory and registers are large enough, all the data will be stored in these two fast memories. Otherwise, they will be placed in the global memory. For the data stored in the global memory, we use two ways to speed up their access. One way is to use texture memory to speed up the access for read-only data, e.g., the parent node indices for each function. The other way is to optimise the data structure to allow a coalesced accessing pattern, e.g., the current state. We explain this in details in Sect. 3.3. This dynamical arrangement of data allows our program to explore the computation ability of a GPU as much as possible, leading to faster speedups for relatively small sparse networks.

### 3.3   Data Optimisation

As mentioned in Sect. 2.2, a GPU usually has a very limited size of fast memory and the latency can vary a lot when accessing the same memory in a different way, e.g., accessing shared memory with or without bank conflict. Therefore, we optimise the data structure of two important data, i.e., the Boolean functions (stored as truth tables) and the states of a PBN, to save space and to maximise the access speed.

**Optimisation on Boolean Functions.** A direct way to store a truth table is to use a Boolean array, which consumes one byte to store each element. Accessing an element of the truth table can be directly made by providing the index of the Boolean array. Instead, we propose to use a primitive 32-bit integer (4 bytes) type to store the truth table. Each bit of an integer stores one entry of the truth table and hence the memory usage can be reduced by 8 in maximum: 4 bytes compared to 32 bytes of a Boolean array. A 32-bit integer can store a truth table of at most 32 elements, corresponding to a Boolean function with max. 5 parent nodes. Since for real biological systems the number of parent nodes is usually small [9], in most cases one integer is enough for storing the truth table of one Boolean function. In the case of a truth table with more than 32 elements, additional integers are needed. In order to save memory and quickly locate a specific truth table, we save the additional integers in a separate array. More precisely, we use a 32-bit integer array $F$ of length $M$ to store the truth tables for all the $M$ Boolean functions and the $i$th ($i \in [0, M - 1]$) element of $F$ stores only the first 32 elements of the $i$th truth table. If the $i$th truth table contains more than 32 elements, the additional integers are stored in an extra integer array $extraF$. In addition, two index arrays $extraFIndex$ and $cumExtraFIndex$ are needed to store the index of the $i$th truth table in $extraF$. Each element of $extraFIndex$ stores one index value of the truth table which requires additional integers. The length of $extraFIndex$ is at most $M$. Each element of $cumExtraFIndex$ stores the cumulative number of additional required integers for all the truth tables whose indices are stored in $extraFIndex$.

As an example, we show how to store a truth table with 128 elements in Fig. 3. We assume that this 128-element truth table is the $i$th one among all $M$ truth tables and that it is the $j$th one among those $m$ truth tables that

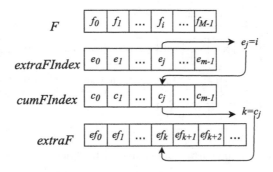

**Fig. 3.** Demonstration of storing Boolean functions in integer arrays.

require additional integers to store. Therefore, its first 32 (0–31th) elements are stored in the $i$th element of $F$ and its index $i$ is stored in the $j$th element of $extraFIndex$, denoted as $e_j$. The $j$th element of $cumExtraFIndex$, denoted as $c_j$, stores the total number of additional integers required to store the $j - 1$ truth tables whose indices are stored in the first $j - 1$ elements of $extraFIndex$. Let $cumExtraFIndex[j] = k$. The $k$th, $(k + 1)$th, and $(k + 2)$th elements of $extraF$ store the 32–127th elements of the $i$th truth table. After storing the truth tables in this way, accessing the $t$th element of the $i$th truth table can be performed in the following way. When $t \in [0, 31]$, $F[i]$ directly provides the information, and when $t \in [32, 127]$, three steps are required: (1) search the array $extraFIndex$ to find the index $j$ such that $extraFIndex[j]$ equals to $i$, (2) fetch the $j$th value of array $cumFIndex$ and let $k = cumFIndex[j]$, (3) the integer $extraF[k + (t - 32)\,\%32]$ contains the required information. Since in most cases the number of parent nodes is very limited, the array $extraFIndex$ is very small. Hence, the search of the index $j$ in the first step can be finished very quickly. In the rare case where the $extraFIndex$ array would be large, e.g., $M$ is large and the length of $extraFIndex$ would be close to $M$, it is preferable to store $extraFindex$ as an array of length $M$ and let $extraFindex[i]$ store the entry in $cumFIndex$ for the $i$th truth table so that the search phase of the first step is eliminated. The required memory for storing this truth table is reduced from 128 bytes (stored as Boolean arrays) to 20 bytes (6 integers to store the truth table and 2 shorts to store the index). In addition to saving memory, the above optimisation can also reduce the chances of bank conflict in shared memory due to the fact that accessing any entry of a truth table is performed by fetching only one integer in array $F$ in most cases. Accessing the elements in $extraFIndex$ requires additional memory fetching; however, as mentioned before, the chance for such cases to happen is very small in a real-life PBN and the gained memory space and improved data fetching pattern can compensate for this penalty.

**Optimisation on PBN States.** The optimisation of the data structure for states is similar to that for Boolean functions, i.e., states are stored as integers and each bit of an integer represent the value of a node. Therefore, a PBN with $n$ nodes requires $\lceil n/32 \rceil$ integers ($4 * \lceil n/32 \rceil$ bytes) to be stored, compared to $n$

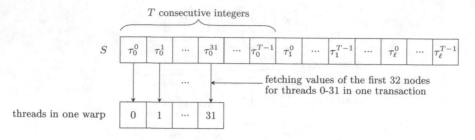

**Fig. 4.** Storing states in one array and coalesced fetching for threads in one warp.

bytes when stored as a Boolean array. During the simulation process, the current state and the next state of a PBN have to be stored. As shown in Table 1, the states are put in registers whenever possible, i.e., when the number of nodes is smaller than 129. In the case of a PBN with nodes number equal to or larger than 129, the global memory has to be used due to the limited register size (shared memory are used to store other data and would not be large enough to store states in this case). To reduce the frequency of accessing global memory, one register (32 bits) is used to cache the integer that stores the values of 32 nodes. Updating of the 32 node values is performed via the register and stored in the global memory with a single access only once after all the 32 node values are updated in the register. Moreover, states for all the threads are stored in one large integer array $S$ in the global memory and we arrange the content of this array to allow for a coalesced accessing pattern. More specifically, starting from the 0th integer, every consecutive $T$ integers store the values of 32 nodes in the $T$ threads (assuming there are $T$ threads in total). Figure 4 shows how to store states of a PBN with $n$ nodes for all the $T$ threads in an integer array $S$ and how the 32 threads in the first warp fetch the first integer in a coalesced pattern. We denote $\tau_i^j$ as the $i$th integer to store values of 32 nodes for thread $j$ and let $\ell = \lceil n/32 \rceil$. For threads in one warp, accessing the values of the same node can be performed via fetching the adjacent integers in the array $S$. This results in a coalesced accessing pattern of the global memory. Hence, all the 32 threads in one warp can fetch the data in a single data transaction.

## 4  Evaluation

We evaluate our GPU-based parallelisation framework for computing steady-state probabilities of PBNs in both randomly generated networks and a real-life biological network. All the experiments are performed on a high performance computing (HPC) machines, each of which contains a CPU of Intel Xeon E5-2680 v3 @ 2.5 GHz and an NVIDIA Tesla K80 Graphic Card with 2496 cores @824 MHz. The program is written in a combination of both Java and C, and the initial and maximum Java virtual machine heap size is set to 4 GB and 11.82 GB, respectively. The C language is used to program operations on GPUs due to the fact that no suitable Java library is currently provided for operations

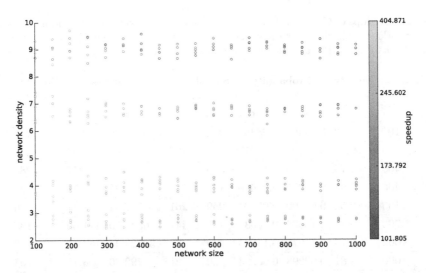

**Fig. 5.** Speedups of GPU-accelerated steady-state computation.

on NVIDIA GPUs. When launching the GPU kernels, the kernel configurations (the number of threads and blocks) are dynamically determined as mentioned in Sect. 3.2.

## 4.1   Randomly Generated Networks

The evaluation on randomly generated networks is performed on 380 PBNs, which are generated using the tool ASSA-PBN [10]. The nodes number of these networks ranges in the set $\{100, 150, 200, 250, 300, 350, 400, 450, 500, 550, 600, 650, 700, 750, 800, 850, 900, 950, 1000\}$. For each of the 380 networks, we compute one steady-state probability using both the sequential two-state Markov chain approach and our GPU-accelerated parallelisation framework. We set the three precision requirements of the two-state Markov chain approach, i.e., the confidence level $s$, the precision $r$, and the steady-state convergence parameter $\epsilon$ to 0.95, $5 \times 10^{-5}$, and $10^{-10}$ respectively. The computation time limit is set to 10 hours. In the end, we obtain 366 pairs of valid results. The 14 invalid pairs are due to that the sequential version two-state Markov chain approach is timed out (the parallel version is not). Among the 366 results, 355 (96.99 %) are comparable, i.e., the differences of computed probabilities satisfy the specified precision requirement. This result meets our confidence level requirement.

We compute the speedups of the GPU-accelerated parallelisation framework with respect to the sequential two-state Markov chain approach for those 366 valid results with the formula $speedup = \frac{s_{pa}/t_{pa}}{s_{se}/t_{se}}$, where $s_{pa}$ and $t_{pa}$ are respectively the sample size and time cost of the parallelisation framework, and $s_{se}$ and $t_{se}$ are respectively the sample size and time cost of the sequential approach. The speedups are plotted in Fig. 5. As can be seen from this figure, we obtain speedups approximately between 102 and 405. There are some small gaps with

**Table 2.** Speedups of GPU-accelerated steady-state computation of 8 randomly generated networks. seq. is short for the sequential two-state Markov chain approach; while par. is short for the GPU-accelerated parallel approach.

| Node # | Density | Probability | | Sample size (million) | | Time (s) | | Speedup |
|---|---|---|---|---|---|---|---|---|
| | | seq. | par. | seq. | par. | seq. | par. | |
| 100 | 2.53 | 0.24409 | 0.24401 | 350 | 367 | 2637.06 | 6.84 | 405 |
| 100 | 9.32 | 0.36221 | 0.36217 | 426 | 427 | 3429.06 | 13.04 | 264 |
| 400 | 2.75 | 0.12003 | 0.12002 | 316 | 318 | 7615.72 | 26.77 | 286 |
| 400 | 8.98 | 0.04657 | 0.04660 | 135 | 137 | 3908.25 | 20.79 | 190 |
| 700 | 2.64 | 0.05800 | 0.05794 | 259 | 261 | 8567.52 | 39.27 | 220 |
| 700 | 9.41 | 0.10632 | 0.10634 | 438 | 441 | 16541.79 | 121.60 | 137 |
| 1000 | 2.73 | 0.14675 | 0.14673 | 838 | 839 | 30626.44 | 184.44 | 166 |
| 1000 | 8.81 | 0.00298 | 0.00293 | 20 | 21 | 792.86 | 8.10 | 103 |

respect to the densities of those networks, e.g., no networks with density between 5 and 6. Those gaps are due to the way how those networks are randomly generated, i.e., one cannot force the ASSA-PBN tool to generate a PBN with a fixed density, but can only provide the following information to affect the density: the number of nodes, the maximum (minimum) number of functions for each node, and the maximum (minimum) number of parent nodes for each function. However, even with the gaps, the tendency of the changes of speedups with respect to densities can be well observed. In fact, this observation is similar to that of the network size. With the network size decreasing and the density decreasing, our GPU-accelerated parallelisation framework gains higher speedups. This is due to our dynamic way of arranging data for different size PBNs: data for relatively small[1] and sparse networks can be arranged in the fastest memory.

To demonstrate the computation details, we select 8 pairs among the 366 results and show in Table 2 the computed probabilities, the sample size (in millions) and the time cost (in seconds) for computing the steady-state probabilities using both the sequential two-state Markov chain approach and the GPU-accelerated parallelisation framework. The two approaches generated comparable results using similar length of samples while our GPU-accelerated parallelisation framework shows speedups of more than two orders of magnitude. All detailed results for the 380 networks can be found at http://satoss.uni.lu/software/ASSA-PBN/benchmark/.

---

[1] In fact all the networks used in this subsection should be called large-size PBNs since the network with the smallest size has already contained $2^{100} \approx 10^{30}$ states.

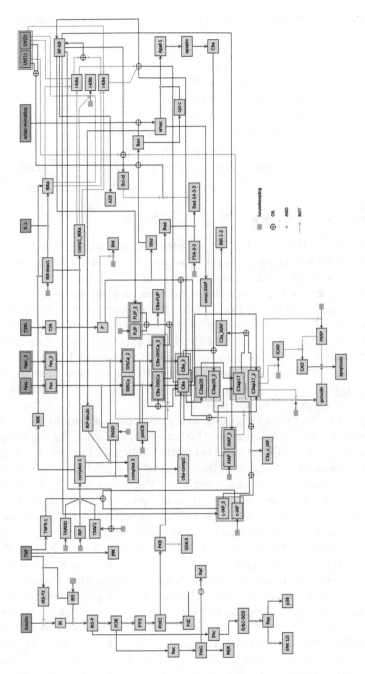

**Fig. 6.** The wiring of the probabilistic Boolean model of apoptosis in [5].

**Table 3.** Speedups of GPU-accelerated steady-state computation of a real-life apoptosis network.

| Steady-state | | | Probability | | Sample size (million) | | Time (s) | | Speedup |
|---|---|---|---|---|---|---|---|---|---|
| R | C | F | seq. | par. | seq. | par. | seq. | par. | |
| 0 | 1 | 1 | 0.003236 | 0.003237 | 589.05 | 590.77 | 3866.04 | 9.28 | 417.81 |
| 1 | 1 | 1 | 0.990053 | 0.990046 | 1809.27 | 1811.71 | 11476.00 | 28.08 | 409.20 |
| 1 | 0 | 1 | 0.005592 | 0.005590 | 1015.95 | 1021.07 | 6662.26 | 15.89 | 421.47 |
| 1 | 1 | 0 | 0.001082 | 0.001080 | 197.80 | 200.12 | 1281.45 | 3.27 | 396.60 |
| * | 1 | 1 | 0.993289 | 0.993288 | 1222.83 | 1241.06 | 7967.42 | 19.30 | 418.99 |
| * | 1 | 0 | 0.001082 | 0.001087 | 197.29 | 206.37 | 1096.90 | 3.36 | 341.62 |
| * | 0 | 1 | 0.005614 | 0.005624 | 1021.87 | 1039.35 | 6725.25 | 16.17 | 422.98 |

## 4.2 An Apoptosis Network

We have analysed a PBN model of an apoptosis network using the sequential two-state Markov chain approach in [6]. The apoptosis network was originally published in [11] as a BN model and cast into the PBN framework in [5]. The PBN model (as shown in Fig. 6) contains 91 nodes and 107 Boolean functions. The selection probabilities of the Boolean functions were fitted to experimental data in [5]. We took the 20 best fitted parameter sets and performed the influence analyses for them. Although we managed to finish this analysis in an affordable amount of time due to an efficient implementation of a sequential PBN simulator, the analysis was still very expensive in terms of computation time since the required trajectories were very long and we needed to compute steady-state probabilities for a number of different states.

In this work, we re-perform part of the influence analyses from [6] using our GPU-accelerated parallel two-state Markov chain approach. In the influence analysis, we consider the PBN with the best fitted values and we aim to compute the *long-term influences* on complex2 from each of its parent nodes: RIP-deubi, complex1, and FADD, in accordance with the definition in [12]. In order to compute this long-term influence, seven different steady-state probabilities are required. We show in the first column of Table 3 the values of the seven steady-states. The three numbers or "*" with two numbers respectively represent the steady-state values of the three genes RIP-deubi, complex1, and FADD: 0 represents active; 1 represents inactive; and "*" represents irrelevant. We compute those seven different steady-state probabilities using both the sequential two-state Markov chain approach and the GPU-accelerated parallelisation framework. We show in Table 3 the computed steady-state probabilities, the sample size (in millions), the time cost (in seconds), and the speedups we obtain for this computation. The confidence level $s$, precision $r$, and the steady-state convergence parameter $\epsilon$ of this computation are set to 0.95, $5 \times 10^{-6}$, and $10^{-10}$ respectively. The density of the network is approximately 1.78. The two approaches compute comparable steady-state probabilities with similartrajectory lengths;

while our GPU-accelerated parallelisation framework reduces the time cost by approximately 400 times. The total time cost for computing the seven probabilities is reduced from about 11 hours to approximately 1.5 min.

## 5    Conclusion and Future Works

In this paper, we have proposed a trajectory-level parallelisation framework to accelerate the computation of steady-state probabilities for large PBNs with the use of GPUs. Our work contributes in three aspects in maximising the performance of a GPU when computing the steady-state probabilities. First, we reduce the time consuming synchronisation cost between GPU cores by allowing each core to simulate all nodes of one trajectory. Secondly, we propose a dynamical data arrangement mechanism for handling different size PBNs with a GPU. This leads to large speedups for handling relatively small-size PBNs. Lastly, we develop a specific way of storing predictor functions of a PBN and the state of the PBN in the GPU memory to save space and to accelerate the memory accessing. We show with experiments that our GPU-based parallelisation gains a speedup of more than two orders of magnitudes. Evaluation on a real-life apoptosis network shows that our GPU-based parallelisation obtains a speedup of approximately 400 times.

There are two directions for our future works. One is to apply our work to analyse large realistic biological models. The other one is to optimise the current structure to better handle very dense and huge networks.

## References

1. Shmulevich, I., Dougherty, E.R.: Probabilistic Boolean Networks: The Modeling and Control of Gene Regulatory Networks. SIAM Press (2010)
2. Trairatphisan, P., Mizera, A., Pang, J., Tantar, A.A., Schneider, J., Sauter, T.: Recent development and biomedical applications of probabilistic Boolean networks. Cell Commun. Signal. **11**, 46 (2013)
3. Kauffman, S.A.: Homeostasis and differentiation in random genetic control networks. Nature **224**, 177–178 (1969)
4. Shmulevich, I., Gluhovsky, I., Hashimoto, R., Dougherty, E., Zhang, W.: Steady-state analysis of genetic regulatory networks modelled by probabilistic Boolean networks. Comp. Funct. Genomics **4**(6), 601–608 (2003)
5. Trairatphisan, P., Mizera, A., Pang, J., Tantar, A.A., Sauter, T.: optPBN: An optimisation toolbox for probabilistic Boolean networks. PLOS ONE **9**(7), e98001 (2014)
6. Mizera, A., Pang, J., Yuan, Q.: Reviving the two-state Markov chain approach (Technical report) (2015). http://arxiv.org/abs/1501.01779
7. Gelman, A., Rubin, D.: Inference from iterative simulation using multiple sequences. Stat. Sci. **7**(4), 457–472 (1992)
8. Mizera, A., Pang, J., Yuan, Q.: Parallel approximate steady-state analysis of large probabilistic Boolean networks. In: Proceedings of 31st ACM Symposium on Applied Computing, pp. 1–8 (2016)

9. Harri, L., Sampsa, H., Ilya, S., Olli, Y.H.: Relationships between probabilistic Boolean networks and dynamic Bayesian networks as models of gene regulatory networks. Signal Process. **86**(4), 814–834 (2006)
10. Mizera, A., Pang, J., Yuan, Q.: ASSA-PBN: An approximate steady-state analyser of probabilistic Boolean networks. In: Finkbeiner, B., Pu, G., Zhang, L. (eds.) ATVA 2015. LNCS, vol. 9364, pp. 214–220. Springer, Heidelberg (2015). doi:10. 1007/978-3-319-24953-7_16
11. Schlatter, R., Schmich, K., Vizcarra, I.A., Scheurich, P., Sauter, T., Borner, C., Ederer, M., Merfort, I., Sawodny, O.: ON/OFF and beyond - a Boolean model of apoptosis. PLOS Comput. Biol. **5**(12), e1000595 (2009)
12. Shmulevich, I., Dougherty, E.R., Kim, S., Zhang, W.: Probabilistic Boolean networks: a rule-based uncertainty model for gene regulatory networks. Bioinformatics **18**(2), 261–274 (2002)

# Behavioural Pseudometrics for Nondeterministic Probabilistic Systems

Wenjie Du[1], Yuxin Deng[2(✉)], and Daniel Gebler[3]

[1] Shanghai Normal University, Shanghai, China
[2] Shanghai Key Laboratory of Trustworthy Computing,
East China Normal University, Shanghai, China
yxdeng@sei.ecnu.edu.cn
[3] VU University Amsterdam, Amsterdam, Netherlands

**Abstract.** For the model of probabilistic labelled transition systems that allow for the co-existence of nondeterminism and probabilities, we present two notions of bisimulation metrics: one is state-based and the other is distribution-based. We provide a sound and complete modal characterisation for each of them, using real-valued modal logics based on Hennessy-Milner logic. The logic for characterising the state-based metric is much simpler than an earlier logic by Desharnais et al. as it uses only two non-expansive operators rather than the general class of non-expansive operators. For the kernels of the two metrics, which correspond to two notions of bisimilarity, we give a comprehensive comparison with some typical distribution-based bisimilarities in the literature.

## 1 Introduction

Bisimulation is an important proof technique for establishing behavioural equivalences of concurrent systems. In probabilistic concurrency theory, there are roughly two kinds of bisimulations: one is state-based that is directly defined over states and then lifted to distributions, and the other is distribution-based as it is a relation between distributions. The former is originally defined in [34] to represent a branching time semantics; the latter, as defined in [13,21,28], represents a linear time semantics.

In correspondence with those bisimulations, there are two notions of behavioural pseudometrics (simply called metrics in the current work). They are more robust ways of formalising behavioural similarity between formal systems than bisimulations because, particularly in the probabilistic setting, bisimulations are too sensitive to probabilities (a very small perturbation of the probabilities would render two systems non-bisimilar). A metric gives a quantitative measure of the distance between two systems and distance 0 usually means that the two systems are bisimilar. A logical characterisation of the state-based bisimulation

Y. Deng—Partially supported by the National Natural Science Foundation of China (61672229, 61261130589), Shanghai Municipal Natural Science Foundation (16ZR1409100), and ANR 12IS02001 PACE.

© Springer International Publishing AG 2016
M. Fränzle et al. (Eds.): SETTA 2016, LNCS 9984, pp. 67–84, 2016.
DOI: 10.1007/978-3-319-47677-3_5

metric for labelled Markov processes is given in [16]. For a more general model of labelled concurrent Markov chains (LCMCs) that allow for the co-existence of nondeterminism and probabilities, a weak bisimulation metric is proposed in [17]. Its logical characterisation uses formulae like $h \circ f$, which is the composition of formula $f$ with any non-expansive operator $h$ on the interval $[0,1]$, i.e. $|h(x)-h(y)| \leq |x-y|$ for any $x, y \in [0,1]$. A natural question then arises: instead of the general class of non-expansive operators, is it possible to use only a few simple non-expansive operators without losing the capability of characterising the bisimulation metric?

In the current work, we give a positive answer to the above question. For simplicity of presentation, we focus on strong bisimulation metrics. But the proof idea can be generalised to the weak case. We work in the framework of probabilistic labelled transition systems (pLTSs) that are essentially the same as LCMCs, so the interplay of nondeterminism and probabilities is allowed. We provide a modal characterisation of the state-based bisimulation metric closely in line with the classical Hennessy-Milner logic (HML) [27]. Our variant of HML makes use of state formulae and distribution formulae, which are formulae evaluated at states and distributions, respectively, and yield success probabilities. We use merely two non-expansive operators: negation ($\neg\phi$) and testing ($\phi \ominus p$). Negation is self-explanatory and the testing operator checks if a state satisfies a property with certain threshold probability. More precisely, if state $s$ satisfies formula $\phi$ with probability $q$, then it satisfies $\neg\phi$ with probability $1 - q$, and satisfies $\phi \ominus p$ with probability $q - p$ if $q > p$ and 0 otherwise. In other words, we do not need the general class of non-expansive operators because negation and testing, together with other modalities inherited from the classical HML, are expressive enough to characterise bisimulation metrics[1]. As regards to the characterisation of distribution-based bisimulation metric, we drop state formulae and use distribution formulae only. In addition, we show that the distribution-based metric is a lower bound of the state-based metric when the latter is lifted to distributions.

The kernels of the two metrics generate two notions of bisimilarities: one is state-based and the other is distribution-based. The state-based bisimilarity is widely accepted by the community of probabilistic concurrency theory, and it admits elegant characterisations from metric, logical, and algorithmic perspectives [10]. On the contrary, there is no general agreement on what is a good notion of distribution-based bisimilarity. We compare the two bisimilarities induced by our metrics with some typical notions of distribution-based bisimilarities proposed in the literature. Our distribution-based bisimilarity turns out to coincide with the one defined in [21] and they constitute the coarsest bisimilarity for distributions.

---

[1] Notice that we do not claim that negation and testing operators, plus some constant functions, suffice to represent all the non-expansive operators on the unit interval. That claim is too strong to be true. For example, the operator $f(x) = \frac{1}{2}x$ cannot be represented by those operators.

The rest of this paper is organised as follows. Section 2 provides some basic concepts on pLTSs. Section 3 defines a two-sorted modal logic that leads to a sound and complete characterisation of the state-based bisimulation metric. Section 4 gives a similar characterisation for the distribution-based bisimulation metric. In Sect. 5 we compare the two metrics discussed in the previous two sections. In Sect. 6 we compare the two bisimilarities generated by the two metrics with some distribution-based bisimilarities appeared in the literature. In Sect. 7 we review some related work. Finally, we conclude in Sect. 8.

## 2 Preliminaries

Let $S$ be a countable set. A *(discrete) probability subdistribution* over $S$ is defined as a function $\Delta : S \to [0,1]$ with $\sum_{s \in S} \Delta(s) \leq 1$. It is a *(full) distribution* if $\sum_{s \in S} \Delta(s) = 1$. Its *support*, written $\lceil \Delta \rceil$, is the set $\{s \in S \mid \Delta(s) > 0\}$. Let $\mathcal{D}_{sub}(S)$ (resp. $\mathcal{D}(S)$) denote the set of all subdistributions (resp. distributions) over $S$. We use $\varepsilon$ to stand for the empty subdistribution, that is $\varepsilon(s) = 0$ for any $s \in S$. We write $\overline{s}$ for the point distribution, satisfying $\overline{s}(t) = 1$ if $t = s$, and 0 otherwise. The *total mass* of subdistribution $\Delta$, written $|\Delta|$, is defined as $\sum_{s \in S} \Delta(s)$. A *weight function* $\omega \in \mathcal{D}(S \times S)$ for $(\Delta, \Theta) \in \mathcal{D}(S) \times \mathcal{D}(S)$ is given if $\sum_{t \in S} \omega(s,t) = \Delta(s)$ and $\sum_{s \in S} \omega(s,t) = \Theta(t)$ for all $s,t \in S$. We denote the set of all weight functions for $(\Delta, \Theta)$ by $\Omega(\Delta, \Theta)$.

A *metric* $d$ over a space $S$ is a distance function $d : S \times S \to \mathbb{R}_{\geq 0}$ satisfying: (i) $d(s,t) = 0$ iff $s = t$ (isolation), (ii) $d(s,t) = d(t,s)$ (symmetry), (iii) $d(s,t) \leq d(s,u) + d(u,t)$ (triangle inequality), for any $s,t,u \in S$. If we replace (i) with $d(s,s) = 0$, we obtain a *pseudometric*. In this paper we are interested in pseudometrics because two distinct states can still be at distance zero if their behaviour is similar. But for simplicity, we often use the term metrics though we really mean pseudometrics. Let $c \in \mathbb{R}_{\geq 0}$ be a positive real number. A metric $d$ over $S$ is $c$-bounded if $d(s,t) \leq c$ for any $s,t \in S$.

Let $d : S \times S \to [0,1]$ be a metric over $S$. We can lift it to be a metric over $\mathcal{D}(S)$ by using the *Kantorowich metric* [31] $K(d) : \mathcal{D}(S) \times \mathcal{D}(S) \to [0,1]$ defined via a linear programming problem as follows:

$$K(d)(\Delta, \Theta) = \min_{\omega \in \Omega(\Delta, \Theta)} \sum_{s,t \in S} d(s,t) \cdot \omega(s,t) \tag{1}$$

for $\Delta, \Theta \in \mathcal{D}(S)$. The dual of the above linear programming problem is the following

$$\max \sum_{s \in S} (\Delta(s) - \Theta(s)) x_s, \text{ subject to } 0 \leq x_s \leq 1 \\ \forall s,t \in S: \ x_s - x_t \leq d(s,t) . \tag{2}$$

The duality theorem in linear programming guarantees that both problems have the same optimal value.

Let $\hat{d}\colon \mathcal{D}(\mathtt{S}) \times \mathcal{D}(\mathtt{S}) \to [0,1]$ be a metric over $\mathcal{D}(\mathtt{S})$. We can lift it to be a metric over the powerset of $\mathcal{D}(\mathtt{S})$, written $\mathcal{P}(\mathcal{D}(\mathtt{S}))$, in the standard way by using the *Hausdorff metric* $H(\hat{d})\colon \mathcal{P}(\mathcal{D}(\mathtt{S})) \times \mathcal{P}(\mathcal{D}(\mathtt{S})) \to [0,1]$ given as follows

$$H(\hat{d})(\Pi_1, \Pi_2) = \max\{ \sup_{\Delta \in \Pi_1} \inf_{\Theta \in \Pi_2} \hat{d}(\Delta, \Theta),\ \sup_{\Theta \in \Pi_2} \inf_{\Delta \in \Pi_1} \hat{d}(\Theta, \Delta)\}$$

for all $\Pi_1, \Pi_2 \subseteq \mathcal{D}(\mathtt{S})$, whereby $\inf \emptyset = 1$ and $\sup \emptyset = 0$.

Probabilistic labelled transition systems (pLTSs) generalize labelled transition systems by allowing for probabilistic choices in the transitions. They are essentially *simple probabilistic automata* [39] without initial states.

**Definition 1.** *A probabilistic labelled transition system is a triple* $(S, A, \to)$, *where* $S$ *is a countable set of states,* $A$ *is a countable set of actions, and the relation* $\to\ \subseteq S \times A \times \mathcal{D}(S)$ *is a transition relation.*

We write $s \xrightarrow{a} \Delta$ for $(s, a, \Delta) \in\ \to$ and $s \xarrownot{a}$ if there is no $\Delta$ satisfying $s \xrightarrow{a} \Delta$. We let $der(s, a) = \{\Delta \mid s \xrightarrow{a} \Delta\}$ be the set of all $a$-successor distributions of $s$. A pLTS is *image-finite* (resp. *deterministic* or *reactive*) if for any state $s$ and action $a$ the set $der(s, a)$ is finite (resp. has at most one element). In the current work, we focus on image-finite pLTSs with finitely many states.

# 3    State-Based Bisimulation Metrics

We consider the complete lattice $([0,1]^{S \times S}, \sqsubseteq)$ defined by $d \sqsubseteq d'$ iff $d(s,t) \le d'(s,t)$, for all $s,t \in S$. For any $D \subseteq [0,1]^{S \times S}$ the least upper bound is given by $(\bigsqcup D)(s,t) = \sup_{d \in D} d(s,t)$, and the greatest lower bound is given by $(\bigsqcap D)(s,t) = \inf_{d \in D} d(s,t)$ for all $s,t \in S$. The bottom element $\mathbf{0}$ is the constant zero function $\mathbf{0}(s,t) = 0$ and the top element $\mathbf{1}$ is the constant one function $\mathbf{1}(s,t) = 1$ for all $s,t \in S$.

**Definition 2.** *A 1-bounded metric $d$ on $S$ is a state-based bisimulation metric if for all $s,t \in S$ with $d(s,t) < 1$, whenever $s \xrightarrow{a} \Delta$ then there exists some $t \xrightarrow{a} \Delta'$ with $K(d)(\Delta, \Delta') \le d(s,t)$.*

The smallest (wrt. $\sqsubseteq$) state-based bisimulation metric, denoted by $\mathbf{d}_s$, is called *state-based bisimilarity metric*. Its kernel is the state-based bisimilarity as defined in [34, 39]. Note that $\mathbf{0}$ does not satisfy Definition 2 for general pLTSs, thus is not a state-based bisimulation metric.

*Example 3.* Let us calculate the distance between states $s$ and $t$ in Fig. 1. Firstly, observe that $\mathbf{d}_s(s_2, t_3) = 0$ because $s_2$ is bisimilar to $t_3$ while $\mathbf{d}_s(s_3, t_3) = 1$ because the two states $s_3$ and $t_3$ perform completely different actions. Secondly, let $\Delta = \frac{1}{2}\overline{s_2} + \frac{1}{2}\overline{s_3}$ and $\Theta = \overline{t_3}$. We see that

$$\begin{aligned}
K(\mathbf{d}_s)(\Delta, \Theta) &= \min_{\omega \in \Omega(\Delta, \Theta)} \mathbf{d}_s(s_2, t_3) \cdot \omega(s_2, t_3) + \mathbf{d}_s(s_3, t_3) \cdot \omega(s_3, t_3) \\
&= \min_{\omega \in \Omega(\Delta, \Theta)} 0 \cdot \omega(s_2, t_3) + 1 \cdot \omega(s_3, t_3) \\
&= 0 \cdot \tfrac{1}{2} + 1 \cdot \tfrac{1}{2} \\
&= \tfrac{1}{2}
\end{aligned}$$

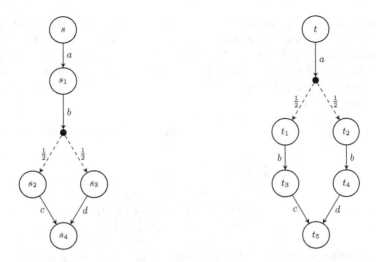

**Fig. 1.** $\mathbf{d}_s(s,t) = \frac{1}{2}$

Here the only legitimate weight function is $\omega$ with $\omega(s_2,t_3) = \omega(s_3,t_3) = \frac{1}{2}$. It follows that $\mathbf{d}_s(s_1,t_1) = \frac{1}{2}$. Similarly, we get $\mathbf{d}_s(s_1,t_2) = \frac{1}{2}$. Then it is not difficult to see that

$$K(\mathbf{d}_s)(\overline{s_1}, \frac{1}{2}\overline{t_1} + \frac{1}{2}\overline{t_2}) = \mathbf{d}_s(s_1,t_1) \cdot \frac{1}{2} + \mathbf{d}_s(s_1,t_2) \cdot \frac{1}{2} = \frac{1}{2}$$

from which we finally obtain $\mathbf{d}_s(s,t) = \frac{1}{2}$.

The above coinductively defined bisimilarity metric can be reformulated as a fixed point of a monotone functional operator. Let us define the functional operator $F_s: [0,1]^{S \times S} \to [0,1]^{S \times S}$ for $d: S \times S \to [0,1]$ and $s,t \in S$ by

$$F_s(d)(s,t) = \sup_{a \in A}\{H(K(d))(der(s,a), der(t,a))\}. \tag{3}$$

It can be shown that $F_s$ is monotone and its least fixed point is given by $\bigsqcup d_i$, where $d_0 = \mathbf{0}$ and $d_{i+1} = F_s(d_i)$ for all $i \in \mathbb{N}$.

**Proposition 4.** $\mathbf{d}_s$ *is the least fixed point of* $F_s$. $\qquad\square$

Essentially the same property as Proposition 4 has appeared in [17].

Now we proceed by defining a real-valued modal logic based on Hennessy-Milner logic [27], called metric HML, to characterize the bisimilarity metric. It is motivated by [4,16,17,30].

**Definition 5.** *Our metric HML is two-sorted and has the following syntax:*

$$\varphi ::= \top \mid \neg\varphi \mid \varphi \ominus p \mid \varphi_1 \wedge \varphi_2 \mid \langle a \rangle \psi$$
$$\psi ::= [\varphi] \mid \neg\psi \mid \psi \ominus p \mid \psi_1 \wedge \psi_2$$

*with* $a \in A$ *and* $p \in [0,1]$.

Let $\mathcal{L}$ denote the set of all metric HML formulae, $\varphi$ range over the set of all *state formulae* $\mathcal{L}^S$, and $\psi$ range over the set of all *distribution formulae* $\mathcal{L}^D$. The two kinds of formulae are defined simultaneously. The operator $\varphi \ominus p$ tests if a state passes $\varphi$ with probability at least $p$. Each state formula $\varphi$ immediately induces a distribution formula $[\varphi]$. Sometimes we abbreviate $\langle a \rangle [\varphi]$ as $\langle a \rangle \varphi$. All other operators are standard.

**Definition 6.** *A state formula $\varphi \in \mathcal{L}^S$ evaluates in $s \in S$ as follows:*

$$\llbracket \top \rrbracket(s) = 1$$
$$\llbracket \neg \varphi \rrbracket(s) = 1 - \llbracket \varphi \rrbracket(s)$$
$$\llbracket \varphi \ominus p \rrbracket(s) = \max(\llbracket \varphi \rrbracket(s) - p, \ 0)$$
$$\llbracket \varphi_1 \wedge \varphi_2 \rrbracket(s) = \min(\llbracket \varphi_1 \rrbracket(s), \llbracket \varphi_2 \rrbracket(s))$$
$$\llbracket \langle a \rangle \psi \rrbracket(s) = \max\nolimits_{s \xrightarrow{a} \Delta} \llbracket \psi \rrbracket(\Delta)$$

*and a distribution formula $\psi \in \mathcal{L}^D$ evaluates in $\Delta \in \mathcal{D}(S)$ as follows:*

$$\llbracket [\varphi] \rrbracket(\Delta) = \sum_{s \in S} \Delta(s) \cdot \llbracket \varphi \rrbracket(s)$$
$$\llbracket \neg \psi \rrbracket(\Delta) = 1 - \llbracket \psi \rrbracket(\Delta)$$
$$\llbracket \psi \ominus p \rrbracket(\Delta) = \max(\llbracket \psi \rrbracket(\Delta) - p, 0)$$
$$\llbracket \psi_1 \wedge \psi_2 \rrbracket(\Delta) = \min(\llbracket \psi_1 \rrbracket(\Delta), \llbracket \psi_2 \rrbracket(\Delta)).$$

We often use constant formulae e.g. $p$ for any $p \in [0, 1]$ with the semantics $\llbracket p \rrbracket(s) = p$, which is derivable in the above logic by letting $p = \top \ominus (1 - p)$. Moreover, we write $\varphi \oplus p$ for $\neg((\neg \varphi) \ominus p)$ with the semantics $\llbracket \varphi \oplus p \rrbracket(s) = \min(\llbracket \varphi \rrbracket(s) + p, \ 1) = 1 - \max(1 - \llbracket \varphi \rrbracket(s) - p, 0)$. In the presence of negation and conjunction we can derive disjunction by letting $\varphi_1 \vee \varphi_2$ be $\neg(\neg \varphi_1 \wedge \neg \varphi_2)$. Intuitively, $\llbracket \varphi \rrbracket(s)$ measures the degree that formula $\varphi$ is satisfied by state $s$; similarly for distribution formulae. Therefore, negation is naturally interpreted as complement, conjunction as minimum and disjunction as maximum[2]. The formula $\langle a \rangle \psi$ specifies the property for a state to perform action $a$ and result in a possible distribution to satisfy $\psi$. Because of nondeterminism, from state $s$ there may be several outgoing transitions labelled by the same action $a$, e.g. $s \xrightarrow{a} \Delta_i$ with $i \in I$. We take the optimal case by taking $\llbracket \langle a \rangle \psi \rrbracket(s)$ to be the maximal $\llbracket \psi \rrbracket(\Delta_i)$ when $i$ ranges over $I$.

The above metric HML induces two natural logical metrics $\mathbf{d}_s^{ls}$ and $\mathbf{d}_s^{ld}$ on states and distributions respectively, by letting

$$\mathbf{d}_s^{ls}(s, t) = \sup\nolimits_{\varphi \in \mathcal{L}^S} |\llbracket \varphi \rrbracket(s) - \llbracket \varphi \rrbracket(t)|$$
$$\mathbf{d}_s^{ld}(\Delta, \Theta) = \sup\nolimits_{\psi \in \mathcal{L}^D} |\llbracket \psi \rrbracket(\Delta) - \llbracket \psi \rrbracket(\Theta)|.$$

---

[2] Since we will compare our logic with that in [17], it is better for our semantic interpretation to be consistent with that in the aforementioned work. In the literature, there are also other ways of interpreting conjunction and disjunction in probabilistic settings, see e.g. [3, 29].

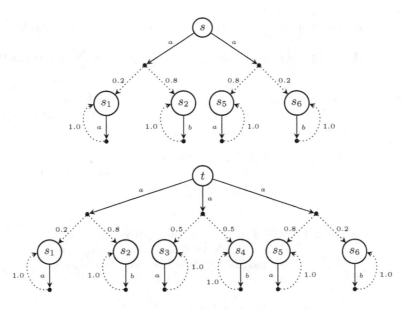

**Fig. 2.** $\mathbf{d}_s^{ls}(s,t) = 0.3$

*Example 7.* Consider the two probabilistic systems depicted in Fig. 2. We have the formula $\varphi = \langle a\rangle\psi$ where $\psi = [\langle a\rangle\top] \wedge [\langle b\rangle\top]$ and would like to know the difference between $s$ and $t$ given by $\varphi$. Let

$$\Delta_1 = 0.2 \cdot \overline{s_1} + 0.8 \cdot \overline{s_2}$$
$$\Delta_2 = 0.8 \cdot \overline{s_5} + 0.2 \cdot \overline{s_6}$$
$$\Delta_3 = 0.5 \cdot \overline{s_3} + 0.5 \cdot \overline{s_4}$$

Note that $[\![\langle a\rangle\top]\!](s_1) = 1$ and $[\![\langle a\rangle\top]\!](s_2) = 0$. Then

$$[\![[\langle a\rangle\top]]\!](\Delta_1) = 0.2 \cdot [\![\langle a\rangle\top]\!](s_1) + 0.8 \cdot [\![\langle a\rangle\top]\!](s_2) = 0.2.$$

Similarly, $[\![[\langle b\rangle\top]]\!](\Delta_1) = 0.8$. It follows that

$$[\![\psi]\!](\Delta_1) = \min([\![[\langle a\rangle\top]]\!](\Delta_1),\ [\![[\langle b\rangle\top]]\!](\Delta_1)) = 0.2.$$

With similar arguments, we see that $[\![\psi]\!](\Delta_2) = 0.2$ and $[\![\psi]\!](\Delta_3) = 0.5$. Therefore, we can calculate that

$$[\![\varphi]\!](s) = \max([\![\psi]\!](\Delta_1), [\![\psi]\!](\Delta_2)) = 0.2$$
$$[\![\varphi]\!](t) = \max([\![\psi]\!](\Delta_1), [\![\psi]\!](\Delta_2), [\![\psi]\!](\Delta_3)) = 0.5.$$

So the difference between $s$ and $t$ with respect to $\varphi$ is $|[\![\varphi]\!](s) - [\![\varphi]\!](t)| = 0.3$. In fact we also have $\mathbf{d}_s^{ls}(s,t) = 0.3$.

In the presence of testing operators in state formulae, one might wonder if the testing operators in distribution formulae can be removed. Unfortunately, this is not the case, as indicated by the following example.

*Example 8.* At first sight the following two equations seem to be sound.

$$[\![\varphi]\!] \ominus p]\!](\Delta) = [\![\varphi \ominus p]\!](\Delta) \quad \text{and} \quad [\![\psi]\!](\sum_i p_i \Delta_i) = \sum_i p_i([\![\psi]\!](\Delta_i))$$

However, in general they do not hold, as witnessed by the counterexamples below. Let $\varphi = \langle b \rangle \top$, $\psi = [\varphi] \ominus 0.5$ and the distribution $\Delta_1$ be the same as in Example 7. Then we have

$$\begin{aligned}
[\![\varphi] \ominus 0.5]\!](\Delta_1) &= \max([\![\varphi]\!](\Delta_1) - 0.5,\ 0) \\
&= \max(0.2[\![\langle b \rangle \top]\!](\overline{s_1}) + 0.8[\![\langle b \rangle \top]\!](\overline{s_2}) - 0.5,\ 0) \\
&= \max(0.2 \cdot 0 + 0.8 \cdot 1 - 0.5,\ 0) \\
&= 0.3
\end{aligned}$$

$$\begin{aligned}
[\![\varphi \ominus 0.5]\!](\Delta_1) &= 0.2[\![\varphi \ominus 0.5]\!](s_1) + 0.8[\![\varphi \ominus 0.5]\!](s_2) \\
&= 0.2 \max([\![\varphi]\!](s_1) - 0.5,\ 0) + 0.8 \max([\![\varphi]\!](s_2) - 0.5,\ 0) \\
&= 0.2 \max(0 - 0.5,\ 0) + 0.8 \max(1 - 0.5,\ 0) \\
&= 0.4
\end{aligned}$$

$$\begin{aligned}
0.2[\![\psi]\!](\overline{s_1}) + 0.8[\![\psi]\!](\overline{s_2}) &= 0.2[\![\varphi] \ominus 0.5]\!](\overline{s_1}) + 0.8[\![\varphi] \ominus 0.5]\!](\overline{s_2}) \\
&= 0.2 \max([\![\varphi]\!](\overline{s_1}) - 0.5,\ 0) + 0.8 \max([\![\varphi]\!](\overline{s_2}) - 0.5,\ 0) \\
&= 0.2 \max(0 - 0.5,\ 0) + 0.8 \max(1 - 0.5,\ 0) \\
&= 0.4
\end{aligned}$$

So we see that $[\![\varphi] \ominus 0.5]\!](\Delta_1) \neq [\![\varphi \ominus 0.5]\!](\Delta_1)$ and $[\![\psi]\!](\Delta_1) \neq 0.2[\![\psi]\!](\overline{s_1}) + 0.8[\![\psi]\!](\overline{s_2})$.

It turns out that the logic $\mathcal{L}$ precisely captures the bisimilarity metric $\mathbf{d}_s$: the metric $\mathbf{d}_s^{ls}$ defined by state formulae coincides with $\mathbf{d}_s$ and the metric $\mathbf{d}_s^{ld}$ defined by distribution formulae coincides with $K(\mathbf{d}_s)$, the lifted form of $\mathbf{d}_s$.

**Theorem 9.** $\mathbf{d}_s = \mathbf{d}_s^{ls}$ *and* $K(\mathbf{d}_s) = \mathbf{d}_s^{ld}$ □

The two properties in Theorem 9 are coupled and should be proved simultaneously because state formulae and distribution formulae are defined reciprocally. The proof is carried out in three steps:

(i) We show $\mathbf{d}_s^{ls} \sqsubseteq \mathbf{d}_s$ and $\mathbf{d}_s^{ld} \sqsubseteq K(\mathbf{d}_s)$ simultaneously by structural induction on formulae.
(ii) We establish $K(\mathbf{d}_s^{ls}) \sqsubseteq \mathbf{d}_s^{ld}$ by exploiting the dual form of the Kantorovich metric in (2). Here it is crucial to require the state space of the pLTS under consideration to be finite in order to use binary conjunctions rather than infinitary conjunctions. The negation and testing operators in state formulae play an important role in the proof.
(iii) We verify that $\mathbf{d}_s^{ls}$ is a state-based bisimulation metric and so obtain $\mathbf{d}_s \sqsubseteq \mathbf{d}_s^{ls}$. This part is based on (ii) and requires the pLTS to be image-finite. Its proof makes use of the negation and testing operators in distribution formulae.

*Remark 10.* For deterministic pLTSs, the proof of Theorem 9 can be greatly simplified. In that case, we can even fold distribution formulae into state formulae and then the state-based bisimilarity metric can be characterised by the following one-sorted metric logic

$$\varphi ::= \top \mid \neg\varphi \mid \varphi \ominus p \mid \varphi_1 \wedge \varphi_2 \mid \langle a \rangle \varphi . \tag{4}$$

Therefore, for deterministic pLTSs, the two-sorted logic in Definition 5 degenerates into the logic considered in [16,23,45], as expected. In the one-sorted logic, the formula $\langle a \rangle (\varphi \ominus p)$ will be interpreted the same as the formula $\langle a \rangle [\varphi \ominus p]$ in $\mathcal{L}^S$, but no formula has the same interpretation as $\langle a \rangle ([\varphi] \ominus p)$ in $\mathcal{L}^S$; the subtlety has already been discussed in Example 8.

In [2,7] a bisimulation metric for game structures is characterised by a quantitative $\mu$-calculus where formulae are evaluated also on states and no distribution formula is needed. This is not surprising because the considered 2-player games are deterministic: at any state $s$, if two players have chosen their moves, say $a_1$ and $a_2$, then there is a unique distribution $\delta(s, a_1, a_2)$ to determine the probabilities of arriving at a set of destination states.

## 4   Distribution-Based Bisimulation Metric

The bisimilarity metric given in Definition 2 measures the distance between two states. Alternatively, it is possible to directly define a metric that measures subdistributions. In order to do so, we first define a transition relation between subdistributions.

**Definition 11.** *With a slight abuse of notation, we also use the notation $\xrightarrow{a}$ to stand for the transition relation between subdistributions, which is the smallest relation satisfying:*

1. *if $s \xrightarrow{a} \Delta$ then $\overline{s} \xrightarrow{a} \Delta$;*
2. *if $s \xrightarrow{a}\!\!\!\!\!/\,$ then $\overline{s} \xrightarrow{a} \varepsilon$;*
3. *if $\Delta_i \xrightarrow{a} \Theta_i$ for all $i \in I$ then $(\sum_{i \in I} p_i \cdot \Delta_i) \xrightarrow{a} (\sum_{i \in I} p_i \cdot \Theta_i)$, where $I$ is a finite index set and $\sum_{i \in I} p_i \leq 1$.*

Note that if $\Delta \xrightarrow{a} \Delta'$ then some (not necessarily all) states in the support of $\Delta$ can perform action $a$. For example, consider the two states $s_2$ and $s_3$ in Fig. 1. Since $s_2 \xrightarrow{c} \overline{s_4}$ and $s_3$ cannot perform action $c$, the distribution $\Delta = \frac{1}{2}\overline{s_2} + \frac{1}{2}\overline{s_3}$ can make the transition $\Delta \xrightarrow{c} \frac{1}{2}\overline{s_4}$ to reach the subdistribution $\frac{1}{2}\overline{s_4}$.

**Definition 12.** *A 1-bounded pseudometric $d$ on $\mathcal{D}_{sub}(S)$ is a distribution-based bisimulation metric if $||\Delta_1| - |\Delta_2|| \leq d(\Delta_1, \Delta_2)$ and for all $\Delta_1, \Delta_2 \in \mathcal{D}_{sub}(S)$ with $d(\Delta_1, \Delta_2) < 1$, whenever $\Delta_1 \xrightarrow{a} \Delta_1'$ then there exists some transition $\Delta_2 \xrightarrow{a} \Delta_2'$ such that $d(\Delta_1', \Delta_2') \leq d(\Delta_1, \Delta_2)$.*

The condition $||\Delta_1| - |\Delta_2|| \leq d(\Delta_1, \Delta_2)$ is introduced to ensure that the distance between two subdistributions should be at least the difference between their total masses. The smallest (wrt. $\sqsubseteq$) distribution-based bisimulation metric, notation $\mathbf{d}_d$, is called *distribution-based bisimilarity metric*. Distribution-based bisimilarity [13] is the kernel of the distribution-based bisimulation metric.

Let $der(\Delta, a) = \{\Delta' \mid \Delta \xrightarrow{a} \Delta'\}$. We define the functional operator

$$F_d \colon [0,1]^{\mathcal{D}_{sub}(S) \times \mathcal{D}_{sub}(S)} \to [0,1]^{\mathcal{D}_{sub}(S) \times \mathcal{D}_{sub}(S)}$$

for $d \colon \mathcal{D}_{sub}(S) \times \mathcal{D}_{sub}(S) \to [0,1]$ and $\Delta, \Theta \in \mathcal{D}_{sub}(S)$ by

$$F_d(d)(\Delta, \Theta) = \max(\sup_{a \in A}\{H(d)(der(\Delta, a), der(\Theta, a))\}, ||\Delta| - |\Theta||). \quad (5)$$

It can be shown that $F_d$ is monotone and its least fixed point is given by $\bigsqcup d_i$, where we set $d_0(\Delta, \Theta) = ||\Delta| - |\Theta||$ for any $\Delta, \Theta \in \mathcal{D}_{sub}(S)$ and $d_{i+1} = F_d(d_i)$ for all $i \in \mathbb{N}$. The property below is analogous to Proposition 4.

**Proposition 13.** $\mathbf{d}_d$ *is the least fixed point of* $F_d$. $\qquad \square$

It is not difficult to see that $\mathbf{d}_s$ is different from $\mathbf{d}_d$, as witnessed by the following example. A more accurate comparison is given in Sect. 5.

*Example 14.* Consider the states in Fig. 1. We first observe that $\mathbf{d}_d(\overline{s_2}, \overline{t_3}) = 0$ because $s_2$ and $t_3$ can match each other's action exactly. Similarly, we have $\mathbf{d}_d(\overline{s_3}, \overline{t_4}) = 0$. Then it is easy to see that $\mathbf{d}_d(\frac{1}{2}\overline{s_2} + \frac{1}{2}\overline{s_3}, \frac{1}{2}\overline{t_3} + \frac{1}{2}\overline{t_4}) = 0$. Since $s_1 \xrightarrow{b} \frac{1}{2}\overline{s_2} + \frac{1}{2}\overline{s_3}$ and $\frac{1}{2}\overline{t_1} + \frac{1}{2}\overline{t_2} \xrightarrow{b} \frac{1}{2}\overline{t_3} + \frac{1}{2}\overline{t_4}$, we infer that $\mathbf{d}_d(\overline{s_1}, \frac{1}{2}\overline{t_1} + \frac{1}{2}\overline{t_2}) = 0$. This, in turn, implies $\mathbf{d}_d(\overline{s}, \overline{t}) = 0$. We have already seen in Example 3 that $\mathbf{d}_s(s, t) = \frac{1}{2}$. Therefore, the two distance functions $\mathbf{d}_s$ and $\mathbf{d}_d$ are indeed different.

We now turn to the logical characterisation of $\mathbf{d}_d$. Consider the metric logic $\mathcal{L}^{D*}$ whose formulae are defined below:

$$\psi ::= \top \mid \neg\psi \mid \psi \ominus p \mid \psi_1 \wedge \psi_2 \mid \langle a \rangle \psi . \quad (6)$$

This logic is the same as that defined in (4) except that now we only have distribution formulae. The semantic interpretation of formulae comes with no surprise.

**Definition 15.** *A formula* $\psi \in \mathcal{L}^{D*}$ *evaluates in* $\Delta \in \mathcal{D}_{sub}(S)$ *as follows:*

$$\begin{aligned}
[\![\top]\!](\Delta) &= |\Delta| \\
[\![\neg\psi]\!](\Delta) &= 1 - [\![\psi]\!](\psi) \\
[\![\psi \ominus p]\!](\Delta) &= \max([\![\psi]\!](\Delta) - p, \ 0) \\
[\![\psi_1 \wedge \psi_2]\!](\Delta) &= \min([\![\psi_1]\!](\Delta), [\![\psi_2]\!](\Delta)) \\
[\![\langle a \rangle \psi]\!](\Delta) &= \max_{\Delta \xrightarrow{a} \Delta'} [\![\psi]\!](\Delta').
\end{aligned}$$

This induces a natural logical metric $\mathbf{d}_d^{\mathrm{ld}}$ over subdistributions defined by

$$\mathbf{d}_d^{\mathrm{ld}}(\Delta, \Theta) = \sup_{\psi \in \mathcal{L}^{D*}} |[\![\psi]\!](\Delta) - [\![\psi]\!](\Theta)|$$

It turns out that $\mathbf{d}_d^{\mathrm{ld}}$ coincides with $\mathbf{d}_d$; the proof is similar to but easier than that of Theorem 9.

**Theorem 16.** $\mathbf{d}_d = \mathbf{d}_d^{\mathrm{ld}}$ $\qquad \square$

# 5   Comparison of the Bisimilarity Metrics

In this section, we compare the state-based bisimilarity metric $\mathbf{d}_s$ with the distribution-based bisimilarity metric $\mathbf{d}_d$. More precisely, we show that $\mathbf{d}_d$ is a lower bound of $K(\mathbf{d}_s)$ when measuring full distributions[3]. The proof makes use of fully enabled pLTSs as a stepping stone. Let us first fix an overall set of actions $Act$ and a special action $\perp \notin Act$. Let $EA(s) = \{a \mid \exists \Delta.\ s \xrightarrow{a} \Delta\}$ be the set of actions that are enabled at state $s$.

**Definition 17.** *A pLTS is* fully enabled *if* $\forall s.\ EA(s) = Act$. *Given any pLTS* $\mathcal{A} = (S, A, \rightarrow)$ *with* $A \subseteq Act$, *we can convert it into a* fully enabled *pLTS* $\mathcal{A}^\perp = (S_\perp, Act \cup \{\perp\}, \rightarrow_\perp)$ *as follows:*

- $S_\perp = S \cup \{\perp\}$
- $\rightarrow_\perp = \rightarrow \cup \{(s, a, \overline{\perp}) \mid s \not\xrightarrow{a} \ and\ a \in Act\} \cup \{(\perp, a, \overline{\perp}) \mid a \in Act \cup \{\perp\}\}$.

Each state $s$ in $\mathcal{A}$ corresponds to a state $s^\perp$ in $\mathcal{A}^\perp$ such that $s^\perp$ keeps all the transitions of $s$ and can evolve into the absorbing state $\perp$ by performing any action in $Act$ not enabled by $s$. As a consequence, each subdistribution $\Delta$ on the states of $\mathcal{A}$ has a corresponding full distribution $\Delta^\perp$ on the states of $\mathcal{A}^\perp$ such that $\Delta^\perp(s^\perp) = \Delta(s)$ and $\Delta^\perp(\perp) = 1 - |\Delta|$.

For any pLTS, let $s, t$ be two states and $\Delta, \Theta$ two subdistributions. It can be shown that $\mathbf{d}_s(s, t) = \mathbf{d}_s(s^\perp, t^\perp)$ and $\mathbf{d}_d(\Delta, \Theta) = \mathbf{d}_d(\Delta^\perp, \Theta^\perp)$. Moreover, for fully enabled pLTSs, the metric $\mathbf{d}_d$ turns out to be a lower bound of $K(\mathbf{d}_s)$ as far as distributions are concerned. Then we arrive at the following theorem.

**Theorem 18.** *Let* $\Delta, \Theta$ *be two distributions on a pLTS. Then* $\mathbf{d}_d(\Delta, \Theta) \leq K(\mathbf{d}_s)(\Delta, \Theta)$. $\qquad\square$

# 6   Bisimulations

The kernel of $\mathbf{d}_s$ (resp. $\mathbf{d}_d$) is the state-based (resp. distribution-based) bisimilarity, denoted by $\sim_s$ (resp. $\sim_d$). They can be defined in a more direct way. The definition of $\sim_s$ requires us to lift a relation on states to be a relation on distributions. There are several different but equivalent formulations of the lifting operation, and they are closely related to the Kantorovich metric; see [10] for more details. The following one is taken from [15].

**Definition 19.** *Let $S$ and $T$ be two sets and $\mathcal{R} \subseteq S \times T$ be a binary relation. The lifted relation $\mathcal{R}^\dagger \subseteq \mathcal{D}_{sub}(S) \times \mathcal{D}_{sub}(T)$ is the smallest relation that satisfies:*

1. *$s \mathcal{R} t$ implies $\overline{s} \mathcal{R}^\dagger \overline{t}$;*
2. *$\Delta_i \mathcal{R}^\dagger \Theta_i$ for all $i \in I$ implies $(\sum_{i \in I} p_i \cdot \Delta_i) \mathcal{R}^\dagger (\sum_{i \in I} p_i \cdot \Theta_i)$, where $I$ is a finite index set and $\sum_{i \in I} p_i \leq 1$.*

---

[3] Although $\mathbf{d}_d$ can measure the distance between two subdistributions, the Kantorovich lifting of $\mathbf{d}_s$ can only measure the distance between full distributions or subdistributions of equal mass, which can easily be normalized to full distributions.

The state-based bisimilarity $\sim_s$ is essentially Larsen and Skou's probabilistic bisimilarity [34], which is originally defined for deterministic systems.

**Definition 20.** *Let* $\sim_s \subseteq S \times S$ *be the largest symmetric relation such that if* $s \sim_s t$ *and* $s \xrightarrow{a} \Delta$ *then there exists some* $t \xrightarrow{a} \Theta$ *with* $\Delta\, (\sim_s)^{\dagger}\, \Theta$.

The distribution-based bisimilarity $\sim_d$ is proposed in [13] as a sound and complete coinductive proof technique for linear contextual equivalence, a natural extensional behavioural equivalence for functional programs.

**Definition 21.** *Let* $\sim_d \subseteq \mathcal{D}_{sub}(S) \times \mathcal{D}_{sub}(S)$ *be the largest symmetric relation such that if* $\Delta \sim_d \Theta$ *then* $|\Delta| = |\Theta|$ *and* $\Delta \xrightarrow{a} \Delta'$ *implies the existence of some* $\Theta'$ *such that* $\Theta \xrightarrow{a} \Theta'$ *and* $\Delta' \sim_d \Theta'$.

It is obvious that $s \sim_s t$ iff $\mathbf{d}_s(s,t) = 0$ iff $[\![\varphi]\!](s) = [\![\varphi]\!](t)$ for any states $s, t$ and formula $\varphi \in \mathcal{L}^S$. Similarly, $\Delta \sim_d \Theta$ iff $\mathbf{d}_d(\Delta, \Theta) = 0$ iff $[\![\psi]\!](\Delta) = [\![\psi]\!](\Theta)$ for any subdistributions $\Delta, \Theta$ and formula $\psi \in \mathcal{L}^{D*}$. Although the state-based bisimilarity is widely accepted, there is no general agreement on what is a good notion of distribution-based bisimilarity. In the literature [14,18,20,21,26,28], several variations of distribution-based bisimulations have been proposed. Some of them are defined for pLTSs with states labelled by atomic propositions. We adapt them to our setting so as to compare with $\sim_d$.

In a pLTS $(S, L, \rightarrow)$, a transition goes from a state to a distribution, e.g. $s \xrightarrow{a} \Delta$. In order to lift $\rightarrow$ to be a relation between distributions, e.g. $\Delta \xrightarrow{a} \Theta$, usually we need to decide whether

(i) to require all the states in the support of $\Delta$ to perform action $a$;
(ii) to combine transitions with the same label, which we explain below.

In [18,20,21] both (i) and (ii) are imposed, while in [28] and also in our definition of $\sim_d$ (i) is not used. The condition (ii) is built in Definition 11 but partially used in [28], as we will see in the sequel. Let $\{s \xrightarrow{a} \Delta_i\}_{i \in I}$ be a collection of transitions, and $\{p_i\}_{i \in I}$ be a collection of probabilities with $\sum_{i \in I} p_i = 1$. Then $s \xrightarrow{a}_C (\sum_{i \in I} p_i \cdot \Delta_i)$ is called a *combined transition* [40]. Let us write $\Delta \xrightarrow{a}_C \Theta$ if $s \xrightarrow{a}_C \Delta_s$ for each $s \in \lceil \Delta \rceil$ and $\Theta = \sum_{s \in \lceil \Delta \rceil} \Delta(s) \cdot \Delta_s$.

*Remark 22.* An equivalent way of defining combined transitions is to use Definition 11. We have that $s \xrightarrow{a}_C \Delta$ iff $\overline{s} \xrightarrow{a} \Delta$ and $|\Delta| = 1$; $\Delta \xrightarrow{a}_C \Theta$ iff $\Delta \xrightarrow{a} \Theta$ and $|\Delta| = |\Theta|$.

Note that a simple way of comparing subdistributions is to lift the state-based bisimilarity and use the relation $(\sim_s)^{\dagger}$. That relation can be slightly weakened by using the combined transition $t \xrightarrow{a}_C \Theta$ in place of $t \xrightarrow{a} \Theta$ in Definition 20 to get a coarser notion of state-based bisimilarity called strong probabilistic bisimulation in [40], written $\sim_s'$, and then lifting it to subdistributions to finally obtain $(\sim_s')^{\dagger}$. This is essentially the relation investigated in [26]. However, most distribution-based bisimilarities proposed in the literature directly compare the transitions between (sub)distributions, so there is no need of defining certain relations on states and then lift them to subdistributions. Below we recall four typical proposals.

Firstly, we adapt the bisimulation of [21] to our setting. Let $(S, A, \rightarrow)$ be a pLTS, we extend it to be a fully enabled pLTS $(S_\perp, Act \cup \{\perp\}, \rightarrow_\perp)$ according to Definition 17.

**Definition 23.** *Let* $\sim_1 \subseteq \mathcal{D}(S_\perp) \times \mathcal{D}(S_\perp)$ *be the largest symmetric relation such that* $\Delta \sim_1 \Theta$ *implies*

1. $\Delta(S) = \Theta(S)$,
2. *for each* $a \in A$, *whenever* $\Delta \overset{a}{\longrightarrow}_C \Delta'$, *there exists* $\Theta'$ *with* $\Theta \overset{a}{\longrightarrow}_C \Theta'$ *and* $\Delta' \sim_1 \Theta'$.

Secondly, we adapt the bisimulation in [14, 26] for subdistributions.

**Definition 24.** *Let* $\sim_2 \subseteq \mathcal{D}_{sub}(S) \times \mathcal{D}_{sub}(S)$ *be the largest symmetric relation such that* $\Delta \sim_2 \Theta$ *implies, for all finite sets of probabilities* $\{p_i \mid i \in I\}$ *satisfying* $\sum_{i \in I} p_i \leq 1$,

1. $|\Delta| = |\Theta|$,
2. *whenever* $\Delta \overset{a}{\longrightarrow}_C \Delta'$, *there exists* $\Theta'$ *with* $\Theta \overset{a}{\longrightarrow}_C \Theta'$ *and* $\Delta' \sim_2 \Theta'$,
3. *whenever* $\Delta = \sum_{i \in I} p_i \cdot \Delta_i$, *for any subdistributions* $\Delta_i$, *there are some subdistributions* $\Theta_i$ *with* $\Theta = \sum_{i \in I} p_i \cdot \Theta_i$, *such that* $\Delta_i \sim_2 \Theta_i$ *for each* $i \in I$.

Thirdly, we adapt the bisimulation in [18] to pLTSs. A subdistribution is *consistent*, if $EA(s) = EA(t)$ for any $s, t \in \lceil \Delta \rceil$. That is, all the states in the support of $\Delta$ have the same set of enabled actions.

**Definition 25.** *Let* $\sim_3 \subseteq \mathcal{D}_{sub}(S) \times \mathcal{D}_{sub}(S)$ *be the largest symmetric relation such that* $\Delta \sim_3 \Theta$ *implies*

1. $|\Delta| = |\Theta|$,
2. *whenever* $\Delta \overset{a}{\longrightarrow}_C \Delta'$, *there exists* $\Theta'$ *with* $\Theta \overset{a}{\longrightarrow}_C \Theta'$ *and* $\Delta' \sim_3 \Theta'$,
3. *if* $\Delta$ *is not consistent, there exist decompositions* $\Delta = \sum_{i \in I} p_i \cdot \Delta_i$ *and* $\Theta = \sum_{i \in I} p_i \cdot \Theta_i$ *such that* $\Delta_i \sim_3 \Theta_i$ *for each* $i \in I$.

Finally, we adapt the bisimulation of [28]. Let $A$ be a set of labels. We write $s \overset{A}{\longrightarrow} \Delta$ if $s \overset{a}{\longrightarrow}_C \Delta$ for some $a \in A$ and denote by $S_A = \{s \mid \exists \Delta. s \overset{A}{\longrightarrow} \Delta\}$ the set of states that can perform some action from $A$. Then we define a transition relation for distributions by letting $\Delta \overset{A}{\longrightarrow} \Theta$ if $s \overset{A}{\longrightarrow} \Delta_s$ for each $s \in S_A \cap \lceil \Delta \rceil$ and $\Theta = \frac{1}{\Delta(S_A)} \sum_{s \in S_A \cap \lceil \Delta \rceil} \Delta(s) \cdot \Delta_s$.

**Definition 26.** *Let* $\sim_4 \subseteq \mathcal{D}_{sub}(S) \times \mathcal{D}_{sub}(S)$ *be the largest symmetric relation such that* $\Delta \sim_4 \Theta$ *implies*

1. $|\Delta| = |\Theta|$ *and* $\Delta(S_A) = \Theta(S_A)$ *for any* $A \subseteq L$,
2. *for each* $A \subseteq L$, *whenever* $\Delta \overset{A}{\longrightarrow} \Delta'$, *there exists* $\Theta'$ *with* $\Theta \overset{A}{\longrightarrow} \Theta'$ *and* $\Delta' \sim_4 \Theta'$.

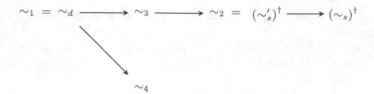

**Fig. 3.** Relationship between the seven bisimilarities for distributions. An arrow pointing from one relation to another means that the former relation is strictly coarser than the latter. Two relations are incomparable if there is no path from one to the other.

**Theorem 27.** *Figure 3 depicts the relationship between the seven bisimilarities for distributions mentioned above.*     □

If we confine ourselves to deterministic pLTSs, then combined transitions add nothing new to ordinary transitions and thus $\sim'_s$ degenerates into $\sim_s$, but the rest of Fig. 3 remains unchanged.

## 7 Other Related Work

Metrics for probabilistic transition systems are first suggested by Giacalone *et al.* [25] to formalize a notion of distance between processes. They are used also in [33,36] to give denotational semantics for deterministic models. De Vink and Rutten [8] show that discrete probabilistic transition systems can be viewed as coalgebras. They consider the category of complete ultrametric spaces. Similar ultrametric spaces are considered by den Hartog in [9]. In [46] Ying proposes a notion of bisimulation index for the usual labelled transition systems, by using ultrametrics on actions instead of using pseudometrics on states. A quantitative linear-time-branching-time spectrum for non-probabilistic systems is given in [19].

Metrics for deterministic systems are extensively studied. Desharnais *et al.* [16] propose a logical pseudometric for labelled Markov chains, which is a deterministic model of probabilistic systems. A similar pseudometric is defined by van Breugel and Worrell [44] via the terminal coalgebra of a functor based on a metric on the space of Borel probability measures. Essentially the same metric is investigated in the setting of continuous Markov decision processes [23]. The metric of [16,23,45] works for continuous probabilistic transition systems, while in this work we concentrate on discrete systems with nondeterminism. In the future it would be interesting to see how to generalise our results to continuous systems. In [43] van Breugel and Worrell present a polynomial-time algorithm to approximate their coalgebraic distances. Furthermore, van Breugel *et al.* propose an algorithm to approximate a behavioural pseudometric without discount [42]. In [22] a sampling algorithm for calculating bisimulation distances in Markov decision processes is shown to have good performance. In [6,7] the probabilistic bisimulation metric on game structures is characterised by a quantitative $\mu$-calculus. Algorithms for game metrics are proposed in [2,38]. A notion of bisimulation distance for distributions is proposed in [21]. It is defined for full distributions only and the definition itself has to be given in terms of fully enabled transition systems. Our distribution-based bisimulation metric generalises it to subdistributions, and allowing transitions between subdistributions has the advantage of allowing our definition to be more direct.

Metrics for nondeterministic probabilistic systems are considered in [17], where Desharnais *et al.* deal with labelled concurrent Markov chains (similar to pLTSs, this model can be captured by the simple probabilistic automata of [39]). They show that the greatest fixed point of a monotonic function on pseudometrics corresponds to the weak probabilistic bisimilarity of [37]. In [24] a notion of uniform continuity is proposed to be an appropriate property of probabilistic processes for compositional reasoning with respect to $d_s$. In [41] a notion of trace metric is proposed for pLTSs and a tool is developed to compute the trace metric. In [1] the boolean-valued logic from [12] is used to characterise state-based bisimulation metrics. It crucially relies on distribution formulae of the form $\bigoplus_{i \in I} p_i \varphi_i$, which is demanding in the sense that if $\Delta$ satisfies that formula then there is some decomposition $\Delta = \sum_{i \in I} p_i \cdot \Delta_i$ such that for each $i \in I$ all the states in the support of $\Delta_i$ must satisfy $\varphi_i$.

Metrics for other quantitative models are also investigated. In [11] a notion of bisimulation metric is proposed that extends the approach of [16,17] to a more general framework called action-labelled quantitative transition systems. In [5] de Alfaro *et al.* consider metric transition systems in which the propositions at each state are interpreted as elements of metric spaces. In that setting, trace equivalence and bisimulation give rise to linear and branching distances that can be characterised by quantitative versions of linear-time temporal logic [35] and the $\mu$-calculus [32].

## 8   Concluding Remarks

We have considered two behavioural pseudometrics for probabilistic labelled transition systems where nondeterminism and probabilities co-exist. They correspond to state-based and distribution-based bisimulations. Our modal characterisation of the state-based bisimulation metric is much simpler than an earlier proposal by Desharnais *et al.* since we only use two non-expansive operators, negation and testing, rather than the general class of non-expansive operators. A similar idea is used to characterise the distribution-based bisimulation metric. The characterisations are shown to be sound and complete. We have also shown that the distribution-based bisimulation metric is a lower bound of the state-based bisimulation metric lifted to distributions. In addition, we have compared the bisimilarities entailed by the two metrics with a few other distribution-based bisimilarities.

In the current work we have not distinguished internal actions from external ones. But it is not difficult to make the distinction and abstract away internal actions so as to introduce weak versions of bisimulation metrics. In a finite-state and finitely branching pLTS, the set of subdistributions reachable from a state by weak transitions may be infinite but can be represented by the convex closure of a finite set [10]. This entails that the logical characterisation of weak bisimulation metrics would be similar to those presented here.

**Acknowledgement.** We thank the anonymous referees for their helpful comments.

# References

1. Castiglioni, V., Gebler, D., Tini, S.: Logical characterization of bisimulation metrics. In: Proceedings of QAPL 2016. EPTCS (2016)
2. Chatterjee, K., De Alfaro, L., Majumdar, R., Raman, V.: Algorithms for game metrics. Log. Methods Comput. Sci. **6**(3:13), 1–27 (2010)
3. Cleaveland, R., Iyer, S.P., Narasimha, M.: Probabilistic temporal logics via the modal mu-calculus. Theor. Comput. Sci. **342**(2–3), 316–350 (2005)
4. D'Argenio, P.R., Sánchez Terraf, P., Wolovick, N.: Bisimulations for non-deterministic labelled Markov processes. Math. Struct. Comput. Sci. **22**(1), 43–68 (2012)
5. De Alfaro, L., Faella, M., Stoelinga, M.: Linear and branching system metrics. IEEE Trans. Softw. Eng. **35**(2), 258–273 (2009)
6. De Alfaro, L., Majumdar, R., Raman, V., Stoelinga, M.: Game relations and metrics. In: Proceedings of LICS 2007, pp. 99–108. IEEE (2007)
7. De Alfaro, L., Majumdar, R., Raman, V., Stoelinga, M.: Game refinement relations and metrics. Log. Methods Comput. Sci. **4**(3:7), 1–28 (2008)
8. de Vink, E.P., Rutten, J.J.M.M.: Bisimulation for probabilistic transition systems: a coalgebraic approach. Theor. Comput. Sci. **221**(1/2), 271–293 (1999)
9. den Hartog, J.I.: Probabilistic Extensions of Semantical Models. Ph.D. thesis, Free University Amsterdam (2002)
10. Deng, Y.: Semantics of Probabilistic Processes: An Operational Approach. Springer, Heidelberg (2015)
11. Deng, Y., Chothia, T., Palamidessi, C., Pang, J.: Metrics for action-labelled quantitative transition systems. ENTCS **153**(2), 79–96 (2006)
12. Deng, Y., Du, W.: Logical, metric, and algorithmic characterisations of probabilistic bisimulation. Technical report CMU-CS-11-110, Carnegie Mellon University, March 2011
13. Deng, Y., Feng, Y., Dal Lago, U.: On coinduction and quantum lambda calculi. In: Proceedings of CONCUR 2015, pp. 427–440. LIPIcs (2015)
14. Deng, Y., Hennessy, M.: On the semantics of Markov automata. Inf. Comput. **222**, 139–168 (2013)
15. Deng, Y., Glabbeek, R., Hennessy, M., Morgan, C.: Testing finitary probabilistic processes. In: Bravetti, M., Zavattaro, G. (eds.) CONCUR 2009. LNCS, vol. 5710, pp. 274–288. Springer, Heidelberg (2009). doi:10.1007/978-3-642-04081-8_19
16. Desharnais, J., Gupta, V., Jagadeesan, R., Panangaden, P.: Metrics for labelled Markov processes. Theor. Comput. Sci. **318**(3), 323–354 (2004)
17. Desharnais, J., Jagadeesan, R., Gupta, V., Panangaden, P.: The metric analogue of weak bisimulation for probabilistic processes. In: Proceedings of LICS 2002, pp. 413–422. IEEE (2002)
18. Eisentraut, C., Godskesen, J.C., Hermanns, H., Song, L., Zhang, L.: Probabilistic bisimulation for realistic schedulers. In: Bjørner, N., de Boer, F. (eds.) FM 2015. LNCS, vol. 9109, pp. 248–264. Springer, Heidelberg (2015). doi:10.1007/978-3-319-19249-9_16

19. Fahrenberg, U., Legay, A.: The quantitative linear-time branching-time spectrum. Theor. Comput. Sci. **538**, 54–69 (2014)
20. Feng, Y., Ying, M.: Toward automatic verification of quantum cryptographic protocols. In: Proceedings of CONCUR 2015. LIPIcs, vol. 42, pp. 441–455 (2015)
21. Feng, Y., Zhang, L.: When equivalence and bisimulation join forces in probabilistic automata. In: Jones, C., Pihlajasaari, P., Sun, J. (eds.) FM 2014. LNCS, vol. 8442, pp. 247–262. Springer, Heidelberg (2014). doi:10.1007/978-3-319-06410-9_18
22. Ferns, N., Panangaden, P., Precup, D.: Bisimulation metrics for continuous Markov decision processes. SIAM J. Comput. **40**(6), 1662–1714 (2011)
23. Ferns, N., Precup, D., Knight, S.: Bisimulation for Markov decision processes through families of functional expressions. In: Breugel, F., Kashefi, E., Palamidessi, C., Rutten, J. (eds.) Horizons of the Mind. A Tribute to Prakash Panangaden: Essays Dedicated to Prakash Panangaden on the Occasion of His 60th Birthday. LNCS, vol. 8464, pp. 319–342. Springer, Heidelberg (2014). doi:10.1007/978-3-319-06880-0_17
24. Gebler, D., Larsen, K.G., Tini, S.: Compositional metric reasoning with probabilistic process calculi. In: Pitts, A. (ed.) FoSSaCS 2015. LNCS, vol. 9034, pp. 230–245. Springer, Heidelberg (2015). doi:10.1007/978-3-662-46678-0_15
25. Giacalone, A., Jou, C., Smolka, S.: Algebraic reasoning for probabilistic concurrent systems. In: Proceedings of IFIP TC2 Working Conference on Programming Concepts and Methods, pp. 443–458 (1990)
26. Hennessy, M.: Exploring probabilistic bisimulations, part I. Formal Aspects Comput. **24**(4–6), 749–768 (2012)
27. Hennessy, M., Milner, R.: Algebraic laws for nondeterminism and concurrency. J. ACM **32**, 137–161 (1985)
28. Hermanns, H., Krčál, J., Křetínský, J.: Probabilistic bisimulation: naturally on distributions. In: Baldan, P., Gorla, D. (eds.) CONCUR 2014. LNCS, vol. 8704, pp. 249–265. Springer, Heidelberg (2014). doi:10.1007/978-3-662-44584-6_18
29. Huth, M., Kwiatkowska, M.Z.: Quantitative analysis and model checking. In: Proceedings of the 12th Annual IEEE Symposium on Logic in Computer Science, pp. 111–122. IEEE Computer Society (1997)
30. Jonsson, B., Larsen, K.G., Yi, W.: Probabilistic extensions of process algebras. In: Handbook of Process Algebra, pp. 685–710. Elsevier, Amsterdam (2001)
31. Kantorovich, L., Rubinshtein, G.: On a space of totally additive functions. Vestn Len. Univ. **13**(7), 52–59 (1958)
32. Kozen, D.: Results on the propositional mu-calculus. Theor. Comput. Sci. **27**, 333–354 (1983)
33. Kwiatkowska, M., Norman, G.: Probabilistic metric semantics for a simple language with recursion. In: Penczek, W., Szałas, A. (eds.) MFCS 1996. LNCS, vol. 1113, pp. 419–430. Springer, Heidelberg (1996). doi:10.1007/3-540-61550-4_167
34. Larsen, K.G., Skou, A.: Bisimulation through probabilistic testing. Inf. Comput. **94**, 1–28 (1991)
35. Manna, Z., Pnueli, A.: The Temporal Logic of Reactive and Concurrent Systems: Specification. Springer, Heidelberg (1991)
36. Norman, G.J.: Metric Semantics for Reactive Probabilistic Systems. Ph.D. thesis, University of Birmingham (1997)
37. Philippou, A., Lee, I., Sokolsky, O.: Weak bisimulation for probabilistic systems. In: Palamidessi, C. (ed.) CONCUR 2000. LNCS, vol. 1877, pp. 334–349. Springer, Heidelberg (2000). doi:10.1007/3-540-44618-4_25
38. Raman, V.: Game Relations, Metrics and Refinements. Ph.D. thesis, University of California (2010)

39. Segala, R.: Modeling and Verification of Randomized Distributed Real-Time Systems. Ph.D. thesis, MIT (1995)
40. Segala, R., Lynch, N.: Probabilistic simulations for probabilistic processes. In: Jonsson, B., Parrow, J. (eds.) CONCUR 1994. LNCS, vol. 836, pp. 481–496. Springer, Heidelberg (1994). doi:10.1007/978-3-540-48654-1_35
41. Song, L., Deng, Y., Cai, X.: Towards automatic measurement of probabilistic processes. In: Proceedings of QSIC 2007, pp. 50–59. IEEE (2007)
42. van Breugel, F., Sharma, B., Worrell, J.: Approximating a behavioural pseudometric without discount for probabilistic systems. Log. Methods Comput. Sci. **4**(2:2), 1–23 (2008)
43. Breugel, F., Worrell, J.: An algorithm for quantitative verification of probabilistic transition systems. In: Larsen, K.G., Nielsen, M. (eds.) CONCUR 2001. LNCS, vol. 2154, pp. 336–350. Springer, Heidelberg (2001). doi:10.1007/3-540-44685-0_23
44. Breugel, F., Worrell, J.: Towards quantitative verification of probabilistic transition systems. In: Orejas, F., Spirakis, P.G., Leeuwen, J. (eds.) ICALP 2001. LNCS, vol. 2076, pp. 421–432. Springer, Heidelberg (2001). doi:10.1007/3-540-48224-5_35
45. van Breugel, F., Worrell, J.: A behavioural pseudometric for probabilistic transition systems. Theor. Comput. Sci. **331**(1), 115–142 (2005)
46. Ying, M.: Bisimulation indexes and their applications. Theor. Comput. Sci. **275**(1/2), 1–68 (2002)

# A Comparison of Time- and Reward-Bounded Probabilistic Model Checking Techniques

Ernst Moritz Hahn[1]($^{(\boxtimes)}$) and Arnd Hartmanns[2]($^{(\boxtimes)}$)

[1] Institute of Software, Chinese Academy of Sciences, Beijing, China
[2] University of Twente, Enschede, The Netherlands
hahn@ios.ac.cn, a.hartmanns@utwente.nl

**Abstract.** In the design of probabilistic timed systems, requirements concerning behaviour that occurs within a given time or energy budget are of central importance. We observe that model-checking such requirements for probabilistic timed automata can be reduced to checking reward-bounded properties on Markov decision processes. This is traditionally implemented by unfolding the model according to the bound, or by solving a sequence of linear programs. Neither scales well to large models. Using value iteration in place of linear programming achieves scalability but accumulates approximation error. In this paper, we correct the value iteration-based scheme, present two new approaches based on scheduler enumeration and state elimination, and compare the practical performance and scalability of all techniques on a number of case studies from the literature. We show that state elimination can significantly reduce runtime for large models or high bounds.

## 1 Introduction

Probabilistic timed automata (PTA, [17]) are a popular formal model for probabilistic real-time systems. They combine nondeterministic choices as in Kripke structures, discrete probabilistic decisions as in Markov chains, and hard real-time behaviour as in timed automata. We are interested in properties of the form "what is the best/worst-case probability to eventually reach a certain system state while accumulating at most $b$ reward", i.e. in calculating reward-bounded reachability probabilities. Rewards can model a wide range of aspects, e.g. the number of retransmissions in a network protocol (accumulating reward 1 for each), energy consumption (accumulating reward at a state-dependent wattage over time), or time itself (accumulating reward at rate 1 everywhere). Reachability probabilities for PTA with rewards can be computed by first turning a PTA into an equivalent Markov decision process (MDP) using the digital clocks semantics [17] and then performing standard probabilistic model checking [3].

The naïve approach to compute specifically *reward-bounded* reachability probabilities is to *unfold* [1] the state space of the model. For the example of

---

This work was supported by the 3TU.BSR project, by CDZ project 1023 (CAP), by the Chinese Academy of Sciences Fellowship for International Young Scientists, and by the National Natural Science Foundation of China (grant no. 61550110506).

M. Fränzle et al. (Eds.): SETTA 2016, LNCS 9984, pp. 85–100, 2016.
DOI: 10.1007/978-3-319-47677-3_6

time-bounded properties, this means adding a new clock variable that is never reset [17]. In the general case on the level of MDP [19], in addition to the current state of the model, one keeps track of the reward accumulated so far, up to $b$. This turns the reward-bounded problem into standard unbounded reachability. Unfolding blows up the model size (the number of states, or the number of variables and constraints in the corresponding linear program) and causes the model checking process to run out of memory even if the original (unbounded) model was of moderate size (cf. Table 1). For PTA, unfolding is the only approach that has been considered so far. A more efficient technique has been developed for MDP, and via the digital clocks semantics it is applicable to PTA just as well:

The probability for bound $i$ depends only on the values for previous bounds $\{i-r, \ldots, i-1\}$ where $r$ is the max. reward in the automaton. We can thus avoid the monolithic unfolding by sequentially computing the values for its "layers" where the accumulated reward is $i = 0$, 1, etc. up to $b$, storing the current layer and the last $r$ result vectors only. This process can be implemented by solving a sequence of $b$ linear programming (LP) problems no larger than the original unbounded model [2]. While it solves the memory problem in principle, LP is known not to scale to large MDP in practice. Consequently, LP has been replaced by value iteration to achieve scalability in the most recent implementation [14]. Value iteration is an approximative numeric technique to compute reachability probabilities up to a predefined error bound $\epsilon$. When used in sequence, this error accumulates, and the final result for bound $b$ may differ from the actual probability by more than $\epsilon$. This has not been taken into account in [14].

In this paper, we first make a small change to the value iteration-based scheme to counteract the error accumulation. We then present two new ways to compute reward-bounded reachability probabilities for MDP (with a particular interest in the application to PTA via digital clocks) without unfolding (Sect. 3). Using either scheduler enumeration or MDP state elimination, they both reduce the model such that a reward of 1 is accumulated on all remaining transitions. A *reward-bounded* property in the original model corresponds to a *step-bounded* property in the reduced model. We use standard step-bounded value iteration [3] to check these properties efficiently and exactly. Observe that we improve the practical efficiency of computing reward-bounded probabilities, but the problem is EXP-complete in general [6]. It can be solved in time polynomial in the size of the MDP and the *value* of $b$, i.e. it is only pseudo-polynomial in $b$. Like all related work, we only present solutions for the case of nonnegative integer rewards.

The unfolding-free techniques also provide the probability for all lower bounds $i < b$. This has been exploited to obtain quantiles [2], and we use it more generally to compute the entire cumulative (sub)distribution function (*cdf* for short) over the bound up to $b$ at no extra cost. We have implemented all techniques in the MCSTA tool (Sect. 4) of the MODEST TOOLSET [10]. It is currently the only publicly available implementation of reward-bounded model checking for PTA and MDP without unfolding. We use it to study the relative performance and scalability of the previous and new techniques on six example models from the literature (Sect. 5). State elimination in particular shows promising performance.

*Other Related Work.* Randour et al. [18] have studied the complexity of computing reward-bounded probabilities (framed as percentile queries) for MDP with *multiple* rewards and reward bounds. They propose an algorithm based on unfolding. For the soft real-time model of Markov automata, which subsumes MDP, reward bounds can be turned into time bounds [13]. Yet this only works for rewards associated to Markovian states, whereas immediate states (i.e. the MDP subset of Markov automata) always implicitly get zero reward.

# 2    Preliminaries

$\mathbb{N}$ is $\{0, 1, \dots\}$, the set of natural numbers. $2^S$ is the powerset of $S$. $\mathrm{Dom}(f)$ is the domain of the function $f$.

**Definition 1.** *A (discrete) probability distribution over a set $\Omega$ is a function* $\mu \in \Omega \to [0, 1]$ *such that* $\mathrm{support}(\mu) \stackrel{\text{def}}{=} \{\omega \in \Omega \mid \mu(\omega) > 0\}$ *is countable and* $\sum_{\omega \in \mathrm{support}(\mu)} \mu(\omega) = 1$. $\mathrm{Dist}(\Omega)$ *is the set of all probability distributions over $\Omega$.* $\mathcal{D}(s)$ *is the* Dirac distribution *for $s$, defined by $\mathcal{D}(s)(s) = 1$.*

**Markov Decision Processes.** To move from one state to another in a Markov decision process, first a transition is chosen nondeterministically; each transition then leads into a probability distribution over rewards and successor states.

**Definition 2.** *A Markov decision process (MDP) is a triple $M = \langle S, T, s_{init}\rangle$ where $S$ is a finite set of states, $T \in S \to 2^{\mathrm{Dist}(\mathbb{N} \times S)}$ is the transition function, and $s_{init} \in S$ is the initial state. For all $s \in S$, we require that $T(s)$ is finite and non-empty. $M$ is a* discrete-time Markov chain *(DTMC) if $\forall s \in S: |T(s)| = 1$.*

We write $s \to_T \mu$ for $\exists \mu \in T(s)$ and call it a *transition.* We write $s \xrightarrow{r}_T s'$ if additionally $\langle r, s'\rangle \in \mathrm{support}(\mu)$. $\langle r, s'\rangle$ is a *branch* with reward $r$. If $T$ is clear from the context, we write just $\to$. Graphically, transitions are lines to an intermediate node from which branches labelled with reward (if not zero) and probability lead to successor states. We may omit the intermediate node and probability 1 for transitions into Dirac distributions, and we may label transitions to refer to them in the text. Figure 1 shows an example MDP $M^e$ with 5 states, 7 (labelled) transitions and 10 branches. Using branch rewards instead of the more standard transition rewards leads to more compact models; in the example, we assign reward 1 to the branches back to $s$ and $t$ to count the number of "failures" before reaching $v$. In practice, high-level formalisms like PRISM's [15] guarded command language are used to specify MDP. They extend MDP with variables over finite domains that can be used in expressions to e.g. enable/disable transitions. This allows to compactly describe very large MDP.

**Definition 3.** *A finite path in $M = \langle S, T, s_{init}\rangle$ is defined as a finite sequence* $\pi_{\mathrm{fin}} = s_0\, \mu_0\, r_0\, s_1\, \mu_1\, r_1\, s_2 \dots \mu_{n-1}\, r_{n-1}\, s_n$ *where $s_i \in S$ for all $i \in \{0, \dots, n\}$ and* $s_i \to \mu_i \wedge \langle r_i, s_{i+1}\rangle \in \mathrm{support}(\mu_i)$ *for all $i \in \{0, \dots, n-1\}$. Let $|\pi_{\mathrm{fin}}| \stackrel{\text{def}}{=} n$,* $\mathrm{last}(\pi_{\mathrm{fin}}) \stackrel{\text{def}}{=} s_n$, *and $\mathrm{reward}(\pi_{\mathrm{fin}}) = \sum_{i=0}^{n-1} r_i$.* $\mathrm{Paths}_{\mathrm{fin}}(M)$ *is the set of all finite*

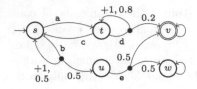

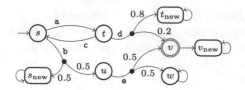

**Fig. 1.** Example MDP $M^e$    **Fig. 2.** Transformed MDP $M^e\downarrow_{F^e}\downarrow_R$

paths starting with $s_{init}$. A path is an infinite sequence $\pi = s_0\,\mu_0\,r_0\,s_1\,\mu_1\,r_1\ldots$ where $s_i \in S$ and $s_i \to \mu_i \wedge \langle r_i, s_{i+1}\rangle \in \text{support}(\mu_i)$ for all $i \in \mathbb{N}$. Paths$(M)$ is the set of all paths starting with $s_{init}$. We define $s \in \pi \overset{\text{def}}{\Leftrightarrow} \exists i\colon s = s_i$.

**Definition 4.** Given $M = \langle S, T, s_{init}\rangle$, $\mathfrak{S} \in \text{Paths}_{\text{fin}}(M) \to \text{Dist}(\text{Dist}(\mathbb{N} \times S))$ is a scheduler for $M$ if $\forall\,\pi_{\text{fin}}\colon \mu \in \text{support}(\mathfrak{S}(\pi_{\text{fin}})) \Rightarrow \text{last}(\pi_{\text{fin}}) \to \mu$. The set of all schedulers of $M$ is Sched$(M)$. $\mathfrak{S}$ is reward-positional if $\text{last}(\pi_1) = \text{last}(\pi_2) \wedge$ reward$(\pi_1) = $ reward$(\pi_2)$ implies $\mathfrak{S}(\pi_1) = \mathfrak{S}(\pi_2)$, positional if $\text{last}(\pi_1) = \text{last}(\pi_2)$ alone implies $\mathfrak{S}(\pi_1) = \mathfrak{S}(\pi_2)$, and deterministic if $|\text{support}(\mathfrak{S}(\pi))| = 1$, for all finite paths $\pi$, $\pi_1$ and $\pi_2$, respectively. A simple scheduler is positional and deterministic. The set of all simple schedulers of $M$ is SSched$(M)$.

Let $M\downarrow_{\mathfrak{S}_s} \overset{\text{def}}{=} \langle S, T', s_{init}\rangle$ with $T'(s) \overset{\text{def}}{=} \{\mu \mid \mathfrak{S}_s(s) = \mathcal{D}(\mu)\}$ for $\mathfrak{S}_s \in$ SSched$(M)$. $M\downarrow_{\mathfrak{S}_s}$ is a DTMC. Using the standard cylinder set construction [3], a scheduler $\mathfrak{S}$ induces a probability measure $\mathcal{P}^{\mathfrak{S}}_M$ on measurable sets of paths starting from $s_{init}$. We define the extremal values $\mathcal{P}^{\max}_M(\Pi) = \sup_{\mathfrak{S}\in\text{Sched}(M)} \mathcal{P}^{\mathfrak{S}}_M(\Pi)$ and $\mathcal{P}^{\min}_M(\Pi) = \inf_{\mathfrak{S}\in\text{Sched}(M)} \mathcal{P}^{\mathfrak{S}}_M(\Pi)$ for measurable $\Pi \subseteq$ Paths$(M)$.

For an MDP $M$ and goal states $F \subseteq S$, we define the unbounded, step-bounded and reward-bounded reachability probabilities for $opt \in \{\max, \min\}$:

- $\text{P}_{opt}(F) \overset{\text{def}}{=} \mathcal{P}^{opt}_M(\{\pi \in \text{Paths}(M) \mid \exists s \in F\colon s \in \pi\})$ is the extremal probability of eventually reaching a state in $F$.
- $\text{P}^{S\le b}_{opt}(F)$ is the extremal probability of reaching a state in $F$ via at most $b \in \mathbb{N}$ transitions, defined as $\mathcal{P}^{opt}_M(\Pi^T_b)$ where $\Pi^T_b$ is the set of paths that have a prefix of length at most $b$ that contains a state in $F$.
- $\text{P}^{R\le b}_{opt}(F)$ is the extremal probability of reaching a state in $F$ with accumulated reward at most $b \in \mathbb{N}$, defined as $\mathcal{P}^{opt}_M(\Pi^R_b)$ where $\Pi^R_b$ is the set of paths that have a prefix $\pi_{\text{fin}}$ containing a state in $F$ with reward$(\pi_{\text{fin}}) \le b$.

**Theorem 1.** For an unbounded property, there exists an optimal simple scheduler, i.e. one that attains the extremal value [3]. For a reward-bounded property, there exists an optimal deterministic reward-positional scheduler [12].

Continuing our example, let $F^e = \{v\}$. We maximise the probability to eventually reach $F^e$ in $M^e$ by always scheduling transition a in $s$ and d in $t$, so $\text{P}_{\max}(F^e) = 1$ with a simple scheduler. We get $\text{P}^{R\le 0}_{\max}(F^e) = 0.25$ by scheduling b in $s$. For higher bound values, simple schedulers are no longer sufficient: we get $\text{P}^{R\le 1}_{\max}(F^e) = 0.4$ by first trying a then d, but falling back to c then b if we return to $t$. We maximise the probability for higher bound values $n$ by trying d until the accumulated reward is $n - 1$ and then falling back to b.

**Probabilistic Timed Automata.** Probabilistic timed automata (PTA [17]) extend MDP with *clocks* and *clock constraints* as in timed automata to model real-time behaviour and requirements. PTA have two kinds of rewards: branch rewards as in MDP and *rate rewards* that accumulate at a certain rate over time. Time itself is a rate reward that is always 1. The digital clocks approach [17] is the only PTA model checking technique that works well with rewards. It works by replacing the clock variables by bounded integers and adding self-loop edges to increment them synchronously as long as time can pass. The reward of a self-loop edge is the current rate reward. The result is (a high-level model of) a finite *digital clocks MDP*. All the algorithms that we develop for MDP in this paper can thus be applied to PTA. While time- and branch reward-bounded properties on PTA are decidable [17], general rate reward-bounded properties are not [4].

**Probabilistic Model Checking.** Probabilistic model checking for MDP (and thus for PTA via the digital clocks semantics) works in two phases: (1) state space *exploration* turns a given high-level model into an in-memory representation of the underlying MDP, then (2) a numerical *analysis* computes the value of the property of interest. In phase 1, the goal states are made absorbing:

**Definition 5.** *Given $M = \langle S, T, s_{init} \rangle$ and $F \subseteq S$, we define the $F$-absorbing MDP as $M{\downarrow}_F = \langle S, T', s_{init} \rangle$ with $T'(s) = \{\mathcal{D}(\langle 1, s \rangle)\}$ for all $s \in F$ and $T'(s) = T(s)$ otherwise. For $s \in S$, we define $M[s] = \langle S, T, s \rangle$.*

An efficient algorithm for phase 2 and unbounded properties is (unbounded) value iteration [3]. We denote a call to a value iteration implementation by $\mathtt{VI}(V, M{\downarrow}_F, opt, \epsilon)$ with initial value vector $V \in S \to [0, 1]$ and $opt \in \{\max, \min\}$. Internally, it iteratively approximates over all states $s$ a (least) solution for

$$V(s) = opt_{\mu \in T(s)} \sum_{\langle r, s' \rangle \in \text{support}(\mu)} \mu(\langle r, s' \rangle) \cdot V(s')$$

up to (relative) error $\epsilon$. Let initially $V = \{s \mapsto 1 \mid s \in F\} \cup \{s \mapsto 0 \mid s \in S \setminus F\}$. Then on termination of $\mathtt{VI}(V, M{\downarrow}_F, opt, \epsilon)$, we have $V(s) \approx_\epsilon \mathrm{P}_{opt}(F)$ in $M[s]$ for all $s \in S$. All current implementations in model checking tools like PRISM [15] use a simple convergence criterion based on $\epsilon$ that in theory only guarantees $V(s) \leq \mathrm{P}_{opt}(F)$, yet in practice delivers $\epsilon$-close results on most, but not all, case studies. Guaranteed $\epsilon$-close results could be achieved at the cost of precomputing and reducing a maximal end component decomposition of the MDP [7]. In this paper, we thus write $\mathtt{VI}$ to refer to an ideal $\epsilon$-correct algorithm, but for the sake of comparison use the standard implementation in our experiments in Sect. 5.

For a step-bounded property, the call $\mathtt{StepBoundedVI}(V = V_0, M{\downarrow}_F, opt, b)$ with bound $b$ can be implemented [3] by computing for all states

$$V_i(s) := opt_{\mu \in T(s)} \sum_{\langle r, s' \rangle \in \text{support}(\mu)} \mu(\langle r, s' \rangle) \cdot V_{i-1}(s')$$

iteratively for $i = 1, \dots, b$. After iteration $i$, we have $V_i(s) = \mathrm{P}_{opt}^{S \leq i}(F)$ in $M[s]$ for all $s \in S$ when starting with $V$ as in the unbounded case above. Note that this algorithm computes exact results (modulo floating-point precision and errors) without any costly preprocessing and is very easy to implement and parallelise.

Reward-bounded properties can naïvely be checked by *unfolding* the model according to the accumulated reward: we add a variable $v$ to the model prior to phase 1, with branch reward $r$ corresponding to an assignment $v := v + r$. To check $P_{opt}^{R \le b}(F)$, phase 1 thus creates an MDP that is up to $b$ times as large as without unfolding. In phase 2, $P_{opt}(F')$ is checked using VI as described above where $F'$ corresponds to the states in $F$ where additionally $v \le b$ holds.

## 3   Reward-Bounded Analysis Techniques

We describe three techniques that allow the computation of reward-bounded reachability probabilities on MDP (and thus PTA) without unfolding. The first one is a reformulation of the value iteration-based variant [14] of the algorithm introduced in [2]. We incorporate a simple fix for the problem that the error accumulation over the sequence of value iterations had not been accounted for and refer to the result as algorithm modvi. We then present two new techniques senum and elim that avoid the issues of unbounded value iteration by transforming the MDP such that step-bounded value iteration can be used instead.

From now on, we assume that all rewards are either zero or one. This simplifies the presentation and is in line with our motivation of improving time-bounded reachability for PTA: in the corresponding digital clocks MDP, all transitions representing the passage of time have reward 1 while the branches of all other transitions have reward 0. Yet it is without loss of generality: for modvi, it is merely a matter of a simplified presentation, and for the two new algorithms, we can preprocess the MDP to replace each branch with reward $r > 1$ by a chain of $r$ Dirac transitions with reward 1. While this may blow up the state space, we found that most models in practice only use rewards 0 and 1 in the first place: among the 15 MDP and DTMC models currently distributed with PRISM [15], only 2 out of the 12 examples that include a reward structure do not satisfy this assumption. It also holds for all case studies that we present in Sect. 5.

For all techniques, we need a transformation $\downarrow_R$ that redirects each reward-one branch to a copy $s'_{new}$ of the branch's original target state $s'$. In effect, this replaces branch rewards by branches to a distinguished category of "new" states:

**Definition 6.** *Given* $M = \langle S, T, s_{init} \rangle$, *we define* $M\downarrow_R$ *as* $\langle S \uplus S_{new}, T^\downarrow, s_{init} \rangle$ *with* $S_{new} = \{ s_{new} \mid s \in S \}$,

$$T^\downarrow(s) = \begin{cases} \{ Conv(s, \mu) \mid \mu \in T(s) \} & \text{if } s \in S \\ \{ \mathcal{D}(\langle 0, s \rangle) \} & \text{if } s \in S_{new} \end{cases}$$

*and* $Conv(s, \mu) \in \text{Dist}(\mathbb{N} \times S \uplus S_{new})$ *is defined by* $Conv(s, \mu)(\langle 0, s' \rangle) = \mu(\langle 0, s' \rangle)$ *and* $Conv(s, \mu)(\langle 1, s'_{new} \rangle) = \mu(\langle 1, s' \rangle)$ *over all* $s' \in S$.

For our example MDP $M^e$ and $F^e = \{v\}$, we show $M^e\downarrow_{F^e}\downarrow_R$ in Fig. 2. Observe that $M^e\downarrow_{F^e}$ is the same as $M^e$, except that the self-loop of goal state $v$

```
1  function ModVI(V, M = ⟨S, T, s_init⟩, F, b, opt, ε)
2  |    for i = 1 to b do
3  |    |    foreach s_new ∈ S_new do V(s_new) := V(s)
4  |    |    VI(V, M↓_F↓_R, opt, ε/(b+1))
```

**Algorithm 1.** Sequential value iterations for reward-bounded reachability

gets reward 1. $M^e\downarrow_{F^e}\downarrow_R$ is then obtained by redirecting the three reward-one branches (originally going to $s$, $t$ and $v$) to new states $s_{new}$, $t_{new}$ and $v_{new}$.

All of the algorithm descriptions we present take a value vector $V$ as input, which they update. $V$ must initially contain the probabilities to reach a goal state in $F$ with zero reward, which can be computed for example via a call to $\text{VI}(V = V_F^0, M\downarrow_F\downarrow_R, opt, \epsilon)$ with sufficiently small $\epsilon$ and

$$V_F^0 \stackrel{\text{def}}{=} \{s \mapsto 1, s_{new} \mapsto 0 \mid s \in F\} \cup \{s \mapsto 0, s_{new} \mapsto 0 \mid s \in S \setminus F\}.$$

### 3.1   Sequential Value Iterations

We recall the technique for model-checking reward-bounded properties of [2] that avoids unfolding. It was originally formulated as a *sequence* of linear programming (LP) problems $LP_i$, each corresponding to bound $i \leq b$. Each $LP_i$ is of the same size as the original (non-unfolded) MDP, representing its state space, but uses the values computed for $LP_{i-r}, \ldots, LP_{i-1}$ with $r$ being the maximal reward that occurs in the MDP. Since LP does not scale to large MDP [7], the technique has been reconsidered using value iteration instead [14]. Using the transformations and assumptions introduced above, we can formulate it as in Algorithm 1. Initially, $V$ contains the probabilities to reach a goal state with zero reward. In contrast to [14], when given an overall error bound $\epsilon$, we use bound $\frac{\epsilon}{b+1}$ for the individual value iteration calls. At the cost of higher runtime, this counteracts the accumulation of error over multiple calls to yield an $\epsilon$-close final result:

Consider $M\downarrow_F\downarrow_R = \langle S, T, s_{init}\rangle$ and $f \in (S \to [0,1]) \to (S \to [0,1])$ with $f = \lim_i f_i$ where for $V \in S \to [0,1]$ it is $f_0(V)(s) = V(s)$ and $f_{i+1}(V)(s) = opt_\mu \sum_{s'} \mu(s')\cdot f_i(V)(s')$, i.e. $f$ corresponds to performing an ideal value iteration with error $\epsilon = 0$. Thus, performing Algorithm 1 using $f$ would result in an error of 0. If we limit the error in each value iteration to $\frac{\epsilon}{b+1}$, then the function we use can be stated as $f' = f_n$ for $n$ large enough such that $||f'(V) - f(V)||_{\max} \leq \frac{\epsilon}{b+1}$ for all $V$ used in the computations. Let $V_0$, $V_0' = V_0 + \delta_0$ be the initial value vectors, $\delta_0 < \frac{\epsilon}{b+1}$. Further, let $V_i, V_i' \in S \to [0,1]$, $i \in \{1, \ldots, b\}$, be the value vectors after the $i$-th call to VI for the case without ($V_i$) and with error ($V_i'$). We can then show by induction that $||V_i - V_i'||_{\max} \leq (i+1)\frac{\epsilon}{b+1}$. Initially, we have $V_0' = V_0 + \delta_0$. Therefore, we have $V_1' = f'(V_0') = f(V_0') + \delta_1 = f(V_0) + \sum_{j=0}^{0} \delta_j + \delta_1$ for some $\delta_1 \in S \to [0,1]$ with $||\delta_1||_{\max} \leq \frac{\epsilon}{b+1}$. Then, we have for some $\delta_j \in S \to [0,1]$, $j \in \{0, \ldots, i\}$, with $||\delta_j||_{\max} \leq \frac{\epsilon}{b+1}$:

$$V_{i+1}' = f'(V_i') = f(V_i') + \delta_{i+1} \stackrel{*}{=} f(V_i) + \sum_{j=0}^{i} \delta_j + \delta_{i+1} = f(V_i) + \sum_{j=0}^{i+1} \delta_j$$

```
1  function SEnum(V, M = ⟨S, T, s_init⟩, F, b, opt)
2      T'' := ∅, M' := M↓_F↓_R = ⟨S ⊎ S_new, T', s_init⟩
3      foreach s ∈ { s_init } ∪ { s'' | ∃ s' : s' →¹_T s'' } do
4          foreach 𝔖 ∈ SSched(M'[s]) do      // enumeration of simple schedulers
5          ⌊ T''(s) := T''(s) ∪ { ComputeProbs(M'[s]↓_𝔖) }
6      T'' := T'' ∪ { ⊥ ↦ { 𝒟(⟨0, ⊥⟩) } }, V(⊥) := 0
7      StepBoundedVI(V, M'' = ⟨Dom(T''), T'', s_init⟩, b, opt)      // step-bounded iter.
8  function ComputeProbs(M = ⟨S ⊎ S_new, ...⟩)                    // M is a DTMC
9      μ := { ⟨0, s⟩ ↦ P_max=min({ s_new }) | s_new ∈ S_new }
10     return μ ∪ { ⟨0, ⊥⟩ ↦ 1 - ∑_{s_new ∈ S_new} μ(⟨0, s⟩) }
```

**Algorithm 2.** Reward-bounded reachability via scheduler enumeration

where $*$ holds by the induction assumption. Finally, $|| \sum_{j=0}^{i} \delta_j ||_{\max} \le (i+1)\frac{\epsilon}{b+1}$, so $||V_i - V_i'||_{\max} \le (i+1)\frac{\epsilon}{b+1}$, which is what had to be proved.

### 3.2 Scheduler Enumeration

Our first new technique, senum, is summarised as Algorithm 2. The idea is to replace the entire sub-MDP between a "relevant" state and the new states (that follow immediately after what was a reward-one branch before the $\downarrow_R$ transformation) by *one direct* transition to a distribution over the new states *for each* simple scheduler. The actual reward-bounded probabilities can be computed on the result MDP $M''$ using the standard step-bounded algorithm (line 7), since one *step* now corresponds to a *reward* of 1.

The relevant states, which remain in the result MDP $M''$, are the initial state plus those states that had an incoming reward-one branch. We iterate over them in line 3. In an inner loop (line 4), we iterate over the simple schedulers for each relevant state. For each scheduler, ComputeProbs determines the distribution $\mu$ s.t. for each new state $s_{new}$, $\mu(s_{new})$ is the probability of reaching it (accumulating 1 reward on the way) and $\mu(\bot)$ is the probability of getting stuck in an end component without being able to accumulate any more reward ever. A transition to preserve $\mu$ in $M''$ is created in line 5. The total number of simple schedulers for $n$ states with max. fan-out $m$ is in $\mathcal{O}(m^n)$, but we expect the number of schedulers that lead to different distributions from one relevant state up to the next reward-one steps to remain manageable (cf. column "avg" in Table 2).

ComputeProbs is implemented either using value iterations, one for each new state, or—since $M'[s]\downarrow_𝔖$ is a DTMC—using DTMC *state elimination* [8]. The latter successively eliminates the non-new states as shown schematically in Fig. 3 while preserving the reachability probabilities, all in one go.

### 3.3 State Elimination

Instead of performing a probability-preserving DTMC state elimination for each scheduler as in senum, technique elim applies a new scheduler- and

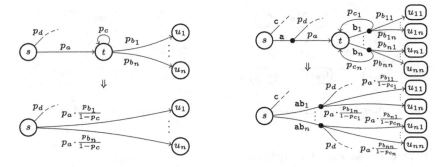

**Fig. 3.** DTMC state elimination [8]        **Fig. 4.** MDP state elimination

```
1  function Elim(V, M = ⟨S, T, s_init⟩, F, b, opt)
2      M' := M↓_F↓_R = ⟨S ⊎ S_new, ...⟩
3      ⟨S ⊎ S_new, T', s_init⟩ := Eliminate(M', S)              // MDP state elimination
4      T'' := { ⊥ ↦ { D(⟨0, ⊥⟩)}}, V(⊥) := 0, μ' := ∅
5      foreach s_new ∈ S_new and μ ∈ T'(s) do                   // state merging
6          μ' := μ' ∪ { ⊥ ↦ ∑_{⟨0,s'⟩∈support(μ)∧s'∈S} μ(⟨0, s'⟩)}
7          μ' := μ' ∪ { s' ↦ μ(⟨0, s'_new⟩)) | ⟨0, s'_new⟩ ∈ support(μ) ∧ s'_new ∈ S_new}
8          T''(s_new) := T''(s_new) ∪ {μ'}, μ' := ∅
9      StepBoundedVI(V, ⟨Dom(T''), T'', s_init⟩, b, opt)        // step-bounded iteration
```

**Algorithm 3.** Reward-bounded reachability via MDP state elimination

probability-preserving state elimination algorithm to the entire MDP. The state elimination algorithm is described by the schema shown in Fig. 4; states with Dirac self-loops will remain. Observe how this elimination process preserves the options that simple schedulers have, and in particular relies on their positional character to be able to redistribute the loop probabilities $p_{c_i}$ onto the same transition only.

elim is shown as Algorithm 3. In line 3, the MDP state elimination procedure is called to eliminate all the regular states in $S$. We can ignore rewards here since they were transformed by $\downarrow_R$ into branches to the distinguished new states. As an extension to the schema of Fig. 4, we also preserve the original outgoing transitions when we eliminate a relevant state (defined as in Sect. 3.2) because we need them in the next step: In the loop starting in line 5, we redirect (1) all branches that go to non-new states to the added bottom state $\bot$ instead because they indicate that we can get stuck in an end component without reward, and (2) all branches that go to new states to the corresponding original states instead. This way, we merge the (absorbing, but not eliminated) new states with the corresponding regular (eliminated from incoming but not outgoing transitions) states. Finally, in line 8, the standard step-bounded value iteration is performed on the eliminated-merged MDP as in senum. Figure 5 shows our example MDP after state elimination, and Fig. 6 shows the subsequent merged MDP. For clarity, transitions to the same successor distributions are shown in a combined way.

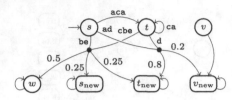

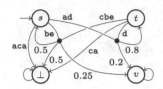

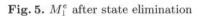

**Fig. 5.** $M_\downarrow^e$ after state elimination     **Fig. 6.** $M_\downarrow^e$ eliminated and merged

## 3.4   Correctness and Complexity

*Correctness.* Let $\mathfrak{S}$ be a deterministic reward-positional scheduler for $M{\downarrow}_F$. It corresponds to a sequence of simple schedulers $\mathfrak{S}_i$ for $M{\downarrow}_F$ where $i \in \{b, \dots, 0\}$ is the remaining reward that can be accumulated before the bound is reached. For each state $s$ of $M{\downarrow}_F$ and $i > 0$, each such $\mathfrak{S}_i$ induces a (potentially substochastic) measure $\mu_s^i$ such that $\mu_s^i(s')$ is the probability to reach $s'$ from $s$ in $M{\downarrow}_F{\downarrow}_{\mathfrak{S}_i}$ over paths whose last step has reward 1. Let $\mu_s^0$ be the induced measure such that $\mu_s^0(s')$ is the probability under $\mathfrak{S}_0$ to reach $s'$ without reward if it is a goal state and 0 otherwise. Using the recursion $\overline{\mu}_s^i(s') \stackrel{\text{def}}{=} \sum_{s'' \in S} \mu_s^i(s'') \cdot \overline{\mu}_{s''}^{i-1}(s')$ with $\overline{\mu}_s^0 \stackrel{\text{def}}{=} \mu_s^0$, the value $\overline{\mu}_s^b(s')$ is the probability to reach goal state $s'$ from $s$ in $M{\downarrow}_F$ under $\mathfrak{S}$. Thus we have $\max_{\mathfrak{S}} \overline{\mu}_s^b(s') = \mathrm{P}_{\max}^{\mathrm{R} \leq b}(F)$ and $\min_{\mathfrak{S}} \overline{\mu}_s^b(s') = \mathrm{P}_{\min}^{\mathrm{R} \leq b}(F)$ by Theorem 1. If we distribute the maximum operation into the recursion, we get

$$\max_{\mathfrak{S}} \overline{\mu}_s^i(s') = \sum_{s'' \in S} \max_{\mathfrak{S}_i} \mu_s^i(s'') \cdot \max_{\mathfrak{S}} \overline{\mu}_{s''}^{i-1}(s') \tag{1}$$

and an analogous formula for the minimum. By computing extremal values w.r.t. simple schedulers for each reward step, we thus compute the value w.r.t. an optimal deterministic reward-positional scheduler for the bounded property overall. The correctness of senum and elim now follows from the fact that they implement precisely the right-hand side of (1): $\overline{\mu}_s^0$ is always given as the initial value of $V$ as described at the very beginning of this section. In senum, we enumerate the relevant measures $\mu_s^\cdot$ induced by all the simple schedulers as one transition each, then choose the optimal transition for each $i$ in the $i$-th iteration inside StepBoundedVI. The argument for elim is the same, the difference being that state elimination is what transforms all the measures into single transitions.

*Complexity.* The problem that we solve is EXP-complete [6]. We make the following observations about senum and elim: Let $n_{\text{new}} \leq n$ be the number of new states, $n_s \leq n$ the max. size of any relevant reward-free sub-MDP (i.e. the max. number of states reachable from the initial or a new state when dropping all reward-one branches), and $s_s \leq m^n$ the max. number of simple schedulers in these sub-MDP. The reduced MDP created by senum and elim have $n_{\text{new}}$ states and up to $n_{\text{new}} \cdot s_s \cdot n_s$ branches. The bounded value iterations thus involve $\mathcal{O}(b \cdot n_{\text{new}} \cdot s_s \cdot n_s)$ arithmetic operations overall. Note that in the worst case, $s_s = m^n$, i.e. it is exponential in the size of the original MDP. To obtain the reduced MDP, senum enumerates $\mathcal{O}(n_{\text{new}} \cdot s_s)$ schedulers; for each, value iteration or DTMC state elimination is done on a sub-MDP of $\mathcal{O}(n_s)$ states. elim needs to eliminate $n - n_{\mathrm{new}}$ states, with each elimination requiring $\mathcal{O}(s_s \cdot n_s)$ operations.

## 4   Implementation

We have implemented the three unfolding-free techniques within MCSTA, the MODEST TOOLSET's model checker for PTA and MDP. When asked to compute $P_{opt}^{R \le b}(\cdot)$, it delivers *all* values $P_{opt}^{R \le i}(\cdot)$ for $i \in \{0, \dots, b\}$ since the algorithms allow doing so at no overhead. Instead of a single value, we thus get the entire (sub-)cdf. Every single value is defined via an individual optimisation over schedulers. However, we have seen in Sect. 3.4 that an optimal scheduler for bound $i$ can be extended to an optimal scheduler for $i + 1$, so there exists an optimal scheduler for all bounds. The max./min. cdf represents the probability distribution induced by that scheduler. We show these functions for the randomised consensus case study [16] in Fig. 7. The top (bottom) curve is the max. (min.) probability for the protocol to terminate within the number of coin tosses given on the x-axis. For comparison, the left and right dashed lines show the means of these distributions. Note that the min. expected value corresponds to the max. bounded probabilities and vice-versa. As mentioned, using the unfolding-free techniques, we compute the curves in the same amount of memory otherwise sufficient for the means only. We also implemented a convergence criterion to detect when the result will no longer increase for higher bounds, i.e. when the unbounded probability has been reached up to $\epsilon$. For the functions in Fig. 7, this happens at 4016 coin tosses for the max. and 5607 for the min. probability.

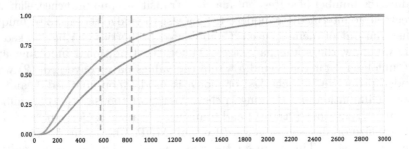

**Fig. 7.** Cdfs and means for the randomised consensus model ($H = 6, K = 4$)

## 5   Experiments

We use six case studies from the literature to evaluate the applicability and performance of the three unfolding-free techniques and their implementation:

- **BEB** [5]: MDP of a bounded exponential backoff procedure with max. back-off value $K = 4$ and $H \in \{5, \dots, 10\}$ parallel hosts. We compute the max. probability of any host seizing the line while all hosts enter backoff $\le b$ times.

- **BRP** [9]: The PTA model of the bounded retransmission protocol with $N \in \{32, 64\}$ frames to transmit, retransmission bound $MAX \in \{6, 12\}$ and transmission delay $TD \in \{2, 4\}$ time units. We compute the max. and min. probability that the sender reports success in $\leq b$ time units.
- **RCONS** [16]: The randomised consensus shared coin protocol MDP as described in Sect. 4 for $N \in \{4, 6\}$ parallel processes and constant $K \in \{2, 4, 8\}$.
- **CSMA** [9]: PTA model of a communication protocol using CSMA/CD, with max. backoff counter $BCMAX \in \{1, \ldots, 4\}$. We compute the min. and max. probability that both stations deliver their packets by deadline $b$ time units.
- **FW** [16]: PTA model ("Impl" variant) of the IEEE 1394 FireWire root contention protocol with either a short or a long cable. We ask for the min. probability that a leader (root) is selected before time bound $b$.
- **IJSS** [14]: MDP model of Israeli and Jalfon's randomised self-stabilising algorithm with $N \in \{18, 19, 20\}$ processes. We compute the min. probability to reach a stable state in $\leq b$ steps of the algorithm (query Q2 in [14]). This is a step-bounded property; we consider IJSS here only to compare with [14].

Experiments were performed on an Intel Core i5-6600T system (2.7 GHz, 4 cores) with 16 GB of memory running 64-bit Windows 10 and a timeout of 30 min.

Looking back at Sect. 3, we see that the only extra states introduced by modvi compared to checking an unbounded probabilistic reachability or expected-reward property are the new states $s_{\text{new}}$. However, this was for the presentation only, and is avoided in the implementation by checking for reward-one branches on-the-fly. The transformations performed in senum and elim, on the other hand, will reduce the number of states, but may add transitions and branches. elim may also create large intermediate models. In contrast to modvi, these two techniques may thus run out of memory even if unbounded properties can be checked. In Table 1, we show the state-space sizes (1) for the traditional unfolding approach ("unfolded") for the bound $b$ where the values have converged, (2) when unbounded properties are checked or modvi is used ("non-unfolded"), and (3) after state elimination and merging in elim. We report thousands (k) or millions (M) of states, transitions ("trans") and branches ("branch"). Column "avg" lists the average size of all relevant reward-free sub-MDP. The values for senum are the same as for elim. Times are for the state-space exploration phase only, so the time for "non-unfolded" will be incurred by all three unfolding-free algorithms. We see that avoiding unfolding is a drastic reduction. In fact, 16 GB of memory are not sufficient for the larger unfolded models, so we used MCSTA's disk-based technique [11]. State elimination leads to an increase in transitions and especially branches, drastically so for BRP, the exceptions being BEB and IJSS. This appears related to the size of the reward-free subgraphs, so state elimination may work best if there are few steps between reward increments.

In Table 2, we report the performance results for all three techniques when run until the values have converged at bound value $b$ (except for IJSS, where we follow [14] and set $b$ to the 99th percentile). For senum, we used the variant based on value iteration since it consistently performed better than the one

**Table 1.** State spaces

| group | model | b | unfolded states | unfolded time | non-unfolded states | trans | branch | time | avg | eliminated states | trans | branch |
|---|---|---|---|---|---|---|---|---|---|---|---|---|
| BEB | 5 | 90 | 1 M | 6 s | 10 k | 12 k | 22 k | 0 s | 3.7 | 3 k | 5 k | 19 k |
| BEB | 6 | 143 | 6 M | 35 s | 45 k | 60 k | 118 k | 0 s | 3.3 | 16 k | 28 k | 104 k |
| BEB | 7 | 229 | 45 M | 273 s | 206 k | 304 k | 638 k | 1 s | 3.0 | 80 k | 149 k | 568 k |
| BEB | 8 | 371 | | | 1.0 M | 1.6 M | 3.4 M | 3 s | 2.8 | 0.4 M | 0.8 M | 3.1 M |
| BEB | 9 | 600 | >30 min | | 4.6 M | 8.3 M | 18.9 M | 26 s | 2.6 | 2.0 M | 4.2 M | 16.6 M |
| BEB | 10 | n/a | | | 22.2 M | 44.0 M | 102.8 M | 138 s | 2.5 | >16 GB | | |
| BRP | 32, 6,2 | 179 | 9 M | 40 s | 0.1 M | 0.1 M | 0.1 M | 0 s | 24.2 | 0.1 M | 0.4 M | 7.1 M |
| BRP | 32, 6,4 | 347 | 50 M | 206 s | 0.2 M | 0.2 M | 0.2 M | 1 s | 22.8 | 0.2 M | 1.0 M | 20.3 M |
| BRP | 32,12,2 | 179 | 22 M | 90 s | 0.2 M | 0.2 M | 0.3 M | 1 s | 22.8 | 0.2 M | 1.1 M | 22.1 M |
| BRP | 32,12,4 | 347 | 122 M | 499 s | 0.6 M | 0.7 M | 0.7 M | 2 s | 21.3 | 0.6 M | 3.2 M | 62.0 M |
| BRP | 64, 6,2 | 322 | 38 M | 157 s | 0.1 M | 0.2 M | 0.2 M | 0 s | 47.1 | 0.1 M | 1.3 M | 53.8 M |
| BRP | 64, 6,4 | 630 | 207 M | 826 s | 0.4 M | 0.4 M | 0.5 M | 1 s | 44.4 | 0.4 M | 3.8 M | 153.7 M |
| BRP | 64,12,2 | 322 | 107 M | 427 s | 0.5 M | 0.5 M | 0.5 M | 1 s | 44.4 | 0.4 M | 4.1 M | 166.0 M |
| BRP | 64,12,4 | 630 | >30 min | | 1.3 M | 1.4 M | 1.5 M | 4 s | 41.3 | >16 GB | | |
| RCONS | 4,4 | 2653 | 54 M | 365 s | 41 k | 113 k | 164 k | 0 s | 4.5 | 35 k | 254 k | 506 k |
| RCONS | 4,8 | 7793 | | | 80 k | 220 k | 323 k | 0 s | 4.1 | 68 k | 499 k | 997 k |
| RCONS | 6,2 | 2175 | >30 min | | 1.2 M | 5.0 M | 7.2 M | 5 s | 11.7 | 1.1 M | 23.6 M | 47.1 M |
| RCONS | 6,4 | 5607 | | | 2.3 M | 9.4 M | 13.9 M | 9 s | 9.1 | 2.2 M | 42.2 M | 84.3 M |
| CSMA | 1 | 2941 | 31 M | 276 s | 13 k | 13 k | 13 k | 0 s | 1.4 | 13 k | 13 k | 15 k |
| CSMA | 2 | 3695 | 191 M | 1097 s | 96 k | 96 k | 97 k | 0 s | 1.3 | 95 k | 95 k | 110 k |
| CSMA | 3 | 5229 | >30 min | | 548 k | 548 k | 551 k | 2 s | 1.2 | 545 k | 545 k | 637 k |
| CSMA | 4 | 8219 | | | 2.7 M | 2.7 M | 2.7 M | 9 s | 1.2 | 2.7 M | 2.7 M | 3.2 M |
| FW | short | 2487 | 9 M | 150 s | 4 k | 6 k | 6 k | 0 s | 4.0 | 4 k | 111 k | 413 k |
| FW | long | 3081 | >30 min | | 0.2 M | 0.5 M | 0.5 M | 1 s | 3.4 | 0.2 M | 2.4 M | 7.7 M |
| IJSS | 18 | 445 | | | 0.3 M | 2.6 M | 5.0 M | 5 s | 1.0 | 0.3 M | 2.5 M | 4.3 M |
| IJSS | 19 | 498 | >30 min | | 0.5 M | 5.5 M | 10.5 M | 10 s | 1.0 | 0.5 M | 5.2 M | 9.0 M |
| IJSS | 20 | 553 | | | 1.0 M | 11.5 M | 22.0 M | 22 s | 1.0 | 1.0 M | 11.0 M | 18.9 M |

using DTMC state elimination. "iter" denotes the time needed for (unbounded or step-bounded) value iteration, while "enum" and "elim" are the times needed for scheduler enumeration resp. state elimination and merging. "#" is the total number of iterations performed over all states inside the calls to VI. "avg" is the average number of schedulers enumerated per relevant state; to get the approx. total number of schedulers enumerated for a model instance, multiply by the number of states for elim in Table 1. "rate" is the number of bound values computed per second, i.e. $b$ divided by the time for value iteration. Memory usage in columns "mem" is MCSTA's peak working set, including state space exploration, reported in mega- (M) or gigabytes (G). MCSTA is garbage-collected, so these values are higher than necessary since full collections only occur when the system runs low on memory. The values related to value iteration for senum are the same as for elim. In general, we see that senum uses less memory than elim,

**Table 2.** Runtime and memory usage

| | model | $b$ | | modvi | | | | senum | | | elim | | | |
|---|---|---|---|---|---|---|---|---|---|---|---|---|---|---|
| | | | iter | # | mem | rate | enum | mem | avg | elim | iter | mem | rate |
| BEB | 5 | 90 | 0 s | 422 | 43 M | $425\frac{1}{s}$ | 0 s | 45 M | 4.2 | 0 s | 0 s | 48 M | $\infty\frac{1}{s}$ |
| | 6 | 143 | 1 s | 666 | 54 M | $168\frac{1}{s}$ | 1 s | 65 M | 6.4 | 0 s | 0 s | 107 M | $1430\frac{1}{s}$ |
| | 7 | 229 | 6 s | 1070 | 132 M | $34\frac{1}{s}$ | 11 s | 210 M | 10.7 | 2 s | 0 s | 409 M | $1145\frac{1}{s}$ |
| | 8 | 371 | 56 s | 1734 | 374 M | $6\frac{1}{s}$ | 88 s | 588 M | 19.9 | 12 s | 2 s | 1.5 G | $247\frac{1}{s}$ |
| | 9 | 600 | 487 s | 2798 | 2.0 G | $1\frac{1}{s}$ | 960 s | 2.6 G | 40.8 | 67 s | 14 s | 6.6 G | $43\frac{1}{s}$ |
| | 10 | n/a | | | > 30 min | | | > 30 min | | | | > 16 GB | |
| BRP | 32, 6, 2 | 179 | 1 s | 837 | 56 M | $209\frac{1}{s}$ | 160 s | 474 M | 6.1 | 4 s | 1 s | 775 M | $164\frac{1}{s}$ |
| | 32, 6, 4 | 347 | 5 s | 1520 | 87 M | $81\frac{1}{s}$ | 498 s | 1.2 G | 5.9 | 12 s | 7 s | 2.6 G | $61\frac{1}{s}$ |
| | 32, 12, 2 | 179 | 3 s | 837 | 92 M | $70\frac{1}{s}$ | 569 s | 1.3 G | 5.9 | 13 s | 4 s | 2.8 G | $56\frac{1}{s}$ |
| | 32, 12, 4 | 347 | 17 s | 1520 | 196 M | $26\frac{1}{s}$ | 1467 s | 3.5 G | 5.5 | 40 s | 22 s | 6.1 G | $20\frac{1}{s}$ |
| | 64, 6, 2 | 322 | 4 s | 1451 | 77 M | $105\frac{1}{s}$ | | | | 31 s | 16 s | 4.7 G | $24\frac{1}{s}$ |
| | 64, 6, 4 | 630 | 20 s | 2735 | 140 M | $38\frac{1}{s}$ | | > 30 min | | 114 s | 91 s | 14.1 G | $8\frac{1}{s}$ |
| | 64, 12, 2 | 322 | 11 s | 1451 | 149 M | $34\frac{1}{s}$ | | | | 132 s | 51 s | 13.9 G | $8\frac{1}{s}$ |
| | 64, 12, 4 | 630 | 62 s | 2735 | 328 M | $12\frac{1}{s}$ | | | | | | > 16 GB | |
| RCONS | 4, 4 | 2653 | 43 s | 21763 | 61 M | $108\frac{1}{s}$ | 2 s | 126 M | 17.5 | 1 s | 3 s | 224 M | $1842\frac{1}{s}$ |
| | 4, 8 | 7793 | 260 s | 66739 | 87 M | $56\frac{1}{s}$ | 4 s | 187 M | 15.8 | 2 s | 16 s | 384 M | $933\frac{1}{s}$ |
| | 6, 2 | 2175 | 1608 s | 19291 | 680 M | $2\frac{1}{s}$ | | > 30 min | | 136 s | 169 s | 11.9 G | $20\frac{1}{s}$ |
| | 6, 4 | 5607 | | | > 30 min | | | | | 275 s | 879 s | 13.4 G | $11\frac{1}{s}$ |
| CSMA | 1 | 2941 | 4 s | 12220 | 45 M | $1363\frac{1}{s}$ | 5 s | 46 M | 3.7 | 0 s | 0 s | 60 M | $\infty\frac{1}{s}$ |
| | 2 | 3695 | 30 s | 15363 | 64 M | $244\frac{1}{s}$ | | | | 1 s | 3 s | 190 M | $2437\frac{1}{s}$ |
| | 3 | 5229 | 226 s | 21780 | 187 M | $46\frac{1}{s}$ | | > 30 min | | 3 s | 24 s | 839 M | $429\frac{1}{s}$ |
| | 4 | 8219 | 1689 s | 34009 | 617 M | $10\frac{1}{s}$ | | | | 19 s | 192 s | 3.8 G | $86\frac{1}{s}$ |
| FW | short | 2487 | 1 s | 6608 | 40 M | $2763\frac{1}{s}$ | 29 s | 79 M | 476.5 | 0 s | 1 s | 113 M | $3109\frac{1}{s}$ |
| | long | 3081 | 60 s | 9610 | 108 M | $51\frac{1}{s}$ | 205 s | 880 M | 82.7 | 13 s | 23 s | 1.4 G | $132\frac{1}{s}$ |
| IJSS | 18 | 445 | 55 s | 891 | 527 M | $8\frac{1}{s}$ | 5 s | 876 M | 10.0 | 16 s | 3 s | 1.7 G | $23\frac{1}{s}$ |
| | 19 | 498 | 126 s | 997 | 947 M | $4\frac{1}{s}$ | 10 s | 1.7 G | 10.5 | 35 s | 7 s | 3.7 G | $12\frac{1}{s}$ |
| | 20 | 553 | 304 s | 1107 | 1.8 G | $2\frac{1}{s}$ | 22 s | 3.5 G | 11.0 | 83 s | 16 s | 7.6 G | $6\frac{1}{s}$ |

but is much slower in all cases except IJSS. If elim works and does not blow up the model too much, it is significantly faster than modvi, making up for the time spent on state elimination with much faster value iteration rates.

# 6 Conclusion

We presented three approaches to model-check reward-bounded properties on MDP without unfolding: a small correction of recent work based on unbounded value iteration [14], and two new techniques that reduce the model such that step-bounded value iteration can be used, which is efficient and exact. We also consider the application to time-bounded properties on PTA and provide the first implementation that is publicly available, within the MODEST TOOLSET at modestchecker.net. By avoiding unfolding and returning the entire probability distribution up to the bound at no extra cost, this could finally make reward- and time-bounded probabilistic timed model checking feasible in practical applications. As we presented the algorithms in this paper, they compute reachability probabilities. However all of them can easily be adapted to compute reward-bounded expected accumulated rewards and instantaneous rewards, too.

*Outlook.* The digital clocks approach for PTA was considered limited in scalability. The presented techniques lift some of its most significant practical limitations. Moreover, time-bounded analysis without unfolding and with computation of the entire distribution in this manner is not feasible for the traditionally more scalable zone-based approaches because zones abstract from concrete timing. We see the possibility to improve the state elimination approach by removing transitions that are linear combinations of others and thus unnecessary. This may reduce the transition and branch blowup on models like the BRP case. Going beyond speeding up simple reward-bounded reachability queries, state elimination also opens up ways towards a more efficient analysis of long-run average and long-run reward-average properties.

# References

1. Andova, S., Hermanns, H., Katoen, J.-P.: Discrete-time rewards model-checked. In: Larsen, K.G., Niebert, P. (eds.) FORMATS 2003. LNCS, vol. 2791, pp. 88–104. Springer, Heidelberg (2004). doi:10.1007/978-3-540-40903-8_8
2. Baier, C., Daum, M., Dubslaff, C., Klein, J., Klüppelholz, S.: Energy-utility quantiles. In: Badger, J.M., Rozier, K.Y. (eds.) NFM 2014. LNCS, vol. 8430, pp. 285–299. Springer, Heidelberg (2014). doi:10.1007/978-3-319-06200-6_24
3. Baier, C., Katoen, J.P.: Principles of Model Checking. MIT Press, Massachusetts (2008)
4. Berendsen, J., Chen, T., Jansen, D.N.: Undecidability of cost-bounded reachability in priced probabilistic timed automata. In: Chen, J., Cooper, S.B. (eds.) TAMC 2009. LNCS, vol. 5532, pp. 128–137. Springer, Heidelberg (2009). doi:10.1007/978-3-642-02017-9_16
5. Giro, S., D'Argenio, P.R., Ferrer Fioriti, L.M.: Partial order reduction for probabilistic systems: a revision for distributed schedulers. In: Bravetti, M., Zavattaro, G. (eds.) CONCUR 2009. LNCS, vol. 5710, pp. 338–353. Springer, Heidelberg (2009). doi:10.1007/978-3-642-04081-8_23
6. Haase, C., Kiefer, S.: The odds of staying on budget. In: Halldórsson, M.M., Iwama, K., Kobayashi, N., Speckmann, B. (eds.) ICALP 2015. LNCS, vol. 9135, pp. 234–246. Springer, Heidelberg (2015). doi:10.1007/978-3-662-47666-6_19
7. Haddad, S., Monmege, B.: Reachability in MDPs: refining convergence of value iteration. In: Ouaknine, J., Potapov, I., Worrell, J. (eds.) RP 2014. LNCS, vol. 8762, pp. 125–137. Springer, Heidelberg (2014). doi:10.1007/978-3-319-11439-2_10
8. Hahn, E.M., Hermanns, H., Zhang, L.: Probabilistic reachability for parametric Markov models. STTT 13(1), 3–19 (2011)
9. Hartmanns, A., Hermanns, H.: A Modest approach to checking probabilistic timed automata. In: QEST, pp. 187–196. IEEE Computer Society (2009)

10. Hartmanns, A., Hermanns, H.: The Modest Toolset: an integrated environment for quantitative modelling and verification. In: Ábrahám, E., Havelund, K. (eds.) TACAS 2014. LNCS, vol. 8413, pp. 593–598. Springer, Heidelberg (2014). doi:10.1007/978-3-642-54862-8_51

11. Hartmanns, A., Hermanns, H.: Explicit model checking of very large MDP using partitioning and secondary storage. In: Finkbeiner, B., Pu, G., Zhang, L. (eds.) ATVA 2015. LNCS, vol. 9364, pp. 131–147. Springer, Heidelberg (2015). doi:10.1007/978-3-319-24953-7_10

12. Hashemi, V., Hermanns, H., Song, L.: Reward-bounded reachability probability for uncertain weighted MDPs. In: Jobstmann, B., Leino, K.R.M. (eds.) VMCAI 2016. LNCS, vol. 9583, pp. 351–371. Springer, Heidelberg (2016). doi:10.1007/978-3-662-49122-5_17

13. Hatefi, H., Braitling, B., Wimmer, R., Fioriti, L.M.F., Hermanns, H., Becker, B.: Cost vs. time in stochastic games and Markov automata. In: Li, X., Liu, Z., Yi, W. (eds.) SETTA 2015. LNCS, vol. 9409, pp. 19–34. Springer, Heidelberg (2015). doi:10.1007/978-3-319-25942-0_2

14. Klein, J., Baier, C., Chrszon, P., Daum, M., Dubslaff, C., Klüppelholz, S., Märcker, S., Müller, D.: Advances in symbolic probabilistic model checking with PRISM. In: Chechik, M., Raskin, J.-F. (eds.) TACAS 2016. LNCS, vol. 9636, pp. 349–366. Springer, Heidelberg (2016). doi:10.1007/978-3-662-49674-9_20

15. Kwiatkowska, M., Norman, G., Parker, D.: PRISM 4.0: verification of probabilistic real-time systems. In: Gopalakrishnan, G., Qadeer, S. (eds.) CAV 2011. LNCS, vol. 6806, pp. 585–591. Springer, Heidelberg (2011). doi:10.1007/978-3-642-22110-1_47

16. Kwiatkowska, M.Z., Norman, G., Parker, D.: The PRISM benchmark suite. In: QEST, pp. 203–204. IEEE Computer Society (2012)

17. Kwiatkowska, M.Z., Norman, G., Parker, D., Sproston, J.: Performance analysis of probabilistic timed automata using digital clocks. FMSD **29**(1), 33–78 (2006)

18. Randour, M., Raskin, J.-F., Sankur, O.: Percentile queries in multi-dimensional Markov decision processes. In: Kroening, D., Păsăreanu, C.S. (eds.) CAV 2015. LNCS, vol. 9206, pp. 123–139. Springer, Heidelberg (2015). doi:10.1007/978-3-319-21690-4_8

19. Ummels, M., Baier, C.: Computing quantiles in Markov reward models.break In: Pfenning, F. (ed.) FoSSaCS 2013. LNCS, vol. 7794, pp. 353–368. Springer, Heidelberg (2013). doi:10.1007/978-3-642-37075-5_23

# Computing Specification-Sensitive Abstractions for Program Verification

Tianhai Liu[1]([✉]), Shmuel Tyszberowicz[3], Mihai Herda[1], Bernhard Beckert[1], Daniel Grahl[1], and Mana Taghdiri[2]

[1] Karlsruhe Institute of Technology, Karlsruhe, Germany
tianhai.liu@kit.edu
[2] Horus Software GmbH, Ettlingen, Germany
[3] The Academic College Tel Aviv Yaffo, Tel Aviv, Israel

**Abstract.** To enable scalability and address the needs of real-world software, deductive verification relies on modularization of the target program and decomposition of its requirement specification. In this paper, we present an approach that, given a Java program and a partial requirement specification written using the Java Modeling Language, constructs a semantic slice. In the slice, the parts of the program irrelevant w.r.t. the partial requirements are replaced by an abstraction. The core idea of our approach is to use bounded program verification techniques to guide the construction of these slices. Our approach not only lessens the burden of writing auxiliary specifications (such as loop invariants) but also reduces the number of proof steps needed for verification.

## 1 Introduction

**Motivation.** The power of deductive program verification has increased considerably over the last decades. To enable scalability and address the needs of real-world software, deductive verification relies on modularization of the target program. This requires annotating sub-procedures with formal auxiliary specifications (method contracts, loop invariants, etc.). To discover useful specifications that are fulfilled by the annotated sub-procedure and also meet the requirements of the calling procedures is, unfortunately, a difficult and error-prone effort (cf. [6, Chap. 5]). To ease the burden, verification engineers routinely break a complex requirement specification into conjunctions of *partial* specifications, i.e., they decompose not only the implementation but also the specification. Then, usually, only parts of the implementation are relevant for proving a partial property, and only partial and less complex auxiliary specifications are needed. To make use of that advantage, the verification engineer needs to identify the slice of the implementation relevant to the partial property. The main contribution of this paper is an automated method for computing such program slices defined by partial specifications.

© Springer International Publishing AG 2016
M. Fränzle et al. (Eds.): SETTA 2016, LNCS 9984, pp. 101–117, 2016.
DOI: 10.1007/978-3-319-47677-3_7

**Our Approach.** Given a Java program and a partial requirement specification, written using the Java Modeling Language (JML) [22], we construct a *semantic slice* (an abstract program). In the slice, the program parts that are irrelevant to the partial requirements are replaced by an abstraction (i.e., they are not completely removed), whereas the rest of the program (i.e., the relevant parts) remains unchanged. (In the rest of the paper we use the terms semantic slice and abstract program interchangeably.) As said above, verifying slices requires fewer auxiliary specifications (as the abstractions have less details), and their correctness—by their construction—implies the correctness of the original program w.r.t. the partial specification under consideration. As a result, our method liberates the verification engineers from finding the relevant slice manually.

Figure 1 illustrates the structure of our novel approach. The core idea is to use bounded program verification techniques to guide the construction of slices. Bounded program verification systems (such as JForge [14], Jalloy [27], and InspectJ [23]) do not require auxiliary specifications. They translate, based on user-provided bounds (that, e.g., limit the number of objects or the number of loop iterations), the analyzed program and its *negated* requirement specification into a satisfiability problem—an SMT [3] formula consisting of a set of constraints, and try to find a solution to that problem. If a solution to that satisfiability problem is found, then that is a counterexample to the correctness of the original program, and no further analysis is required. If no solution is found, the partial property holds—but only w.r.t. the bounds, thus a deductive verification—an unbounded program verification, is still needed.

Before continuing with the deductive verification we compute the slice of the program relevant to the partial requirements. The computation is based on the *unsatisfiable core* (unsat core)—a subset of constraints that is unsatisfiable, obtained during the unsatisfiability proof for the bounded problem. Then we minimize the unsat core to ensure that the proof requires all its elements. The Java program statements that are related to the constraints in the unsat core (by the construction of constraints from the Java code) are known to be relevant for the bounded proof of the requirement specification. We generate a semantic slice by over-approximating the behaviors of the other statements.

Finally, if the semantic slice can be verified using deductive program verification, which requires auxiliary specifications, the original program satisfies the specification as well (by the construction of the slice). Otherwise, we use counterexample-guided refinement to refine the abstraction and repeat the deductive verification.

The semantic slice is generated based on a particular bounded proof. Therefore it (i) may be too abstract, and thus deductive verification is not possible, and (ii) may exclude unnecessary, yet helpful, details, hence deductive verification may require more effort. But, as our evaluation shows, in practice the slice is sufficiently precise.

Our approach not only lessens the burden of writing auxiliary specifications but also eases the deductive verification: less proof steps are needed. Besides, by the *small-scope hypothesis* [20], if the program does not satisfy its specification,

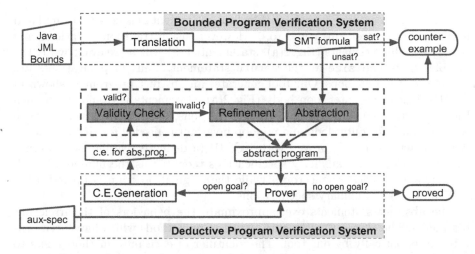

**Fig. 1.** Structure of our approach.

in many cases that will be detected during the bounded-verification phase of our approach, avoiding unnecessary attempts at deductive verification.

We have built a prototype tool, *AbstractJ*, that implements the abstraction as well as the refinement and validity check, and we have performed several experiments to evaluate the benefits of our approach.

## 2   Motivating Examples

We use two examples (Figs. 2 and 3) to demonstrate our approach. To specify the Java modules we employ JML, a behavioral interface specification language. We shortly explain the JML clauses used in the examples; for more details see [22]. The specification is written between /*@ and */. The `ensures` clause specifies properties that are guaranteed to hold at the end of the method call, and `\result` refers to the value returned by the method. (Both clauses refer to the case that the method terminated normally.) The `diverges` clause is used to specify when a method may either loop forever or not return normally to its caller. Writing `diverges true` means that non-termination is allowed for the method. The `assignable` clause provides the locations that can be assigned to during the execution of the method (frame conditions). The clause `assignable \strictly_nothing` denotes that the relevant methods neither modify heap locations nor allocate objects, whereas `assignable \nothing` allows object allocations; `assignable \everything` enables the method both to modify any heap location and to allocate objects.

The program in Fig. 2(a) computes the number of prime numbers between two given integers x and y (exclusive). The first line denotes that if the number of prime numbers is larger than 0, then x < y. Carefully inspecting the code, a verification engineer will notice that the `ensures` clause becomes false only when x >= y. In that case, the outer loop (Fig. 2(a), statement 2) is never executed and the variable size remains equal to 0. However, using traditional static slicing techniques, all statements (Fig. 2(a), statements 1–9) will be relevant w.r.t. the variables x, y, and size at the return statement. Thus loop invariants are required for the two loops. Our tool generates an abstract program as shown in Fig. 2(b) where the outer loop body (Fig. 2(a), statements 3–9) and the branch (Fig. 2(a), statements 10–11) are abstracted. Thus, it becomes easier to write loop invariants for outer loop and no loop invariant is needed for the inner loop. The abstract program is proved with KeY [1] (a deductive verification system) using 646 rules and 8 auxiliary specifications (counted as the number of JML constructs and logical connectors), while the original program is proved using 5802 rules and 26 auxiliary specifications.

The abstract statements over-approximate the behaviors of the irrelevant statements. The abstract statements invoke pure methods which have been generated automatically by our tool. The identifier of each pure method refers to the type of original statement. Each native method returns an unspecified value of the appropriate type.

```
/*@ ensures \result>0==>x<y;
  @ diverges true;
  @ assignable \everything;*/
  int numberOfPrime(int x, int y){
1   int size = 0;
2   for(int i=x; i<y; i++){
3     boolean isPrime = true;
4     for(int j=2; j<i; j++)
5       if (i%j==0){
6         isPrime=false;
7         break;
        }
8     if (isPrime)
9       size++;
    }
10  if(size > 0){
11    int[] a = new int[y-x];
    }
12  return size;
  }
```

(a) Original program

```
/*@ ensures \result>0==>x<y;
  @ diverges true;
  @ assignable \everything;*/
  int numberOfPrime(int x, int y){
1   int size = 0;
2   for(int i=x; i<y; i++){
3     size = pure_int();



    }
4   pure_allocArrayInt();

5   return size;
  }
  //@ assignable \strictly_nothing;
  native int pure_int();
  //@ assignable \nothing;
  native int[] pure_intArray();
```

(b) Abstract program

**Fig. 2.** A data-structure-poor program to compute the number of prime numbers between two integers. The empty lines in the abstract program are left deliberately for an intuitive comparison.

Figure 3(a) shows the second example. It provides a map data type implemented using associative arrays. Keys and values are recorded in separate arrays, `keys` and `values`, respectively, and have the same index in the arrays. The method `put(k,v)` invokes the method `getIndexOf` to check whether k already exists in the map. If it exists, the old value is replaced by v; otherwise, the methods `addKey` and `addValue` reallocate the arrays `keys` and `values`, respectively, and add k and v to the new arrays. The `ensures` clause guarantees that the value v is in this map (the `\exists` quantifier). By default, referenced variables are not null, thus the

```
class Key {} class Value {}
class Map {
  /*@ nullable */ Key[] keys;
  /*@ nullable */ Value[] values;
  /*@ ensures (\exists int i;0<=i&&
   @ i<values.length;values[i]==v);
   @ diverges true;
   @ assignable \everything; */
  void put(Key k, Value v){
1    int pos = getIndexOf(k);
2    if (pos>=0)
3      values[pos] = v;
     else {
4      addKey(k);
5      addValue(v);
     }
  }
  int getIndexOf(Key k){
6    int r = -1;
7    for(int i=0;i<keys.length;i++)
8      if (keys[i] == k)
9        r = i;
10   return r;
  }
  void addKey(Key k){
11   Key[] oldKs = keys;
12   keys = new Key[keys.length+1];
13   keys[keys.length - 1] = k;
14   for (int i=0;i<oldKs.length;i++)
15     keys[i] = oldKs[i];
  }
  void addValue(Value v){
16   Value[] oldVs = values;
17   values=new Value[values.length+1];
18   values[values.length - 1] = v;
19   for (int i=0;i<oldVs.length;i++)
20     values[i] = oldVs[i];
  }
}
```

(a) Original program

```
class Key {} class Value {}
class Map {
  /*@ nullable */ Key[] keys;
  /*@ nullable */ Value[] values;
  /*@ ensures (\exists int i;0<=i&&
   @ i<values.length;values[i]==v);
   @ diverges true;
   @ assignable \everything; */
  void put(Key k, Value v){
1    int pos = pure_int(k);
2    if(pure_boolean())
3      values[pos] = v;
     else {
4      impure_keys(k);
5      addValue(v);
     }
  }
  void addValue(Value v){
6    Value[] oldVs = values;
7    values=new Value[values.length+1];
8    values[values.length - 1] = v;
9    for (int i=0;i<oldVs.length;i++)
10     values[i] = pure_Value();
  }
  //@ assignable \strictly_nothing;
  native int pure_int();
  //@ assignable \strictly_nothing;
  native boolean pure_boolean();
  //@ assignable this.keys;
  native void impure_keys();
  //@ assignable \strictly_nothing;
  native /*@nullable*/ Value
        pure_Value();
}
```

(b) Abstract program

**Fig. 3.** A data-structure-rich program to put a key and a value to a map.

nullable clause enables also the null value. The abstract program is shown in Fig. 3(b), where the methods getIndexOf and addKey are irrelevant to the property of interest, thus their call sites (Fig. 3(a), statements 4, 5) are abstracted, such that this fact is directly exposed to the verification engineers. The abstract method put contains half of the statements of the original program, relieving the user from the burden of writing some auxiliary specifications. The methods getIndexOf and addKey are abstracted, thus loop invariants are not needed for them. It is difficult to discover this fact without non-trivial efforts. Besides, the abstraction indicates that the loop (Fig. 3(a), statements 19–20) only modifies locations $[0 \ldots (values.length - 2)]$ of the values array, whereas the concrete behaviors to modify the other slots can be left out completely in the loop invariants. The abstract program has been proved with KeY using 3879 rules and 4 auxiliary specifications, while the original program requires 14684 rules and 14 auxiliary specifications.

## 3  Techniques

We now explain the principle techniques of our approach. We describe: program translation, program abstraction, validity check of counterexamples and refinement of abstract programs, and runtime exception handling. See Fig. 1.

We focus on analyzing object-oriented programs, and currently support a basic subset of Java that does not include floating point numbers, concurrency, and user-defined exceptions. We support a class hierarchy definition without interfaces and abstract classes. A detailed program syntax can be found in a previous work (cf. [23, Sect. 2]). We currently support a basic subset of JML that does not include model fields and exceptional behaviors.

### 3.1  Translation

We explain the translation techniques of bounded program verification (the "Translation" box in Fig. 1). Based on user-provided bounds we translate a Java program and its JML requirement specifications into an SMT formula. Some code transformations are performed on the analyzed program before the translation: Loops are unrolled the number of times defined in bounds; methods are inlined into their call sites; constructors are split into object allocation and initialization; and all variables and fields are renamed such that they are assigned at most once. The preprocessed program (called *bounded program* in the paper) is represented using a *computation graph* [27], a directed acyclic graph that has a single entry node and a single exit node. The nodes of the graph represent the control points in the bounded program, and the edges represent the state transitions. Figure 4 provides a simple example. The computation graph of the program in Fig. 4(a) is shown in Fig. 4(b), where variable names are indexes; the initial index is 0, and the index is incremented every time the variable is updated. Figure 4(c) gives the SMT constraints encoding the control flow. An SMT formula consists of logical conjunction-connected SMT constraints (enclosed in the

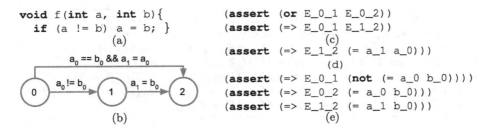

```
void f(int a, int b){
    if (a != b) a = b; }
         (a)
```

(assert (or E_0_1 E_0_2))
(assert (=> E_0_1 E_1_2))
               (c)

(assert (=> E_1_2 (= a_1 a_0)))
               (d)

(assert (=> E_0_1 (not (= a_0 b_0))))
(assert (=> E_0_2 (= a_0 b_0)))
(assert (=> E_1_2 (= a_1 b_0)))
               (e)

**Fig. 4.** Encoding: (a) a sample Java code, (b) computation graph, (c) control constraints, (d) frame conditions, (e) data constraints.

assert command). Basic constraint are combined using the boolean operators and, or, not, and => (implies). We introduce a boolean variable E_i_j to represent an edge from node i to node j; the data constraints in Fig. 4(e) provide the correct semantics for state transitions; the frame condition in Fig. 4(d) explicitly prevents variables to be unspecified. Variables (fields) in JML expressions are replaced by the appropriate variables (fields) in the pre-/post-state of the bounded program. More details can be found in a previous work [23].

## 3.2   Abstraction

When an SMT formula is unsatisfiable, an SMT solver capable of generating proofs is used to find a proof of invalidity, i.e., an *unsat core*. Minimization is performed on the core returned by an SMT solver to ensure the core is locally minimal: removing any single constraint from the core renders it satisfiable. (The algorithm is presented later in this section.) Let the set $C$ denote the inconsistent constraints extracted from the SMT formula; i.e., $C$ encodes the reason that no post-state violates the requirement specification. To discover which statements are responsible for a constraint in the unsat core, we maintain a *constraint map* $M := \{C \mapsto S\}$ to store the connection between the constraints $C$ and statements $S$. When generating data constraints (e.g., Fig. 4(e)), the mapping from a data constraint $c \in C$ to the statement $s \in S$ (where the constraint is generated) is added to the constraint map $M$. Figure 5 presents the rules used for updating the constraint map $M$. Rules $R_1$ and $R_2$ shows data constraints are directly mapped to the simple assignment statements. We translate the assignments e.f = e to two constraints: $e.f' = e$ and $\forall T o, o \neq e \Rightarrow o.f' = o.f$ ($T$ represents the type of e), where only the former is used to update the constraint map $M$ ($R_3$). The translations of the create statement and the array update statement are handled in the same way ($R_4$ and $R_5$ respectively). The rule $R_6$ shows that the constraints translated from branch conditions are mapped to the branch (or loop) statement. The loop condition is negated after the last iteration, the rule $R_7$ maps the negation of the loop condition to the loop statement. The statements mapped by the constraints of $C$ are the relevant statements w.r.t. the property under consideration for user-provided bounds.

| $R_1$: | $\mathcal{T}_d[\![\texttt{T v = e;}, E]\!]$ | $\rightarrow$ | $M' = M \cup \{(E \Rightarrow (v_0 = \mathcal{T}_e[\![e]\!])) \mapsto S\}$ |
|--------|---------------------------------------------|---------------|-----------------------------------------------------------------------------|
| $R_2$: | $\mathcal{T}_d[\![\texttt{v = e;}, E]\!]$ | $\rightarrow$ | $M' = M \cup \{(E \Rightarrow (v' = \mathcal{T}_e[\![e]\!])) \mapsto S\}$ |
| $R_3$: | $\mathcal{T}_d[\![\texttt{e.f = e;}, E]\!]$ | $\rightarrow$ | $M' = M \cup \{(E \Rightarrow (e.f' = \mathcal{T}_e[\![e]\!])) \mapsto S\}$ |
| $R_4$: | $\mathcal{T}_d[\![\texttt{e = new T;}, E]\!]$ | $\rightarrow$ | $M' = M \cup \{(E \Rightarrow (e' = \mathcal{T}_e[\![newT]\!])) \mapsto S\}$ |
| $R_5$: | $\mathcal{T}_d[\![\texttt{e[e] = e;}, E]\!]$ | $\rightarrow$ | $M' = M \cup \{(E \Rightarrow (e[\mathcal{T}_e[\![e]\!]] = \mathcal{T}_e[\![e]\!])) \mapsto S\}$ |
| $R_6$: | $\mathcal{T}_d[\![\texttt{if (e)}, E_t, E_f]\!]$ | $\rightarrow$ | $M' = M \cup \{(E_t \Rightarrow \mathcal{T}_e[\![e]\!]) \mapsto S, (E_f \Rightarrow \mathcal{T}_e[\![!e]\!]) \mapsto S\}$ |
| $R_7$: | $\mathcal{T}_d[\![\texttt{assume (e)}, E]\!]$ | $\rightarrow$ | $M' = M \cup \{(E \Rightarrow \mathcal{T}_e[\![e]\!]) \mapsto S\}$ |

**Fig. 5.** The rules for updating the constraint map $M$ to $M'$. The new variables (or fields) are marked with apostrophes. $\mathcal{T}_d$ and $\mathcal{T}_e$ represent the translation of program statements and expressions respectively. $S$ denotes a program statement and $E$ represents the edge for the statement. $E_t$ and $E_f$ denote the outgoing edges of branch statements.

| $R_1$: | $\mathcal{A}[\![\texttt{T v = e;}]\!]$ | $:=$ | `T v = pure_T();` |
|--------|-----------------------------------------|------|-------------------|
| $R_2$: | $\mathcal{A}[\![\texttt{v = e;}]\!]$ | $:=$ | `v = pure_T();` |
| $R_3$: | $\mathcal{A}[\![\texttt{e.f = e;}]\!]$ | $:=$ | `e.f = pure_T();` |
| $R_4$: | $\mathcal{A}[\![\texttt{if (e)}]\!]$ | $:=$ | `if (pure_T())` |
| $R_5$: | $\mathcal{A}[\![\texttt{while (e)}]\!]$ | $:=$ | `while (pure_T())` |
| $R_6$: | $\mathcal{A}[\![\texttt{return e;}]\!]$ | $:=$ | `return pure_T();` |

**Fig. 6.** The rules for transforming original statements to abstract statements. The transformation is denoted by $\mathcal{A}$. The concrete Java statements on the left are replaced by the abstract statements on the right. The pure_T methods return unspecified values.

These relevant statements are marked as *mustHave* statements and will not be abstracted. The other statements in the bounded program, that are named *mayHave* statements, are not necessary for bounded verification, but may be helpful in the deductive program verification. We generate an abstract program by over-approximating the behaviors of the statements in the original program when their transformed statements are mayHave statements—each transformed statement gets the location of its original statement. Thus all feasible executions of the original program are feasible in the abstract program, but not vice versa. Abstract programs are generated using the abstraction rules in Fig. 6. The original statement (on the left of Fig. 6), from which the mayHave statement has been transformed, is replaced with a statement (on the right of Fig. 6) that calls a JML-annotated *pure_T* method /*@ assignable \strictly_nothing;*/ native /*@ nullable */ T pure_T();. The JML assignable clause ensures that no memory location is changed by the pure method and that distinct unspecified values will be returned by the pure method, T represents an appropriate type required by the original statement, and the pure method returns an unspecified value of T which includes null as well. The Java keyword native is used to avoid implementations of the pure methods.

Using the rules in Fig. 6, the generated abstract programs provide to the verification engineers the information which statements are necessary for the properties under consideration. Thus, writing auxiliary specifications could be easier. However, it may increase the proof complexity compared to the concrete programs. Typically, a deductive program verification system (e.g., KeY [1]) symbolic executes a program and applies various calculus rules to make a proof. During symbolic execution, the symbolic states of the original program are very likely more concrete than those of the abstract program. Therefore, the symbolic execution paths which are invalid for the concrete programs are traversed when proving the abstract programs. Furthermore, symbolic execution of an abstract statement may require more rules than its original statement.

We optimize the abstract program. In order for the abstract programs to have appropriate concrete states, statements that are unnecessary for bounded program verification, yet helpful for the deductive program verification, are marked as mustHave statements. For example, assignment statements where the expression on the right-hand side is an object allocation, constant, etc., and that their defined variables are used in some mustHave statements. When possible, we abstract a set S of mayHave statements into one single statement, thus reducing the number of abstract statements. This is available for any two nodes m and n in the computation graph g (see, e.g., Fig. 4(b)), where m dominates n, n post-dominates m, and *all* the statements in S whose edges are in the paths from m to n are mayHave statements. The new abstract statement calls an impure method /*@ assignable loc;*/ native T impure_T();, where the JML specification denotes the memory locations modified by the statements. We compute the modifiable locations loc as a collection of the fields and variables that are updated in the mayHave statements.

**Minimization of an Unsat Core.** A locally minimal unsat core is useful for computing optimal abstractions. To the best of our knowledge, none of the SMT solvers guarantees its unsat core is locally minimal. We present an algorithm (Algorithm 1) that minimizes an unsat core by exhaustively checking whether a constraint is necessary for the unsatisfiability of the SMT formula. If the formula remains unsatisfiable when deactivating (negating) the constraints of a program statement, the constraints are not needed and their statement is a mayHave statement. The new unsat core returned by the SMT solver is the input for

---

**Algorithm 1.** Minimize an unsat core

---

**Input:** $\mathcal{C}$: unsatisfiable SMT constraints; $S \leftarrow \emptyset$: unnecessary constraints;
$muc \leftarrow \emptyset$: locally minimal unsat core.
**for** $c \in \mathcal{C}$ **do**
    **if** $c \notin S$ **then**
        **if** *(check-sat $((\mathcal{C} - c) \setminus S))$ is UNSAT* **then**
            $muc \leftarrow getUnsatCoreFromSolver()$;
            $S \leftarrow S \cup ((\mathcal{C} - c) \setminus muc)$
**return** $muc$

---

the next check. Otherwise, we reactivate the constraints and check the other constraints till all constraints are flipped.

### 3.3    Validity Check and Refinement

To check the validity of the counterexamples, new bounds are always required since the abstract program fulfills the property w.r.t. the old bounds. For a counterexample ce, the new bound of class C is $CPre_{ce}+max(CPath_1,\ldots,CPath_n)$, where $CPre_{ce}$ is the number of C instances in ce, $CPath_n$ is the number of allocations of C instances on the n-th program path, and function max returns the maximum. To compute new loop unrolls, we transform a while(cond){stmts;} loop into if(cond){stmts; if(!cond) var=var;}, where var is a variable that is modifiable in stmts. This transformation prevents unrolling the loops that are irrelevant for the program correctness.

We compute a new SMT formula that is the conjunction of the translation of the counterexample and the translation of the original program for the new bounds. When the formula is satisfiable, then either the counterexample is valid, or the loop requires further iterations if the loop condition is still true after traversing the last iteration. In the latter case, we double the loop bounds and repeat the validity check. If the formula is unsatisfiable, we find the statements w.r.t. the counterexample using the techniques as shown in Sect. 3.2. In the bounded program, we highlight a mayHave statement as a mustHave statement when the statement is in the newly found statements. Finally, using the technique shown in Sect. 3.2, we generate a new abstract program for deductive verification.

### 3.4    Runtime Exceptions

For each property to be proved, verification systems also prove that no runtime exception is thrown. When more than one functional property has to be verified, the same proof steps for checking runtime exceptions are redone. Our approach separates the verification of functional properties from checking runtime exceptions: usually the statement o = o.f, e.g., is translated into $(o \neq null \Rightarrow o' = o.f) \vee (o = null \Rightarrow exc)$, where $exc$ denotes runtime exception, whereas we translate it into $o \neq null \wedge o' = o.f$. To check that there are no runtime exceptions, we also inject guards into the code, such that if a guard passes an exception is thrown. We treat the possible exception types separately.

Figure 7 presents the code from Fig. 2(a) with one guard injected. We insert a guard (statements 11–13) which sets to true the flag $NASE$ in the class $RTE$ if a NegativeArraySizeException is about to be thrown (statement 14). Thus, when the program in Fig. 7 preserves the value of the exception flag (it is false when calling the method and when returning from it[1]), no NegativeArraySizeException is thrown in the original program, as the guard is checking the statement at line 14. All program parts not relevant to whether

---

[1] The requires clause specifies the method's precondition.

```
        /*@ requires RTE.NASE = false;
         @ ensures RTE.NASE = false;
         @ diverges true;
         @ assignable \everything;*/
        int numberOfPrime(int x, int y) {
            ...// statements 1-9 are omitted to save space.
10      if (size>0){
11          if (y-x < 0) {
12              RTE.NASE = true;
13              return;
            }
14          this.a = new int[y-x];
        }
15      return size;}
```

**Fig. 7.** The example from Fig. 2(a) with an injected guard.

the exception is thrown are abstracted. In our approach, when there is no run-time exception and the functional properties have been fulfilled by the analyzed abstract programs, the original program is also verified.

## 4   Evaluation

The approach that we have presented (i) liberates verification engineers from finding the relevant program slices manually, and (ii) reduces the proof complexity especially for partial properties, for which most of the program slices are irrelevant.

We have implemented the techniques introduced in the paper in a proto-type tool, *AbstractJ*. We use InspectJ [23] as the bounded verification tool and KeY [1] as the deductive verification tool. The KeY system performs *symbolic execution* [21] of sequential Java programs, using various calculus rules. Program verification with KeY is usually done in *auto-active* style: the user interacts with the system only through provided auxiliary specifications, while the proof result is obtained automatically. The number of rule applications is our primary measure of proof complexity. We have used 5 benchmark programs, all taken from the related program verification literature and from the KeY repository. Each program has 2 to 6 partial properties to be verified. We have considered also two other approaches to evaluate the effectiveness of our approach (*abstraction*) in program verification. One approach, *baseline*, proves the original programs using KeY as usual. The other approach, *highlight*, is similar to the *abstraction* approach, but it only highlights the relevant program statements and retains the irrelevant statements rather than abstracting them. We have completed 21 verification tasks using each approach, and in total we have completed 63 (= 21 ∗ 3) verification tasks in our experiments. We have written the auxiliary specifications as compact as possible and measured the auxiliary specifications as the

**Table 1.** Evaluation results

| method | properties | origin stmts | baseline specs | baseline rules | highlight specs | highlight rules | abstraction stmts | abstraction specs | abstraction rules |
|---|---|---|---|---|---|---|---|---|---|
| List. | nullPointer | 27 | 22 | 3578 | 12 | 3046 | 4 | 0 | 196 |
| merge(list) | indexBounds | 43 | 59 | 4641 | 46 | 4434 | 33 | 46 | 3717 |
| | negSize | 31 | 13 | 4316 | 14 | 2723 | 16 | 6 | 1188 |
| | leElems | 22 | 14 | 2962 | 14 | 2962 | 13 | 6 | 1715 |
| | subset | 22 | 82 | 6299 | 56 | 5715 | 15 | 52 | 4404 |
| Map. | nullPointer | 32 | 28 | 4485 | 14 | 3780 | 9 | 0 | 512 |
| put(key,value) | indexBounds | 48 | 61 | 6154 | 54 | 5557 | 48 | 54 | 5488 |
| | negSize | 32 | 17 | 4084 | 12 | 3753 | 16 | 0 | 654 |
| | oldKey | 26 | 30 | 4295 | 30 | 4295 | 11 | 22 | 1725 |
| | sameValues | 26 | 27 | 9823 | 34 | 8494 | 12 | 26 | 4647 |
| | kvMatched | 26 | 50 | 7327 | 50 | 7327 | 26 | 50 | 8814 |
| LRS. | nullPointer | 39 | 11 | 3022 | 8 | 2818 | 13 | 0 | 753 |
| doLRS() | indexBounds | 43 | 44 | 5006 | 14 | 4545 | 30 | 14 | 4502 |
| | foundOrNot | 26 | 32 | 4155 | 14 | 2908 | 17 | 10 | 1255 |
| Set. | nullPointer | 48 | 23 | 10937 | 18 | 10226 | 25 | 6 | 5505 |
| intersect(set) | negSize | 38 | 17 | 14555 | 14 | 9963 | 23 | 10 | 4586 |
| | indexBounds | 58 | 57 | 19715 | 33 | 12287 | 51 | 33 | 6714 |
| | emptySet | 33 | 94 | 64807 | 46 | 13557 | 16 | 38 | 3875 |
| | subset | 33 | 142 | RO | 60 | 136225 | 16 | 52 | 11211 |
| Graph. | sameNodes | 54 | 78 | RO | 60 | 14985 | 13 | 39 | 3923 |
| remove(nodes) | sameEdges | 54 | 119 | RO | 83 | RO | 18 | 67 | 12334 |

number of the operands of JML expressions, JML constructs, and logical connectors, e.g., `loop_invariant`, `assignable`, `forall`, `&&`, etc.[2] We used the SMT solver Z3 [25] to compute the unsat cores. For the experiments described in this paper, we have used the default minimal bounds of InspectJ—at most 3 objects and at most 3 loop iterations. All experiments[3] have been performed on an Intel Core i5-2520M CPU with 2.50 GHz running on a 64-bit Linux.

To evaluate the effect of the *abstraction* approach on reducing the complexity of programs, we have compared the number of Java statements of original and abstract programs. The results are shown in Table 1. The column *method* shows the Java class and its method to be verified; the verified properties are listed in the column *properties*. The *nullPointer*, *indexBounds*, and *negSize* represent the runtime exceptions `NullPointerException`, `ArrayIndexOutOfBoundsException`, and `NegativeArraySizeException`, respectively. The *orgStmts* column displays

---

[2] Different engineers may write different auxiliary specifications for the same programs. We have asked an experienced KeY engineer to prove the original programs and a relatively inexperienced KeY user to prove the abstract programs. They carefully inspected and ensured that the annotations are compact enough w.r.t. the requirement specifications.

[3] The complete experiments can be found at http://asa.iti.kit.edu/458.php.

the number of the original program statements.[4] The column *stmts* shows the number of the program statements that have been generated by the *abstraction* approach. On average, 49.5 % (median 50 %, maximum 85.2 %) of statements in the original programs have been abstracted by the *abstraction* approach. There are 2 properties (indexBounds, and kvMatched for the method put) for which the approach *abstraction* seems has no effect. A careful inspection reveals that one single concrete statement is abstracted. From the results, the abstract programs contain less, yet enough details for partial properties. The more partial the verified property, the fewer details the abstract programs have. Conservatively speaking, even in the case where the abstract programs are identical to the original programs, the *abstraction* approach assists verification engineers at exploring the relevant statements—all program statements that have not been abstracted are necessary for the properties under consideration. The *highlight* approach shows the relevant program statements to verification engineers, while the *abstraction* approach provides additional benefits: (i) automatic generation of auxiliary specifications for the irrelevant program statements, and (ii) *possible* reduction of proof complexity for partial properties. Besides, the *abstraction* can increase users confidence in the correctness of their programs, before starting deductive verification.

For a fair comparison of the amount of manually written auxiliary specifications, the *highlight* approach reused the auxiliary specifications that have been written manually in the *abstraction* approach (shown in the column *specs* of the column *abstraction* in Table 1). The *abstraction* approach generates annotations for the unnecessary program slices, for which the verification engineers need to write annotations using the *highlight* approach. On average, 37.2 % (median 26.7 %) of annotations for the highlighted programs have been automatically generated by the *abstraction* approach.

All properties in Table 1 have been proved using the *abstraction* approach. When using the approaches *highlight* and *baseline*, several properties are unprovable. The column *rules* provides the number of rule applications. Any rule application beyond our threshold of 2000000[5] is denoted by *RO*. For 18 properties that have been proved by all approaches, the *abstraction* approach needed only 50.1 % (median 55.2 %) of the rules required by the *highlight* approach. It is not guaranteed that the *abstraction* approach requires less rule applications than the other two approaches for arbitrary properties. Besides of the reasons talked in Sect. 3.2, KeY creates branches for each abstract statement, to check its pre-/post-conditions.[6] When the rule cost introduced by the abstract statements is lower than the cost of symbolic execution of the irrelevant original statements, only then the *abstraction* approach requires fewer rules than other approaches, by assuming they use same auxiliary specifications. In other words, the more

---

[4] The injected guard statements are treated as original statements when handling runtime exceptions.

[5] The time cost and memory consumption grow exponentially w.r.t. the rule applications. It required ∼30 min and more than 4 GB memory for 2000000 rules.

[6] The trivial pre-/post-conditions of each abstract statement requires ∼20–100 rules.

partial the verified property, the less proof complexity of the abstract programs. The property *kvMatched* is an example for less partial property.

Although we used small bounds for InspectJ in the experiments, there are no refinement cases in Table 1. On the other hand, when the refinement of an abstract program is needed, the abstract program will contain much less details, thus it is easy to find the relevant program statements. The verification engineers are free to provide even higher bounds for InspectJ. Given the same input formula, Z3 may find an unsat core that is different from the core found by other SMT solvers. AbstractJ may generate different abstract programs using other SMT solver, but the abstract programs will still contain less details than the concrete programs if the analyzed property is partial enough.

## 5  Related Work

Several methods have been proposed to split the program under analysis with respect to particular concerns. Traditional program slicing techniques (e.g., static/dynamic slicing) generate a group of accessible statement (a slice) w.r.t. variables of interest at particular locations. Due to the complexity of the specification expressions and various data structures in the analyzed programs, it is very difficult to find specification-sensitive slices correctly.

Conditioned slicing techniques [4,8,9,12,13,17] have been widely applied to simplify programs with respect to the specifications. Comuzzi et al. [12] introduced predicates as a slicing criterion; the slice contains the statements affecting the predicates. That idea has been extended by introducing preconditions [9], symbolic execution [4], and program verification [13] into conditioned slicing techniques. Typically, conditioned slicing produces a group of all accessible statements w.r.t. the specification by symbolic execution with the inputs generated by a solver. The pre/postcondition (generally formulas of first-order logic) are expressed in terms of the (input) variables at program locations of interest. However, intensive human interaction is required to guide the symbolic execution by choosing a suitable criterion. GamaSlicer [13] verifies the program w.r.t. specifications before generating semantic-based slices. Nevertheless, it may not terminate with a conclusive result, since it targets an undecidable logic. Our approach ensures that the soundness of the proof depends only on the deductive verification.

The following three approaches tried to improve the verification process using bounded analysis. Bormer et al. [6] claim that verifying programs using the bounded model checker LLBMC [24] facilitates proving with VCC [11]. Annotations written in VCCs specification language are translated into assertions that can be checked by LLBMC. El Ghazi et al. [16] try to verify Alloy problems using deductive verification, after the Alloy analyzer [20]—based on bounded analysis, fails in finding a counterexample in bounds dictated by the machine. Kroening et al. [15] combine $k$-induction and inductive invariant method to facilitate program verification using significantly weaker annotations. These approaches do not aim to reduce the overhead of writing specifications. However, the $k$-induction frequently allows using weaker loop invariants than are required by the inductive invariant approach. Our approach can reduce the burden of specifications not only for loops.

Using unsat core is not new in bounded program verification. The authors of [26] used the unsat core to refine the method summaries in program verification. In [14], a code coverage metric is constructed by the program statements that are mapped from the unsat core.

Counterexample-guided abstraction refinement (CEGAR) has been widely used in program verification. To the best of our knowledge, the abstractions have been constructed mostly at the predicate level [2,5,7,10,18] and rarely at the function level [26]. Our approach constructs the abstractions at the levels including the ones mentioned above and statement level.

## 6  Conclusion and Future Work

We presented a novel method to compute specification-sensitive abstractions for program verification. The abstractions are constructed with the help of bounded program verification. The counterexample-guided refinement framework has been used to refine the abstractions. We exploited the characteristics of the unsat core to discover irrelevant statements. The novelty of our approach is to abstract the program statements that are irrelevant for the properties of interest, to help verification engineers to write auxiliary specifications. We described how to: encode programs, map program statements to constraints, generate abstractions based on abstraction rules, and refine the abstractions with new bounds computation. We evaluated our experiments on 5 programs that were already used in related papers and in the KeY repository. Initial results show that our approach generates suitable abstract programs for verification, and all abstract programs have been proved for all 21 properties, while the original programs have been proved for 18 properties. Our tool took off 50 % of the user's workload in writing auxiliary specifications. Only about half of the proof rules used to prove the original program are needed for proving the abstract program.

We plan to apply our approach to larger programs, and investigate incorporating loop invariant generators, e.g., Invgen [19], to improve the automation of the approach.

**Acknowledgement.** This work has been partially supported by GIF (grant No. 1131-9.6/2011) and by DFG under project "DeduSec" within SPP 1496 "RS3" and by BMBF under project FIfAKS within the Software Campus program.

## References

1. Ahrendt, W., et al.: The KeY platform for verification and analysis of Java programs. In: Giannakopoulou, D., Kroening, D. (eds.) VSTTE 2014. LNCS, vol. 8471, pp. 55–71. Springer, Heidelberg (2014). doi:10.1007/978-3-319-12154-3_4
2. Ball, T., Cook, B., Levin, V., Rajamani, S.K.: SLAM and static driver verifier: technology transfer of formal methods inside Microsoft. In: Boiten, E.A., Derrick, J., Smith, G. (eds.) IFM 2004. LNCS, vol. 2999, pp. 1–20. Springer, Heidelberg (2004). doi:10.1007/978-3-540-24756-2_1

3. Barrett, C., Fontaine, P., Tinelli, C.: The SMT-LIB standard: Version 2.5. Technical report, The University of Iowa (2015)
4. Barros, J.B., da Cruz, C.D., Henriques, P.R., Pinto, J.S.: Assertion-based slicing and slice graphs. Formal Asp. Comput. **24**(2), 217–248 (2012)
5. Beyer, D., Henzinger, T.A., Jhala, R., Majumdar, R.: The software model checker Blast. STTT **9**(5–6), 505–525 (2007)
6. Bormer, T.: Advancing Deductive Program-Level Verification for Real-World Application: Lessons Learned from an Industrial Case Study. Ph.D. thesis, KIT (2014)
7. Chaki, S., Clarke, E.M., Groce, A., Jha, S., Veith, H.: Modular verification of software components in C. IEEE Trans. Softw. Eng. **30**(6), 388–402 (2004)
8. Chebaro, O., Kosmatov, N., Giorgetti, A., Julliand, J.: Program slicing enhances a verification technique combining static and dynamic analysis. In: SAC, pp. 1284–1291. ACM (2012)
9. Chung, I.S.: Program slicing based on specification. In: SAC, pp. 605–609. ACM (2001)
10. Clarke, E., Kroening, D., Sharygina, N., Yorav, K.: SATABS: SAT-based predicate abstraction for ANSI-C. In: Halbwachs, N., Zuck, L.D. (eds.) TACAS 2005. LNCS, vol. 3440, pp. 570–574. Springer, Heidelberg (2005). doi:10.1007/978-3-540-31980-1_40
11. Cohen, E., Dahlweid, M., Hillebrand, M., Leinenbach, D., Moskal, M., Santen, T., Schulte, W., Tobies, S.: VCC: a practical system for verifying concurrent C. In: Berghofer, S., Nipkow, T., Urban, C., Wenzel, M. (eds.) TPHOLs 2009. LNCS, vol. 5674, pp. 23–42. Springer, Heidelberg (2009). doi:10.1007/978-3-642-03359-9_2
12. Comuzzi, J.J., Hart, J.M.: Program slicing using weakest preconditions. In: Gaudel, M.-C., Woodcock, J. (eds.) FME 1996. LNCS, vol. 1051, pp. 557–575. Springer, Heidelberg (1996). doi:10.1007/3-540-60973-3_107
13. da Cruz, D.C., Henriques, P.R., Pinto, J.S.: GamaSlicer: an online laboratory for program verification and analysis. In: LDTA. ACM (2010)
14. Dennis, G.D.: A Relational Framework for Bounded Program Verification. Ph.D. thesis, MIT (2009)
15. Donaldson, A.F., Haller, L., Kroening, D., Rümmer, P.: Software verification using $k$-induction. In: Yahav, E. (ed.) SAS 2011. LNCS, vol. 6887, pp. 351–368. Springer, Heidelberg (2011). doi:10.1007/978-3-642-23702-7_26
16. Ghazi, A.A., Ulbrich, M., Gladisch, C., Tyszberowicz, S., Taghdiri, M.: JKelloy: a proof assistant for relational specifications of Java programs. In: Badger, J.M., Rozier, K.Y. (eds.) NFM 2014. LNCS, vol. 8430, pp. 173–187. Springer, Heidelberg (2014). doi:10.1007/978-3-319-06200-6_13
17. Fox, C., Danicic, S., Harman, M., Hierons, R.M.: ConSIT: a fully automated conditioned program slicer. Softw. Pract. Experience **34**(1), 15–46 (2004)
18. Gupta, A., Popeea, C., Rybalchenko, A.: Predicate abstraction and refinement for verifying multi-threaded programs. ACM SIGPLAN Not. **46**(1), 331–344 (2011)
19. Gupta, A., Rybalchenko, A.: InvGen: an efficient invariant generator. In: Bouajjani, A., Maler, O. (eds.) CAV 2009. LNCS, vol. 5643, pp. 634–640. Springer, Heidelberg (2009). doi:10.1007/978-3-642-02658-4_48
20. Jackson, D.: Software Abstractions: Logic, Language, and Analysis. The MIT Press, Cambridge (2012)
21. King, J.C.: Symbolic execution and program testing. CACM **19**(7), 385–394 (1976)
22. Leavens, G.T., Baker, A.L., Ruby, C.: Preliminary design of JML: a behavioral interface specification language for Java. ACM SIGSOFT SEN **31**(3), 1–38 (2006)

23. Liu, T., Nagel, M., Taghdiri, M.: Bounded program verification using an SMT solver: a case study. In: ICST, pp. 101–110. IEEE (2012)
24. Merz, F., Falke, S., Sinz, C.: LLBMC: bounded model checking of C and C++ programs using a compiler IR. In: Joshi, R., Müller, P., Podelski, A. (eds.) VSTTE 2012. LNCS, vol. 7152, pp. 146–161. Springer, Heidelberg (2012). doi:10.1007/978-3-642-27705-4_12
25. Moura, L., Bjørner, N.: Z3: an efficient SMT solver. In: Ramakrishnan, C.R., Rehof, J. (eds.) TACAS 2008. LNCS, vol. 4963, pp. 337–340. Springer, Heidelberg (2008). doi:10.1007/978-3-540-78800-3_24
26. Taghdiri, M., Jackson, D.: Inferring specifications to detect errors in code. Autom. Softw. Eng. **14**(1), 87–121 (2007)
27. Vaziri, M.: Finding Bugs in Software with a Constraint Solver. Ph.D. thesis, MIT (2004)

# Reducing State Explosion for Software Model Checking with Relaxed Memory Consistency Models

Tatsuya Abe[1]([✉]), Tomoharu Ugawa[2], Toshiyuki Maeda[1],
and Kousuke Matsumoto[2]

[1] STAIR Lab, Chiba Institute of Technology, Narashino, Japan
{abet,tosh}@stair.center
[2] Kochi University of Technology, Kami, Japan
{ugawa,matsumoto}@plas.info.kochi-tech.ac.jp

**Abstract.** Software model checking suffers from the so-called state explosion problem, and relaxed memory consistency models even worsen this situation. What is worse, parameterizing model checking by memory consistency models, that is, to make the model checker as flexible as we can supply definitions of memory consistency models as an input, intensifies state explosion. This paper explores specific reasons for state explosion in model checking with multiple memory consistency models, provides some optimizations intended to mitigate the problem, and applies them to McSPIN, a model checker for memory consistency models that we are developing. The effects of the optimizations and the usefulness of McSPIN are demonstrated experimentally by verifying copying protocols of concurrent copying garbage collection algorithms. To the best of our knowledge, this is the first model checking of the concurrent copying protocols under relaxed memory consistency models.

**Keywords:** Software model checking · Relaxed memory consistency models · State explosion · Reordering of instructions · Integration of states · Concurrent copying garbage collection

## 1 Introduction

Modern computing systems are based on concurrent/parallel processing designs for their performance advantages, and programs therefore must also be written to exploit these designs. However, writing such programs is quite difficult and error-prone, because humans cannot exhaustively consider the behaviors of computers very well. One approach to this problem is to use software model checking, in which all possible states that can be reached during a program's execution are explored. Many such model checkers have been developed (e.g., [7,8,13,19,26,27]).

However, most existing model checkers adopt *strict consistency* as a Memory Consistency Model (MCM) on shared memories, which only allows interleaving of instruction execution, and ignore more *relaxed* MCMs than strict consistency,

© Springer International Publishing AG 2016
M. Fränzle et al. (Eds.): SETTA 2016, LNCS 9984, pp. 118–135, 2016.
DOI: 10.1007/978-3-319-47677-3_8

which allow reorderings of instructions. This is not realistic because many modern computer architectures such as IA64, SPARC, and POWER [21,23,38] have adopted relaxed MCMs. Relaxed MCMs facilitate the performance of parallel-processing implementations because instructions may be reordered and multiple threads may observe distinct views on shared memory while strict consistency, which requires synchronization at each memory operation, is prohibitively expensive to be implemented on computer architectures.

As interest in MCMs has grown, some model checkers have introduced support for them [25,28–30]. However, these have been specific to certain MCMs, such as Total Store Ordering (TSO) and Partial Store Ordering (PSO) [11]. We are in the process of developing a model checker, McSPIN [9], that can handle multiple MCMs [1–3,5]. McSPIN can take an MCM as an input with a program to be verified. It has a specification language that covers various MCMs including TSO, PSO, Relaxed Memory Ordering (RMO), acquire and release consistency [24], Itanium MCM [22], and UPC MCM [40]. By using McSPIN, we can easily model check a *fixed* program under *various* MCMs.

However, software model checking suffers from the *state explosion problem*, and relaxed MCMs even worsen this, because the reordering of instructions allowed under relaxed MCMs enormously increases the number of reachable states. What is worse, parameterizing model checking by MCMs, that is, to make the model checker as flexible as we can supply definitions of MCMs as an input intensifies the state explosion.

This paper explains how model checking with multiple MCMs increases the number of reachable states, and clarifies the reasons for state explosion specific to model checking with multiple MCMs. In addition, some optimizations are provided that reduce state explosion, and their effects are demonstrated through experiments. The ideas behind the optimizations are simple: Pruning traces, partial order reduction, and predicate abstraction are well known to reduce state explosion in conventional model checking [18]. In our former paper [3], we arranged pruning traces and partial order reduction for model checking with relaxed MCMs. In this paper, we arrange predicate abstraction, and propose *stages*, which are integrations of states under relaxed MCMs.

Although the optimization in our earlier work have enabled verification of non-toy programs such as Dekker's mutual exclusion algorithm [3], it was difficult to apply McSPIN to larger problems such as verifications of copying protocols of Concurrent Copying Garbage Collection algorithms (CCGCs), due to the state explosion. In this paper, we demonstrated the optimizations above enables McSPIN to verify larger programs; we checked if a desirable property of CCGCs, "in a single thread program, what the program reads is what it has most recently written", are held or not for several CCGCs on multiple MCMs. Though we used verifications of GCs as examples in this paper, safety of GC is an important issue in the field (e.g., [15,17]), and this achievement is a positive development. To the best of our knowledge, this is the first model checking of copying protocols of CCGCs with relaxed MCMs.

The rest of this paper is organized as follows: Sect. 2 describes McSPIN with exploring the reasons for state explosion specific to model checking with MCMs, and Sect. 3 describes the relevant optimizations we have applied in McSPIN. Section 4 presents experimental results using McSPIN on different CCGCs and shows the effectiveness of the optimizations. Section 5 discusses related work, and the conclusions and directions for future work are presented in Sect. 6.

## 2    McSPIN

We first briefly review our earlier work [1–3] on constructing a general model checking framework with relaxed MCMs and developing and implementation. In the following, we do not distinguish the framework from its implementation and refer to both as McSPIN. In McSPIN, threads on computers with shared memory are uniformly regarded as processes that have their own memories. Therefore, we formally call threads (in the usual sense) *processes* (in McSPIN), while we refer to them as "threads" when informally explaining behavior on shared-memory systems.

### 2.1    Syntax

A program is an $N$-tuple of sequences of instructions defined as follows:

| | |
|---|---|
| (Instruction) | $i ::= \langle L, A, \iota \rangle$, |
| (Raw Instruction) | $\iota ::= \mathsf{Move}\, r\, t \mid \mathsf{Load}\, r\, x \mid \mathsf{Store}\, x\, t \mid \mathsf{Jump}\, L\, \mathsf{if}\, t \mid \mathsf{Nop}$, |
| (Term) | $t ::= v \mid r \mid t + t \mid t - t \mid \cdots$, |
| (Attributes) | $A ::= \{a, \ldots, a\}$, |

where $N$ is the number of processes. An instruction $i$ is a triple of a label, attributes, and a raw instruction. A label $L$ designates an instruction in a program. An attribute $a \in A$ denotes an additional label for a raw instruction, has no effect itself, and are used to describe constraints specified by an MCM.

Here $r$ is variable local to a process and $x, y, \ldots$, are shared variables. The raw instruction $\mathsf{Move}\, r\, t$ denotes the assignment of an evaluated value of a term $t$ to a process-local variable $r$, which does not affect other processes. The term $v$ denotes an immediate value. The terms $t_0 + t_1, t_0 - t_1, \ldots$, denote standard arithmetic expressions. $\mathsf{Load}\, r\, x$ represents loading $x$ from its own memory and assigning its value to $r$. $\mathsf{Store}\, x\, t$ denotes storing an evaluated value of $t$ to $x$ on its own memory. $\mathsf{Jump}\, L\, \mathsf{if}\, t$ denotes a conditional jump to $L$ depending on the evaluated value of $t$. Note that $t$ contains no shared values; to jump to $L$ depending on $x$, it is necessary to perform $\mathsf{Load}\, r\, x$ in advance. $\mathsf{Nop}$ denotes the usual no-operation.

Careful readers may wonder why no synchronization instructions such as *memory fence* and *compare-and-swap* instructions appear. In McSPIN, a memory fence is represented as a $\mathsf{Nop}$ with attribute **fence**, and its effect is defined at each input MCM, that is, multiple types of fences can be defined. This flexibility enables verification of a *fixed* program with *different* MCMs as explained in more detail in Sect. 2.3. Compare-and-swap (usually an instruction on a computer architecture) is also represented by compound statements, which can be seen in [10].

Programs (inputs to McSPIN) have to be written in the assembly-like modeling language. Such low-level languages are suitable for handling MCMs that require one to carefully take into account effects on specific computer architectures. However, these languages may not be practical for writing programs. McSPIN has a C-like modeling language to facilitate programming, but this is beyond the scope of the present paper.

## 2.2 Semantics

McSPIN adopts trace semantics with states. Execution traces are sequences of *operations*, defined as follows:

$$\text{(Operation)} \qquad o ::= \text{Fe}_q^j \, p \, i \mid \text{Is}_q^j \, p \, i \mid \text{Ex}_q^j \, p \, i \, \ell \, v \mid \text{Re}_q^j \, [p \Rightarrow p] \, i \, \ell \, v.$$

One key point in handling different MCMs is to consider at most four kinds of operations for an instruction. For any instruction, its fetch and issue operations are considered. Load and store instructions have execution operations. Store instructions have reflect operations. An effect of each operation is formally defined in our former paper [5]. In this paper, we roughly explain why such operations are introduced.

Under very relaxed MCMs such as C++ [24] and UPC MCM [40], distinct threads can exhibit different program behaviors; that is, each thread has its own execution trace. To represent these in one trace, we add a process identifier $q$, denoting an observer process as a subscript of an operation. In addition, McSPIN can handle programs with loops. To distinguish multiple operations corresponding to an instruction, an operation has a branch counter $j$ that designates the $j$th iteration within a loop.

We explain the four kinds of operations by example. $\text{Fe}_q^j \, p \, i$ denotes fetching an instruction $i$ from a process $p$, which enables the issuance of $i$. By default, this also increments the program counter of $p$ if $i$'s raw instruction is not Jump. If so, the program counter is not changed and will be changed when the Jump is issued. McSPIN is equipped with a *branch prediction* mode that can be switched on or off. In branch prediction mode, the program counter is non-deterministicly incremented or set to $l$, when Jump $l$ if $t$ is fetched. Thus, in order to handle branch prediction, fetch has to be distinguished from issue.

Although branch prediction is often ignored in specifications of MCMs, note that no branch prediction implicitly prohibits some kinds of reorderings across conditionals. For example, no branch prediction on the process-model that McSPIN adopts cannot perform the so-called *out-of-thin-air* read [31] in the program in Table 17.6 of Java language specification [33], although *legal* executions under Java MCM are specified by not using a *total* order of operations on such process-model but consistency between *partial* orders of operations on threads.

Operation $\mathsf{Is}_q^j\, p\, i$ denotes the issuing of an instruction $i$ to a process $p$. Effects that complete inside the register on $p$ (not $p$'s own memory) are performed. For example, while issuing $\mathsf{Move}\, r\, t$ indicates assignment of an evaluated value of $t$ to $r$, $\mathsf{Store}\, x\, t$ implies evaluation of $t$ only. In branch prediction mode, a predicted execution trace in fetching $\mathsf{Jump}$ is checked.

Operation $\mathsf{Ex}_q^j\, p\, i\, \ell\, v$ denotes execution of an instruction $i$ on a process $p$. Effects that complete inside $p$ are performed. For example, while the execution of $\mathsf{Load}\, r\, x$ means that $v$ is loaded from $x$ (at location $\ell$) and assigned to $r$, $\mathsf{Store}\, x\, t$ represents storing an evaluated value $v$ of $t$ to $x$ (at location $\ell$) in $p$'s own memory. While an instruction is issued, its (intra-process) effect may not have occur yet. Itanium MCM allows such situation, by distinguishing issues from executions of instructions.

Operation $\mathsf{Re}_q^j\, [p_0 \Rightarrow p_1]\, i\, \ell\, v$ denotes reflects of an instruction $i$ from process $p_0$ to $p_1$. The reflect of $\mathsf{Store}\, x\, t$ means storing an evaluated value $v$ of $t$ to $x$ at $\ell$ in $p_1$'s own memory. While a store instruction is executed, that is, its effect is reflected to its store buffer, its (inter-process) effect may not be reflected to other processes yet. One reflect may be immediately passed, and another reflect may be delayed. Moreover, processes can observe distinct views a.k.a. *the IRIW test* [12]. Our definition covers such situations.

While the distinction enables delicate handling effects of instructions, it intensifies state explosion since the number of interleavings of operations increases.

To handle more relaxed MCMs, it is also necessary to distinguish multiple operations that are generated from an instruction in a loop statement, whereas this is unnecessary when queues can be used to handle specific MCMs such as TSO and PSO. For example, in a code ($\mathsf{Store}\, x\, r_0$; $\mathsf{Move}\, r_0\, r_0 + 1$; $\mathsf{Jump}\, 0\, \mathtt{if}\, 1$) $\|$ $\mathsf{Load}\, r_0\, x$, the second fetch of the $\mathsf{Store}$ on the former process may follow the fetch of the $\mathsf{Load}$ on the latter process, while the first fetch of the $\mathsf{Store}$ on the former process may precede it. To the best of our knowledge, no existing method can handle such low-level jumps (across which instructions may be reordered) in a detailed fashion, which is necessary for verification of CCGCs.

### 2.3    Formalized Memory Consistency Models

MCMs are sets of constraints that control program behaviors on the very relaxed semantics that McSPIN adopts and are formally defined as a first-order formula as follows:

$$\varphi ::= \mathrm{x}_\tau = \mathrm{x}'_\tau \mid \mathrm{x}_\tau < \mathrm{x}'_\tau \mid \neg \varphi \mid \varphi \supset \varphi' \mid \forall \mathrm{x}_\tau.\varphi(\mathrm{x}_\tau),$$

where $\tau$ denotes one of Variable, Location, Label, Value, Instruction, Raw Instruction, Attribute, Branch Counter, and Operation. Here $\mathrm{x}_\tau$ represents metavariables in the syntax of McSPIN. For example, $\mathrm{x}_{\text{Location}} < \mathrm{x}'_{\text{Location}}$ can be read as $\ell < \ell'$. In addition, $<$ with respect to Operation identifies the order of execution between operations. We use standard notation such as $\wedge$, $\vee$, and $\exists$ and assign higher precedence to $\neg$, $\wedge$, $\vee$, and $\supset$.

Example constraints can be seen in [1–3], and Itanium and UPC MCMs are fully formalized in their journal version [5]. Here we focus on only two. In Sect. 2.1, we stated that the effect of a memory fence can be flexibly defined by an input MCM. A memory fence forces evaluation of all the reflects of store instructions that are fetched before the memory fence. This is represented as follows:

$$\mathsf{Fe}_q^{j_0} \; p \; i_0 < \mathsf{Fe}_q^{j_1} \; p \; (L_1, A_1, \mathsf{Nop}) \supset \mathsf{Re}_q^{j_0} \; [p {\Rightarrow} p_0] \; i_0 \; \ell_0 \; v_0 < \mathsf{Is}_q^{j_1} \; p \; (L_1, A_1, \mathsf{Nop}) \,,$$

where $\mathsf{fence} \in A_1$, $i_0$'s raw instruction is $\mathsf{Store}$, and all free variables are universally quantified. Meanwhile, we can consider another operation that forces $\mathsf{Load}$ only:

$$\mathsf{Fe}_q^{j_0} \; p \; i_0 < \mathsf{Fe}_q^{j_1} \; p \; (L_1, A_1, \mathsf{Nop}) \supset \mathsf{Ex}_q^{j_0} \; p \; i_0 \; \ell_0 \; v_0 < \mathsf{Is}_q^{j_1} \; p \; (L_1, A_1, \mathsf{Nop}) \,,$$

where $\mathsf{fence} \in A_1$ and $i_0$'s raw instruction is $\mathsf{Load}$.

One constraint that differentiates TSO from PSO with *multiple-copy-atomicity* [37], which prohibits two threads from observing different behaviors of write operations that the two threads do not perform, is whether reflects of store instructions are *atomically* performed *in program order*. This can be represented as follows:

$$\mathsf{Fe}_q^{j_0} \; p \; i_0 < \mathsf{Fe}_q^{j_1} \; p \; i_1 \supset \mathsf{Re}_q^{j_0} \; [p {\Rightarrow} p_0] \; i_0 \; \ell_0 \; v_0 < \mathsf{Re}_q^{j_1} \; [p {\Rightarrow} p_1] \; i_1 \; \ell_1 \; v_1 \,,$$

where $i_0$'s and $i_1$'s raw instructions are $\mathsf{Store}$ instructions. This constraint causes *every* reflect of $i_1$ to await completion of *all* reflects of $i_0$. Full constraints of TSO, PSO, and other relaxed MCMs are formalized in McSPIN's public repository [9].

## 2.4  Translation into PROMELA

McSPIN uses the model checker SPIN as an engine and translates programs written in our modeling language into PROMELA, the modeling language of SPIN. The underlying idea is quite simple. McSPIN translates sequential compositions of statements $i_0; i_1; \ldots$ written in our modeling language into PROMELA loop statements as follows:

```
do
:: (guard0,0) -> (operation of Fe of i0); (epilogue0,0);
:: (guard0,1) -> (operation of Is of i0); (epilogue0,1);
:: (guard0,2) -> (operation of Ex of i0); (epilogue0,2);
:: (guard0,3) -> (operation of Re of i0 to p0); (epilogue0,3);
:: ...
:: (guard0,(N-1)+3) -> (operation of Re of i0 to pN-1); (epilogue0,(N-1)+3);
:: (guard1,0) -> (operation of Fe of i1); (epilogue1,0);
:: ...
:: else -> break;
od;
```

A PROMELA loop statement has multiple clauses with guards. One of those clauses whose guards are satisfied is non-deterministicly chosen and processed. Let $\mathsf{clock}$ be a time counter. Each clause corresponds to performing an operation as follows:

```
end_o==0 -> o; end_o=clock; clock++;
```

where the positiveness of **end_**$o$ denotes that $o$ has already performed.

Although such a PROMELA code may admit very relaxed behavior that does not satisfy an input MCM, McSPIN appropriately removes such execution traces. Assertions can be written not only at the end of a program, but also at any place within. This is important for CCGC verification, because we would like to confirm data consistency at a certain place and moment. McSPIN modifies assertion statements to follow the input MCM. Let $\varphi$ be an assertion that we wish to verify. McSPIN adds (formalized) constraints that an input MCM obligates to $\varphi$ as a conjunct. For example, the constraint that differentiates TSO and PSO, as explained in Sect. 2.3, is translated into

$$!(\text{end\_}\{Fe_q^{j0}\ p\ i_0\}<\text{end\_}\{Fe_q^{j1}\ p\ i_1\})||\text{end\_}\{Re_q^{j0}\ [p\Rightarrow p_0]\ i_0\ \ell_0\ v_0\}<\text{end\_}\{Re_q^{j1}\ [p\Rightarrow p_1]\ i_1\ \ell_1\ v_1\}$$

and added to the assertion $\varphi$ as a conjunct, where ! and || represent negation and disjunction in PROMELA, respectively. Thus, execution traces that violate the MCM are removed when assertions are checked.

# 3    Optimizations

Here we provide MCM-sensitive optimization techniques to reduce the problem specific to model checking with multiple MCMs. The optimizations described in Sects. 3.1 and 3.2 were introduced in [3]; we briefly review them here in order to make it easy to understand an optimization introduced in Sect. 3.3.

## 3.1    Enhanced Guards: Pruning Inadmissible Execution Traces

As explained in Sect. 2.4, McSPIN explores all execution traces and removes traces that are inadmissible under an input MCM in checking assertions. This is obviously redundant. A straightforward method to prune inadmissible execution traces is to enhance guards for clauses corresponding to operations. A guard that is uniformly generated as **end_**$o$==0 from an operation $o$ in Sect. 2.4 is enhanced by an input MCM (details are provided in [3]). We explain this using the constraint that differentiates TSO and PSO, as set out in Sect. 2.3. The constraint claims that all reflects of $i_1$ must wait for all reflects of $i_0$, where $i_0$ precedes $i_1$ in program order. McSPIN adds a condition

$$!(\text{end\_}\{Fe_q^{j0}\ p\ i_0\}<\text{end\_}\{Fe_q^{j1}\ p\ i_1\})||\text{end\_}\{Re_q^{j0}\ [p\Rightarrow p_0]\ i_0\ \ell_0\ v_0\}>0$$

corresponding to this claim to the guard of the reflect of $i_1$.

## 3.2    Defining Predicates: Promoting Partial Order Reduction

As explained in Sect. 2.4, it is necessary to judge whether an execution trace is admissible to a given MCM. This means that it is also necessary to remember orders between operations in the execution trace. The most straightforward method is to use a time counter; that is, to substitute a variable $\mathsf{end\_}o$ (defined at each operation) with the time at which operation $o$ was performed. However, time counters are too concrete to reduce state explosion. For example, consider four operations $o_0$, $o_1$, $o_2$, $o_3$ under the constraint $o_0 < o_1 \supset o_2 < o_3$. If times are substituted for the variables $\mathsf{end\_}o_k$ $(0 \leq k < 4$, then the number of combinations $\langle \mathsf{end\_}o_0, \mathsf{end\_}o_1, \mathsf{end\_}o_2, \mathsf{end\_}o_3 \rangle$ is 24 (=4!), which distinguishes states more concretely than the constraint requires.

When considering the constraint rule, it suffices to remember the order of $o_0$ and $o_1$ and of $o_2$ and $o_3$, because nothing else is used to define the constraint. We introduce new variables $\mathsf{ord\_}o_0\_o_1$ and $\mathsf{ord\_}o_2\_o_3$, and call them *defining predicates* of the constraint or, formally, atomic formulas consisting of the predicate symbol $<$ (or $\leq$) between operations that occur in the constraint. Because the defining predicates preserve the order of times at which the operations are performed, we change $\mathsf{end\_}o_k$ to boolean variables that denote whether the operation has been performed. After all the operations have been performed (that is, $\mathsf{end\_}o_k = 1$ $(0 \leq k < 4)$), the possible states are $\langle \mathsf{ord\_}o_0\_o_1, \mathsf{ord\_}o_2\_o_3 \rangle = \{\langle 0,0 \rangle, \langle 0,1 \rangle, \langle 1,1 \rangle\}$, of cardinality 3.

## 3.3    Stage: Abstracting Programs by MCM-Deriving Predicates

Predicate abstraction [18] is one promising method to reduce state explosion in model checking. In this subsection, we show that predicates exist that are determined by an input MCM. Such predicates integrate states that do not have to be separated with respect to an input MCM. Therefore, the predicate abstractions have no omission of checking.

To handle the effects of instructions more delicately, McSPIN has at most four kinds of operations, $\mathsf{Fe}$, $\mathsf{Is}$, $\mathsf{Ex}$, and $\mathsf{Re}$ for one instruction. However, some MCMs do not require complete distinction. Assume that an input MCM has the constraint $\mathsf{Is}_q^j\,p\,i < o \supset \mathsf{Ex}_q^j\,p\,i\,\ell_1\,v_1 < o$ as called *integration* in [2,3], which indicates that no operation can interleave two operations $\mathsf{Is}_q^j\,p\,i$ and $\mathsf{Ex}_q^j\,p\,i\,\ell_1\,v_1$. In an earlier version, McSPIN generated clauses that had guards waiting for $\mathsf{Ex}_q^j\,p\,i\,\ell_1\,v_1$ when $\mathsf{Is}_q^j\,p\,i$ was performed. Such guards control program behaviors in accordance with an input MCM.

In this paper, we promote integration to state level rather than execution-trace level. In earlier versions, McSPIN generated one clause at each operation; that is, at most $3+N$, the cardinality of $\{\mathsf{Fe}, \mathsf{Is}, \mathsf{Ex}\} \cup \{\mathsf{Re}\,k \mid 0 \leq k < N\}$, clauses at each instruction, where $N$ is again the number of processes, and $\mathsf{Re}\,k$ denotes a reflect to $k$. In the current version, McSPIN can accept additional input *stages* $S = \{s_0, s_1, \ldots, s_{M-1}\}$ for an input MCM. Formally, stages are partitions of $\{\mathsf{Fe}, \mathsf{Is}, \mathsf{Ex}\} \cup \{\mathsf{Re}\,k \mid 0 \leq k < N\}$. We write $f_S$ for the induced mapping from the stages. McSPIN generates $M$ clauses at each instruction $i$, where $M$ is the number of stages of $i$ as follows:

```
do
:: (guards₀,fₛ(s₀)) -> (operation of fₛ(s₀) of i₀); (epilogue₀,fₛ(s₀));
:: (guards₀,fₛ(s₁)) -> (operation of fₛ(s₁) of i₀); (epilogue₀,fₛ(s₁));
:: ...
:: (guards₀,fₛ(sₘ₋₁)) -> (operation of fₛ(sₘ₋₁) of i₀ to pₙ₋₁); (epilogue₀,fₛ(sₘ₋₁));
:: ...
:: else -> break;
od;
```

This optimized translation reduces checking space and time. By loading such a PROMELA code, SPIN remembers not unintegrated states themselves but stages. This implies that state-vector on SPIN is kept small. Memory is not, therefore, consumed so much. This optimization also saves time to check whether clauses are executable since the number of clauses is smaller.

Let us see example stages for TSO and PSO with neither *branch prediction* nor *multiple-copy-atomicity* [37], which prohibits two threads from observing different behaviors of write operations that the two threads do not perform. Since these MCMs allow Loads to overtake (inter-process) effects of Stores, each member of $\{\mathrm{Re}\,k \mid 0 \leq k < N\}$ has to be separated from Ex. However, Fe, Is, and Ex do not have to be separated. Also, Re $k$ does not have to be distinguished from Re $k'$ ($k' \neq k$) by multiple-copy-atomicity. We can therefore introduce the following stages:

$$S = \{s_0, s_1\} \qquad f_S(s) = \begin{cases} \{\mathrm{Fe, Is, Ex}\} & \text{if } s = s_0 \\ \{\mathrm{Re}\,k \mid 0 \leq k < N\} & \text{if } s = s_1. \end{cases}$$

Given a stage $S$ (and its mapping $f_S$), McSPIN automatically returns PROMELA code in which clauses are integrated; in particular, guards and epilogues are appropriately generated from an input MCM.

## 4  Experiments

In this section, we demonstrate the effects of the optimizations introduced in Sect. 3. The figure to the right shows our experimental environment, with ample memory.

| CPU: | Intel Xeon E5-2670 2.6GHz |
|---|---|
| Memory: | DDR3-1066 1.5TB |
| SPIN: | 6.4.5 |
| GCC: | 5.3.0 |

The optimizations described in Sects. 3.1 and 3.2 have enabled verification of relatively large programs such as Dekker's algorithm [3]. Here we demonstrate that the optimization described in Sect. 3.3 enables verification of genuinely large programs.

### 4.1  Experimental Setting

We chose CCGCs as examples of large programs. In this subsection, we briefly explain the CCGCs we used.

Garbage collection (GC) is a basic service of modern programming languages. Its role is to find garbage, that is, data objects that are no longer in use by the application, and to reclaim the memory that those objects occupy. Copying GC accomplishes this by copying live objects, i.e., those that may be used in the future, to a separate space and then releasing the old space that contains the copied objects and garbage. Concurrent GC, as the name suggests, runs concurrently with the application. What is difficult in designing CCGC algorithms is that the garbage collector thread and an application thread may race; the application thread may change the contents of an object that is being copied by the garbage collector. This may be the case even with an single thread application. Because an application thread changes, or mutates, the object, we call it a *mutator*. If a mutator writes to the object that is being copied, the collector may copy a stale value, which means that the latest value gets lost. Various copying protocols have been proposed to provide application programmers with reasonable MCMs, all of which require the mutators to do some work on every read (*read barrier*) or write (*write barrier*) operation or both, in which the mutator synchronizes with the collector.

Because such barriers incur overhead for every read or write operation, one goal of CCGC algorithms is to design barriers that are as lightweight as possible. Thus, synchronizations such as compare-and-swap should be minimized. With relaxed MCMs, memory barriers should also be minimized. Unfortunately, the synchronizations required for safety depend on the given MCM; it is often the case that those synchronizations that are redundant for one MCM are mandatory for another.

**Model.** We experimentally checked the safety of concurrent copying protocols, *in a single thread program, what the mutator reads is what it has most recently written*. This property is expected to be held in any reasonable MCMs such as the happens-before consistency of Java [33]. The complete McSPIN models for checking this property can be found in [10] or the McSPIN public repository [9]. Here, we briefly explain the model.

In our model checking, we made some assumptions. We assume that there is a single mutator thread, i.e., the application is a single thread program. Remark that even if there is a single mutator thread, there is another thread, the collector thread, and they may race. We also assume that there is only a single object with a single integer slot in the heap.

The mutator has a pointer to the object and repeatedly reads from and write to the object through the pointer. On write operations, it remembers the value it wrote.On read operations, it checks if the read value is equal to the value it lastly wrote. Meanwhile, the collector copies the object following to the copying protocol of each algorithm. Once it successfully copied, the collector rewrites the mutator's pointer to the object so that the pointer points at the copy.

To cooperate with the collector, the mutator uses the read and write barriers required by the copying protocol on its read and write operations. For some algorithms, the mutator also performs so called the checkpoint operation between object accessing operations, where the mutator polls and answers collector's requests. Some collectors request the mutator to answer the handshake by setting a per-mutator handshake request flag. The checkpoint operation clears the flag to let the collector know the mutator has observed the flag set. In TSO, if a mutator observes the flag is set, all stores preceding the store setting the flag are guaranteed to be visible to the mutator.

We created McSPIN models for each CCGC algorithms we describe below. In the models, the mutator has an infinite loop, where it reads or writes once per an iteration. It also performs a checkpoint operation before and after each read or write. Thus, the supremum of loop iterations on the mutator limits the number of mutator's memory accesses.

**GC Algorithm.** In this paper, we checked three GC algorithms: Chicken [35], Staccato [32] and Stopless [34]. The details of these algorithms can be found in their papers. Here, we briefly explain their features.

Chicken and Staccato were basically the same algorithm though they are developed independently. The only difference is their target MCMs; Chicken is designed for the MCMs of Intel CPU such as IA64 [23], while Staccato's main target seems to be POWER MCM [21]. These algorithms use compare-and-swap operations to resolve races between the collector and a mutator. In the IA64 MCM, the compare-and-swap is usually realized by the instruction sequence `lock cmpxchg`. This sequence implies memory fences. As for POWER MCM, the manual [21] shows a sample implementation of the compare-and-swap operation that does not imply memory fences.

Stopless is a different algorithm from those two. It uses compare-and-swap operations that implies memory fence excessively, hence chances of reordering are fewer.

### 4.2   Effect of Optimization

In this subsection, we reveal the effectness of the *stage* optimization described in Sect. 3.3. We verified the models created in Sect. 4.1 by using McSPIN with and without the optimization. We fixed the supremum of iteration on the collector to 1 and varied that on the mutator from 1 to 2.

Table 1 shows the results of the verification. Note that any PROMELA code produced by McSPIN consumed around 170 MB of memory as constant overhead. As Table 1 shows, the amounts of memory consumed and elapsed times are greatly reduced in all algorithms compared with those without the optimization. In particular, when the supremum of loop iterations on the mutator was set to 2, McSPIN without the optimization often required around 1 TB of memory, which is far from reasonable. However, a single iteration could not detect any error even for the algorithms that actually work incorrectly with PSO, i.e., Chicken and Stopless.

Table 1 also suggests that the more instructions the model had, the more effective the stage optimization was. For example, in Chicken and Staccato with a single iteration, the net memory consumption was reduced to 3.9–8.8 %, while, in Stopless, it was reduced to 3.3 and 4.4 %. This is because the stage optimization reduced the number of units that are subject to reordering, or clauses of the do-loop in the PROMELA code.

**Table 1.** Effects of optimization: In TSO, a compare-and-swap instruction implies memory fences. In PSO, it does not. Columns labeled with "col." and "mut." list the num ber of instructions of the collector and the mutator, respectively. Column labeled with "loop" lists the supremum of loop iterations on the mutator. For verification either with or without stages, the first column shows the results; ✓ means no error was found and × means a violation was found. The following columns are the number of state transitions, the amount of memory consumed, and elapsed times for verification, respectively. Columns labeled with "mem. ratio" and "time ratio" list the ratios of memory and time consumption for verification with/without stages. The column labeled with "net mem. ratio" lists those that do not count constant overhead.

| MCM | Algorithm | col. | mut. | loop | Without stages | | | With stages | | | mem. ratio (%) | net mem. ratio (%) | time ratio (%) |
|---|---|---|---|---|---|---|---|---|---|---|---|---|---|
| | | | | | State (K) | Memory (MB) | Time (sec.) | State (K) | Memory (MB) | Time (sec.) | | | |
| TSO | chicken | 24 | 42 | 1 | ✓ 108 | 8,595 | 132 | ✓ 23 | 908 | 8 | 10.6 | 8.8 | 6.2 |
| | | | | 2 | ✓ 2,506 | 546,038 | 8,637 | ✓ 534 | 24,960 | 433 | 4.6 | 4.5 | 5.0 |
| | staccato | 32 | 46 | 1 | ✓ 141 | 14,918 | 236 | ✓ 26 | 1,032 | 11 | 6.9 | 5.8 | 4.7 |
| | | | | 2 | ✓ 3,888 | 1,097,491 | 16,022 | ✓ 733 | 43,432 | 735 | 4.0 | 3.9 | 4.6 |
| | stopless | 33 | 87 | 1 | ✓ 90 | 28,183 | 378 | ✓ 14 | 1,404 | 19 | 5.0 | 4.4 | 5.0 |
| | | | | 2 | ✓ 564 | 564,635 | 7,705 | ✓ 87 | 18,885 | 585 | 3.3 | 3.3 | 7.6 |
| PSO | chicken | 24 | 42 | 1 | ✓ 308 | 25,208 | 430 | ✓ 65 | 1,652 | 28 | 6.6 | 5.9 | 6.6 |
| | | | | 2 | × 1,136 | 264,857 | 4,248 | × 237 | 12,190 | 227 | 4.6 | 4.5 | 5.3 |
| | staccato | 32 | 46 | 1 | ✓ 143 | 15,166 | 243 | ✓ 26 | 1,032 | 11 | 6.8 | 5.7 | 4.7 |
| | | | | 2 | ✓ 4,020 | 1,144,602 | 16,975 | ✓ 751 | 44,920 | 768 | 3.9 | 3.9 | 4.5 |
| | stopless | 33 | 87 | 1 | ✓ 177 | 55,210 | 833 | ✓ 30 | 2,520 | 46 | 4.6 | 4.3 | 5.5 |
| | | | | 2 | × 45 | 45,416 | 630 | × 8 | 2,148 | 41 | 4.7 | 4.4 | 6.6 |

To the best of our knowledge, this is the first model checking of these algorithms with PSO, due to the optimizations given in this paper.

## 4.3  Reducing Memory Fences of Staccato

Because Staccato is designed for the relaxed MCM of POWER, some memory fences are redundant on a stricter MCM. Thus, we designed and verified a variant of Staccato for a PSO MCM with a compare-and-swap that does not imply memory fences. In addition, we created an incorrect variant that lacks mandatory fences for the PSO MCM. These variants are labeled staccato_pso and staccato_bug.

The result of verification is shown in Table 2. The verification is conducted with the stage optimization. The result of staccato_bug shows that McSPIN detected an error if we reduced fences too much.

The variants of Staccato demonstrate the usefulness of McSPIN. When we modify a GC algorithm for a machine with some MCM that is different from the one that the GC is originally designed for, we add or remove some synchronizations. However, the modified model often lacks synchronizations. McSPIN can detect such errors in the variant with a reasonable memory consumption. This enables us to check the GC when we are performing modifications.

**Table 2.** Variants of Staccato

| Algorithm | col. | mut. | loop | TSO | | | | | PSO | | | |
|-----------|------|------|------|-----|---|---------------|--------------|---------------|---|---------------|--------------|---------------|
| | | | | | State (K) | Memory (MB) | Time (sec.) | | | State (K) | Memory (MB) | Time (sec.) |
| staccato | 32 | 46 | 1 | ✓ | 26 | 1,032 | 11 | | ✓ | 26 | 1,032 | 11 |
| | | | 2 | ✓ | 733 | 43,432 | 735 | | ✓ | 751 | 44,920 | 768 |
| staccato_pso | 31 | 44 | 1 | ✓ | 25 | 908 | 10 | | ✓ | 25 | 1,032 | 10 |
| | | | 2 | ✓ | 719 | 38,721 | 637 | | ✓ | 755 | 40,953 | 703 |
| staccato_bug | 30 | 44 | 1 | ✓ | 28 | 1,032 | 11 | | ✓ | 35 | 1,156 | 14 |
| | | | 2 | ✓ | 819 | 43,184 | 726 | | × | 235 | 12,810 | 217 |

## 4.4  McSPIN vs. Hand-Coding

In this subsection, we compare PROMELA codes generated by McSPIN with codes written by hand and confirm how close McSPIN is to an ideal implementation.

Whereas McSPIN generates uniform PROMELA codes that contain variables to remember orders between operations, etc., to support different MCMs, some variables are essentially unnecessary for verifications specific to TSO and PSO. Because TSO never reorders store instructions, queues (for all shared variables) at each thread to buffer effects of write instructions suffice for verifications under TSO as shown in Fig. 1. The two WRITEs put $\langle x0, 0 \rangle$ and $\langle x1, 1 \rangle$ into the queue in order. Reflects from the queue to shared memory are performed by COMMIT_WRITEs on a process mem. We omit the implementation details. For PSO, one queue at *each* shared variable is enough to reorder the effects of write instructions to distinct shared variables.

Table 3 compares PROMELA codes generated by McSPIN with those written by hand where the constant overhead is removed. The programs are simple, consisting of multiple store instructions (without loops). Verified properties are fixed to be true. Each column is similar to Tab. 1. The digits in the names of the codes denote the number of store instructions at each thread, respectively. The number of states almost coincides. Slight differences appear to derive from

```
active proctype main() {
 run mem();
 run proc0();
 ...
}
proctype proc0() {
 WRITE(x0,0);
 WRITE(x1,1);
}
...
```

```
proctype mem() {
endmem:
 do
 ::atomic{COMMIT_WRITE(queue_proc0);}
 ::atomic{COMMIT_WRITE(queue_proc1);}
 ...
 od;
}
inline WRITE (var,val) {...}
inline COMMIT_WRITE (queue) {...}
```

**Fig. 1.** Hand-written code

**Table 3.** Comparison between McSPIN and hand-coding

| | TSO | | | | | | PSO | | | | | |
|---|---|---|---|---|---|---|---|---|---|---|---|---|
| | McSPIN | | | Hand-written | | | McSPIN | | | Hand-written | | |
| | State | Memory (MB) | Time (sec.) | State | Memory (MB) | Time (sec.) | State | Memory (MB) | Time (sec.) | State | Memory (MB) | Time (sec.) |
| 1 | 25 | 0.006 | 0.01 | 19 | 0.002 | 0.01 | 25 | 0.006 | 0.02 | 19 | 0.003 | 0.01 |
| 2 | 52 | 0.017 | 0.02 | 60 | 0.009 | 0.00 | 65 | 0.021 | 0.02 | 79 | 0.017 | 0.01 |
| 3 | 116 | 0.053 | 0.02 | 149 | 0.026 | 0.00 | 241 | 0.110 | 0.05 | 337 | 0.095 | 0.01 |
| 4 | 241 | 0.153 | 0.04 | 313 | 0.064 | 0.02 | 977 | 0.619 | 0.20 | 1,405 | 0.504 | 0.02 |
| 5 | 457 | 0.391 | 0.08 | 585 | 0.143 | 0.01 | 3,985 | 3.405 | 0.98 | 5,749 | 2.500 | 0.07 |
| 6 | 800 | 0.897 | 0.18 | 1,004 | 0.276 | 0.01 | 16,145 | 18.107 | 6.13 | 23,269 | 11.894 | 0.31 |
| 7 | 1,312 | 1.882 | 0.35 | 1,615 | 0.493 | 0.01 | 65,041 | 93.290 | 34.06 | 93,637 | 54.294 | 1.66 |
| 8 | 2,041 | 3.659 | 0.61 | 2,469 | 0.829 | 0.02 | 261,137 | 468.195 | 171.10 | 375,685 | 246.497 | 8.28 |

the current implementation of SPIN, because we observe that SPIN returns fewer states for a PROMELA code with a loop statement and control variables (such as code generated by McSPIN) than another PROMELA code with a sequential composition of statements (like hand-written code). However, we have not investigated this in detail.

McSPIN consumes more memory and time. This is a result of the sizes of the state vectors and is inevitable, because McSPIN defines more variables to determine program structures than hand-written codes, as explained in the beginning of this subsection.

## 5 Related Work

There exists no work, which is directly compared with our work, of model checking to take multiple MCMs in a uniform way. Therefore, we can find no work for its optimization has been studied.

Jonsson's seminal work discovered the potential of SPIN for program translation toward model checking with relaxed MCMs [25]. However, he could not conduct a large number of experiments, because his program translation was not completely automatic and optimized. This paper has addressed the problems that he left open. McSPIN supports various MCMs and takes an MCM as an input, and its program translation is automatic. McSPIN is greatly optimized and enables verification of larger concurrent algorithms such as copying protocols of CCGCs.

Linden et al. [28–30] tackled the state explosion problem by representing store buffers as automata. However, they handled relatively strict relaxed MCMs such as TSO and PSO, unlike McSPIN. It is an open issue to extend their representation so as to handle more relaxed MCMs and apply it to McSPIN.

Modex [20], a model extractor of SPIN that is guided by a user-defined test harness, translates C codes into PROMELA codes. However, Modex ignores relaxed MCMs. Although revising Modex so as to handle relaxed MCMs is surely one approach, we have developed McSPIN in order to show the potential of program translation toward model checking with relaxed MCMs with no restriction derived from the existing tool.

Travkin et al. [41] developed a similar tool that translates programs into PROMELA codes and uses SPIN as the engine for model checking, demonstrated verifications of linearizability of concurrent algorithms under TSO, and planned to tackle PSO. However, their translator, which generates codes that are similar to hand-written PROMELA code as introduced in Sect. 4.4, cannot be immediately applied to relaxed MCMs beyond PSO. Unlike their approach, ours supports relaxed MCMs by virtue of constructing a base that allows such relaxed behaviors and then defines MCMs as constraints on the base. Although an issue of our approach is addressing the state explosion problem, this paper has presented optimizations for the problem.

Dan et al. [14] reported high utility of predicate abstractions in model checking with relaxed MCMs by verifying some programs with predicate abstractions under TSO and PSO. They proposed the notion of predicate *extrapolation* to abstract a boolean program for an input program. Although the stages introduced in this paper can be regarded as predicate abstractions, there is a difference in usage: McSPIN considers at most four kinds of multiple states at one instruction to support various MCMs beyond TSO/PSO. Although it is necessary to handle the worst case under the most relaxed MCM, this is not always the case. Stages are states that are integrated by predicates that are uniformly generated by an input MCM. Therefore, abstractions by the predicates never leak out of checking. Dan et al.'s technique of extrapolating predicates seems to be compatible with stages, and its combination with stages is an open issue.

Theorem proving in program logic is also one promising approach to program verification with relaxed MCMs [4,6,16,36,42]. Formal verifications of GC algorithms with relaxed MCMs using theorem provers have recently appeared [17]. However, fully automated verification by model checking is usually preferable to manual (or semi-automatic) construction of proofs in theorem proving.

# 6    Conclusion and Future Work

We have explained the reasons for the state explosion problem specific to model checking with multiple MCMs, presented optimizations modified from pruning execution traces, partial order reduction, and predicate abstraction, and applied them to McSPIN, our model checker with MCMs. We have also shown the effectiveness of the optimizations through experiments of verifications of copying protocols of CCGCs.

There are four future directions for this work. Although we verified copying protocols of CCGCs as examples of large programs, a verification of GC algorithm is itself subject of our interest. The first is verifications of wide range of GC algorithms and other properties such as wait-freedom for Chicken, which the authors designed as a wait-free CCGC [35]. These verifications may require more complicated settings including pointers and/or multiple mutators, which need still larger models. The second is to show a verification of concurrent copying protocols with MCMs that are more relaxed than PSO. An advantage of McSPIN is its ability to support various MCMs. The third is to show more realistic benchmark programs, e.g., SV-COMP benchmarks [39]. The fourth is further optimization of McSPIN to verify even larger programs.

**Acknowledgments.** The authors thank the anonymous reviewers for several comments to improve the final version of the paper. This research partly used computational resources under Collaborative Research Program for Young Scientists provided by Academic Center for Computing and Media Studies, Kyoto University. This work was supported by JSPS KAKENHI Grant Numbers 25871113, 25330080, and 16K21335.

# References

1. Abe, T., Maeda, T.: Model checking with user-definable memory consistency models. In: Proceedings of PGAS, short paper, pp. 225–230 (2013)
2. Abe, T., Maeda, T.: A general model checking framework for various memory consistency models. In: Proceedings of HIPS, pp. 332–341 (2014)
3. Abe, T., Maeda, T.: Optimization of a general model checking framework for various memory consistency models. In: Proceedings of PGAS (2014)
4. Abe, T., Maeda, T.: Concurrent program logic for relaxed memory consistency models with dependencies across loop iterations. J. Inform. Process. (2016, to appear)
5. Abe, T., Maeda, T.: A general model checking framework for various memory consistency models. Int. J. Softw. Tools Technol. Transf. (2016, to appear)
6. Abe, T., Maeda, T.: Observation-based concurrent program logic for relaxed memory consistency models. In: Proceedings of APLAS (2016, to appear)
7. Abe, T., Maeda, T., Sato, M.: Model checking with user-definable abstraction for partitioned global address space languages. In: Proceedings of PGAS (2012)
8. Abe, T., Maeda, T., Sato, M.: Model checking stencil computations written in a partitioned global address space language. In: Proceedings of HIPS, pp. 365–374 (2013)
9. Abe, T., Maeda, T., Ugawa, T.: McSPIN. https://bitbucket.org/abet/mcspin/
10. Abe, T., Ugawa, T., Maeda, T., Matsumoto, K.: Reducing state explosion for software model checking with relaxed memory consistency models (2016). arXiv:1608.05893
11. Adve, S., Gharachorloo, K.: Shared memory consistency models: a tutorial. Computer **29**(12), 66–76 (1996)
12. Boehm, H.J., Adve, S.V.: Foundations of the C++ concurrency memory model. In: Proceedings of PLDI, pp. 68–78 (2008)
13. Cavada, R., Cimatti, A., Jochim, C.A., Olivetti, G.K.E., Pistore, M., Roveri, M., Tchaltsev, A.: NuSMV User Manual. 2.5 edn. (2002)
14. Dan, A.M., Meshman, Y., Vechev, M., Yahav, E.: Predicate abstraction for relaxed memory models. In: Logozzo, F., Fähndrich, M. (eds.) SAS 2013. LNCS, vol. 7935, pp. 84–104. Springer, Heidelberg (2013). doi:10.1007/978-3-642-38856-9_7
15. Doligez, D., Gonthier, G.: Portable, unobtrusive garbage collection for multiprocessor systems. In: Proceedings of POPL, pp. 70–83 (1994)
16. Ferreira, R., Feng, X., Shao, Z.: Parameterized memory models and concurrent separation logic. In: Gordon, A.D. (ed.) ESOP 2010. LNCS, vol. 6012, pp. 267–286. Springer, Heidelberg (2010). doi:10.1007/978-3-642-11957-6_15

17. Gammie, P., Hosking, T., Engelhardt, K.: Relaxing safely: verified on-the-fly garbage collection for x86-TSO. In: Proceedings of PLDI, pp. 99–109 (2015)
18. Graf, S., Saidi, H.: Construction of abstract state graphs with PVS. In: Grumberg, O. (ed.) CAV 1997. LNCS, vol. 1254, pp. 72–83. Springer, Heidelberg (1997). doi:10. 1007/3-540-63166-6_10
19. Holzmann, G.J.: The SPIN Model Checker. Addison-Wesley, Reading (2003)
20. Holzmann, G.J., Smith, M.H.: An automated verification method for distributed systems software based on model extraction. IEEE Trans. Softw. Eng. **28**(4), 364–377 (2002)
21. IBM: PowerPC Architechture Book, Version 2.02 (2005)
22. Intel: A Formal Specification of Intel Itanium Processor Family Memory Ordering (2002)
23. Intel: Intel 64 and IA-32 Architectures Software Developer's Manual (2016)
24. ISO/IEC 14882: 2011: Programming Language C++ (2011)
25. Jonsson, B.: State-space exploration for concurrent algorithms under weak memory orderings: (preliminary version). SIGARCH Comput. Archit. News **36**(5), 65–71 (2008)
26. Kroening, D., Tautschnig, M.: CBMC – C bounded model checker (Competition Contribution). In: Ábrahám, E., Havelund, K. (eds.) TACAS 2014. LNCS, vol. 8413, pp. 389–391. Springer, Heidelberg (2014). doi:10.1007/978-3-642-54862-8_26
27. Kwiatkowska, M., Norman, G., Parker, D.: PRISM 4.0: verification of probabilistic real-time systems. In: Gopalakrishnan, G., Qadeer, S. (eds.) CAV 2011. LNCS, vol. 6806, pp. 585–591. Springer, Heidelberg (2011). doi:10.1007/978-3-642-22110-1_47
28. Linden, A., Wolper, P.: An automata-based symbolic approach for verifying programs on relaxed memory models. In: Pol, J., Weber, M. (eds.) SPIN 2010. LNCS, vol. 6349, pp. 212–226. Springer, Heidelberg (2010). doi:10.1007/978-3-642-16164-3_16
29. Linden, A., Wolper, P.: A verification-based approach to memory fence insertion in relaxed memory systems. In: Groce, A., Musuvathi, M. (eds.) SPIN 2011. LNCS, vol. 6823, pp. 144–160. Springer, Heidelberg (2011). doi:10.1007/978-3-642-22306-8_10
30. Linden, A., Wolper, P.: A verification-based approach to memory fence insertion in PSO memory systems. In: Piterman, N., Smolka, S.A. (eds.) TACAS 2013. LNCS, vol. 7795, pp. 339–353. Springer, Heidelberg (2013). doi:10.1007/978-3-642-36742-7_24
31. Manson, J., Pugh, W., Adve, S.V.: The Java memory model. In: Proceedings of POPL, pp. 378–391 (2005)
32. McCloskey, B., Bacon, D.F., Cheng, P., Grove, D.: Staccato: a parallel and concurrent real-time compacting garbage collector for multiprocessors. Report RC24504, IBM (2008)
33. Oracle: The Java Language Specification. Java SE 8 edn. (2015)
34. Pizlo, F., Frampton, D., Petrank, E., Steensgaard, B.: Stopless: a real-time garbage collector for multiprocessors. In: Proceedings of ISMM, pp. 159–172 (2007)
35. Pizlo, F., Petrank, E., Steensgaard, B.: A study of concurrent real-time garbage collectors. In: Proceedings of PLDI, pp. 33–44 (2008)
36. Ridge, T.: A rely-guarantee proof system for x86-TSO. In: Leavens, G.T., O'Hearn, P., Rajamani, S.K. (eds.) VSTTE 2010. LNCS, vol. 6217, pp. 55–70. Springer, Heidelberg (2010). doi:10.1007/978-3-642-15057-9_4
37. Sarkar, S., Sewell, P., Alglave, J., Maranget, L., Williams, D.: Understanding POWER multiprocessors. In: Proceedings of PLDI, pp. 175–186 (2011)

38. SPARC International Inc: The SPARC Architecture Manual, Version 9 (1994)
39. SV-COMP: Competition on Software Verification. https://sv-comp.sosy-lab.org/
40. The UPC Consortium: UPC Language Specifications Version 1.3 (2013)
41. Travkin, O., Mütze, A., Wehrheim, H.: SPIN as a linearizability checker under weak memory models. In: Bertacco, V., Legay, A. (eds.) HVC 2013. LNCS, vol. 8244, pp. 311–326. Springer, Heidelberg (2013). doi:10.1007/978-3-319-03077-7_21
42. Vafeiadis, V., Narayan, C.: Relaxed separation logic: a program logic for C11 concurrency. In: Proceedings of OOPSLA, pp. 867–884 (2013)

# Identifying XML Schema Constraints Using Temporal Logic

Ruifang Zhao[1]([⊠]), Ke Liu[1], Hongli Yang[1], and Zongyan Qiu[2]

[1] Beijing University of Technology, Beijing 10000, China
zrf@emails.bjut.edu.cn, liuke1985@163.com, yhl@bjut.edu.cn
[2] Peking University, Beijing 10000, China

**Abstract.** Twig pattern minimization is an important aspect of XML query optimization. During the minimizing process, it usually needs to take advantage of the constraints of XML Schema. The traditional methods for identifying constraints is to develop corresponding algorithms based on the type of constraints. It is inflexible because the constraints may be changed as new Twig pattern optimizing rules are found. Since the constraints of XML Schema mainly depict the sequence relationship of nodes, it is natural to be described by temporal logic. Based on the recognition, this paper proposes a method of identifying XML Schema constraints using temporal logic. Concretely, an XML Schema is modeled as a graph. In order to easily represent constraints related to parent and ancestor nodes, we made some modifications to Computational Tree Logic(CTL) with backward temporal operators, and developed model checking algorithms for automatically identifying XML Schema constraints. Compared with traditional methods, our method is more flexibility.

**Keywords:** XML Schema constraint · Model checking · Temporal logic

## 1 Introduction

XML has been widely used in the Internet environment. With the emergence of XML, XPath and XQuery as its query languages, have received a lot of attention. The core query pattern in these standard XML query languages is a tree-like structure, which is often referred to as a Twig pattern [1]. In particular, an XPath query is normally modeled as a Twig pattern query. As an example, let us consider the following XPath query:

$$\text{doc(``bib.xml'')}//\text{book[publisher]/name}$$

It returns all *name* nodes that are children nodes of those *book* nodes with a *publisher* child in the *bib* document. Figure 1 shows its corresponding Twig pattern. Note that single lines represent the Parent-Child relationship between nodes, double lines represent the Ancestor-Descendant relationship.

© Springer International Publishing AG 2016
M. Fränzle et al. (Eds.): SETTA 2016, LNCS 9984, pp. 136–146, 2016.
DOI: 10.1007/978-3-319-47677-3_9

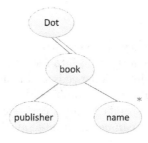

**Fig. 1.** A twig pattern example

The returned node is marked with symbol $*$. The node labeled as *Dot* is the starting node of a Twig pattern.

In order to improve the efficiency of query, there are lots of work on the Twig pattern minimization [1–4]. Those work are broadly divided into constraint-based approaches [1, 2, 4] and unconstrained ones [2, 3]. Generally, the constraints include Required Parent-Child (RPC), Required Ancestor-Descendant (RAD), etc. These constraints are usually defined in corresponding XML Schema and DTD.

One of the key problems for constraint-based Twig pattern minimization is to identify whether those constraints hold in corresponding XML Schema. The traditional methods [6] for identifying constraints is to develop corresponding algorithms based on the type of constraints. Since the constraints may changed as new Twig pattern optimizing rules are found, the search algorithms need to be changed frequently in order to identify new constraints. Moreover, it is difficult to guarantee the correctness of such modified algorithms. Thus, a formal and precise method for XML Schema constraints identification would be very useful.

Considering the constraints of XML Schema mainly depict the sequence relationship of nodes, it is nature to describe the constraints by temporal logic. So we propose a method of identifying XML Schema constraints using temporal logic [7] and model checking. In our approach, XML Schema is modeled as state transition graphs. CTL [7], a well-understood and widely used temporal logic, is used to express XML Schema constraints. In order to easily represent parent and ancestor nodes constraints such as: whether a node $X$ has a required parent $Y$ or not, we made some modifications to CTL with backward temporal operators, and developed model checking algorithm for automatically identifying XML Schema constraints. Compared with traditional approach, it avoids frequent modification of the search algorithm, and only requires representing the new constraint as temporal logic formula for automatic checking. Thus, our approach has more flexibility.

In the rest of the paper, Sect. 2 surveys some related work. Section 3 introduces XML Schema constraints and models. Section 4 presents CTL representation of constraints and develops identifying algorithms. Section 5 concludes.

## 2    Related Work

In order to improve the efficiency of Twig pattern matching in XML documents, researchers have put a lot of efforts [3–5] into reducing the size of Twig patterns. Typically, Twig pattern minimization needs to identify constraints from XML Schema, DTD etc. The limitation of traditional approaches is that a specific algorithm only works for one specific Schema constraint. Those methods are not only time consuming, but also can not guarantee all the cases are covered by the algorithms.

As XML documents are naturally represented as labeled transition systems, connections between XML querying and temporal logics were discovered, thus opening a possibility of using efficient model-checking algorithms in XML querying. Libkin and other coauthors have done much work [11–13] in this area. [11] studies static analysis of XML specifications and transformations. The authors observe that many properties of interest in the XML context are related to navigation, and can be formulated in temporal logics for trees, which can be translated into unranked tree automata and are convenient for reasoning about unary node-selecting queries. [13] presents a technique for combining temporal logics to capture unary XML queries expressible in two yardstick languages: FO and MSO. Moreover, [9] uses temporal logic formulas to express XPath expressions, and transforms XML documents into the input model of NuSMV for evaluating XPath expressions. [10] describes how to interpret various integrity constraints as sequenced constraints defined in XML Schema. There are few works on using temporal logic in the process of Twig pattern minimization.

There are a lot of works on CTL and its extensions. Standard temporal logics only refer to the future of the current time such as LTL,CTL and $CTL^*$. In order to make specifications easier to write and more natural, standard temporal logics are extended to logics combining past [21] and future modalities. The most simple linear-past extension is $PCTL^*$ [16], obtained by adding the past counterparts of standard linear-time modalities 'next' and 'until'. The usage of past-time modalities is very limited. A more interesting and meaningful linear past extension of CTL$^*$ is the logic $CTL_{lp}^*$[17]. $CTL_{lp}^*$ is as expressive as CTL$^*$, but the translation of $CTL_{lp}^*$ into CTL$^*$ is of non-elementary complexity [18]. Kupferman and Vardi [20] introduce a memoryful variant of CTL$^*$ called mCTL$^*$, which unifies CTL$^*$ and the Pistore-Vardi logic [19]. The unique difference is the adding of a special proposition 'present' which is needed to emulate the ability of CTL$^*$ to talk about the 'present'.

## 3    XML Schema Constraints and Models

This section will firstly introduce XML Schema and its constraints. Then, the model MCSG is defined.

## 3.1   XML Schema

XML Schema defines a kind of valid format of XML document. It defines various constraints such as what elements are (and are not) allowed and the times of an element occurrence. Particularly, it limits the relations between element nodes.

```
⟨?xml version="1.0" encoding="utf-8"⟩
⟨xs:schema id="G" xmlns:xs="http://www.w3.org/2001/XML Schema"⟩
 ⟨xs: element name="D1"⟩
  ⟨xs:complexType⟩
   ⟨xs:choice⟩
    ⟨xs:element minOccurs="0" name="E1" type="xs:string"⟩
    ⟨xs:element name="F1"⟩
     ⟨xs:complexType⟩
      ⟨xs:sequence⟩
       ⟨xs:element name="G1" type="xs:string"⟩
        ⟨xs:element minOccurs="0" maxOccurs="unbounded" ref="D1"⟩
      ⟨/xs:sequence⟩
     ⟨/xs:complexType⟩
    ⟨/xs:element⟩
   ⟨/xs:choice⟩
  ⟨/xs:complexType⟩
 ⟨/xs:element⟩
 ⟨xs:element name="A1"⟩
  ⟨xs:complexType⟩
   ⟨xs:sequence⟩
    ⟨xs:element name="B1" type="xs:string"/⟩
    ⟨xs:element name="C1" type="xs:string"/⟩
    ⟨xs:element ref="D1"/⟩
   ⟨/xs:sequence⟩
  ⟨/xs:complexType⟩
 ⟨/xs:element⟩
⟨/xs:schema⟩
```

**Fig. 2.** An XML Schema example

Figure 2 is an example of XML Schema. It defines two element nodes $D1$ and $A1$. The element $A1$ contains three child elements $B1$, $C1$ and $D1$. Furthermore, elements *sequence* (indicated as *SEQ*), *all* (indicated as *ALL*) and *choice* (indicated as *CHOICE*) represent order indicators. Node *SEQ* represents that all of its child nodes must occur in order. *ALL* defines its child nodes occur in any order. *CHOICE* means that only one child may occur. *minOccurs* and *maxOccurs* represent minimum and maximum occurrences of a node.

## 3.2   XML Schema Constraints

XML Schema constraints refer to the required relationships among nodes in an XML Schema. Currently, we only care about the constraints that are related to structural relationships and the number of times nodes may occur. These constraints are related to the order and occurrence indicators in XML Schema, and can be used to minimize Twig patterns. In the following, we will introduce some main constraints of XML Schema.

- Required Parent-Child ($RPC$): For a given XML Schema, if node $A$ has a required child node $B$ (short as RPC(A, B)), $A$ and $B$ must satisfy constraints: (1) node $A$ is defined as a complex type and contains a child $B$; (2) The order indicator between nodes $A$ and $B$ must be $SEQ$ or $ALL$, and the value of $minOccurs$ of $B$ is not less than one. For example, in Fig. 2, XML Schema has constraints $RPC(A1, C1)$, $RPC(A1, B1)$, $RPC(A1, D1)$ and $RPC(F1, G1)$.
- Required Ancestor-Descendant (RAD): For a given XML Schema, if $RPC(B_0, B_1)$, $RPC(B_1, B_2)$, $\cdots$, $RPC(B_{n-1}, B_n)$ hold, we can say $B_i$ has a descendant $B_n$ (short as $RAD(B_i, B_n)$) where i $\in \{0,1,\cdots,n-1\}$.
- Backward Required Parent-Child ($BRPC$): For a given XML Schema, if node $A$ and $B$ satisfy constraint: node $A$ has certain parent node $B$, then we say that XML Schema has backward required parent-child constraint $BRPC(A, B)$.
- Backward Required Ancestor-Descendant ($BRAD$): For a given XML Schema, if $BRPC(B_0, B_1)$, $BRPC(B_1, B_2)$, $\cdots$, $BRPC(B_{n-1}, B_n)$ hold, we can say that XML Schema has backward required ancestor-descendant constraint $BRAD(B_i, B_n)$, where i $\in \{0,1,\cdots,n-1\}$.

### 3.3   MCSG Model

Considering the use of model checking techniques for identifying constraints, we define model $MCSG$ (Model Checking Schema Graph) in Definition 2.

**Definition 2.** A MCSG model $M$ is a four-tuple: $M = (S, L, R, F)$, where

- $S$ is a set of integers, that are identities of nodes in the underlying XML Schema graph $G$.
- $L$ is a set of strings, which are names of all the nodes in XML Schema graph $G$. There are three new node names $STAR$, $OPT$ and $PLUS$ added. They stand for the times of occurrence $*, ?, +$ in graph $G$.
- $R$ is a mapping from $S$ to $S$, which stand for the edges of the corresponding graph $G$.
- $F$ is a mapping from $S$ to $L$.

Figure 3 is the corresponding MCSG model of XML Schema in Fig. 2. According to the MCSG definition we have:

$S = \{0, 1, 2, 3, 4, 5, 6, 7, 8, 9, 10, 11, 12\}$

$L = \{SCHEMA, A1, D1, SEQ, CHOICE, B1, C1, OPT, F1, E1, SEQ, G1, STAR\}$

$R = \{0 \rightarrow 1, 0 \rightarrow 2, 1 \rightarrow 3, 2 \rightarrow 4, 3 \rightarrow 5, 3 \rightarrow 6, 4 \rightarrow 7, 4 \rightarrow 8, 7 \rightarrow 9, 8 \rightarrow 10, 10 \rightarrow 11,$
$\quad\quad 11 \rightarrow 12\}$

$F = \{0 \rightarrow SCHEMA, 1 \rightarrow A1, 2 \rightarrow D1, 3 \rightarrow SEQ, 4 \rightarrow CHOICE, 5 \rightarrow B1, 6 \rightarrow C1,$
$\quad\quad 7 \rightarrow OPT, 8 \rightarrow F1, 9 \rightarrow E1, 10 \rightarrow SEQ, 11 \rightarrow G1, 12 \rightarrow STAR\}$

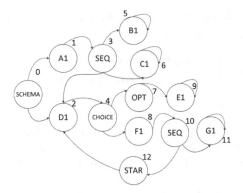

**Fig. 3.** An MCSG model

# 4    Identifying Constraints

This section will presents CTL representation of constraints and develops identifying algorithms.

## 4.1    Constraints Representation

We give the standard CTL definition [7] as follows.

$$\Phi ::= \top \mid \bot \mid p \mid \neg\Phi \mid \Phi \vee \Phi \mid \Phi \wedge \Phi \mid \Phi \to \Phi \mid AX(\Phi) \mid EX(\Phi) \mid AF(\Phi)$$
$$\mid EF(\Phi) \mid AG(\Phi) \mid EG(\Phi) \mid A\,[\,\Phi U \Phi\,] \mid E\,[\,\Phi U \Phi\,]$$

(1)

For constraint $RPC(A, B)$, which means node $A$ has a certain child node $B$, it can be expressed as CTL formula (2).

$$A \to EX(\,E\,[(\,SEQ \vee PLUS \vee ALL\,)\,UB\,])$$

(2)

It shows that for node $A$, there must exist a path whose next nodes may be *SEQ*, *PLUS* or *ALL* until node $B$ is reached.

Similarly, constraint $RAD(A, B)$ denotes node $A$ has certain descendant node $B$, can be expressed as CTL formula (3).

$$A \to EX(\,E\,[(\neg STAR \wedge \neg OPT \wedge \neg CHOICE)UB\,])$$

(3)

It indicates that for node $A$, there must exist a path whose next nodes must not be *STAR*, *OPT*, *CHOICE* until node $B$ is reached.

We made some modifications to CTL in order to make it easier to express backward constraints in XML Schema, such as past temporal operators $\overleftarrow{E}$, $\overleftarrow{A}$, $\overleftarrow{EX}$, and $\overleftarrow{EG}$. This allows one to intuitively specify backward constraints. Similar extensions for backward operators have been used in program analyses [8,15].

In order to simplify the expression of backward constraints, we give definitions of sub-formulas shown as (4), (5) and (6).

$$orpc = SEQ \vee ALL \vee PLUS \tag{4}$$
$$orad = (\neg OPT) \wedge (\neg CHOICE) \wedge (\neg STAR) \tag{5}$$
$$ornoelem = SEQ \vee PLUS \vee ALL \vee CHOICE \vee STAR \vee OPT \tag{6}$$

The formula (4) means that the node is either *SEQ*, *ALL* or *PLUS*. The formula (5) indicates that the node is neither *OPT*, *CHOICE* nor *STAR*. While the formula (6) means that the node must be indicator node, no element node is permitted.

The constraint $BRPC(A, B)$ is defined as:

$$A \rightarrow \overleftarrow{EX}( \overleftarrow{E}[\, ornoelem U B\,]) \tag{7}$$

The constraint $BRAD(A, B)$ is defined as:

$$A \rightarrow \overleftarrow{EX}( \overleftarrow{E}[\, true U B\,]) \tag{8}$$

## 4.2  Main Algorithm and Auxiliary Procedure

Algorithm $Identify(M, F)$ takes an MCSG model $M$ and a CTL formula $F$ as input, and output *true* or *false*. The main idea is to dispatch sub-formula checking tasks to the corresponding procedures. The algorithm does not consider *EG*, *AG*, *AF* and *EF* operators because current XML Schema constraint representations have not involved these temporal operators. It is easy to add the corresponding procedure to process these operators.

The line 1 in Algorithm $Identify(M, F)$ declares a two-dimensional table $Result[subformula][node]$. Its rows are the numbers of sub-formulas. The numbering rule of the sub-formula in the syntax tree is: (1) the number of next-sibling is bigger than the preceding-sibling; (2) parent is greater than descendant. Its columns are all the nodes in the state set $S$ of MCSG model $M$. For any sub-formula $f$, node $n$, $Result[f][n]$ is a boolean value to indicate whether node $n$ satisfies the sub-formula $f$.

The lines $2-6$ are initialization of the table. The function $getithformu(i)$ in line 8 obtains the $i$-th sub-formula in CTL syntax tree. The lines $7-33$ process each sub-formula from bottom to up of the syntax tree. Based on the deferent operator type of $Currentformu$, the algorithm will call the corresponding procedure. For instance, in order to calculate satisfied node set of the $i$-th sub-formula in model $M$, the algorithm call procedure $processEX(M, i-1)$. Here $i-1$ means $i-1$-th sub-formula, which is the child of $Currentformu$. The lines $34-38$ checks if the formula $F$ (it is called root formula of the syntax tree) is satisfied by the Model $M$.

We present two auxiliary procedures for processing $EX$ and corresponding backward temporal operator $BEX$. Other procedures are omitted here.

---

**Algorithm 1.** Identify(M,F)

---

**Input:**
    MCSG model M;
    CTL formula F;
**Output:**
    whether M satisfies F: true or false;
1: Result[F.size][M.size]
2: **for** $i = 0$ to F.size-1 **do**
3:    **for** every node in M.S **do**
4:       Result[i][node]=false
5:    **end for**
6: **end for**
7: **for** $i = 0$ to F.size-1 **do**
8:    Currentformu = getithformu(i)
9:    **switch** (C)urrentformu.operator
10:    **case** *basic_predicate*:
11:      **for** every node satisfied Currentformu **do**
12:        Result[i][node]=true
13:      **end for**
14:    **case** $EX$:
15:      processEX(M, i-1)
16:    **case** $AX$:
17:      processAX(M, i-1)
18:    **case** $EU$:
19:      processEU(M, leftchild(i), i-1)
20:    **case** $AU$:
21:      processAU(M, leftchild(i), i-1)
22:    **case** $BEX$:
23:      processBEX(M, i-1)
24:    **case** $BAX$:
25:      processBAX(M, i-1)
26:    **case** $BEU$:
27:      processBEU(M, leftchild(i), i-1)
28:    **case** $BAU$:
29:      processBAU(M, leftchild(i), i-1)
30:    **default:**
31:      Error
32:    **end switch**
33: **end for**
34: **for** every node in M.S **do**
35:    **if** Result[F.size-1][node]==false **then**
36:      **return** false
37:    **end if**
38: **end for**
39: **return** true

---

| | |
|---|---|
| 1 | *Procedure processEX(M, i)* |
| 2 | **For** *every node in M.S* |
| 3 | **If**(*Result*[*i*][*node*] == *true*) |
| 4 | **For** *every prenode of node* |
| 5 | *Result*[*i* + 1][*prenode*] = *true*; |
| 6 | **End For** |
| 7 | **End If** |
| 8 | **End For** |

| | |
|---|---|
| 1 | *Procedure processBEX(M, i)* |
| 2 | **For** *every node in M.S* |
| 3 | **If**(*Result*[*i*][*node*] == *true*) |
| 4 | **For** *every postnode of node* |
| 5 | *Result*[*i* + 1][*postnode*] = *true*; |
| 6 | **End For** |
| 7 | **End If** |
| 8 | **End For** |

**Fig. 4.** Procedure *processEX*          **Fig. 5.** Procedure *processBEX*

The procedure for *EX* is defined in Fig. 4. It takes MCSG model $M$ and the $i$-th sub-formula as parameters. It firstly obtains the node set that satisfy the $i$-th sub-formula by checking the table $Result[i][node]$, then update $Result[i + 1][prenode]$ as *true*. The node *prenode* in $Result[i + 1][prenode]$ is the parent of the *node* in $Result[i][node]$.

Similarly for procedure *processBEX* defined in Fig. 5.

Considering time complexity, let us assume the total number of the logical connectors and the temporal operators of CTL formula $F$ be $N$, the vertex number of the MCSG model $M$ be $V$, and the number of edges be $E$. Time complexity of each procedure is calculated as follows: O(processEX)=E, O(processAX)=E, O(processEU)=V*E, O(processAU)=V*E

The time complexity of the algorithm *Identify* is: O(*Identify*)=f*V*E and the space complexity is: f*V

The complexity of our algorithm is polynomial. It is succinct compared with the complexity of pure-future temporal logic model checking [22], which is exponential.

## 5   Conclusion

The main goal of this work is to bring techniques developed in the temporal logic community into the field of Twig pattern minimization. In order to efficiently solve XML Schema constraint identification issues, we begin with an observation that XML Schema constraints can be represented as temporal logic formulas, and XML Schema documents can be naturally represented as labeled transition systems. We develop algorithms for identifying constraints automatically.

The main contributions of this work lie in that: it tries to use model checking approach for resolving Twig pattern minimization issue, particularly, for identifying XML Schema constraints. Compared with traditional approach, it is more flexible and easy to guaranty the correctness. If there is a new constraint need to be identified, as long as the constraint can be expressed by temporal logic formula, it can be automatically identified, without to develop a new algorithm.

For the future work, on the one hand, although we have modified CTL for expressing backward constraints, there still exist some constraints that can not be easily expressed, such as unique path between two elements [14]. So we will make modifications on CTL to support more constraints. On the other hand, we will progressively use temporal logic techniques to automatically generate optimization action of Twig patterns.

# References

1. Jagadish, H.V., Lakshmanan, L.V.S., Srivastava, D., Thompson, K.: TAX: a tree algebra for XML. In: Ghelli, G., Grahne, G. (eds.) DBPL 2001. LNCS, vol. 2397, pp. 149–164. Springer, Heidelberg (2002). doi:10.1007/3-540-46093-4_9
2. Amer-Yahia, S., Cho, S.R., Lakshmanan, L.V.S., et al.: Tree pattern query minimization. VLDB J. Int. J. Very Large Data Bases **11**(4), 315–331 (2002)
3. Che, D.: An efficient algorithm for tree pattern query minimization under broad integrity constraints. Int. J. Web Inform. Syst. **3**(3), 231–256 (2007)
4. Chen, D., Chan, C.Y.: Minimization of tree pattern queries with constraints. In: Proceedings of the 2008 ACM SIGMOD International Conference on Management of Data, pp. 609–622 (2008)
5. Lee, K.H., Whang, K.Y., Han, W.S.: Xmin: minimizing tree pattern queries with minimality guarantee. World Wide Web **13**(3), 343–371 (2010)
6. Li, H., Liao, H.S., Su, H.: Optimize twig query pattern based on XML schema. J. Softw. **8**(6), 1479–1486 (2013)
7. Huth, M., Ryan, M.: Logic in Computer Science, 2nd edn., pp. 207–216. Cambridge University Press, Cambridge (2005)
8. Kalvala, S., Warburton, R., Lacey, D.: Program transformations using temporal logic side conditions. ACM Trans. Program. Lang. Syst. (TOPLAS) **31**(4), 824–833 (2009)
9. Afanasiev, L., Franceschet, M., Marx, M., et al.: Ctl model checking for processing simple xpath queries. Proc. Temp. Represent. Reasoning **152**(6), 117–124 (2004)
10. Currim, F.A., Currim, S.A., Dyreson, C.E., et al.: Adding temporal constraints to XML schema. IEEE Trans. Knowl. Data Eng. **24**(8), 1361–1377 (2012)
11. Libkin, L., Sirangelo, C.: Reasoning about XML with temporal logics and automata. In: Cervesato, I., Veith, H., Voronkov, A. (eds.) LPAR 2008. LNCS (LNAI), vol. 5330, pp. 97–112. Springer, Heidelberg (2008). doi:10.1007/978-3-540-89439-1_7
12. Libkin, L.: Logics for unranked trees: an overview. In: Caires, L., Italiano, G.F., Monteiro, L., Palamidessi, C., Yung, M. (eds.) ICALP 2005. LNCS, vol. 3580, pp. 35–50. Springer, Heidelberg (2005). doi:10.1007/11523468_4
13. Arenas, M., Barceló, P., Libkin, L.: Combining temporal logics for querying XML documents. In: Schwentick, T., Suciu, D. (eds.) ICDT 2007. LNCS, vol. 4353, pp. 359–373. Springer, Heidelberg (2006). doi:10.1007/11965893_25
14. Zhang, J.M., Tao, S.Q., Liang, J.Y.: Minimization of path expression under structural integrity constraints for XML. J. Softw. **20**(11), 2977–2987 (2009)
15. Steffen, B.: Data flow analysis as model checking. In: Ito, T., Meyer, A.R. (eds.) TACS 1991. LNCS, vol. 526, pp. 346–364. Springer, Heidelberg (1991). doi:10.1007/3-540-54415-1_54
16. Hafer, T., Thomas, W.: Computation tree logic CTL* and path quantifiers in the monadic theory of the binary tree. In: Ottmann, T. (ed.) ICALP 1987. LNCS, vol. 267, pp. 269–279. Springer, Heidelberg (1987). doi:10.1007/3-540-18088-5_22
17. Kupferman, O., Pnueli, A.: Once and for all. In: Proceedings of the 10th LICS, pp. 25–35

18. Bozzelli, L.: The complexity of CTL* + linear past. In: Amadio, R. (ed.) FoSSaCS 2008. LNCS, vol. 4962, pp. 186–200. Springer, Heidelberg (2008). doi:10.1007/978-3-540-78499-9_14
19. Pistore, M., Vardi, M.Y.: The planning spectrum - one, two, three, infinity. In: Proceedings of the 18th LICS, pp. 234–243
20. Kupferman, O., Vardi, M.Y.: Memoryful branching-time logic. In: Proceedings of the 21th LICS, pp. 265–274
21. http://www.lsv.ens-cachan.fr/~markey/PLTL.php
22. Laroussinie, F., Markey, N., Schnoebelen, P.: Temporal logic with forgettable past. In: Proceedings of the 17th IEEE Symposium on Logic in Computer Science (LICS 2002), Copenhagen, Denmark, July 2002, pp. 383–392. IEEE Comp. Soc. Press (2002)

# Schedulability Analysis of Timed Regular Tasks by Under-Approximation on WCET

Bingbing Fang[1], Guoqiang Li[1(✉)], Daniel Sun[2], and Hongming Cai[1]

[1] School of Software, Shanghai Jiao Tong University, Shanghai, China
{fke_htj,li.g,hmcai}@sjtu.edu.cn
[2] Data61, CSIRO, New South Wales, Australia
daniel.sun@data61.csiro.au

**Abstract.** *Schedulability analysis* is one of the most important issues in developing and analyzing real-time systems. Given a task system where each task is characterized by a *worst-case execution time (WCET)* and a *relative deadline*, the schedulability analysis is decidable. However in reality, it is difficult to calculate the WCET of a complex task, even after it is abstracted to a formal model, e.g., *timed automata (TAs)*. This paper proposes a schedulability analysis method without the information of the WCET, by introducing a model named *timed regular task automata (TRTAs)*. Each task is described by a TA, a starting point with a clock valuation, a status and a relative deadline. A test is performed on each TA for an under-approximation of the WCET. The system may still be unschedulable under the approximation. A further schedulability checking is then performed by encoding to the reachability problem of *nested timed automata (NeTAs)*. The methodology is thus sound and complete.

## 1 Introduction

*Real-time systems* are playing a crucial role in the society, and in the past decades, there has been an explosive growth in the number of real-time systems being used in our daily lives and in industry production. Schedulability analysis is one of the most important issues in developing and analyzing real-time systems. Given a task system, each task is usually characterized by a *worst-case execution time (WCET)* and a *relative deadline*. The schedulability analysis is decidable under a singleton processor [1,2]. However in reality, it is difficult to calculate the WCET of a complex task [3], even after it is abstracted and modeled by some formal frameworks, such as, *timed automata (TAs)*.

This paper proposes a schedulability analysis methodology without the information of the WCET. We introduce a formal model named *timed regular task automata (TRTAs)*, extended from *task automata* [2,4]. A TRTA is a TA where each control location is assigned to a task. Each task is characterized by a TA to describe the behavior of the task, a starting control location of the task, an initial clock valuation, a status and a relative deadline. Similar to that of task

© Springer International Publishing AG 2016
M. Fränzle et al. (Eds.): SETTA 2016, LNCS 9984, pp. 147–162, 2016.
DOI: 10.1007/978-3-319-47677-3_10

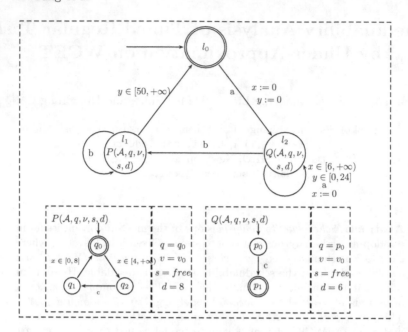

**Fig. 1.** Example of a TRTA

automata, a scheduling queue is prepared for the execution and release of tasks. In the system, we assume the task system is preemptive and no feedback [2].

Instead of calculating the WCET directly, a test is performed to each TA for an under-approximation of the WCET of the respective task, which is used to bound the schedulable queue. The system may still be unschedulable, after the queue is bounded by the approximation. A schedulability checking is further performed to detect unschedulable cases, by encoding to the reachability problem of *nested timed automata (NeTAs)* [5–7], which is known to be decidable. The methodology is thus sound and complete, preserving the schedulability analysis.

We will illustrate the method by the following example. Figure 1 is a TRTA, with three locations, $l_0$, $l_1$, $l_2$ and two tasks $P$, $Q$. The behavior of task $P$ and $Q$ is described as TAs shown in lower left and lower right of Fig. 1, respectively. When the system switches to the location $l_1$ and $l_2$, task $P$ and $Q$ will be triggered and a task instance will be inserted into the task queue waiting for executing, respectively. In location $l_2$, at most four copies of task $Q$ can be created and every instance will reach its final state within deadline due to the constraints $x \in [6, +\infty)$, $y \in [0, 24]$. In $l_1$, the system can create any numbers of task $P$ instance since there is no constraints labelled on the transition from anywhere to it.

We can see the task $P$ described in the lower left of Fig. 1, although it is modeled by a quite simple timed automaton, it is still difficult to get the exact WCET of the task. In this paper, a test will be performed to each TA to get the under-approximation of the WCET of the respective task, which is used to check if the length of the queue meets the condition of a schedulable queue. If the queue is not bounded by the approximation execution time, the system is non-schedulable. However, if it is bounded, the system may still be non-schedulable. For this kind of situation, we encoded scheduling strategy to a NeTA and perform the scheduling checking problem by reachability checking problem of the NeTA, which we will introduce in the following sections in detail.

**Related Work.** A variety of methods have been published in schedulability analysis for real-time systems. Real time tasks may be periodic, sporadic, pre-emptive or non-preemptive. For systems with only periodic tasks [8], the *rate monotonic scheduling* algorithm and efficient methods for schedulability checking are widely used [1]. For systems with non-periodic tasks, the *controller synthesis* approach [9] has been adopted, which is to achieve schedulability by constructions. Timed automata are regarded as a formal model for analysis to solve non-preemptive scheduling problems, mainly for job-shop scheduling [10, 11]. For preemptive scheduling problems, stopwatch automata are used with the restriction on the assumption that task preemptions occur only at integer points. Task automata [2, 12] for real time systems with non-uniformly recurring computation tasks describe tasks that are generated non-deterministically according to timing constraints and may have a BCET and a WCET. A decidable class is identified to solve scheduling problems algorithmically without assuming that preemptions occur only at integer points. These works are all associated with the BCET and WCET of a task which is difficult to calculate. In this paper, we propose a schedulability analysis methodology without the information of the BCET and WCET by introducing TRTAs. Recently, a group of researches reveal the decidability results of time sensitive pushdown systems under different types of clocks, such as *recursive timed automata (RTAs)* [13], *timed recursive state machines (TRSMs)* [14,15], and NeTAs [5,6], which provide us the backbone analysis techniques.

**Paper Organization.** The rest of this paper is organized as follows: Sect. 2 briefly reviews the TAs, NeTAs and their parallel compositions. Section 3 introduces timed regular tasks, and the syntax and semantics of timed regular task automata. The schedulability problem for timed regular task automata is proposed in Sect. 4. Section 5 gives the methodology of the under-approximation for WCET and contributes to encode schedulability analysis to the reachability problem of NeTAs. Section 6 concludes the paper and introduces the future works.

## 2   Preliminaries

Let $\mathbb{R}^{\geq 0}$ and $\mathbb{N}$ denote the sets of non-negative real numbers and natural numbers, respectively. We define $\mathbb{N}^{\omega} := \mathbb{N} \cup \{\omega\}$, where $\omega$ is the first limit ordinal. Let $\mathcal{I}$ denotes the set of *intervals*. An interval is a set of numbers, written as $(a, b)$,

$[a, b]$, $[a, b)$ or $(a, b]$, where $a \in \mathbb{N}$ and $b \in \mathbb{N}^\omega$. We say an interval is *bounded* if $a, b \in \mathbb{N}$. For a number $r \in \mathbb{R}^{\geq 0}$ and an interval $I \in \mathcal{I}$, we use $r \in I$ to denote that $r$ belongs to $I$.

Let $X = \{x_1, \ldots, x_n\}$ be a finite set of *clocks*. The set of *clock constraints*, $\Phi(X)$, over $X$ is defined by

$$\phi ::= \top \mid x \in I \mid \phi \wedge \phi$$

where $x \in X$ and $I \in \mathcal{I}$. A *clock valuation* $\nu : X \to \mathbb{R}^{\geq 0}$, assigns a value to each clock $x \in X$. $\nu_0$ represents all clocks in $X$ assigned to zero. For a clock valuation $\nu$ and a clock constraint $\phi$, we write $\nu \models \phi$ to denote that $\nu$ satisfies the constraint $\phi$. Given a clock valuation $\nu$ and a time $t \in \mathbb{R}^{\geq 0}$, $(\nu + t)(x) = \nu(x) + t$, for $x \in X$. Given a set of clocks $\lambda \subseteq X$ and a clock valuation $\nu$, let a *clock reset function* $\nu[\lambda]$ be a clock valuation, defined as follows:

$$(\nu[\lambda])(x) = \begin{cases} 0 & \text{if } x \in \lambda \\ \nu(x) & \text{otherwise} \end{cases}$$

### 2.1 Timed Automata

A timed automaton is an automaton augmented with a finite set of clocks [16,17]. Time can elapse in a location, while switches are instantaneous.

**Definition 1 (Timed Automata).** *A timed automaton is a tuple* $\mathcal{A} = (\Sigma, Q, q_0, F, X, I, \Delta) \in \mathscr{A}$, *where*

- $\Sigma$ *is a finite set of input symbols.*
- $Q$ *is a finite set of control locations.*
- $q_0 \in Q$ *is the initial location.*
- $F \in Q$ *is the finite set of final locations.*
- $X$ *is a finite set of clocks.*
- $I : Q \to \Phi(X)$ *is a function assigning each location with a clock constraint, called an* invariant.
- $\Delta \subseteq Q \times \Sigma \times \Phi(X) \times 2^X \times Q$.

When $\langle q_1, a, \phi, \lambda, q_2 \rangle \in \Delta$, we write $q_1 \xrightarrow{a, \phi, \lambda} q_2$.

Given a timed automaton $\mathcal{A} \in \mathscr{A}$, we use $\Sigma(\mathcal{A})$, $Q(\mathcal{A})$, $q_0(\mathcal{A})$, $F(\mathcal{A})$ and $X(\mathcal{A})$ to represent its set of input symbols, control locations, initial location, final locations and set of clocks, respectively. We will use similar notations for other automata.

The semantics of timed automata includes progress transitions, for time elapsing within one control location, and discrete transitions, for transference between two control locations [16].

**Definition 2 (Semantics of Timed Automata).** *A configuration of a TA is a pair* $(q, \nu)$ *of a control location* $q \in Q$, *and a clock valuation* $\nu$ *on* $X$. *The labelled transition system (LTS) of timed automata is represented as follows,*

– Progress transition: $(q, \nu) \xrightarrow{t}_{\mathscr{A}} (q, \nu + t)$, where $t \in \mathbb{R}^{\geq 0}$ and $(\nu + t) \models I(q)$.
– Discrete transition: $(q_1, \nu) \xrightarrow{a}_{\mathscr{A}} (q_2, \nu[\lambda])$, if $q_1 \xrightarrow{a,\phi,\lambda} q_2$, and $\nu \models \phi$, and $\nu[\lambda] \models I(q_2)$.

The initial configuration is $(q_0, \nu_0)$. The transition relation is $\rightarrow$ and we define $\rightarrow = \xrightarrow{t}_{\mathscr{A}} \cup \xrightarrow{\phi}_{\mathscr{A}}$, and define $\rightarrow^*$ to be the reflexive and transitive closure of $\rightarrow$.

*Remark 1.* For conciseness of proof, the Definition 1 is slightly different from that in [5], by allowing testing and resetting value of clocks in one transition rule, following the definition style in [16]. This can be encoded with an extra clock by resetting it to 0 and checking it still 0 after transitions, and introducing fresh control locations in the original definition.

Although general verification problems, such as the language inclusion problem, are undecidable on timed automata, the reachability problem for real-time systems [16,17] is decidable.

**Fact 1.** *The reachability problem of timed automata is decidable [16,17].*

Furthermore, we define $BCET(\mathcal{A})$ and $WCET(\mathcal{A})$ for the best-case execution time and worst-case execution time of $\mathcal{A}$, respectively. Say, beginning with the $q_0(\mathcal{A})$, the shortest time passage and longest time passage when a final control location is met.

## 2.2 Nested Timed Automata

*Nested timed automata (NeTAs)* [5,6] extend TAs with the recursive structure, which allow clocks of TAs in the stack elapse simultaneously with the current running clocks during time passage.

**Definition 3.** *[Nested Timed automata] A nested timed automaton(NeTA) is a tuple $\mathcal{N} = (T, \mathcal{A}_0, X, C, \Delta)$, where*

– $T$ is a finite set of timed automata $\{\mathcal{A}_0, \mathcal{A}_1, \ldots, \mathcal{A}_k\}$, with the initial timed automaton $\mathcal{A}_0 \in T$.
– $X$ is the finite set of $k$ local clocks.
– $C$ is the finite set of global clocks.
– $\Delta \subseteq Q \times (Q \cup \{\varepsilon\}) \times Actions^+ \times Q \times (Q \cup \{\varepsilon\})$ describes transition rules below, where $Q = \cup_{\mathcal{A}_i \in T} Q(\mathcal{A}_i)$.

*A transition rule is described by a sequence of Actions* $- \{internal, push, pop, c \in I, c \leftarrow I, c \leftarrow x\}$, *where* $c \in C$, $x \in X$ *and* $I \in \mathcal{I}$.

– **Internal** $(q, \varepsilon, internal, q', \varepsilon)$, which describes an internal transition in the working TA (placed at a control location) with $q, q' \in Q(\mathcal{A}_i)$, the internal transition with $q, q'$ is the same as the Definition 1.
– **Push** $(q, \varepsilon, push, q_0(\mathcal{A}_i'), q)$, which interrupts the currently working TA $\mathcal{A}_i$ at $q \in Q(\mathcal{A}_i)$, then a TA $\mathcal{A}_i'$ newly starts.

- **Pop** $(q, q', pop, q', \varepsilon)$ which restarts $\mathcal{A}'_i$ in the stack from $q' \in Q(\mathcal{A}'_i)$ after $\mathcal{A}_i$ has finished at $q \in Q(\mathcal{A}_i)$.
- **Global-test** $(q, \varepsilon, c \in I?, q', \varepsilon)$, which tests whether the value of a global clock $c$ is in $I$.
- **Global-assign** $(q, \varepsilon, c \leftarrow I, q', \varepsilon)$, which assigns a value $r \in I$ to a global clock c.
- **Global-store** $(q, \varepsilon, c \leftarrow x, q', \varepsilon)$, which assign the value of a local clock $x \in X$ of the working TA to a global clock $c$.

**Definition 4 (Semantics of NeTAs).** *Given a NeTA* $(T, \mathcal{A}_0, X, C, \Delta)$, *the current control state is referred by $q$. Let* $\mathcal{V}al_X = \{\nu : X \to \mathbb{R}^{\geq 0}\}$ *and* $\mathcal{V}al_C = \{\mu : C \to \mathbb{R}^{\geq 0}\}$. *A configuration of a NeTA is an element in* $(Q \times \mathcal{V}al_X \times \mathcal{V}al_C, (Q \times \mathcal{V}al_X)^*)$. $(Q \times \mathcal{V}al_X)^*$ *is a stack content denoted by $s$. Let* $s = \langle q_1, \nu_1 \rangle \langle q_2, \nu_2 \rangle ... \langle q_n, \nu_n \rangle$, *then $s + t$ will be* $\langle q_1, \nu_1 + t \rangle \langle q_2, \nu_2 + t \rangle ... \langle q_n, \nu_n + t \rangle$. *The transition rules of NeTAs are presented as follows:*

- *Progress transitions:* $(\langle q, \nu, \mu \rangle, s) \xrightarrow{t} (\langle q, \nu + t, \mu + t \rangle, s + t)$.
- *Discrete transitions:* $\kappa \xrightarrow{\phi} \kappa'$ *is defined as follows.*
  - **Internal-action** $(\langle q, \nu, \mu \rangle, s) \xrightarrow{\phi} (\langle q', \nu[\lambda], \mu \rangle, s)$, *if* $q \xrightarrow{a, \phi, \lambda} q' \in \Delta(\mathcal{A})$, *where $\mathcal{A}$ is the current running TA, $a \subseteq \Sigma(\mathcal{A})$, $\lambda \subseteq X(\mathcal{A})$, and $\nu \models \phi$, and $\nu[\lambda] \models I(q')$.*
  - **Push** $(\langle q, \nu, \mu \rangle, s) \xrightarrow{\phi} (\langle q_0(\mathcal{A}'), \nu'_0, \mu \rangle, \langle q, \nu \rangle s)$.
  - **Pop** $(\langle q, \nu, \mu \rangle, \langle q', \nu' \rangle s) \xrightarrow{pop} (\langle q', \nu', \mu \rangle, s)$.
  - **Global-test** $(\langle q, \nu, \mu \rangle, s) \xrightarrow{c \in I?} (\langle q', \nu, \mu \rangle, s)$, *if* $\mu(c) \in I$.
  - **Global-assign** $(\langle q, \nu, \mu \rangle, s) \xrightarrow{c \leftarrow I} (\langle q', \nu, \mu[c \leftarrow r] \rangle, s)$ *for* $r \in I$.
  - **Global-store** $(\langle q, \nu, \mu \rangle, s) \xrightarrow{c \leftarrow x} (\langle q', \nu, \mu[c \leftarrow \nu[x]] \rangle, s)$.

According to [5], we know that the reachability problem for nested timed automata is decidable by encoding NeTAs to dense timed pushdown automata [18]. And the global clocks do not affect the safety property of an NeTA. We have the following fact.

**Fact 2.** *The state reachability problem of NeTAs is decidable [5].*

### 2.3 Parallel Composition of TA and NeTA

Given a TA $\mathcal{A}$ and a NeTA $\mathcal{N}$, we construct a parallel composition automaton $\mathcal{A} \| \mathcal{N}$. A formal definition of the parallel composition between $\mathcal{A}$ and $\mathcal{N}$ is defined as follows.

Assuming a TA $\mathcal{A} = (Q, q_0, F, X, I, \Delta)$ and a NeTA $\mathcal{N} = (T, \mathcal{A}_0, X, C, \Delta)$ are running concurrently over a shared finite set of actions $\Sigma$. $\Sigma^\tau = \Sigma \cup \{\tau\}$, where $\tau$ stands for a silent action. Transition of $\mathcal{N}$ is redefined by $\Delta(\mathcal{N}) \subseteq Q \times O \times \Sigma^\tau \times Action \times Q \cup \{\varepsilon\}$, where $Action \in \{push, pop\}$. A rule $(p, \phi, a, \Phi, p') \in \Delta(\mathcal{N})$ is written as $p \xrightarrow{\phi, a, \Phi} p'$. Transition of the TA is defined by $\Delta(\mathcal{A}) \subseteq Q \times \Sigma^\tau \times O \times Q$. A rule $(p, a, \phi, p') \in \Delta(\mathcal{A})$ is written as $p \xrightarrow{a, \phi} p'$. We usually omit $a$ when $a = \tau$.

**Definition 5 (Semantics of Parallel Composition of NeTA and TA).**
*Given a nested timed automaton $\mathcal{N} = (T, \mathcal{A}_0, X, C, \Delta)$ and a timed automaton $\mathcal{A} = (Q, q_0, F, X, I, \Delta)$, a finite set of actions $\Sigma$, a configuration of a parallel composition $\mathcal{N}\|\mathcal{A}$ is a tuple $(q_\mathcal{A}, \nu_\mathcal{A}, (\langle q, \nu, \mu \rangle, s))$, where $q_\mathcal{A} \in Q(\mathcal{A})$, $\nu_\mathcal{A}$ is a clock valuation on $X(\mathcal{A})$, $s$ is the stack belongs to configuration of $\mathcal{N}$, and $(q, \nu, \mu)$ is the configuration of a TA is executing now which is in the top of stack $s$. The transition rules of $\mathcal{N}\|\mathcal{A}$ is defined as follows:*

- *Progress transitions:* $(q_\mathcal{A}, \nu_\mathcal{A}, (\langle q, \nu, \mu \rangle, s)) \xrightarrow{t} (q_\mathcal{A}, \nu_\mathcal{A}+t, (\langle q, \nu+t, \mu+t \rangle, s+t))$.
- *Discrete transitions:* $(q_\mathcal{A}, \nu_\mathcal{A}, (\langle q, \nu, \mu \rangle, s)) \xrightarrow{a, \phi, \phi_\mathcal{N}} (q'_\mathcal{A}, \nu'_\mathcal{A}, (\langle q', \nu', \mu' \rangle, s'))$ *is defined as a union of the following transition rules:*
  - *TA-movement* $(q_\mathcal{A}, \nu_\mathcal{A}, (\langle q, \nu, \mu \rangle, s)) \xrightarrow{\tau, \phi} (q'_\mathcal{A}, \nu'_\mathcal{A}, (\langle q, \nu, \mu \rangle, s))$, *if* $(q_\mathcal{A}, \nu_\mathcal{A}) \xrightarrow{\tau, \phi}_\mathcal{A} (q'_\mathcal{A}, \nu'_\mathcal{A})$.
  - *NeTA-intra-movement* $(q_\mathcal{A}, \nu_\mathcal{A}, (\langle q, \nu, \mu \rangle, s)) \xrightarrow{\tau, \phi} (q_\mathcal{A}, \nu_\mathcal{A}, (\langle q', \nu', \mu \rangle, s))$, *if* $(q, \nu, \mu) \xrightarrow{\tau, \phi}_\mathcal{N} (q', \nu', \mu)$.
  - *NeTA-pop-movement* $(q_\mathcal{A}, \nu_\mathcal{A}, (\langle q, \nu, \mu \rangle, s)) \xrightarrow{\phi_\mathcal{N}, pop} (q_\mathcal{A}, \nu_\mathcal{A}, (\langle q', \nu', \mu' \rangle, s'))$, *if* $(q, \nu, \mu) \xrightarrow{\phi_\mathcal{N}, pop}_\mathcal{N} (q', \nu', \mu')$.
  - *Push-synchronization* $(q_\mathcal{A}, \nu_\mathcal{A}, (\langle q, \nu, \mu \rangle, s)) \xrightarrow{\tau, \phi, \phi_\mathcal{N}, push,} (q'_\mathcal{A}, \nu'_\mathcal{A}, (\langle q_0(\mathcal{A}'), \nu'_0, \mu' \rangle, s'))$, *if* $(q_\mathcal{A}, \nu_\mathcal{A}) \xrightarrow{a, \phi}_\mathcal{A} (q', \nu')$ *and* $(\langle q, \nu, \mu \rangle, s) \xrightarrow{a, push, \phi_\mathcal{N}}_\mathcal{N} (\langle q_0(\mathcal{A}'), \nu'_0, \mu' \rangle, s')$.

The initial configuration of $\mathcal{N}\|\mathcal{A}$ is $(q_{0\mathcal{A}}, \nu_{0\mathcal{A}}, (\langle q_0(\mathcal{A}_0), \nu_0, \mu_0 \rangle, \varepsilon))$, where $\mathcal{A}_0 \in T(\mathcal{N})$, $\nu_{0\mathcal{A}}(x) = 0$ for $x \in X(\mathcal{A})$, $\nu_0(x) = 0$ for $x \in X(\mathcal{A}_0)$ and $\mu_0(c) = 0$ for $c \in C(\mathcal{N})$.

Note that, the parallel composition of a TA and a NeTA is essentially a NeTA with global clocks [6], which does not increase expressiveness of the model we defined in Definition 3.

## 3   Task Automata with Timed Regular Behaviours

### 3.1   Timed Regular Tasks

Let $\mathcal{P}$ be a set of *task types*, or *tasks* ranged over by $P, Q, R, \ldots$. A task type is a tuple $(P, \mathscr{A}, q_\mathscr{A}, \nu_\mathscr{A}, S, D)$, written by $P(\mathscr{A}, q_\mathscr{A}, \nu_\mathscr{A}, S, D)$, where $P$ is the task name, $\mathscr{A}$ is the timed automaton, $q_\mathscr{A} \in Q(\mathscr{A})$ is the location of $\mathscr{A}$ to describe the current running control location of the task, $\nu_\mathscr{A}$ is a clock valuation on $X(\mathscr{A})$, $S$ is the status of the timed automaton and $D$ is the relative deadline. $S = \{free, released, preempted, running\}$, where *free* denotes that a task is not triggered, *released* denotes that a task is triggered but not started yet, *preempted* means that a task is started but not running now and *running* means that a task is running on the processor. The status of task is initialized to *free*. Whenever a task is triggered, its status is set to *released*. A task type may have several task instances. A task instance is a tuple $P(\mathcal{A}, q, \nu, s, d)$, where $\mathcal{A} \in \mathscr{A}$ is a timed automaton describing the behaviour of the task, $q \in Q(\mathcal{A})$ is a control location of $\mathcal{A}$ to describe the current running control location of $\mathcal{A}$, $\nu$ is a clock valuation on $X(\mathcal{A})$, $s \in S$ is the current status of $\mathcal{A}$ and $d \in \mathbb{R}^{\geq 0}$ is a relative deadline. We shall use $p_i$ to denote a task instance, and $p_i$'s task type will be understood as $P_i(\mathscr{A}, q_{\mathscr{A}}, \nu_{\mathscr{A}}, S, D)$.

We define a *task queue* as a list of task instances, denoted as $[P_1(\mathcal{A}_1, q_1, \nu_1, s_1, d_1), P_2(\mathcal{A}_2, q_2, \nu_2, s_2, d_2), \ldots, P_n(\mathcal{A}_n, q_n, \nu_n, s_n, d_n)]$. A set of task queues containing instances of the task types from $\mathcal{P}$ is denoted $\mathcal{Q}_{\mathcal{P}}$. We use P, Q to represent task queues, and $[\,]$ is used to represent an empty queue.

We assumed that there is only one processor, the task executed on the processor is the first element of the *task queue* and the others are waiting for executing. Whenever a task is released, it is inserted into the *task queue* according to a certain scheduling strategy, e.g., *fixed priority strategy (FPS)*, *rate monotone strategy (RMS)*, or *earliest deadline first (EDF)*. A scheduling strategy is a function Sch : $\mathcal{P} \times \mathcal{Q}_{\mathcal{P}} \to \mathcal{Q}_{\mathcal{P}}$, which given a task instance from $\mathcal{P}$ and a task queue then returns a task queue with the task instance inserted into the proper position of the queue according to some parameters, e.g., deadline or priority. For example, $EDF(P(\mathcal{A}_1, q_1, \nu_1, released, 10), [Q(\mathcal{A}_2, q_2, \nu_2, s_2, 9), R(\mathcal{A}_3, q_3, \nu_3, s_3, 13)]) = [Q(\mathcal{A}_2, q_2, \nu_2, s_2, 9), P(\mathcal{A}_1, q_1, \nu_1, released, 10), R(\mathcal{A}_3, q_3, \nu_3, s_3, 13)]$.

Given a task queue $\mathbf{Q} = [P_1(\mathcal{A}_1, q_1, \nu_1, s_1, d_1), P_2(\mathcal{A}_2, q_2, \nu_2, s_2, d_2), \ldots, P_n(\mathcal{A}_n, q_n, \nu_n, s_n, d_n)]$, when $t$ time units passage happens, the task instance $P_1$ in the first position of the task queue will be executed on the processor with $t$ time units. If $s_1 \in \{released, preempted\}$, the status of $P_1$ will be reset to $s_1 := running$; otherwise $s_1$ remains to $running$. The current running control location $q \in Q(\mathcal{A}_1)$ may be changed to other control locations due to transitions of $\mathcal{A}_1$. The clock valuations on clock $X(\mathcal{A}_1)$ will be changed accordingly and clock valuations on clock $X(\mathcal{A}_i)$, where $i > 1$ will be changed if the status of a task is *preempted*. Each relative deadline of task instances in the queue will decease with $t$.

To describe the above intuition, we introduce a function Exec : $\mathcal{Q}_{\mathcal{P}} \times \mathbb{R}^{\geq 0} \to \mathcal{Q}_{\mathcal{P}}$ that given a task queue and a real number $t$ then returns a task queue executing $t$ time units on a processor. Given a task queue $\mathbf{Q} = [P_1(\mathcal{A}_1, q_1, \nu_1, s_1, d_1), P_2(\mathcal{A}_2, q_2, \nu_2, s_2, d_2), \ldots, P_n(\mathcal{A}_n, q_n, \nu_n, s_n, d_n)]$, the result of Exec$(\mathbf{Q}, t) =$ can be defined inductively as follows:

- Exec$(\mathbf{Q}, 0) = \mathbf{Q}$
- Exec$(\mathbf{Q}, t) = [P_1(\mathcal{A}_1, q_1', \nu_1', running, d_1'), P_2(\mathcal{A}_2, q_2, \nu_2', s_2, d_2'), \ldots, P_n(\mathcal{A}_n, q_n, \nu_n', s_n, d_n')]$, if $q_1' \notin F(\mathcal{A})$, where
  - $\nu_i' = \nu_i + t$ if $s_i = preempted$ and $\nu_i' = \nu_i$ otherwise for $i > 1$,
  - $d_i' = d_i - t$, and
  - $(q_1, \nu_1) \to^* (q_1', \nu_1')$.
  If $d_i' < 0$, then we say that the task queue is *non-schedulable* which will be introduced later in detail.
- Exec$(\mathbf{Q}, t) = $ Exec$([P_2(\mathcal{A}_2, q_2, \nu_2', s_2, d_2'), \ldots, P_n(\mathcal{A}_n, q_n, \nu_n', s_n, d_n')], t - t')$, if after $t'$ time units $(q_1, \nu_1) \to^* (q_1', \nu_1')$, and $q_1' \in F(\mathcal{A})$, where
  - $\nu_i' = \nu_i + t'$ if $s_i = preempted$ and $\nu_i' = \nu_i$ otherwise for $i > 1$,
  - $d_i' = d_i - t'$

*Example 1.* Given a task queue $\mathbb{Q} = [P_1(\mathcal{A}_1, q_1, \nu_1, released, 6), P_2(\mathcal{A}_2, q_2, \nu_2,$ $preempted, 4), P_3(\mathcal{A}_3, q_3, \nu_3, released, 8)]$, we assumed that

- when 2 time units passaged, $\mathcal{A}_1$ reaches its final states, then $\text{Exec}(\mathbb{Q}, 3) =$ $\text{Exec}([P_2(\mathcal{A}_2, q_2, \nu_2', running, 2), P_3(\mathcal{A}_3, q_3, \nu_3, released, 6)], 1)$.
- when 3 time units passaged, $\mathcal{A}_1$ have not reached its final states yet, then $\text{Exec}(\mathbb{Q}, 3) = [P_1(\mathcal{A}_1, q_1', \nu_1', running, 3), P_2(\mathcal{A}_2, q_2, \nu_2', preempted, 1), P_3(\mathcal{A}_3, q_3, \nu_3, released, 5)]$.
- when 5 time units passaged, $\mathcal{A}_1$ have not reached its final states yet, then we say that $\mathbb{Q}$ is non-schedulable, as the deadline of $\mathcal{A}_2$ is less than zero after 5 time units.

### 3.2   Timed Regular Task Automata

**Definition 6 (Timed Regular Task Automata).** *A timed regular task automaton (TRTA) over actions $\mathcal{A}ct$ and task types $\mathcal{P}$ is a tuple $\mathcal{R} = (S, s_0, \mathcal{C}, \mathcal{I}, \mathcal{M}, \Delta) \in \mathcal{R}$, where*

- *$S$ is a finite set of states.*
- *$s_0 \in S$ is the initial state.*
- *$\mathcal{C}$ is a finite set of clocks.*
- *$\mathcal{I} : S \to \Phi(\mathcal{C})$ is a function assigning each state with an invariant.*
- *$\mathcal{M} : S \hookrightarrow \mathcal{P}$ is a partial function assigning states with task types.*
- *$\Delta \subseteq S \times \Phi(\mathcal{C}) \times \mathcal{A}ct \times 2^{\mathcal{C}} \times S$.*

*When $(s_1, \phi, a, \lambda, s_2) \in \Delta$, we write $s_1 \xrightarrow{\phi, a, \lambda} s_2$.*

### 3.3   Operational Semantics

**Definition 7 (Semantics of TRTA).** *Given a TRTA $\mathcal{R} = (S, s_0, \mathcal{C}, I, M, \Delta)$, a configuration is a tuple $(s, \mu, q, \nu, \mathbb{Q})$, where,*

- *$s \in S$ is a state.*
- *$\mu$ is a clock valuation of $\mathcal{C}$.*
- *$\mathbb{Q}$ is the current task queue.*
- *$q$ is a control location of the TA in the head of $\mathbb{Q}$. If $\mathbb{Q}$ is empty, $q$ is denoted by __.*
- *$\nu$ is a clock valuation on the clocks of the TA in the head of $\mathbb{Q}$. If $\mathbb{Q}$ is empty, $\nu$ is also denoted by __.*

*Given a scheduling strategy $\text{Sch}$, the semantics of $\mathcal{R}$ is defined by the LTS with an initial state $(s_0, \mu_0, __, __, [])$, and transitions as the following rules,*

- Progress transitions:
  - $(s, \mu, __, __, []) \xrightarrow{t}_{Sch} (s, \mu + t, __, __, [])$, where $t \in \mathbb{R}^{\geq 0}$ and $(\mu + t) \models \mathcal{I}(s)$.
  - $(s, \mu, q, \nu, \mathbb{Q}) \xrightarrow{t}_{Sch} (s, \mu + t, q, \nu + t, \text{Exec}(\mathbb{Q}, t))$, where $t \in \mathbb{R}^{\geq 0}$, $(\mu + t) \models \mathcal{I}(s)$, and $(\nu + t) \models I(q)$.
- Discrete transitions:
  - $(s, \mu, q, \nu, \mathbb{Q}) \xrightarrow{a}_{Sch} (s', \mu[\lambda], q, \nu, \text{Sch}(\mathcal{M}(s'), \mathbb{Q}))$, if $s \xrightarrow{\phi, a, \lambda} s'$, $\mu \models \mathcal{I}(s)$, and $\mu[\lambda] \models \mathcal{I}(s')$.

## 4    Schedulability Analysis

In this section, we study the verification problems of TRTAs presented in the previous section. One of the most interesting properties is schedulability.

**Definition 8 (Schedulability).** *A timed regular task automaton $\mathcal{R}$ with the initial state $(s_0, \mu_0, \text{---}, \text{---}, [\,])$ is non-schedulable under a scheduling policy* Sch, *iff* $(s_0, \mu_0, \text{---}, \text{---}, [\,]) \rightarrow^*_{Sch} (s, \mu, q, \nu, \text{Error})$ *for some $s$, $\mu$, $q$, and $\nu$, where $(s, \mu, q, \nu, \text{Error})$ is a failure state $(s, \mu, q, \nu, \mathbb{Q})$ with $\mathbb{Q} = [P_1(\mathcal{A}_1, q_1, \nu_1, s_1, d_1), P_2(\mathcal{A}_2, q_2, \nu_2, s_2, d_2), \ldots, P_n(\mathcal{A}_n, q_n, \nu_n, s_n, d_n)]$ and there exists $i$ such that $d_i < 0$. Otherwise, we say that $\mathcal{R}$ is* schedulable *with* Sch.

In general, The queue of the model is unbounded, which leads to Turing-completeness [19]. However, there is an important observation that a schedulable queue is bounded [2]. Firstly, a task instance that has been started cannot be preempted by another instance of the same task type, which means that others must wait for executing unless the first one reaches its final states and is removed from the task queue. Thus the number of instances of each task type $P_i(\mathcal{A}_i, q, \nu, S_i, D_i) \in \mathcal{P}$ in a schedulable queue is bounded by $\lceil \frac{D_i}{WECT(\mathcal{A}_i)} \rceil$, and the size of a schedulable queue is bounded by $\sum_{P_i(\mathcal{A}_i, q, \nu, S_i, D_i) \in \mathcal{P}} \lceil \frac{D_i}{WECT(\mathcal{A}_i)} \rceil$. Hence, if we abstract each task instance as an atomic task and assume each WCET is pre-known, a TRTA thus becomes a task automaton, and as the proof given in [2], we immediately have the following results.

**Theorem 1.** *The schedulability analysis of TRTAs is decidable.*

In order for the later analysis, we further discuss the non-schedulable queue.

**Definition 9.** *A queue* $\mathbb{Q} = [P_1(\mathcal{A}_1, q_1, \nu_1, s_1, d_1), P_2(\mathcal{A}_2, q_2, \nu_2, s_2, d_2), \ldots, P_n(\mathcal{A}_n, q_n, \nu_n, s_n, d_n)]$ *with* Sch *is* non-schedulable *if $(\sum_{i \leq k} WECT(\mathcal{A}_i)) > d_k$ for some $k \leq n$.*

According to Definitions 8 and 9, we conclude that a task queue is *non-schedulable* if it meets one of the following situations:

- **Deadline-Missing:** the task queue contains a task instance whose deadline $d_i < 0$ for some $i$ which means that some instances already miss deadline.
- **Overflow:** The task queue contains more than $\lceil \frac{D_i}{WECT(\mathcal{A}_i)} \rceil$ instances of $P_i$ for some $i$.
- **Error-Queue:** The task queue will inevitably be an *error-queue* as defined in Definition 8.

If a task queue belongs to the first and second situation, we can conclude that it is *non-schedulable* immediately. For the last situation, all reachable states of the system $\mathcal{R}$ should be checked. Unlike that in task automata, all three situations can be checked through one reachability checking of a TA, we have to check them separately through our methods in the following section.

# 5    Under-Approximation of WCET

It is well-known that it is difficult to get BCET and WCET of a complex task. We firstly run each task of a TRTA, and calculate the time a task used to reach one of its final state, denoted as $et_i$, as an under-approximation of the WCET. By the under-approximation, the system may still be unschedulable, we further translate the TRTA into a NeTA, and the schedulability is checked by the reachability of the NeTA.

In detail, given a TRTA $\mathcal{R}$ and a preemptive scheduling strategy Sch, the goal is to check if the $\mathcal{R}$ is schedulable with the Sch. We construct a product automaton $E(\mathcal{R})\|E(\text{Sch})$, and check pre-defined error-states in the product automaton. $E(\mathcal{R})$ will be constructed the same as a TA described in [4], and $E(\text{Sch})$ will be constructed as a NeTA. The product automaton is essentially an NeTA [5].

## 5.1    A Testing as an Under-Approximation

We list three situations to conclude that a task queue is *non-schedulable*, in which the second one "The task queue contains more than $\lceil \frac{D_i}{WECT(\mathcal{A}_i)} \rceil$ instances of $P_i$ for some $i$" depends on the $WECT(\mathcal{A}_i)$. According to the obvious observation that $et_i \leq WCET(\mathcal{A}_i)$, we change the second situation to "The task queue contains more than $\lceil \frac{D_i}{at_i} \rceil$ instances of $P_i$ for some $i$, where $at_i = \frac{\sum_{i \leq num} et_i}{num}$, where $num$ is the sum of instances of task type $P_i$". Note that the under-approximation of the WCET does not guarantee soundness with respect to schedulability of the system, however, it allows us to erase most unschedulable cases and make the restricted system decidable for further analysis.

## 5.2    Deadline-Missing as a Guarded Automaton

A non-schedulable queue caused by the deadline-missing may not be checking by predefined error states in the model of a scheduling policy, thus we have to translate the TA of each task to its respective *guarded automaton*, in which an error state is predefined as to checking the deadline missing of the task.

Given a task type $P(\mathcal{A}, q_{\mathcal{A}}, \nu_{\mathcal{A}}, S, D) \in \mathcal{P}$, Guarded : $\mathcal{P} \to \mathscr{A}$ is defined by Guarded$(P(\mathcal{A}, q_{\mathcal{A}}, \nu_{\mathcal{A}}, S, D)) = (Q^G, q_0^G, F^G, X^G, \delta^G)$, where

- $Q^G = Q(\mathcal{A}) \cup Q_\Delta \cup q_{err}$, where $Q_\Delta = \{q_\delta \mid \text{for each } \delta \in \Delta(\mathcal{A})\}$.
- $q_0^G = q_0(\mathcal{A})$, and $F^G = F(\mathcal{A})$.
- $X^G = X(\mathcal{A}) \cup \{x_{sch}\}$.
- $\Delta^G = \Delta_{sch} \cup \Delta_{err}$, where
  - $\Delta_{sch} = \{q \xrightarrow{a} q_\delta, q_\delta \xrightarrow{x_{sch} \in [0,D]?} q' \mid \delta = (q, 0, q') \in \Delta\}$
  - $\Delta_{err} = \{q_\delta \xrightarrow{x_{sch} \in [D,\infty]?} q_{err} \mid q \in Q(\mathcal{A}) \cup Q_\Delta\}$.

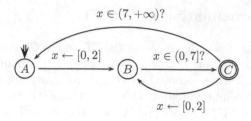

**Fig. 2.** An example of TA

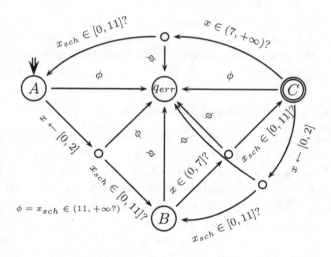

**Fig. 3.** The TA with error location transformed from Fig. 2

*Example 2.* Figure 2 is a TA used to described the behavior of a task $P$, and the relative deadline of $P$ is 11. According to the function Guarded : $P \rightarrow \mathscr{A}$, we transform it to a TA with error locations shown in Fig. 3, once the clock $x_{sch} \in (11, \infty)$ which misses the relative deadline, the system reaches the error location.

## 5.3   Encoding a Scheduler as a Nested Timed Automaton

For each task type $P_i(\mathcal{A}, q_{\mathcal{A}}, \nu_{\mathcal{A}}, S, D)$, we use $P_{ij}$ to denote the $j$th instance of the task type $P_i$. For each $P_{ij}$, we need one deadline clock $x(i, j)$ used to remember the deadline and reset to 0 when $P_{ij}$ is released.

Given a finite set of tasks $\mathcal{P}$, and a priority policy is described by a relation $\prec$ on $\mathcal{P}$. For Example, EDF(earliest deadline first), where $\prec$ can be coded as constraints over the deadline clock. If we say $P_{ij} \prec P_{mn}$ that $P_{mn}$ has the shorter deadline than $P_{ij}$, then it can coded as a constraint that $D(m) - x(m, n) \leq D(i) - x(i, j)$. For EDF scheduling strategy on tasks $\mathcal{P}$, Sch($\mathcal{P}$) is defined by a nested timed automaton $(T, \mathcal{A}_0, X, C, \Delta)$ over a set of input symbols $\Sigma$ where,

- $\Sigma = \{released_P \mid \text{for each } P \in \mathcal{P}\}$.
- $T = \{\texttt{Guarded}(P) \mid \text{for each } P \in \mathcal{P}\} \cup \{\mathcal{A}_{idle}\}$, where $\mathcal{A}_{idle}$ is a singleton timed automaton without any transitions.
- $X = \cup_{\mathcal{A}_i \in T} X(\mathcal{A}_i)$.
- $C = \{c\}$.
- $\mathcal{A}_0 = \mathcal{A}_{idle}$.
- $\Delta$ is defined by $\Delta_{idle} \cup \Delta_{push} \cup \Delta_{pop}$:
  - $\Delta_{idle} = \{q_{idle} \xrightarrow{c \leftarrow [0,0], released_P, push} q_0(\mathcal{A}) \mid \forall \mathcal{A} \in T, q_{idle} \in Q(\mathcal{A}_{idle})\}$.
  - $\Delta_{push} = \{q \xrightarrow{c \in [0, D_p - D_{p'})?, released_{P'}, push, c \leftarrow [0,0]} q_0(\mathcal{A}') \mid \forall \mathcal{A}, \mathcal{A}' \in T, q \in Q(\mathcal{A})\}$.
  - $\Delta_{pop} = \{q \xrightarrow{c \leftarrow x_{sch}, pop} q' \mid \forall \mathcal{A}, \mathcal{A}' \in T, q \in Q(\mathcal{A}), q' \in Q(\mathcal{A}')$, the stack is not empty, $\mathcal{A}'$ is in the stack next to $\mathcal{A}$, $x_{sch} \in X(\mathcal{A}')$ which is the deadline clock$\}$.
  - $\Delta_{pop} = \{q \xrightarrow{c \leftarrow [0,0], pop} \varepsilon \mid \forall \mathcal{A} \in T, q \in Q(\mathcal{A})$, the stack is empty$\}$.

*Example 3.* Figure 4 is a NeTA with three TAs, showing that the EDF scheduling strategy on two tasks $P$ and $Q$. $\mathcal{A}_{idle}$ is an empty TA for the idle state. $\mathcal{A}_p$ and $\mathcal{A}_q$ model the behavior of task P and task Q, respectively. Note that the $\Delta_{pop}$ transition rules are not explicitly represented in the figure, only when the running TA reaches its final states, can $\Delta_{pop}$ be applied that the running TA will be popped. The NeTA starts from the timed automaton $\mathcal{A}_{idle}$ may move to $\mathcal{A}_p$ or $\mathcal{A}_q$ by action $released_P$ or $released_Q$, then the corresponding task will be pushed into the stack. A task instance may be preempted if at some point the system meets the conditions labelled in the push transition.

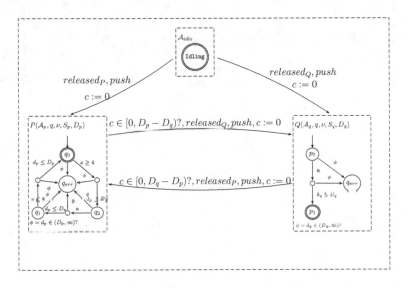

**Fig. 4.** Encoding the scheduling strategy into a NeTA

## 5.4    Construct the Product Automaton

The last step is to construct the product automaton $E(\mathcal{R})\|E(\text{Sch})$ in which the $E(\mathcal{R})$ is a timed automaton and the $E(\text{Sch})$ is a nested timed automaton. $E(\mathcal{R})$ and $E(\text{Sch})$ can only synchronize on identical action symbols.

An example is shown below.

*Example 4.* Figure 5 is a TA $\mathcal{A}_{environment}$ which removes tasks assigned in a TRTA shown in Fig. 1, it is used as an environment. Figure 4 is a nested timed automaton $\mathcal{N}_{\text{sch}}$ which encodes an EDF scheduler. $\mathcal{N}_{\text{sch}}$ and $\mathcal{A}_{environment}$ start from $A_{idle}$ and $l_0$ respectively, and they are only synchronized on the same actions $released_P$ and $released_Q$.

Now, we show that the product automaton is bounded.

**Lemma 1.** *Let $\mathcal{R}$ be a timed regular task automaton and Sch a scheduling strategy, they are encoded into a TA $\mathcal{A}$ and a NeTA $\mathcal{N}$. Assumed that $(s_0, \mu_0, q_0, \nu_0, \mathbf{Q}_0)$ and $(s_0, \mu_0, (\langle q_0, \nu_0, \mu_{0N}\rangle, c_0))$ are the initial states of $\mathcal{R}$ and the product automaton $\mathcal{A}\|\mathcal{N}$ respectively, where $s_0$ is the initial state of $\mathcal{R}$, $\mu_0$ and $\mu_{0N}$ are clock assignments assigning all clocks with 0 for clocks $X(\mathcal{R})$ and $C(\mathcal{N})$ respectively, $q_0$ and $\nu_0$ is the initial control location and clock assignment of TA in the head of $\mathbf{Q}_0$. As $\mathbf{Q}_0$ is the empty task queue, $q_0$ and $\nu_0$ are not exist, $c_0$ is the initial stack for NeTA which is empty. Then for all $s$, $\mu$, $q$, $\nu$, $\mu_N$, $c$ and the predefined error state $q_{err}$: $(s_0, \mu_0, q_0, \nu_0, \mathbf{Q}_0) \rightarrow^* (s, \mu, q, \nu, \text{Error})$ iff $(s_0, \mu_0, (\langle q_0, \nu_0, \mu_{0N}\rangle, c_0)) \rightarrow^* (s, \mu, (\langle q_{err}, \nu, \mu_N\rangle, c)).$*

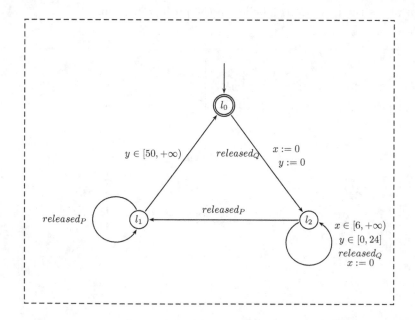

**Fig. 5.** Encoding the TRTA into a TA as an environment

*Proof.* It is by induction on the length of transition sequence.

The above lemma states that that the schedulability analysis problem can be solved by reachability analysis for timed regular task automata. Due to Fact 2, the reachability problem is decidable, our result stated in Theorem 1 is proved.

# 6  Conclusion

This paper investigates the schedulability analysis on complex real-time task systems, in which each task is described as a timed automaton. Without any information of worst case execution time of each task, a test is then performed to give an under-approximation. If the task queue is not bound with the approximation, it is certainly unschedulable. If bounded, a schedulability checking is further performed by encoding to the reachability of nested timed automata. The whole method gains soundness and completeness.

We will consider schedulability analysis on soft real-time systems. That is, the relative deadline may change due to environment and given conditions. Following our methodology, it is required that each TA in a NeTA contains an updatable clock to record the relative deadline. The model is thus named nested updatable timed automata with one updatable clock (NeUTA1). The properties, such as reachability, as well as boundedness and termination, are under our consideration.

**Acknowledgements.** This work is supported by National Natural Science Foundation of China with grant No. 61472240, 61672340, 61472238, and the NSFC-JSPS bilateral joint research project with grant No. 61511140100.

# References

1. Buttazzo, G.C.: Hard Real-Time Computing Systems: Predictable Scheduling Algorithms and Applications. Springer, New York (2004)
2. Fersman, E., Krcal, P., Pettersson, P., Wang, Y.: Task automata: schedulability, decidability and undecidability. Inform. Comput. **205**(8), 1149–1172 (2007)
3. Wilhelm, R., Engblom, J., Ermedahl, A., Holsti, N., Thesing, S., Whalley, D.B., Bernat, G., Ferdinand, C., Heckmann, R., Mitra, T., Mueller, F., Puaut, I., Puschner, P.P., Staschulat, J., Stenström, P.: The worst-case execution-time problem - overview of methods and survey of tools. ACM Trans. Embed. Comput. Syst. **7**(3), 1–53 (2008)
4. Ericsson, C., Wall, A., Wang, Y.: Timed automata as task models for event-driven systems. In: Proceedings of the 6th International Conference on Real-Time Computing Systems and Applications (RTCSA 1999), pp. 182–189. IEEE Computer Society (1999)
5. Li, G., Cai, X., Ogawa, M., Yuen, S.: Nested timed automata. In: Braberman, V., Fribourg, L. (eds.) FORMATS 2013. LNCS, vol. 8053, pp. 168–182. Springer, Heidelberg (2013). doi:10.1007/978-3-642-40229-6_12

6. Li, G., Ogawa, M., Yuen, S.: Nested timed automata with frozen clocks. In: Sankaranarayanan, S., Vicario, E. (eds.) FORMATS 2015. LNCS, vol. 9268, pp. 189–205. Springer, Heidelberg (2015). doi:10.1007/978-3-319-22975-1_13

7. Wang, Y., Li, G., Yuen, S.: Nested timed automata with various clocks. Sci. Found. China **24**(2), 51–68 (2016)

8. Choffrut, C., Goldwurm, M.: Timed automata with periodic clock constraints. J. Autom. Lang. Comb. **5**(4), 371–404 (2000)

9. Altisen, K., Gössler, G., Pnueli, A., Sifakis, J., Tripakis, S., Yovine, S.: A framework for scheduler synthesis. In: Proceedings of the 20th IEEE Real-Time Systems Symposium (RTSS 1999), pp. 154–163. IEEE Computer Society (1999)

10. Abdeddam, Y., Maler, O.: Job-Shop Scheduling Using Timed Automata? Springer, Berlin (2001)

11. Fehnker, A.: Scheduling a steel plant with timed automata. In: Proceedings of the Sixth International Conference on Real-Time Computing Systems and Applications, pp. 280–286 (1999)

12. Fersman, E., Pettersson, P., Yi, W.: Timed automata with asynchronous processes: schedulability and decidability. In: Katoen, J.-P., Stevens, P. (eds.) TACAS 2002. LNCS, vol. 2280, pp. 67–82. Springer, Heidelberg (2002). doi:10.1007/3-540-46002-0_6

13. Trivedi, A., Wojtczak, D.: Recursive timed automata. In: Bouajjani, A., Chin, W.-N. (eds.) ATVA 2010. LNCS, vol. 6252, pp. 306–324. Springer, Heidelberg (2010). doi:10.1007/978-3-642-15643-4_23

14. Benerecetti, M., Minopoli, S., Peron, A.: Analysis of timed recursive state machines. In: Proceedings of the 17th International Symposium on Temporal Representation and Reasoning (TIME 2010), pp. 61–68. IEEE Computer Society (2010)

15. Benerecetti, M., Peron, A.: Timed recursive state machines: expressiveness and complexity. Theor. Comput. Sci. **625**, 85–124 (2016)

16. Alur, R., Dill, D.L.: A theory of timed automata. Theor. Comput. Sci. **126**(2), 183–235 (1994)

17. Henzinger, T.A., Nicollin, X., Sifakis, J., Yovine, S.: Symbolic model checking for real-time systems. Inform. Comput. **111**(2), 193–244 (1994)

18. Abdulla, P.A., Atig, M.F., Stenman, J.: Dense-timed pushdown automata. In: Proceedings of the 27th Annual IEEE Symposium on Logic in Computer Science (LICS 2012), pp. 35–44. IEEE Computer Society (2012)

19. Kozen, D.C.: Automata and Computability. Springer, New York (1951)

# Importance Sampling for Stochastic Timed Automata

Cyrille Jegourel[1], Kim G. Larsen[2], Axel Legay[2,3], Marius Mikučionis[2(✉)],
Danny Bøgsted Poulsen[2], and Sean Sedwards[3]

[1] National University of Singapore, Singapore, Singapore
[2] Department of Computer Science, Aalborg University, Aalborg, Denmark
marius@cs.aau.dk
[3] Inria Rennes – Bretagne Atlantique, Rennes, France
sean.sedwards@inria.fr

**Abstract.** We present an importance sampling framework that combines symbolic analysis and simulation to estimate the probability of rare reachability properties in stochastic timed automata. By means of symbolic exploration, our framework first identifies states that cannot reach the goal. A state-wise change of measure is then applied on-the-fly during simulations, ensuring that dead ends are never reached. The change of measure is guaranteed by construction to reduce the variance of the estimator with respect to crude Monte Carlo, while experimental results demonstrate that we can achieve substantial computational gains.

## 1 Introduction

Stochastic Timed Automata [7] extend Timed Automata [1] to reason on the stochastic performance of real time systems. Non-deterministic time delays are refined by stochastic choices and discrete non-deterministic choices are refined by probabilistic choices. The semantics of stochastic timed automata is given in terms of nested integrals over products of uniform and exponential distributions. Abstracting from the stochasticity of the model, it is possible to find the symbolic paths reaching a set of goal states, but solving the integrals to calculate the probability of a property becomes rapidly intractable. Using a similar abstraction it is possible to bound the maximum and minimum probabilities of a property, but this can lead to results such as *the system could work or fail with high probability*. Our goal is to quantify the *expectation* of rare behaviour with specific distributions.

A series of works [3–5,7] has developed methods for analysing Stochastic Timed Automata using Statistical Model Checking (SMC) [18]. SMC includes a collection of Monte Carlo techniques that use simulation to avoid "state space explosion" and other intractabilities encountered by model checking. It is typically easy to generate sample executions of a system, while the confidence of estimates increases with the number of independently generated samples. Properties with low probability (rare properties) nevertheless pose a challenge for

© Springer International Publishing AG 2016
M. Fränzle et al. (Eds.): SETTA 2016, LNCS 9984, pp. 163–178, 2016.
DOI: 10.1007/978-3-319-47677-3_11

SMC because the relative error of estimates scales inversely with rarity. A number of standard variance reduction techniques to address this have been known since the early days of simulation [11]. The approach we present here makes use of *importance sampling* [11,15], which works by performing Monte Carlo simulations under a probabilistic *measure* that makes the rare event more likely to occur. An unbiased estimate is achieved by compensating for the *change of measure* during simulation.

Our model may include rarity arising from explicit Markovian transitions, but our main contribution is addressing the more challenging rarity that results from the intersection of timing constraints and continuous distributions of time. To gain an intuition of the problem, consider the example in Fig. 1. The automaton first chooses a delay uniformly at random in $[0, 10^6]$ and then selects to either go to A or B. Since the edge to A is only enabled in the interval $[10^6 - 1, 10^6]$, reaching A constitutes a rare event with probability $\int_{10^6-1}^{10^6} 10^{-6} \cdot \frac{1}{2} \, dt = \frac{1}{2} \cdot 10^{-6}$.

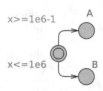

**Fig. 1.** A rare event of reaching A due to timing constraints.

The probability theory relating to our model has been considered in the framework of generalised semi Markov processes, with related work done in the context of queueing networks. Theory can only provide tractable analytical solutions for special cases, however. Of particular relevance to our model, [17] proposes the use of state classes to model stochastic distributions over dense time, but calculations for the closely related *Duration Probabilistic Automata* [14] do not scale well [12]. Monte Carlo approaches provide an approximative alternative to analysis, but incur the problem of rare events. Researchers have thus turned to importance sampling. In [19] the authors consider rare event verification of a model of stochastic hybrid automata that shares a number of features in common with our own model. They suggest using the *cross-entropy method* [16] to refine a parametrised change of measure for importance sampling, but do not provide a means by which this can be applied to arbitrary hybrid systems.

Our contribution is an automated importance sampling framework that is integrated into UPPAAL SMC and applicable to arbitrary time-divergent priced timed automata [7]. By means of symbolic analysis we first construct an exhaustive *zone*-based reachability graph of the model and property, thus identifying all "dead end" states that cannot reach a satisfying state. Using this graph we generate simulation traces that always avoid dead ends and satisfy the property, applying importance sampling to compensate estimates for the loss of the dead ends. In each concrete state we integrate over the feasible times of enabled actions to calculate their total probabilities, which we then use to choose an action at random. We then choose a new concrete vector of clock values at random from the feasible times of the chosen action, using the appropriately composed distribution. All simulated traces reach satisfying states, while our change of measure is guaranteed by construction to reduce the variance of estimates with respect to crude Monte Carlo. Our experimental results demonstrate substantial reductions of variance and overall computational effort.

The remainder of the paper is as follows. Sections 2 and 3 provide background: Sect. 2 recalls the basic notions of importance sampling and Sect. 3 describes Stochastic Timed Automata in terms of Stochastic Timed Transition Systems. We explain the basis of our importance sampling technique in Sect. 4 and describe how we realise it for Stochastic Timed Automata in Sect. 5. In Sect. 6 we present experimental results using our prototype implementation in UPPAAL SMC and then briefly summarise our achievements and future work in Sect. 7.

## 2   Variance Reduction

Let $F$ be a probability measure over the measurable set of all possible executions $\omega \in \Omega$. The expected probability $p_\varphi$ of property $\varphi$ is defined by

$$p_\varphi = \int_\Omega \mathbf{1}_\varphi \, \mathrm{d}F, \tag{1}$$

where the indicator function $\mathbf{1}_\varphi : \Omega \to \{0,1\}$ returns 1 iff $\omega$ satisfies $\varphi$. This leads to the standard ("crude") unbiased Monte Carlo estimator used by SMC:

$$p_\varphi \approx \frac{1}{N} \sum_{i=1}^{N} \mathbf{1}_\varphi(\omega_i), \tag{2}$$

where each $\omega_i \in \Omega$ is selected at random and distributed according to $F$, denoted $\omega_i \sim F$. The variance of the random variable sampled in (2) is given by

$$\sigma^2_{crude} = \int_\Omega (\mathbf{1}_\varphi - p_\varphi)^2 \, \mathrm{d}F = \int_\Omega \mathbf{1}_\varphi \, \mathrm{d}F - (p_\varphi)^2 \tag{3}$$

The variance of an $N$-sample average of i.i.d. samples is the variance of a single sample divided by $N$. Hence the variance of the crude Monte Carlo estimator (2) is $\sigma^2_{crude}/N$ and it is possible to obtain more confident estimates of $p_\varphi$ by increasing $N$. However, when $p_\varphi \approx 0$, i.e., $\varphi$ is a rare property, standard concentration inequalities require infeasibly large numbers of samples to bound the *relative* error.

In this work we use importance sampling to reduce the variance of the random variable from which we sample, which then reduces the number of simulations necessary to estimate the probability of rare properties. Referring to the same probability space and property used in (1), importance sampling is based on the integral

$$p_\varphi = \int_\Omega \mathbf{1}_\varphi \frac{\mathrm{d}F}{\mathrm{d}G} \, \mathrm{d}G, \tag{4}$$

where $G$ is another probability measure over $\Omega$ and $\mathrm{d}F/\mathrm{d}G$ is called the *likelihood ratio*, with $\mathbf{1}_\varphi F$ *absolutely continuous* with respect to $G$. Informally, this means that $\forall \omega \in \Omega, \mathrm{d}G(\omega) = 0 \implies \mathbf{1}_\varphi \mathrm{d}F(\omega) = 0$. Hence $\mathbf{1}_\varphi(\omega)\mathrm{d}F(\omega)/\mathrm{d}G(\omega) > 0$ for all realisable paths under $F$ that satisfy $\varphi$ and is equal to 0 otherwise.

The integral (4) leads to the unbiased importance sampling estimator

$$p_\varphi \approx \frac{1}{N} \sum_{i=1}^{N} \mathbf{1}_\varphi(\omega_i) \frac{\mathrm{d}F(\omega_i)}{\mathrm{d}G(\omega_i)}, \quad \omega_i \sim G. \tag{5}$$

In practice, a simulation is performed under measure $G$ and if the resulting trace satisfies $\varphi$, its contribution is compensated by the likelihood ratio, which is calculated on the fly. To reduce variance, the intuition is that $G$ is constructed to make traces that satisfy $\varphi$ more likely to occur in simulations.

The variance $\sigma_{is}^2$ of the random variable sampled by the importance sampling estimator (4) is given by

$$\sigma_{is}^2 = \int_\Omega \left( \mathbf{1}_\varphi \frac{\mathrm{d}F}{\mathrm{d}G} - p_\varphi \right)^2 \mathrm{d}G = \int_\Omega \mathbf{1}_\varphi \left( \frac{\mathrm{d}F}{\mathrm{d}G} \right)^2 \mathrm{d}G - (p_\varphi)^2 \tag{6}$$

If $F = G$, the likelihood ratio of realisable paths is uniformly equal to 1, (4) reduces to (1) and (6) reduces to (3). To ensure that the variance of (5) is less than the variance of (2) it is necessary to make $\sigma_{is}^2 < \sigma_{crude}^2$, for which it is sufficient to make $\mathrm{d}F/\mathrm{d}G < 1, \forall \omega \in \Omega$.

**Lemma 1.** *Let $F, G$ be probability measures over the measurable space $\Omega$, let $\mathbf{1}_\varphi : \Omega \to \{0, 1\}$ be an indicator function and let $\mathbf{1}_\varphi F$ be absolutely continuous with respect to $G$. If for all $\omega \in \Omega$, $\mathbf{1}_\varphi(\omega) \cdot \frac{\mathrm{d}F(\omega)}{\mathrm{d}G(\omega)} \leq 1$ then $\sigma_{is}^2 \leq \sigma_{crude}^2$.*

*Proof.* From the definitions of $\sigma_{crude}^2$ (3) and $\sigma_{is}^2$ (6), we have

$$\sigma_{is}^2 \leq \sigma_{crude}^2 \iff \int_\Omega \mathbf{1}_\varphi \left( \frac{\mathrm{d}F}{\mathrm{d}G} \right)^2 \mathrm{d}G - (p_\varphi)^2 \leq \int_\Omega \mathbf{1}_\varphi \mathrm{d}F - (p_\varphi)^2,$$

where $p_\varphi$ is the expectation of $\mathbf{1}_\varphi F$. Noting that $(p_\varphi)^2$ is outside the integrals and common to both sides of the inequality, we conclude

$$\sigma_{is}^2 \leq \sigma_{crude}^2 \iff \int_\Omega \mathbf{1}_\varphi \frac{\mathrm{d}F}{\mathrm{d}G} \mathrm{d}F \leq \int_\Omega \mathbf{1}_\varphi \mathrm{d}F.$$

Hence, given $\mathbf{1}_\varphi \in \{0, 1\}$, to ensure $\sigma_{is}^2 \leq \sigma_{crude}^2$ it is sufficient that $\mathbf{1}_\varphi(\omega) \cdot \frac{\mathrm{d}F(\omega)}{\mathrm{d}G(\omega)} \leq 1, \forall \omega \in \Omega$.

## 3   Timed Systems

The modelling formalism we consider in this paper is a stochastic extension of Timed Automata [1] in which non-deterministic time delays are refined by stochastic choices and non-deterministic discrete choices are refined by probabilistic choices. Let $\Sigma = \Sigma_! \cup \Sigma_?$ be a set of actions split into output ($\Sigma_!$) and input ($\Sigma_?$). As usual we assume there is a one-to-one-mapping between input actions and output actions. We adopt the scheme that $a!$ is an output action and $a?$ is the corresponding input action.

**Definition 1 (Timed Transition System).** *A timed transition system over actions $\Sigma$ split into input actions $\Sigma_?$ and output actions $\Sigma_!$ is a tuple $\mathcal{L} = (S, s^0, \rightarrow, AP, P)$ where 1) $S$ is a set of states, 2) $s^0$ is the initial state, 3) $\rightarrow \subseteq S \times (\Sigma \cup \mathbb{R}_{\geq 0}) \times S$ is the transition relation, 4) $AP$ is a set of propositions and 5) $P : S \rightarrow 2^{\overline{AP}}$ maps states to propositions.* □

For shorthand we write $s \xrightarrow{a} s'$ whenever $(s, a, s') \in \rightarrow$. Following the compositional framework laid out by David et al. [6] we expect timed transition systems to be action-deterministic i.e. if $s \xrightarrow{a} s'$ and $s \xrightarrow{a} s''$ then $s' = s''$ and we expect them to be *input-enabled* meaning for all input actions $a? \in \Sigma_?$ and all states $s$ there exists $s'$ such that $s \xrightarrow{a?} s'$. Let $s, s' \in S$ be two states then we write $s \rightarrow^* s'$ if there exists a sequence of transitions such that $s'$ is reachable and we write $s \not\rightarrow^* s'$ if $s'$ is not reachable from $s$. Generalising this to a set of states $\mathsf{G} \subseteq S$, we write $s \rightarrow^* \mathsf{G}$ if there exists $s' \in \mathsf{G}$ such that $s \rightarrow^* s'$ and $s \not\rightarrow^* \mathsf{G}$ if for all $s' \in \mathsf{G}$, $s \not\rightarrow^* s'$.

A run over a timed transition system $\mathcal{L} = (S, s^0, \rightarrow, AP, P)$ is an alternating sequence of states, reals and output actions, $s_0 d_0 a_0! s_1 d_1 a_1! \ldots$ such that $s_i \xrightarrow{d_i} \xrightarrow{a_i!} s_{i+1}$. We denote by $\Omega(\mathcal{L})$ the entire set of runs over $\mathcal{L}$. The set of propositional runs is the set $\Omega^{AP}(\mathcal{L}) = \{P(s_0) d_0, \ldots | s_0 d_0 a_0! \in \Omega(\mathcal{L})\}$.

Several Timed Transition Systems $\mathcal{L}_1 \ldots \mathcal{L}_n$, $\mathcal{L}_i = (S_i, s_i^0, \rightarrow_i, AP_i, P_i)$, may be composed in the usual manner. We denote this by $\mathcal{L} = \mathcal{L}_1 | \mathcal{L}_2 | \ldots | \mathcal{L}_n$ and for a state $\boldsymbol{s} = (s_1, s_2, \ldots, s_n)$ of $\mathcal{L}$ we let $\boldsymbol{s}[i] = s_i$.

*Timed Automata.* Let $X$ be a finite set of variables called *clocks*. A valuation over a set of clocks is a function $v : X \rightarrow \mathbb{R}_{\geq 0}$ assigning a value to each clock. We denote by $V(X)$ all valuations over $X$. Let $v \in V(X)$ and $Y \subseteq X$ then we denote by $v[Y]$ the valuation assigning 0 whenever $x \in Y$ and $v(x)$ whenever $x \notin Y$. For a value $d \in \mathbb{R}_{\geq 0}$ we let $(v + d)$ be the valuation assigning $v(x) + d$ for all $x \in X$. An *upper bound* (*lower bound*) over a set of clocks is an element $x \triangleleft n$ $(x \triangleright n)$ where $x \in X$, $n \in \mathbb{N}$ and $\triangleleft \in \{<, \leq\}$ $(\triangleright \in \{>, \geq\})$. We denote the set of finite conjunctions of upper bounds (lower bounds) over $X$ by $\mathcal{B}^\triangleleft(X)$ $(\mathcal{B}^\triangleright(X))$ and the set of finite conjunctions over upper and lower bounds by $\mathcal{B}(X)$. We write $v \models g$ whenever $v \in V(X)$ satisfies an element $g \in \mathcal{B}(X)$. We let $v_0 \in V(X)$ be the valuation that assigns zero to all clocks.

**Definition 2.** *A Timed Automaton over output actions $\Sigma_!$ and input actions $\Sigma_?$ is a tuple $(L, \ell_0, X, E, Inv)$ where 1) $L$ is a set of control locations, 2) $\ell_0$ is the initial location, 3) $X$ is a finite set of clocks, 4) $E \subseteq L \times \mathcal{B}^\triangleright(X) \times (\Sigma_! \cup \Sigma_?) \times 2^X \times L$ is a finite set of edges 5) $Inv : L \rightarrow \mathcal{B}^\triangleleft(X)$ assigns an invariant to locations.* □

The semantics of a timed automaton $\mathcal{A} = (L, \ell_0, X, E, Inv)$ is a timed transition system $\mathcal{L} = (S, s^0, \rightarrow, L, P)$ where 1) $S = L \times V(X)$, 2) $s^0 = (\ell_0, v_0)$, 3) $(\ell, v) \xrightarrow{d} (\ell, (v + d))$ if $(v + d) \models Inv(\ell)$, 4) $(\ell, v) \xrightarrow{a} (\ell', v')$ if there exists $(\ell, g, a, r, \ell') \in E$ such that $v \models g$, $v' = v[r]$ and $v' \models Inv(\ell')$ and 5) $P((\ell, v)) = \{\ell\}$.

## 3.1   Stochastic Timed Transition System

A stochastic timed transition system (STTS) is a pair $(\mathcal{L}, \nu)$ where $\mathcal{L}$ is a timed transition system defining allowed behaviour and $\nu$ gives for each state a density-function, that assigns densities to possible successors. Hereby some behaviours may, in theory, be possible for $\mathcal{L}$ but rendered improbable by $\nu$.

**Definition 3 (Stochastic Timed Transition System).** *Let $\mathcal{L} = (S, s^0, \rightarrow, AP, P)$ be a timed transition system with output actions $\Sigma_!$ and input actions $\Sigma_?$. A stochastic timed transition system over $\mathcal{L}$ is a tuple $(\mathcal{L}, \nu)$ where $\nu : S \rightarrow \mathbb{R}_{\geq 0} \times \Sigma_! \rightarrow \mathbb{R}_{\geq 0}$ assigns a joint-delay-action density where for all states $s \in S$, (1) $\sum_{a! \in \Sigma_!}(\int_{\mathbb{R}_{\geq 0}} \nu(s)(t, a!)\, dt) = 1$ and (2) $\nu(s)(t, a!) \neq 0$ implies $s \xrightarrow{t} \xrightarrow{a!}$.*   □

(1) captures that $\nu$ is a probability density and (2) demands that if $\nu$ assigns a non-zero density to a delay-action pair then the underlying timed transition system should be able to perform that pair. Note that (2) is not a bi-implication, reflecting that $\nu$ is allowed to exclude possible successors of $\mathcal{L}$.

Forming the core of a stochastic semantics for a stochastic timed transition system $\mathcal{T} = ((S, s^0, \rightarrow, \mathrm{AP}, P), \nu)$, let $\pi = p_0 I_0 p_1 I_1 p_2 \ldots I_{n-1} p_n$ be a cylinder construction where for all $i$, $I_i$ is an interval with rational end points and $p_i \subseteq$ AP. For a finite run $\omega = p_1' d_1 p_2' \ldots d_{n-1} p_n$ we write $\omega \models \pi$ if for all $i$, $d_i \in I_i$ and $p_i' = p_i$. The set of runs within $\pi$ is then $C(\pi) = \{\omega\omega' \in \Omega^{\mathrm{AP}}(\mathcal{T}) \mid \omega \models \pi\}$. Using the joint density, we define the measure of runs in $C(\pi)$ from $s$ recursively:

$$F_s(\pi) = (p_0 = P(s)) \cdot \int_{t \in I_0} \sum_{a! \in \Sigma_!} \left( \nu(s)(t, a!) \cdot F_{[[s]^d]^{a!}}(\pi^1) \right) dt,$$

where $\pi^1 = p_1 I_1 \ldots p_{n-1} p_n$, base case $F_s(p) = (P_{\mathcal{T}}(s) = (p))$ and $[s]^a$ is the uniquely defined $s'$ such that $s \xrightarrow{a} s'$. With the cylinder construction above and the probability measure $F$, the set of runs reaching a certain proposition $p$ within a time limit $t$, denoted as $\Diamond_{\leq t}\, p$, is measurable.

*Stochastic Timed Automata (STA).* Following [7], we associate to each state, $s$, of timed automaton $\mathcal{A} = (L, \ell_0, X, E, \mathrm{Inv})$ a delay density $\delta$ and a probability mass function $\gamma$ that assigns a probability to output actions. The delay density is either a uniform distribution between the minimal delay $(d_{min})$ before a guard is satisfied and maximal delay $(d_{max})$ while the invariant is still satisfied, or an exponential distribution shifts $d_{min}$ time units in case no invariant exists. The $\gamma$ function is a simple discrete uniform choice of all possible actions. Formally, let $d_{min}(s) = \min\{d \mid s \xrightarrow{d} \xrightarrow{a!}\ \text{ for some } a!\}$ and $d_{max}(s) = \sup\{d \mid s \xrightarrow{d}\}$ then $\delta(s)(t) =$

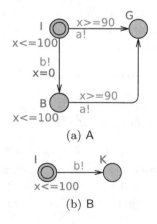

(a) A

(b) B

**Fig. 2.** Two stochastic timed automata.

$\frac{1}{d_{max}(s) - d_{min}(s)}$ if $d_{max}(s) \neq \infty$ and $\delta(s)(t) = \lambda \cdot e^{-\lambda(t - d_{min}(s))}$, for a user spec-ified $\lambda$, if $d_{max}(s) = \infty$. Regarding output actions, let $\mathsf{Act}(s) = \{a! | s \xrightarrow{a!}\}$, then $\gamma(s)(a!) = \frac{1}{|\mathsf{Act}|}$ for $a! \in \mathsf{Act}$. With $\delta$ and $\gamma$ at hand we define the sto-chastic timed transition system for $\mathcal{A}$ with underlying timed transition sys-tem $\mathcal{L}$ as $\mathcal{T}^{\mathcal{A}} = (\mathcal{L}, \delta \bullet \gamma)$, where $\delta \bullet \gamma$ is a composed joint-delay-density function and $(\delta \bullet \gamma)(s)(t, a!) = \delta(s)(t) \cdot \gamma([s]^t)(a!)$. Notice that for any $t$, $\sum_{a! \in \Sigma_!} \nu(s)(t, a!) = \delta(s)(t)$. In the remainder we will often write a stochastic timed transition as $(\mathcal{L}, \delta \bullet \gamma)$. Also we will write $\gamma(s)(t)(a!)$ in lieu of $\gamma([s]^t)(a!)$.

*Example 1.* Consider Fig. 2 and the definition of $\delta_{\mathsf{A}}$ and $\gamma_{\mathsf{A}}$ in the initial state (A.I, x=0). By definition we have $\gamma_{\mathsf{A}}((\mathsf{I}, \mathsf{x}{=}0)(t)(a!) = 1$ for $t \in [90, 100]$ and $\delta_{\mathsf{A}}(\mathsf{I}, \mathsf{x}{=}0)(t) = \frac{1}{100 - 90}$ for $t \in [90, 100]$. Similarly for the B component in the state (B.I, x=0) we have $\gamma_{\mathsf{B}}((\mathsf{I}, \mathsf{x}{=}0)(t)(b!) = 1$ if $t \in [0, 100]$ and zero otherwise and $\delta_{\mathsf{B}}(\mathsf{I}, \mathsf{x}{=}0)(t) = \frac{1}{100 - 0}$ if $t \in [0, 100]$ and zero otherwise.

## 3.2   Composition of Stochastic Timed Transitions Systems

Following [7], the semantics of STTS is race based, in the sense that each com-ponent first chooses a delay, then the component with the smallest delay wins the race and finally selects an output to perform. For the remainder we fix $\mathcal{T}_i = (\mathcal{L}_i, \nu_i)$ where $\mathcal{L}_i = (S_i, s_i^0, \to_i, \mathsf{AP}_i, P_i)$ is over the output actions $\Sigma_!^i$ and the common input actions $\Sigma_?$.

**Definition 4.** *Let $\mathcal{T}_1, \mathcal{T}_2, \ldots, \mathcal{T}_n$ be stochastic timed transition systems with dis-joint output actions. The composition of these is a stochastic timed transition system $\mathcal{J} = (\mathcal{L}_1 | \mathcal{L}_2 | \ldots | \mathcal{L}_n, \nu)$ where*

$$\nu(\boldsymbol{s})(t, a!) = \nu_k(\boldsymbol{s}[k])(t, a!) \cdot \prod_{j \neq k} \left( \int_{\tau > t} \sum_{b! \in \Sigma_!^j} \nu_j(\boldsymbol{s}[j])(\tau, b!) \, d\tau \right) \quad \text{for } a! \in \Sigma_!^k.$$

The race based semantics is apparent from $\nu$ in Definition 4, where the $k^{th}$ com-ponent chooses a delay $t$ and action $a!$ and each of the other components inde-pendently select a $\tau > t$ and an output action $b!$. For a composition of Stochas-tic Timed Automata we abstract from the losing components' output actions and just integrate over $\delta$, as the following shows. Let $\mathcal{J} = (\mathcal{L}, \nu)$ be a com-position of stochastic timed transition systems, $\mathcal{T}_1, \mathcal{T}_2, \ldots, \mathcal{T}_n$, and let for all $i$, $\mathcal{T}_i = (\mathcal{L}_i, \delta_i \bullet \gamma_i)$ originate from a timed automaton. Let $a! \in \Sigma_!^k$, then $\nu(\boldsymbol{s})(t, a!)$ is given by

$$\delta_k(\boldsymbol{s}[k])(t) \cdot \gamma_k(s_k)(t)(a!) \cdot \prod_{j \neq k} \left( \int_{\tau > t} \sum_{b! \in \Sigma_!^j} \delta_j(\boldsymbol{s}[j])(\tau) \gamma_j(\boldsymbol{s}[j])(\tau)(b!) \, d\tau \right)$$

$$= \delta_k(\boldsymbol{s}[k])(t) \cdot \prod_{j \neq k} \left( \int_{\tau > t} \delta_j(\boldsymbol{s}[j])(\tau) \, d\tau \right) \cdot \gamma_k(\boldsymbol{s}[k])(t)(a!).$$

The term $\delta_k(s[k])(t) \cdot \prod_{j \neq k} \left( \int_{\tau > t} \delta_j(s[j])(\tau) \, d\tau \right)$ is essentially the density of the $k^{th}$ component winning the race with a delay of $t$. In the sequel we let $\kappa_k^{\delta}(t) = \delta_k(s[k])(t) \cdot \prod_{j \neq k} \left( \int_{\tau > t} \delta_j(s[j])(\tau) \, d\tau \right)$.

*Example 2.* Returning to our running example of Fig. 2, we consider the joint-delay density of the composition in the initial state $s = (s_A, s_B)$, where $s_A = (1, x = 0)$ and $s_B = (1, x = 0)$. Applying the definition of composition we see that

$$\nu_{A|B}(s)(t, c!) = \begin{cases} \frac{1}{100-90} \cdot \frac{100-t}{100-0} \cdot 1 & \text{if } t \in [90, 100] \text{ and } c! = a! \\ \frac{1}{100-0} \cdot 1 & \text{if } t \in [0, 90[ \\ \frac{1}{100-0} \cdot \frac{100-t}{100-90} \cdot 1 & \text{if } t \in [90, 100] \end{cases} \right\} \text{ and } c! = b!.$$

## 4   Variance Reduction for STTS

For a stochastic timed transition system $\mathcal{T} = ((S, s^0, \rightarrow, \mathrm{AP}, P_{\mathcal{L}}), \nu)$ and a set of goal states $G \subseteq S$ we split the state space into dead ends ($\frown_G$) and good ends ($\smile_G$) i.e. states that can never reach $G$ and those that can. Formally,

$$\frown_G = \{s \in S \mid s \not\rightarrow^* G\} \text{ and } \smile_G = \{s \in S \mid s \rightarrow^* G\}.$$

For a state $s$, let $\mathrm{Act}_{G,t}(s) = \{a! \mid [[s]^t]^{a!} \in \smile_G\}$ and $\mathrm{Del}_F(s) = \{d \mid [s]^d \in \smile_G \wedge \mathrm{Act}_{G,d}(s) \neq \emptyset\}$. Informally, $\mathrm{Act}_{G,t}(s)$ extracts all the output actions that after a delay of $t$ will ensure having a chance to reach $G$. Similarly, $\mathrm{Del}_G(s)$ finds all the possible delays after which an action can be performed that ensures staying in good ends.

**Definition 5 (Dead End Avoidance).** *For a stochastic timed transition system $\mathcal{T} = ((S, s^0, \rightarrow, AP, P), \nu)$ and goal states $G$, we define an alternative dead end-avoiding stochastic timed transition system as any stochastic timed transition system $\acute{\mathcal{T}} = ((S, s^0, \rightarrow, AP, P), \acute{\nu})$ where if $\acute{\nu}(s)(t, a!) \neq 0$ then $a! \in \mathrm{Act}_{G,t}(s)$.*   □

Recall from Lemma 1 in Sect. 2 that in order to guarantee a variance reduction, the likelihood ratio should be less than 1. Let $\mathcal{T} = ((S, s^0, \rightarrow, \mathrm{AP}, P_{\mathcal{L}}), \nu)$ be a stochastic timed transition system, let $G \subseteq S$ be a set of goal states and let $\acute{\mathcal{T}} = ((S, s^0, \rightarrow, \mathrm{AP}, P_{\mathcal{L}}), \acute{\nu})$ be a dead end-avoiding alternative. Let $\omega = s_0, d_0, a_0! s_1, \ldots d_{n-1} a_{n-1}! s_n$ be a time bounded run, then the likelihood ratio of $\omega$ sampled under $\acute{\mathcal{T}}$ is

$$\frac{d\mathcal{T}(\omega)}{d\acute{\mathcal{T}}(\omega)} = \frac{\nu(s_0)(d_0, a_0!)}{\acute{\nu}(s_0)(d_0, a_0!)} \cdot \frac{\nu(s_1)(d_1, a_1!)}{\acute{\nu}(s_1)(d_1, a_1!)} \cdots \frac{\nu(s_{n-1})(d_{n-1}, a_{n-1}!)}{\acute{\nu}(s_{n-1})(d_{n-1}, a_{n-1}!)}$$

Clearly, if for all $i$, $\nu(s_i)(d_i, a_i!) \leq \acute{\nu}(s_i)(d_i, a_i!)$ then $\frac{d\mathcal{T}(\omega)}{d\acute{\mathcal{T}}(\omega)} \leq 1$. For a stochastic timed transition system $(\mathcal{L}, \nu = \delta \bullet \gamma)$ originating from a stochastic timed automaton we achieve this by proportionalising $\delta$ and $\gamma$ with respect to good ends, i.e. we use $\tilde{\nu} = \tilde{\delta} \bullet \tilde{\gamma}$ where

$$\tilde{\delta}(s)(t) = \frac{\delta(s)(t)}{\int_{\mathrm{Del}_G(s)} \delta(s)(\tau) \, d\tau} \text{ and } \tilde{\gamma}(s)(t)(a!) = \frac{\gamma(s)(t)(a!)}{\sum_{b! \in \mathrm{Act}_{G,t}(s)} \gamma(s)(t)(b!)}.$$

**Lemma 2.** *Let* $\mathcal{T} = (\mathcal{L}, \nu = \delta \bullet \gamma)$ *be a stochastic timed transition system from a stochastic timed automata, let* $\mathsf{G}$ *be a set of goal states and let* $\tilde{\mathcal{T}} = (\mathcal{L}, \tilde{\nu})$ *be a dead end avoiding alternative where* $\tilde{\nu}(s)(t, a!) = \tilde{\delta}(s)(t) \bullet \tilde{\gamma}(s)(t)(a!)$. *Also, let* $\mathbf{1}_{\mathsf{G}}$ *be an indicator function for* $\mathsf{G}$. *Then for any finite* $\omega \in \Omega(\mathcal{T})$, $\mathbf{1}_{\mathsf{G}}(\omega) \cdot \frac{\mathrm{d}\mathcal{T}(\omega)}{\mathrm{d}\tilde{\mathcal{T}}(\omega)} \leq 1$    □

For a composition, $\mathcal{J} = (\mathcal{L}, \nu)$ of stochastic timed transition systems $\mathcal{T}_1, \mathcal{T}_2, \ldots,$ $\mathcal{T}_n$ where for all $i$, $\mathcal{T}_i = (\mathcal{L}_i, \delta_i \bullet \gamma_i)$, we define a dead end avoiding stochastic timed transition system for $\mathsf{G}$ as $\acute{\mathcal{J}} = (\mathcal{L}, \tilde{\nu}^*)$ where

$$\tilde{\nu}^*(s)(t, a!) = \begin{cases} 0 & \text{if } t \notin \mathtt{Del}_{\mathsf{G},k}(s) \\ \frac{\kappa_k^\delta(s[k])(t)}{\sum_{i=1}^n (\int_{t' \in \mathtt{Del}_{\mathsf{G},i}(s)} \kappa_i^\delta(s[i])(t')\, \mathrm{d}t')} \cdot \kappa_k^\gamma(s[k])(t, a!) & \text{otherwise} \end{cases}$$

where $\mathtt{Del}_{\mathsf{G},k}(s) = \{d \mid (\mathtt{Act}_{\mathsf{G},d}(s)) \cap \Sigma_!^k \neq \emptyset\}$ and

$$\kappa_k^\gamma(s[k])(t, a!) = \begin{cases} \frac{\gamma_k(s[k])(t)(a!)}{\sum_{b! \in (\mathtt{Act}_{\mathsf{G},t}(s) \cap \Sigma_!^k)} \gamma_k(s[k])(t)(b!)} & \text{if } a! \in \mathtt{Act}_{\mathsf{G},t}(s) \cap \Sigma_!^k \\ 0 & \text{otherwise} \end{cases}$$

First the density of the $k^{th}$ component winning $(\kappa_k^\delta)$ is proportionalised with respect to all components winning delays. Afterwards, the probability mass of the actions leading to good ends for the $k^{th}$ component is proportionalised as well $(\kappa_k^\gamma)$.

**Lemma 3.** *Let* $\mathcal{J} = (\mathcal{L}, \nu)$ *be a stochastic timed transition system for a composition of stochastic timed transitions* $\mathcal{T}_1, \mathcal{T}_2, \ldots, \mathcal{T}_n$, *where for all* $i$, $\mathcal{T}_i = (\mathcal{L}_i, \delta_i \bullet \gamma_i)$ *originates from a stochastic timed automaton. Let* $\mathsf{G}$ *be a set of goal states and let* $\acute{\mathcal{J}} = (\mathcal{L}, \tilde{\nu}^*)$, *where* $\tilde{\nu}^*$ *is defined as above. Also, let* $\mathbf{1}_{\mathsf{G}}$ *be an indicator function for* $\mathsf{G}$. *Then for any finite* $\omega \in \Omega(\mathcal{J})$, $\mathbf{1}_{\mathsf{G}}(\omega) \cdot \frac{\mathrm{d}\mathcal{J}(\omega)}{\mathrm{d}\acute{\mathcal{J}}(\omega)} \leq 1$.    □

*Example 3.* For our running example let us consider $\tilde{\nu}_{\mathsf{A}|\mathsf{B}}^*$ as defined above:

$$\tilde{\nu}_{\mathsf{A}|\mathsf{B}}^*(s)(t, c!) = \begin{cases} \frac{\frac{1}{10} \cdot \frac{100-t}{100}}{\int_{90}^{100} \frac{1}{10} \cdot \frac{100-\tau}{100}\, \mathrm{d}\tau + \int_0^{10} \frac{1}{100}\, \mathrm{d}\tau} \cdot \frac{1}{1} & \text{if } c! = a! \text{ and } t \in [90, 100] \\ \frac{\frac{1}{100}}{\int_{90}^{100} \frac{1}{10} \cdot \frac{100-\tau}{100}\, \mathrm{d}\tau + \int_0^{10} \frac{1}{100}\, \mathrm{d}\tau} \cdot \frac{1}{1} & \text{if } c! = b! \text{ and } t \in [0, 10] \end{cases}$$
$$= \begin{cases} \frac{20}{30} \cdot \frac{100-t}{100} \cdot 1 & \text{if } c! = a! \text{ and } t \in [90, 100] \\ \frac{20}{300} \cdot 1 & \text{if } c! = b! \text{ and } t \in [0, 10] \end{cases}$$

## 5   Realising Proportional Dead End Avoidance for STA

In this section we focus on how to obtain the modified stochastic timed transition $\acute{\mathcal{T}} = (\mathcal{L}, \tilde{\nu})$ for a stochastic timed transition system $\mathcal{T} = (\mathcal{L}, \nu)$ originating from a stochastic timed automaton $\mathcal{A}$ and how to realise $\acute{\mathcal{T}} = (\mathcal{L}, \tilde{\nu}^*)$ for a composition of stochastic timed automata. In both cases the practical realisation consists of two steps: first the sets $\smile_G$ and $\frown_G$ are located by a modified algorithm of UPPAAL TIGA [2]. The result of running this algorithm is a reachability graph annotated with what actions to perform in certain states to ensure staying in good ends. On top of this reachability graph the sets $\mathtt{Del}_{G,k}(s)$ and $\mathtt{Act}_{G,k}(s)$ can be extracted.

## 5.1   Identifying Good Ends

Let $X$ be a set of clocks, a *zone* is a convex subset of $V(X)$ described by a conjunction of integer bounds on individual clocks and clock differences. We let $\mathcal{Z}_M(X)$ denote all sets of zones, where the integers bounds do not exceed $M$.

For $\mathcal{A} = (L, \ell_0, X, E, \mathrm{Inv})$ we call elements $(\ell, Z)$ of $L \times \mathcal{Z}_M(X)$, where $M$ is the maximal integer occuring in $\mathcal{A}$, for symbolic states and write $(\ell, v) \in (\ell, Z)$ if $v \in Z$. An element of $2^{\mathcal{Z}_M(X)}$ is called a federation of zones and we denote all federations by $\mathcal{F}_M(X)$. For a valuation $v$ and federation $F$ we write $v \in F$ if there exists a zone $Z \in F$ such that $v \in Z$.

Zones may be effectively represented using Difference Bound Matrices (DBM) [8]. Furthermore, DBMs allow for efficient symbolic exploration of the reachable state space of timed automata as implemented in the tool UPPAAL [13]. In particular, a forward symbolic search will result in a finite set $\mathcal{R}$ of symbolic states:

$$\mathcal{R} = \{(\ell_0, Z_0), \dots, (\ell_n, Z_n)\} \tag{7}$$

such that whenever $v_i \in Z_i$, then the state $(\ell_i, v_i)$ is reachable from the initial state $(\ell_i, v_0)$ (where $v_0(x) = 0$ for all clocks $x$). Dually, for any reachable state $(\ell, v)$ there is a zone $Z$ such that $v \in Z$ and $(\ell, Z) \in \mathcal{R}$.

To capture the good ends, i.e. the subset of $\mathcal{R}$ which may actually reach a state in the goal-set $G$, we have implemented a simplified version of the backwards propagation algorithm of UPPAAL TIGA [2] resulting in a *strategy* $\mathcal{S}$ "refining" $\mathcal{R}$:

$$\mathcal{S} = \{(\ell_0, Z_0, F_0, a_0!), \dots, (\ell_k, Z_k, F_k, a_k!)\} \tag{8}$$

where $F_i \subseteq Z_i$ and whenever $v_i \in F_i$ then $(\ell_i, v_i) \xrightarrow{a_i!} \to^* G$. Also, $(\ell_i, Z_i) \in \mathcal{R}$ whenever $(\ell_i, Z_i, F_i, a_i!) \in \mathcal{S}$. Thus, the union of the symbolic states $(\ell_i, F_i)$ appearing in quadruples of $\mathcal{S}^1$ identifies exactly the reachable states from which a discrete action $a_i!$ guarantees to enter a good end of the timed automaton (or network). Figure 3 depicts the reachability set $\mathcal{R}$ (grey area) and strategy set $\mathcal{S}$ (blue area) of our running example.

Given the strategy set $\mathcal{S}$ (8) and a state $s = (\ell, v)$, the set of possible delays after which an output action $a!$ leads to a good end is given by

$$\mathtt{Del}_{a!}(s) = \{ d \mid \exists (\ell_i, Z_i, F_i, a!) \in \mathcal{S} \text{ s.t. } [s]^d \in F_i) \}.$$

---

[1] One symbolic state may appear in several quadruples.

For a single stochastic timed automaton $\mathtt{Del_G}(s) = \bigcup_{a!\in\Sigma_!}\mathtt{Del}_{a!}(s)$ and for a network $\mathtt{Del_{G,i}}(s) = \bigcup_{a!\in\Sigma_!^i}\mathtt{Del}_{a!}(s)$. Importantly, note that $\mathtt{Del}_{a!}((\ell,v))$ – and thus also $\mathtt{Del_G}((\ell,v))$ – can be represented as a finite union of disjoint intervals. Given a closed zone $Z$ and a valuation $v$, the UPPAAL DBM library[2] provides functions that return the minimal delay $(d_{min})$ for entering a zone as well as the maximal delay for leaving it again $(d_{max})$. Due to convexity of zones then $\{(v+d) \mid d_{min} \leq d \leq d_{max}\} \subseteq Z$ and thus the possible delays to stay in $Z$ from $v$ is equal to the interval $[d_{min}, d_{max}]$. For the remainder of this paper we write

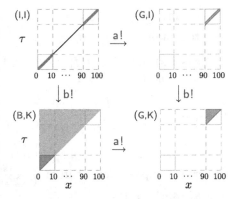

**Fig. 3.** Running example reachability set $\mathcal{R}$ (grey) and strategy set $\mathcal{S}$ (blue) for goal $\mathsf{G} \wedge \tau \leq 100$. (Color figure online)

$\{I_1, I_2 \ldots, I_n\} = \mathtt{Del_G}(s)$ where $I_1, I_2 \ldots I_n$ are the intervals making up $\mathtt{Del_G}(s)$.

Extracting the possible actions from a state $s = (\ell_i, v_i)$ after a delay of $d$ is simply a matter of iterating over all elements $(\ell_i, Z_i, F_i, a_i!)$ in $\mathcal{S}$ and checking whether $[s]^d \in F_i$. Formally, given a state $s = (\ell_i, v_i)$, $\mathtt{Act_{G,d}}((s)) = \{a_i! \in \Sigma_! \mid \exists (\ell_i, Z_i, F_i, a_i!) \in \mathcal{S}$ s.t. $[s]^d \in F_i\}$.

## 5.2   On-the-Fly State-Wise Change of Measure

Having found methods for extracting the sets $\mathtt{Act_{G,d}}(s)$ and $\mathtt{Del_{G,k}}(s)$, we focus on how to perform the state-wise change of measure.

**Single Stochastic Timed Automaton.** In the following let $\mathcal{A}$ be a timed automaton, $\mathcal{T}_\mathcal{A} = (\mathcal{L}, \delta_\mathcal{A} \bullet \gamma_\mathcal{A})$ be its stochastic timed transition system and let $\mathsf{G}$ be the goal states. For a fixed $t$ obtaining samples $\tilde{\gamma}_\mathcal{A}(s)(t)$ is a straightforward normalised weighted choice. Sampling delays from $\tilde{\delta}_\mathcal{A}(s)$ requires a bit more work: let $\mathcal{I} = \{I_1, I_2, \ldots I_n\} = \mathtt{Del_G}(s)$ and let $t \in I_j$, for some $j$ then

$$\tilde{\delta}_\mathcal{A}(s)(t) = \frac{\delta_\mathcal{A}(s)(t)}{\int_{\tau\in\mathtt{Del_G}(s)}\delta_\mathcal{A}(s)(t)\,\mathrm{d}\tau} = \frac{\int_{\tau\in I_j}\delta_\mathcal{A}(s)(t)\,\mathrm{d}\tau}{\sum_{I\in\mathcal{I}}\int_{\tau\in I}\delta_\mathcal{A}(s)(\tau)\,\mathrm{d}\tau} \cdot \frac{\delta_\mathcal{A}(s)(t)}{\int_{\tau\in I_j}\delta_\mathcal{A}(s)(t)\,\mathrm{d}\tau}.$$

Thus, to sample a delay we first choose an interval–weighted by its probability– and then sample from the conditional probability distribution of being inside that interval. Since $\delta_\mathcal{A}(s)$ is either an exponential distribution or uniform, integrating over it is easy and sampling from the conditional distribution is straightforward.

**Network of Stochastic Timed Automata.** In the following let $\mathcal{A}_1, \mathcal{A}_2, \ldots, \mathcal{A}_n$ be timed automata, $\mathcal{T}_i = (\mathcal{L}_i, \delta_i \bullet \gamma_i)$ be their stochastic timed transition

[2] http://people.cs.aau.dk/~adavid/UDBM/index.html.

systems and let $\mathcal{J} = (\mathcal{J}, \nu)$ be their composition. Recall from previously we wish to obtain $\tilde{v}^*(s)(t, a!) = \frac{\kappa_k^\delta(s[k])(t)}{\sum_{i=1}^n (\int_{t' \in \text{Del}_{G,i}(s)} \kappa_i^\delta(s[i])(t') \, dt')} \cdot \kappa_k^\gamma(s[k])(t, a!)$ for $t \in \text{Del}_{G,k}(s)$. Let $\mathcal{I}^i = \{I_1^i, I_2^i, \ldots, I_k^i\} = \text{Del}_{G,i}(s)$ for all $1 \le i \le n$ and let $t \in I$ for some $I \in \mathcal{I}^w$ then

$$\tilde{v}^*(s)(t, a!) = \frac{\int_I \kappa_w^\delta(s[w])(\tau) \, d\tau}{\sum_{i=1}^n \sum_{I' \in \mathcal{I}^i} \int_{I'} \kappa_i^\delta(s[i])(\tau) \, d\tau} \frac{\kappa_w^\delta(s[w])(t)}{\int_I \kappa_w^\delta(s[w])(\tau) \, d\tau} \cdot \kappa_w^\gamma(s[w])(t, a!),$$

when $t \in I$ and thus sampling from $\tilde{v}^*$ reduces to selecting an interval $I$ and winner $w$, sample a delay $t$ from $\frac{\kappa_w^\delta(s[m])(t)}{\int_I \kappa_w^\delta(s[w])(\tau) \, d\tau}$ and finally sample an action $a!$ from $\kappa_w^\gamma(s[w])(t, a!)$.

Algorithm 1 is our importance sampling algorithm for a composition of STA. In line 5 the delay densities of components according to standard semantics is extracted, and the win-densities $(\kappa_i^\delta)$ of each component winning is defined in line 6. In line 7 the delay intervals we should alter the distributions of $\kappa_i^\delta$ into is found. Lines 8 and 9 find a winning component ,$w$, and an interval in which it won, and then lines 10 and 11 sample a delay from that interval according to $\kappa_w^\delta$. After sampling a delay, lines 13 and 14 sample an action. Afterwards the current state and likelihood ratio $(L)$ is updated. The sampling in line 14 is a standard weighted choice, likewise is the sampling in line 9 - provided we have first calculated the integrals over $\kappa_i^\delta$ for all $i$. In line 11 the sampling from the conditional distribution is performed by *Inverse Transform Sampling*, requiring integration of $\kappa_k^\delta(s[k])$.

A recurring requirement for the algorithm is thus that $\kappa_k^\delta(s[k])$ is integrable: In the following we assume, without loss of generality, that $s$ is a state where there exists some $k$ such that for all $i \le k$, $\delta_i(s[i])$ is a uniform distribution between $a_i$ and $b_i$ and for all $i > k$ that $\delta_i(s[i])(t) = \lambda_i e^{-\lambda_i(t-d_i)}$ for $t > d_i$ i.e. $\delta_i(s[i])$ is a shifted exponential distribution. For any $i \le k$ we can now derive that $\kappa_i^\delta(s[i])(t) = \delta_i(s[i])(t) \prod_{j \ne i} \left( \int_{\tau > t} \delta_j(s[j])(\tau \, d\tau) \right)$ is

$$\frac{1}{b_i - a_i} \prod_{j \le k, j \ne i} \left( \begin{cases} \frac{b_j - t}{b_j - a_j} & \text{if } a_j \le t \le b_j \\ 1 & \text{if } t < a_j \\ 0 & \text{if } b_j < t \text{ or } t < a_i \\ & \text{or } t > a_i \end{cases} \right) \cdot \prod_{j > k} \left( \begin{cases} e^{-\lambda_i(t-d_i)} & \text{if } t > d_i \\ 1 & \text{else} \end{cases} \right)$$

and in general it can be seen that $\kappa_i^\delta(s[i])(t) = \begin{cases} \mathcal{P}_0(t) \cdot \mathcal{E}_0(t) & \text{if } t \in I_0 \\ \mathcal{P}_1(t) \cdot \mathcal{E}_1(t) & \text{if } t \in I_1 \\ \quad \vdots \\ \mathcal{P}_k(t) \cdot \mathcal{E}_k(t) & \text{if } t \in I_l \end{cases}$

where $I_0, I_1 \ldots I_l$ are disjoint intervals covering $[a_i, b_i]$, and for all $j$, $\mathcal{P}_j(t)$ is a polynomial constructed by the multiplication of uniform distribution and $\mathcal{E}_j(t)$ is an exponential function of the form $e^{\alpha \cdot t + \beta}$ constructed by multiplying shifted

---

**Algorithm 1.** Importance Sampling for Composition of STA

---

**Data:** Stochastic Timed Automata: $\mathcal{A}_1|\mathcal{A}_2|\ldots|\mathcal{A}_n$
**Data:** Goal States: $\mathsf{G}$

1  Let $(\mathcal{L}, \delta_{\mathcal{A}_i} \bullet \gamma_{\mathcal{A}_1}) = \mathcal{T}^{\mathcal{A}_i}$ for all $i$;
2  $s_c =$ initial state ;
3  $L = 1$ ;
4  **while** $s_c \notin \mathsf{G}$ **do**
5  $\quad$ Let $\delta_i = \delta_{\mathcal{A}_i}(s_c[i])$ for all $i$;
6  $\quad$ Let $\kappa_i^\delta(t) = \delta_i(t) \cdot \prod_{j \neq i} \left( \int_{\tau > t} \delta_j(\tau) \, d\tau \right)$ for all $i$;
7  $\quad$ Let $\mathcal{I}^i = \{I_1, I_2, \ldots, I_n\} = \mathsf{Del}_{\mathsf{G},i}(s_c)$ for all $i$;
8  $\quad K(I, w) = \dfrac{\int_I \kappa_w^\delta(t) \, dt}{\sum_{i=1}^n \sum_{I' \in \mathcal{I}^i} \int_{I'} \kappa_i^\delta(\tau) \, d\tau}$ for $I \in \mathcal{I}^m$ ;
9  $\quad (I, w) \sim K$ ;
10 $\quad d(T) = \dfrac{\kappa_w^\delta(T)}{\int_I \kappa_w^\delta(\tau) \, d\tau}$ for $T \in I$;
11 $\quad t \sim d$;
12 $\quad \gamma = \gamma_{\mathcal{A}_w}(s)(t)$;
13 $\quad m(a!) = \dfrac{\gamma(a!)}{\sum_{b! \in \mathsf{Act}_{\mathsf{G},t}(s) \cap \Sigma_!^i} \gamma(b!)}$ for $a! \in \mathsf{Act}_{\mathsf{G},t}(s) \cap \Sigma_!^i$;
14 $\quad a! \sim m$;
15 $\quad s_c = [[s_c]^t]^{a!}$;
16 $\quad L = L \cdot \dfrac{\delta_i(t)}{K(I,w) \cdot d(t)} \cdot \dfrac{\gamma(a!)}{m(a!)}$;
17 **return** $L$;

---

exponential distributions. Notice that although we assumed $i \leq k$, the above generalisation also holds for $i > k$. As a result we need to show for any polynomial, $\mathcal{P}(t)$, that $\mathcal{P}(t) \cdot e^{\alpha \cdot t + \beta}$ is integrable.

**Lemma 4** ([3]). *Let $\mathcal{P}(t) = \sum_{i=0}^n a_i t^i$ be a polynomial and let $\mathcal{E}_0(t) = e^{\alpha \cdot t + \beta}$ be an exponential function with $\alpha, \beta \in \mathbb{R}_{\geq 0}$. Then $\int \mathcal{P}(t) \cdot \mathcal{E}(t) \, dt = \hat{\mathcal{P}}(t) \cdot \mathcal{E}(t)$, with $\hat{\mathcal{P}}(t) = \sum_{i=0}^{n+1} b_i t^i$, where $b_{n+1} = 0$ and $b_i = \frac{a_i - b_{i+1}(i+1)}{\alpha}$.* $\square$

## 6  Experiments

In this section we compare the variance of our importance sampling (IS) estimator with that of standard SMC. We include a variety of scalable models, with both rare and not-so-rare properties to compare performance. *Running* is a parametrised version of our running example. *Race* is based on a simple race between automata. Both *Running* and *Race* are parametrised by scale, which affects the bounds in guards and invariants. *Race* considers the property of reaching goal location Obs.G within a fixed time, denoted $\Diamond_{\leq 1}$ Obs.G. *Running* considers the property of reaching goal location A.G within a time related to scale,

---

[3] Details can be found at http://people.cs.aau.dk/~marius/stratego/rare.html.

denoted $\Diamond_{\leq \mathsf{scale}\cdot 100}$ A.G. The DPA (Duration Probabilistic Automata) are job-scheduling models [7] that have proven challenging to analyse in other contexts [12]. We consider the probability of all processes completing their tasks within parametrised time limit $\tau$. The property has the form $\Diamond_{\leq \tau}$ DPA1.G $\wedge \cdots \wedge$ DPA$n$.G, where DPA1.G, ..., DPA$n$.G are the goal states of the $n$ components that comprise the model.

The variance of our IS estimator is typically lower than that of SMC, but SMC simulations are generally quicker. IS also incurs additional set-up and initial analysis with respect to SMC. To make a valid comparison we therefore consider the amount of variance reduction after a fixed amount of CPU time. For each approach and model instance we calculate results based on 30 CPU-time-bounded experiments, where individual experiments estimate the probability of the property and the variance of the estimator after 1 s. This time is sufficient to generate good estimates with IS, while SMC may produce no successful traces. Using an estimate of the expected number of SMC traces after 1 s, we are nevertheless able to make a valid comparison, as described below.

Our results are given in Table 1. The Gain column gives an estimate of the true performance improvement of IS by approximating the ratio (*variance of*

**Table 1.** Experimental results. $\mathbb{P}$ is exact probability, when available. SMC (IS) indicates crude Monte Carlo (importance sampling). $\widehat{\sigma_{\mathrm{is}}^2}$ is empirical variance of likelihood ratio. Gain estimates true improvement of IS at 1 s CPU time. Mem. reports memory use. Model DPA$x$S$y$ contains $x$ processes and $y$ tasks.

| Model | Param. | $\mathbb{P}$ | Estimated Prob. | | IS | | Mem. (MB) | |
|---|---|---|---|---|---|---|---|---|
| | scale / $\tau$ | | SMC | IS | $\widehat{\sigma_{\mathrm{is}}^2}$ | Gain | SMC | IS |
| Running | scale = 1 | 1.0e−1 | 1.0e−1 | 1.0e−1 | 2.5e−3 | 1.6 | 10.2 | 14.0 |
| | 10 | 1.0e−2 | 1.0e−2 | 1.0e−2 | 2.5e−5 | 1.7e1 | 10.4 | 13.6 |
| | 100 | 1.0e−3 | 1.0e−3 | 9.6e−4 | 2.5e−7 | 1.4e2 | 10.2 | 13.7 |
| Race | 1 | 1.0e−5 | 1.3e−5 | 1.6e−5 | 2.1e−11 | 8.3e3 | 10.0 | 13.5 |
| | 2 | 3.0e−6 | 1.6e−6 | 3.2e−6 | 1.3e−12 | 2.0e4 | 10.0 | 13.3 |
| | 3 | 1.0e−6 | 0 | 1.4e−6 | 2.5e−13 | 6.8e4 | 10.0 | 13.6 |
| | 4 | 8.0e−7 | 0 | 8.0e−7 | 8.1e−14 | 1.5e4 | 11.5 | 14.4 |
| DPA1S3 | $\tau = 200$ | n/a | 1.4e−1 | 1.4e−1 | 2.8e−2 | 1.1 | 10.2 | 11.9 |
| | 40 | n/a | 3.6e−4 | 3.5e−4 | 1.7e−7 | 2.8e2 | 10.2 | 12.0 |
| | 16 | n/a | 0 | 2.2e−8 | 6.9e−16 | 2.7e6 | 10.2 | 11.8 |
| DPA2S6 | 423 | n/a | 1.0e−5 | 2.9e−5 | 3.7e−7 | 0.9 | 10.2 | 13.5 |
| | 400 | n/a | 2.7e−5 | 7.6e−6 | 2.8e−8 | 3.0 | 10.3 | 13.5 |
| | 350 | n/a | 0 | 5.5e−8 | 4.9e−13 | 1.1e3 | 10.3 | 13.3 |
| DPA4S3 | 395 | n/a | 7.0e−5 | 1.9e−5 | 4.9e−8 | 4.4 | 10.3 | 53.1 |
| | 350 | n/a | 1.4e−5 | 9.0e−6 | 1.3e−8 | 7.1 | 10.5 | 53.1 |
| | 300 | n/a | 0 | 2.7e−7 | 2.3e−11 | 1.1e2 | 10.4 | 53.0 |

SMC *estimator*)/(*variance of* IS *estimator*) after 1 s of CPU time. The variance of the IS estimator is estimated directly from the empirical distribution of all IS samples. The variance of the SMC estimator is estimated by $\hat{p}(1 - \hat{p})/N_{\mathrm{SMC}}$, where $\hat{p}$ is our best estimate of the true probability (the true value itself or the mean of the IS estimates) and $N_{\mathrm{SMC}}$ is the mean number of standard SMC simulations observed after 1 s.

For each model type we see that computational gain increases with rarity and that real gains of several orders of magnitude are possible. We also include model instances where the gain is approximately 1, to give an idea of the largest probability where IS is worthwhile. Memory usage is approximately constant over all models for SMC and approximately constant within a model type for IS. The memory required for the reachability graph is a potential limiting factor of our IS approach, although this is not evident in our chosen examples.

# 7 Conclusion

Our approach is guaranteed to reduce estimator variance, but it incurs additional storage and simulation costs. We have nevertheless demonstrated that our framework can make substantial real reductions in computational effort when estimating the probability of rare properties of stochastic timed automata. Computational gain tends to increase with rarity, hence we observe marginal cases where the performance of IS and SMC are similar. We hypothesise that it may be possible to make further improvements in performance by applying cross-entropy optimisation to the discrete transitions of the reachability graph, along the lines of [9], making our techniques more efficient and useful for less-rare properties.

Importance *splitting* [11,15] is an alternative variance reduction technique with potential advantages for SMC [10]. We therefore intend to compare our current framework with an implementation of importance splitting for stochastic timed automata, applying both to substantial case studies.

**Acknowledgements.** This research has received funding from the Sino-Danish Basic Research Centre, IDEA4CPS, funded by the Danish National Research Foundation and the National Science Foundation, China, the Innovation Fund Denmark centre DiCyPS, as well as the ERC Advanced Grant LASSO. Other funding has been provided by the Self Energy-Supporting Autonomous Computation (SENSATION) and Collective Adaptive Systems Synthesis with Non-zero-sum Games (CASSTING) European FP7-ICT projects, and the Embedded Multi-Core systems for Mixed Criticality (EMC$^2$) ARTEMIS project.

# References

1. Alur, R., Dill, D.L.: A theory of timed automata. Theoret. Comput. Sci. **126**(2), 183–235 (1994)
2. Behrmann, G., Cougnard, A., David, A., Fleury, E., Larsen, K.G., Lime, D.: Uppaal-Tiga: time for playing games!. In: Damm, W., Hermanns, H. (eds.) CAV 2007. LNCS, vol. 4590, pp. 121–125. Springer, Heidelberg (2007). doi:10.1007/978-3-540-73368-3_14

3. Bulychev, P., David, A., Larsen, K.G., Legay, A., Li, G., Poulsen, D.B.: Rewrite-based statistical model checking of WMTL. In: Qadeer, S., Tasiran, S. (eds.) RV 2012. LNCS, vol. 7687, pp. 260–275. Springer, Heidelberg (2013). doi:10.1007/978-3-642-35632-2_25

4. Bulychev, P., David, A., Larsen, K.G., Legay, A., Li, G., Bøgsted Poulsen, D., Stainer, A.: Monitor-based statistical model checking for weighted metric temporal logic. In: Bjørner, N., Voronkov, A. (eds.) LPAR 2012. LNCS, vol. 7180, pp. 168–182. Springer, Heidelberg (2012). doi:10.1007/978-3-642-28717-6_15

5. Bulychev, P., David, A., Larsen, K.G., Legay, A., Mikučionis, M., Bøgsted Poulsen, D.: Checking and distributing statistical model checking. In: Goodloe, A.E., Person, S. (eds.) NFM 2012. LNCS, vol. 7226, pp. 449–463. Springer, Heidelberg (2012). doi:10.1007/978-3-642-28891-3_39

6. David, A., Larsen, K.G., Legay, A., Nyman, U., Wasowski, A.: Timed i/o automata: a complete specification theory for real-time systems. In: HSCC, pp. 91–100. ACM (2010)

7. David, A., Larsen, K.G., Legay, A., Mikučionis, M., Poulsen, D.B., Vliet, J., Wang, Z.: Statistical model checking for networks of priced timed automata. In: Fahrenberg, U., Tripakis, S. (eds.) FORMATS 2011. LNCS, vol. 6919, pp. 80–96. Springer, Heidelberg (2011). doi:10.1007/978-3-642-24310-3_7

8. Dill, D.L.: Timing assumptions and verification of finite-state concurrent systems. In: Proceedings of Automatic Verification Methods for Finite State Systems, International Workshop, Grenoble, France, June 12–14, 1989, pp. 197–212 (1989)

9. Jegourel, C., Legay, A., Sedwards, S.: Cross-entropy optimisation of importance sampling parameters for statistical model checking. In: Madhusudan, P., Seshia, S.A. (eds.) CAV 2012. LNCS, vol. 7358, pp. 327–342. Springer, Heidelberg (2012). doi:10.1007/978-3-642-31424-7_26

10. Jegourel, C., Legay, A., Sedwards, S.: Importance splitting for statistical model checking rare properties. In: Sharygina, N., Veith, H. (eds.) CAV 2013. LNCS, vol. 8044, pp. 576–591. Springer, Heidelberg (2013). doi:10.1007/978-3-642-39799-8_38

11. Kahn, H.: Use of different monte carlo sampling techniques. Technical report P-766, Rand Corporation, November 1955

12. Kempf, J.-F., Bozga, M., Maler, O.: Performance evaluation of schedulers in a probabilistic setting. In: Fahrenberg, U., Tripakis, S. (eds.) FORMATS 2011. LNCS, vol. 6919, pp. 1–17. Springer, Heidelberg (2011). doi:10.1007/978-3-642-24310-3_1

13. Larsen, K.G., Pettersson, P., Yi, W.: Uppaal in a nutshell. STTT **1**(1–2), 134–152 (1997)

14. Maler, O., Larsen, K.G., Krogh, B.H.: On zone-based analysis of duration probabilistic automata. In: Proceedings 12th International Workshop on Verification of Infinite-State Systems (INFINITY), pp. 33–46 (2010)

15. Rubino, G., Tuffin, B.: Rare Event Simulation using Monte Carlo Methods. Wiley, Hoboken (2009)

16. Rubinstein, R.: The cross-entropy method for combinatorial and continuous optimization. Meth. Comput. Appl. Probab. **1**(2), 127–190 (1999)

17. Vicario, E., Sassoli, L., Carnevali, L.: Using stochastic state classes in quantitative evaluation of dense-time reactive systems. IEEE Trans. Softw. Eng. **35**(5), 703–719 (2009)

18. Younes, H.L.S.: Verification and planning for stochastic processes with asynchronous events. Ph.D. thesis, Carnegie Mellon University (2005)

19. Zuliani, P., Baier, C., Clarke, E.M.: Rare-event verification for stochastic hybrid systems. In: Proceedings of the 15th ACM International Conference on Hybrid Systems: Computation and Control, HSCC, pp. 217–226 (2012)

# Semipositivity in Separation Logic with Two Variables

Zhilin Wu[⊠]

State Key Laboratory of Computer Science, Institute of Software,
Chinese Academy of Sciences, Beijing, People's Republic of China
wuzl@ios.ac.cn

**Abstract.** In a recent work by Demri and Deters (CSL-LICS 2014), first-order separation logic restricted to two variables and separating implication was shown undecidable, where it was shown that even with only two variables, if the use of negations is unrestricted, then they can be nested with separating implication in a complex way to get the undecidability result. In this paper, we revisit the decidability and complexity issues of first-order separation logic with two variables, and proposes *semi-positive* separation logic with two-variables (SPSL2), where the use of negations is restricted in the sense that negations can only occur in front of atomic formulae. We prove that satisfiability of the fragment of SPSL2 where neither separating conjunction nor septraction (the dual operator of separating implication) occurs in the scope of universal quantifiers, is NEXPTIME-complete. As a byproduct of the proof, we show that the finite satisfiability problem of first-order logic with two variables and a bounded number of function symbols is NEXPTIME-complete (the lower bound holds even with only one function symbol and without unary predicates), which may be of independent interest beyond separation logic community.

## 1 Introduction

**Decidability and Separation Logics.** Separation logic is a prominent logical formalism to verify programs with pointers and it comes in different flavours and many fragments and extensions exist. The decidability status of first-order separation logic with two record fields has been answered negatively quite early in [5] thanks to Trakhtenbrot's Theorem [23]: finitary satisfiability for predicate logic restricted to a single binary predicate symbol is undecidable and not recursively enumerable. The undecidability of first-order separation logic with a single record field was then established in [4] and a bit later in [8] with the further restriction that only two individual variables are permitted. Undecidability can be established in various ways: in [9] by reduction from the halting problem for Minsky machines [15] or from the satisfiability problem for FO2 on data words with a linear ordering on data [3]. Despite these negative results, many

Partially supported by the NSFC grants (Nos. 61100062, 61272135, 61472474, and 61572478).

M. Fränzle et al. (Eds.): SETTA 2016, LNCS 9984, pp. 179–196, 2016.
DOI: 10.1007/978-3-319-47677-3_12

fragments of separation logic are known to be decidable and used in practice, mainly thanks to the absence of the separation implication, see e.g. [1,7,17,20]. For instance, the symbolic-heap fragment is free of separating implication and the propositional fragment of separation logic can be decided in polynomial space [5]. Semi-decision procedures for fragments with separating implication can be found in [22]. First-order separation logic with all separating connectives but with a single variable is shown in PSPACE in [10].

**Our Motivations.** Undecidable fragments of separation logic allow still too much freedom whereas the decidable fragments with relatively low complexity are still poorly expressive. The real question is how to reduce this gap by introducing restrictions based on negations (see also the related work [21] about restrictions on negations). In this paper, we consider semi-positive separation logic with two variables, denoted by SPSL2, where negation symbols only occur in front of the atomic formulae. Our goal is to understand the influence of the restricted use of negations on the decidability/complexity of the satisfiability problem.

We know that $\mathrm{SPSL2}\left((\overset{f}{\hookrightarrow})_{f\in\mathbb{F}}, *\right)$ (where $\mathbb{F}$ is a finite set of fields, $\overset{f}{\hookrightarrow}$ and $*$ represent the "points-to" and "separating conjunction" modality respectively), the fragment of SPSL2 where separating implication does not occur, admits a decidable satisfiability problem if $\mathbb{F} = \{f\}$ (that is, there is exactly one field), since $\mathrm{SL2}\left(\overset{f}{\hookrightarrow}, *\right)$, the smallest extension of $\mathrm{SPSL2}\left(\overset{f}{\hookrightarrow}, *\right)$ closed under negations, is decidable with a non-elementary computational complexity (cf. [4], the lower bound with only two variables was shown in [9]). Nevertheless, to the best of our knowledge, the decidability and complexity of various fragments of SPSL2 are still largely open.

**Our contributions.** As a starting point towards a complete decidability/complexity charaterization of SPSL2, we show that the satisfiability of the following fragments of SPSL2 is NEXPTIME-complete (cf. Sect. 2 for the definition of these fragments).

1. $\mathrm{SPSL2}\left((\overset{f}{\hookrightarrow})_{f\in\mathbb{F}}\right)$ and $\mathrm{SPSL2}\left((\overset{f}{\hookrightarrow})_{f\in\mathbb{F}}, \mathcal{P}\right)$, where separating operators do not occur, and $\mathcal{P}$ denotes a finite set of unary predicates. The NEXPTIME lower bound holds even if there is only one field, that is, $\mathbb{F} = \{f\}$. The upper bound proof is obtained by a reduction to the finite satisfiability of first-order logic with two variables and counting quantifiers, which is NEXPTIME-complete [18,19]. The lower bound is shown by encoding the solutions of a given exponential-size tiling problem into a formula in $\mathrm{SPSL2}\left(\overset{f}{\hookrightarrow}\right)$, where only one function symbol $f$ is used and no unary predicates are needed.

2. $\mathrm{ESPSL2}\left((\overset{f}{\hookrightarrow})_{f\in\mathbb{F}}, *, \overset{\rightharpoonup}{*}\right)$, the extension of $\mathrm{SPSL2}\left((\overset{f}{\hookrightarrow})_{f\in\mathbb{F}}\right)$ with separating conjunction $*$ and septraction $\overset{\rightharpoonup}{*}$ (the dual operator of magic wand $\mathrel{-\!*}$), where neither $*$ nor $\overset{\rightharpoonup}{*}$ occurs in the scope of universal quantifiers. The result is obtained by a reduction to the satisfiability of $\mathrm{SPSL2}\left((\overset{f}{\hookrightarrow})_{f\in\mathbb{F}}, \mathcal{P}\right)$ formulae.

Related work. Logics with two variables are a classical topic in mathematical logic and theoretical computer science. Over arbitrary relational structures, first-order logic with two variables and its extensions have been investigated intensively, see [12–14, 16, 18, 19] (to cite a few). In [18, 19], it is well-known that the satisfiability and finite satisfiability problem of $\mathcal{C}^2$, first-order logic with two variables and counting quantifiers, are NEXPTIME-complete. Since function symbols can be encoded by relation symbols with the help of counting quantifiers, it follows that the satisfiability problem of $\mathrm{SPSL2}\left( (\xrightarrow{\mathfrak{f}})_{\mathfrak{f} \in \mathbb{F}}, \mathcal{P} \right)$ is in NEXPTIME. In addition, it was shown that the (finite) satisfiability problem of first-order logic with two variables and unary predicates is already NEXPTIME-hard. The NEXPTIME lower bound we obtained is novel in the sense that in our reduction, only one function symbol but no unary predicates is used. First-order logic with two variables on special classes of structures, e.g. words and trees, has also been investigated [2, 11]. In [6], first-order logic with two variables and deterministic transitive closure over one binary relation was considered and its the satisfiability problem was shown to be EXPSPACE-complete.

Outline of the paper. Preliminaries are given in Sect. 2. In Sect. 3, the satisfiability of $\mathrm{SPSL2}\left( (\xrightarrow{\mathfrak{f}})_{\mathfrak{f} \in \mathbb{F}} \right)$ and $\mathrm{SPSL2}\left( (\xrightarrow{\mathfrak{f}})_{\mathfrak{f} \in \mathbb{F}}, \mathcal{P} \right)$ is shown to be NEXPTIME-complete. Section 4 is devoted to $\mathrm{ESPSL2}\left( (\xrightarrow{\mathfrak{f}})_{\mathfrak{f} \in \mathbb{F}}, *, \xrightarrow{\twoheadrightarrow} \right)$.

## 2  Preliminaries

For $n \in \mathbb{N}$, let $[n]$ denote the set $\{0, \ldots, n-1\}$. For $n, m \in \mathbb{N}$ such that $n \leq m$, let $[n, m] = \{n, n+1, \ldots, m\}$. Let $\mathbb{F}$ denote a finite set of *fields*. A *heap* $\mathfrak{h}$ over $\mathbb{F}$ is a collection of partial functions $(\mathfrak{h}_{\mathfrak{f}})_{\mathfrak{f} \in \mathbb{F}} : \mathbb{N} \rightharpoonup \mathbb{N}$ such that each of them has a finite domain. We write $\mathrm{dom}(\mathfrak{h}_{\mathfrak{f}})$ to denote the *domain* of $\mathfrak{h}_{\mathfrak{f}}$ and $\mathrm{ran}(\mathfrak{h}_{\mathfrak{f}})$ to denote its *range*. In addition, we use $\mathrm{loc}(\mathfrak{h}_{\mathfrak{f}})$ to denote $\mathrm{dom}(\mathfrak{h}_{\mathfrak{f}}) \cup \mathrm{ran}(\mathfrak{h}_{\mathfrak{f}})$. We also use $\mathrm{dom}(\mathfrak{h})$ to denote $\bigcup_{\mathfrak{f} \in \mathbb{F}} \mathrm{dom}(\mathfrak{h}_{\mathfrak{f}})$, $\mathrm{ran}(\mathfrak{h})$ to denote $\bigcup_{\mathfrak{f} \in \mathbb{F}} \mathrm{ran}(\mathfrak{h}_{\mathfrak{f}})$, and $\mathrm{loc}(\mathfrak{h})$ to denote $\bigcup_{\mathfrak{f} \in \mathbb{F}} \mathrm{loc}(\mathfrak{h}_{\mathfrak{f}})$. Two heaps $\mathfrak{h}_1 = (\mathfrak{h}_{1,\mathfrak{f}})_{\mathfrak{f} \in \mathbb{F}}$ and $\mathfrak{h}_2 = (\mathfrak{h}_{2,\mathfrak{f}})_{\mathfrak{f} \in \mathbb{F}}$ are said to be *disjoint*, denoted $\mathfrak{h}_1 \perp \mathfrak{h}_2$, if $\mathrm{dom}(\mathfrak{h}_1) \cap \mathrm{dom}(\mathfrak{h}_2) = \emptyset$; when this holds, we write $\mathfrak{h}_1 \uplus \mathfrak{h}_2$ to denote the collection of partial functions $(\mathfrak{h}_{\mathfrak{f}})_{\mathfrak{f} \in \mathbb{F}}$ such that $\mathfrak{h}_{\mathfrak{f}}$ is obtained from $\mathfrak{h}_{1,\mathfrak{f}}$ and $\mathfrak{h}_{2,\mathfrak{f}}$ by taking their disjoint union.

We introduce some graph-theoretical notations for heaps. Let $\mathfrak{h} = (\mathfrak{h}_{\mathfrak{f}})_{\mathfrak{f} \in \mathbb{F}}$ and $\mathfrak{f} \in \mathbb{F}$.

- Let $\mathfrak{l}$ and $\mathfrak{l}'$ be two locations. If $\mathfrak{h}_{\mathfrak{f}}(\mathfrak{l}) = \mathfrak{l}'$, then $\mathfrak{l}'$ is said to be the $\mathfrak{f}$-*successor* of $\mathfrak{l}$ (resp. $\mathfrak{l}$ is said to be an $\mathfrak{f}$-*predecessor* of $\mathfrak{l}'$) in $\mathfrak{h}$.
- An $\mathfrak{f}$-*path* in $\mathfrak{h}$ is a sequence of locations, say $\mathfrak{l}_0 \mathfrak{l}_1 \ldots \mathfrak{l}_k$ (where $k \geq 0$), such that for each $j : 0 \leq j < k$, $\mathfrak{l}_{j+1}$ is the $\mathfrak{f}$-successor of $\mathfrak{l}_j$. The location $\mathfrak{l}_0$ and $\mathfrak{l}_k$ are called the *start location* and *end location* of the $\mathfrak{f}$-path respectively. In addition, $k$ is called the *length* of the $\mathfrak{f}$-path. An $\mathfrak{f}$-path $\mathfrak{l}_0 \mathfrak{l}_1 \ldots \mathfrak{l}_k$ is called an

$\mathfrak{f}$-*cycle* if $k \geq 1$ and $\mathfrak{l}_0 = \mathfrak{l}_k$. If $\mathfrak{l} \in \mathbb{N}$ is the end location of an $\mathfrak{f}$-path, then we also call the $\mathfrak{f}$-path as a *backward* $\mathfrak{f}$-*path* of $\mathfrak{l}$.

- Let $\mathcal{G}[\mathfrak{h}_\mathfrak{f}]$ be the directed graph corresponding to $\mathfrak{h}_\mathfrak{f}$, that is, the graph where the set of nodes is $\mathtt{loc}(\mathfrak{h}_\mathfrak{f})$, and for each pair of locations $\mathfrak{l}, \mathfrak{l}' \in \mathtt{loc}(\mathfrak{h}_\mathfrak{f})$, there is an arc from $\mathfrak{l}$ to $\mathfrak{l}'$ iff $\mathfrak{h}_\mathfrak{f}(\mathfrak{l}) = \mathfrak{l}'$. Note that $\mathcal{G}[\mathfrak{h}_\mathfrak{f}]$ has a special structure in the sense that each node has at most one successor (as a result of the fact that $\mathfrak{h}_\mathfrak{f}$ is a partial function). Suppose that $\mathcal{C}$ is a connected component of $\mathcal{G}[\mathfrak{h}_\mathfrak{f}]$, then the partial function $\mathfrak{h}'_\mathfrak{f}$ such that $\mathcal{G}[\mathfrak{h}'_\mathfrak{f}] = \mathcal{C}$ (this means that $\mathtt{loc}(\mathfrak{h}'_\mathfrak{f})$ is the set of nodes in $\mathcal{C}$, and $\mathfrak{h}'_\mathfrak{f}(\mathfrak{l}) = \mathfrak{l}'$ iff there is an arc from $\mathfrak{l}$ to $\mathfrak{l}'$ in $\mathcal{C}$ iff $\mathfrak{h}_\mathfrak{f}(\mathfrak{l}) = \mathfrak{l}'$), is called a *connected component* of $\mathfrak{h}$.

Formulae in $\mathrm{SPSL2}\left((\overset{\mathfrak{f}}{\hookrightarrow})_{\mathfrak{f}\in\mathbb{F}}, *, -\!\!*\right)$, *semi-positive separation logic with two variables*, are defined by the following rules:

$$\mathbf{v} ::= \mathbf{x} \mid \mathbf{y},$$
$$\alpha ::= \mathbf{v} = \mathbf{v} \mid \mathbf{v} \overset{\mathfrak{f}}{\hookrightarrow} \mathbf{v},$$
$$\phi ::= \alpha \mid \neg\alpha \mid \phi \vee \phi \mid \phi \wedge \phi \mid \exists \mathbf{v}.\phi \mid \forall \mathbf{v}.\phi \mid \phi * \phi \mid \phi -\!\!* \phi,$$

where $\mathbf{x}$ and $\mathbf{y}$ are two distinguished first-order variables, and $\mathfrak{f} \in \mathbb{F}$.

We write $\mathrm{SPSL2}\left((\overset{\mathfrak{f}}{\hookrightarrow})_{\mathfrak{f}\in\mathbb{F}}\right)$ to denote the fragment of $\mathrm{SPSL2}\left((\overset{\mathfrak{f}}{\hookrightarrow})_{\mathfrak{f}\in\mathbb{F}}, *, -\!\!*\right)$ without separating connectives (remove the last two rules in the definition of $\phi$). Note that $\mathrm{SPSL2}\left((\overset{\mathfrak{f}}{\hookrightarrow})_{\mathfrak{f}\in\mathbb{F}}\right)$ can be seen as first-order logic with two variables and $|\mathbb{F}|$ function symbols (where $|\mathbb{F}|$ denote the cardinality of $\mathbb{F}$). Similarly, we write $\mathrm{SPSL2}\left((\overset{\mathfrak{f}}{\hookrightarrow})_{\mathfrak{f}\in\mathbb{F}}, *\right)$ to denote the fragment of $\mathrm{SPSL2}\left((\overset{\mathfrak{f}}{\hookrightarrow})_{\mathfrak{f}\in\mathbb{F}}, *, -\!\!*\right)$ without separating implication (remove the last rule in the definition of $\phi$). We use $|\phi|$ to denote the size of $\phi$. In addition, the set of subformulae of $\phi$, denoted by $\mathtt{Sub}(\phi)$, can be defined in a standard way.

An *assignment* is a map $\mathfrak{m} : \{\mathbf{x}, \mathbf{y}\} \to \mathbb{N}$. For an assignment $\mathfrak{m}$ and $\mathfrak{l} \in \mathbb{N}$, we use $\mathfrak{m}[\mathbf{v} \mapsto \mathfrak{l}]$ denote the assignment that is the same as $\mathfrak{m}$, except that it maps $\mathbf{v}$ to $\mathfrak{l}$ (where $\mathbf{v} = \mathbf{x}, \mathbf{y}$). The satisfaction relation $\models$ is parameterised by assignments and defined as follows (clauses are omitted when these can be obtained by permuting the two variables below):

- $\mathfrak{h} \models_\mathfrak{m} \mathbf{v}_1 = \mathbf{v}_2 \overset{\mathrm{def}}{\Leftrightarrow} \mathfrak{m}(\mathbf{v}_1) = \mathfrak{m}(\mathbf{v}_2)$.
- $\mathfrak{h} \models_\mathfrak{m} \neg(\mathbf{v}_1 = \mathbf{v}_2) \overset{\mathrm{def}}{\Leftrightarrow} \mathfrak{m}(\mathbf{v}_1) \neq \mathfrak{m}(\mathbf{v}_2)$.
- $\mathfrak{h} \models_\mathfrak{m} \mathbf{v}_1 \overset{\mathfrak{f}}{\hookrightarrow} \mathbf{v}_2 \overset{\mathrm{def}}{\Leftrightarrow} \mathfrak{m}(\mathbf{v}_1) \in \mathtt{dom}(\mathfrak{h}_\mathfrak{f})$ and $\mathfrak{h}_\mathfrak{f}(\mathfrak{m}(\mathbf{v}_1)) = \mathfrak{m}(\mathbf{v}_2)$.
- $\mathfrak{h} \models_\mathfrak{m} \neg(\mathbf{v}_1 \overset{\mathfrak{f}}{\hookrightarrow} \mathbf{v}_2) \overset{\mathrm{def}}{\Leftrightarrow} \mathfrak{m}(\mathbf{v}_1) \notin \mathtt{dom}(\mathfrak{h}_\mathfrak{f})$ or otherwise $\mathfrak{h}_\mathfrak{f}(\mathfrak{m}(\mathbf{v}_1)) \neq \mathfrak{m}(\mathbf{v}_2)$.
- $\mathfrak{h} \models_\mathfrak{m} \phi_1 \wedge \phi_2 \overset{\mathrm{def}}{\Leftrightarrow} \mathfrak{h} \models_\mathfrak{m} \phi_1$ and $\mathfrak{h} \models_\mathfrak{m} \phi_2$.
- $\mathfrak{h} \models_\mathfrak{m} \phi_1 \vee \phi_2 \overset{\mathrm{def}}{\Leftrightarrow} \mathfrak{h} \models_\mathfrak{m} \phi_1$ or $\mathfrak{h} \models_\mathfrak{m} \phi_2$.
- $\mathfrak{h} \models_\mathfrak{m} \phi_1 * \phi_2 \overset{\mathrm{def}}{\Leftrightarrow}$ there exist $\mathfrak{h}_1, \mathfrak{h}_2$ such that $\mathfrak{h}_1 \perp \mathfrak{h}_2$, $\mathfrak{h} = \mathfrak{h}_1 \uplus \mathfrak{h}_2$, $\mathfrak{h}_1 \models_\mathfrak{m} \phi_1$ and $\mathfrak{h}_2 \models_\mathfrak{m} \phi_2$.

- $\mathfrak{h} \models_{\mathfrak{m}} \phi_1 \mathbin{-\!\!*} \phi_2 \overset{\text{def}}{\Leftrightarrow}$ for all $\mathfrak{h}'$, if $\mathfrak{h} \perp \mathfrak{h}'$ and $\mathfrak{h}' \models_{\mathfrak{m}} \phi_1$ then $\mathfrak{h} \uplus \mathfrak{h}' \models_{\mathfrak{m}} \phi_2$.
- $\mathfrak{h} \models_{\mathfrak{m}} \exists v.\phi \overset{\text{def}}{\Leftrightarrow}$ there is $\mathfrak{l} \in \mathbb{N}$ such that $\mathfrak{h} \models_{\mathfrak{m}[v \mapsto \mathfrak{l}]} \phi$.
- $\mathfrak{h} \models_{\mathfrak{m}} \forall v.\phi \overset{\text{def}}{\Leftrightarrow}$ for every $\mathfrak{l} \in \mathbb{N}$, $\mathfrak{h} \models_{\mathfrak{m}[v \mapsto \mathfrak{l}]} \phi$.

If $\phi$ is a sentence, we also omit $\mathfrak{m}$ and we write $\mathfrak{h} \models \phi$ since $\mathfrak{m}$ is irrelevant in this case. A formula $\phi$ is *satisfiable* if there is a pair $(\mathfrak{h}, \mathfrak{m})$ such that $\mathfrak{h} \models_{\mathfrak{m}} \phi$. The satisfiability problem asks whether $\phi$ is satisfiable, given a formula $\phi$.

We are also interested another separating operator $\overset{\rightharpoonup}{-\!*}$, called "septraction", which is the dual operator of $-\!\!*$, that is, $\phi_1 \overset{\rightharpoonup}{-\!*} \phi_2 \equiv \neg(\phi_1 \mathbin{-\!\!*} \neg\phi_2)$. More specifically, $\mathfrak{h} \models_{\mathfrak{m}} \phi_1 \overset{\rightharpoonup}{-\!*} \phi_2 \overset{\text{def}}{\Leftrightarrow}$ there is $\mathfrak{h}'$ such that $\mathfrak{h} \perp \mathfrak{h}'$, $\mathfrak{h}' \models_{\mathfrak{m}} \phi_1$, and $\mathfrak{h} \uplus \mathfrak{h}' \models_{\mathfrak{m}} \phi_2$. Then we can define the logic $\mathrm{SPSL2}\left((\overset{f}{\hookrightarrow})_{f \in \mathbb{F}}, *, \overset{\rightharpoonup}{-\!*}\right)$ by replacing the rule $\phi \mathbin{-\!\!*} \phi$ in the definition of $\mathrm{SPSL2}\left((\overset{f}{\hookrightarrow})_{f \in \mathbb{F}}, *, -\!\!*\right)$ with $\phi \overset{\rightharpoonup}{-\!*} \phi$. Note that if the rule $\phi \overset{\rightharpoonup}{-\!*} \phi$ was *added* to the definition of $\mathrm{SPSL2}\left((\overset{f}{\hookrightarrow})_{f \in \mathbb{F}}, *, -\!\!*\right)$, then the resulting logic would become undecidable ([8]).

In this paper, we also consider an additional fragment whose definition is presented below. Let $\mathrm{ESPSL2}\left((\overset{f}{\hookrightarrow})_{f \in \mathbb{F}}, *, \overset{\rightharpoonup}{-\!*}\right)$ denote existential SPSL2 $\left((\overset{f}{\hookrightarrow})_{f \in \mathbb{F}}, *, \overset{\rightharpoonup}{-\!*}\right)$, which is the extension of $\mathrm{SPSL2}\left((\overset{f}{\hookrightarrow})_{f \in \mathbb{F}}\right)$ with $*$ and $\overset{\rightharpoonup}{-\!*}$ such that no occurrences of $*$ and $\overset{\rightharpoonup}{-\!*}$ are in the scope of universal quantifiers. More precisely, $\mathrm{ESPSL2}\left((\overset{f}{\hookrightarrow})_{f \in \mathbb{F}}, *, \overset{\rightharpoonup}{-\!*}\right)$ formulae $\psi$ are defined by the following rules,

$$\psi ::= \phi \mid \psi \vee \psi \mid \psi \wedge \psi \mid \exists v.\ \psi \mid \psi * \psi \mid \psi \overset{\rightharpoonup}{-\!*} \psi,$$

where $\phi$ is an $\mathrm{SPSL2}\left((\overset{f}{\hookrightarrow})_{f \in \mathbb{F}}\right)$ formula. Note that since $\mathrm{SPSL2}\left((\overset{f}{\hookrightarrow})_{f \in \mathbb{F}}\right)$ formulae may contain universal quantifiers, we notice that universal quantifiers may still occur in the $\mathrm{ESPSL2}\left((\overset{f}{\hookrightarrow})_{f \in \mathbb{F}}, *, \overset{\rightharpoonup}{-\!*}\right)$ formulae.

*Example 1.* Let $\psi \overset{\text{def}}{=} (\neg\, x = y) \wedge (\psi_1' \overset{\rightharpoonup}{-\!*} \psi_2') \overset{\rightharpoonup}{-\!*} (\psi_3' \overset{\rightharpoonup}{-\!*} \psi_4')$, where

- $\psi_1' \overset{\text{def}}{=} x \overset{f}{\hookrightarrow} y$ expresses that $y$ is the $f$-successor of $x$,
- $\psi_2' \overset{\text{def}}{=} (\exists y.\ x \overset{f}{\hookrightarrow} y) \wedge (\exists y.\ y \overset{f}{\hookrightarrow} x)$ expresses that the $f$-successor of $x$ exists and there is an $f$-predecessor of $x$,
- $\psi_3' \overset{\text{def}}{=} \neg\, x \overset{f}{\hookrightarrow} y \wedge \exists y.\ (x \overset{f}{\hookrightarrow} y \wedge \forall x.\ \neg y \overset{f}{\hookrightarrow} x)$ expresses that $y$ is not the $f$-successor of $x$, but the $f$-successor of $x$ exists, and the $f$-successor of $x$ has no $f$-successor, and
- $\psi_4' \overset{\text{def}}{=} ((\exists y.\ y \overset{f}{\hookrightarrow} x) * (\exists y.\ y \overset{f}{\hookrightarrow} x)) \wedge \exists y.\ (x \overset{f}{\hookrightarrow} y \wedge \exists x.\ (y \overset{f}{\hookrightarrow} x))$ expresses that there are two distinct $f$-predecessors of $x$ and a path of length at least two starting from $x$.

Then $\psi$ is an $\mathrm{ESPSL2}\left((\xrightarrow{\mathsf{f}})_{\mathsf{f}\in\mathbb{F}}, *, \xrightarrow{\twoheadrightarrow}\right)$ formula. It is not hard to see that $\mathfrak{h} \models_\mathfrak{m}$ $\psi'_1 \xrightarrow{\twoheadrightarrow} \psi'_2$ iff $\mathfrak{m}(x)$ has an $\mathsf{f}$-predecessor in $\mathfrak{h}$, and $\mathfrak{h} \models_\mathfrak{m} \psi'_3 \xrightarrow{\twoheadrightarrow} \psi'_4$ iff $\mathfrak{m}(x)$ has two $\mathsf{f}$-predecessors in $\mathfrak{h}$ and there is a location $\mathfrak{l} \neq \mathfrak{m}(y)$ such that the $\mathsf{f}$-successor of $\mathfrak{l}$ exists in $\mathfrak{h}$. Therefore, $\mathfrak{h} \models_\mathfrak{m} \psi$ iff $\mathfrak{m}(x) \neq \mathfrak{m}(y)$, there is an $\mathsf{f}$-predecessor of $\mathfrak{m}(x)$ in $\mathfrak{h}$, and there is a location $\mathfrak{l} \neq \mathfrak{m}(y)$ such that the $\mathsf{f}$-successor of $\mathfrak{l}$ exists in $\mathfrak{h}$. $\square$

For the presentation of the decision procedure for $\mathrm{ESPSL2}\left((\xrightarrow{\mathsf{f}})_{\mathsf{f}\in\mathbb{F}}, *, \xrightarrow{\twoheadrightarrow}\right)$ in Sect. 4, the extension of $\mathrm{SPSL2}\left((\xrightarrow{\mathsf{f}})_{\mathsf{f}\in\mathbb{F}}\right)$ with *unary predicates* is also relevant.

Let $\mathcal{P}$ be a finite set of unary predicates. The extension of $\mathrm{SPSL2}\left((\xrightarrow{\mathsf{f}})_{\mathsf{f}\in\mathbb{F}}\right)$ with unary predicates from $\mathcal{P}$, denoted by $\mathrm{SPSL2}\left((\xrightarrow{\mathsf{f}})_{\mathsf{f}\in\mathbb{F}}, \mathcal{P}\right)$, is defined by the syntax rules of $\mathrm{SPSL2}\left((\xrightarrow{\mathsf{f}})_{\mathsf{f}\in\mathbb{F}}\right)$, plus two new rules $\phi ::= P(v) \mid \neg P(v)$, where $P \in \mathcal{P}$.

The semantics of $\mathrm{SPSL2}\left((\xrightarrow{\mathsf{f}})_{\mathsf{f}\in\mathbb{F}}, \mathcal{P}\right)$ formulae are defined as a relation $(\mathfrak{h}, \mathfrak{I}) \models_\mathfrak{m} \phi$, where $\mathfrak{h}, \mathfrak{m}$ are as before, and $\mathfrak{I} : \mathbb{N} \to 2^\mathcal{P}$ is a function such that $\mathrm{dom}(\mathfrak{I}) = \{\mathfrak{l} \in \mathbb{N} \mid \mathfrak{I}(\mathfrak{l}) \neq \emptyset\}$ is finite. The relation $(\mathfrak{h}, \mathfrak{I}) \models_\mathfrak{m} \phi$ is a natural extension of the relation $\mathfrak{h} \models_\mathfrak{m} \phi$ defined above, where $(\mathfrak{h}, \mathfrak{I}) \models_\mathfrak{m} P(v) \overset{\text{def}}{\Leftrightarrow} P \in \mathfrak{I}(\mathfrak{m}(v))$, and $(\mathfrak{h}, \mathfrak{I}) \models_\mathfrak{m} \neg P(v) \overset{\text{def}}{\Leftrightarrow} P \notin \mathfrak{I}(\mathfrak{m}(v))$. The function $\mathfrak{I}$ can also be seen in another way: It assigns each $P \in \mathcal{P}$ a finite subset of $\mathbb{N}$, that is, the set $\{\mathfrak{l} \in \mathbb{N} \mid P \in \mathfrak{I}(\mathfrak{l})\}$. The pairs $(\mathfrak{h}, \mathfrak{I})$ are called *labeled heaps*. Note that in a labeled heap $(\mathfrak{h}, \mathfrak{I})$, there may exist $\mathfrak{l} \notin \mathrm{loc}(\mathfrak{h})$ such that $\mathfrak{I}(\mathfrak{l}) \neq \emptyset$, in other words, $\mathrm{dom}(\mathfrak{I})$ may not necessarily be a subset of $\mathrm{loc}(\mathfrak{h})$.

# 3    $\mathrm{SPSL2}\left((\xrightarrow{\mathsf{f}})_{\mathsf{f}\in\mathbb{F}}\right)$ and $\mathrm{SPSL2}\left((\xrightarrow{\mathsf{f}})_{\mathsf{f}\in\mathbb{F}}, \mathcal{P}\right)$

This section is devoted to the proof of the following result.

**Theorem 1.** *The satisfiability of $SPSL2\left((\xrightarrow{\mathsf{f}})_{\mathsf{f}\in\mathbb{F}}\right)$ and $SPSL2\left((\xrightarrow{\mathsf{f}})_{\mathsf{f}\in\mathbb{F}}, \mathcal{P}\right)$ is* NEXPTIME-*complete.*

The rest of this section is devoted to the proof of Theorem 1. We consider the lower bound first, then the upper bound.

## 3.1    Lower Bound

We show the lower bound for the special case that $\mathbb{F} = \{\mathsf{f}\}$ and $\mathrm{SPSL2}(\xrightarrow{\mathsf{f}})$, that is, the satisfiability problem is NEXPTIME-hard even if there is only one field and there are no unary predicates. Since $\mathbb{F} = \{f\}$, in the following, for brevity, we will omit $f$ in the proof of the lower bound. The lower bound is obtained by a reduction from the exponential-size tiling problem. The problem is defined as follows: Given a tuple $(D, H, V, u)$, where $D = \{\mathfrak{d}_1, \ldots, \mathfrak{d}_s\}$ is a finite set of *tiles*, $H, V \subseteq D \times D$, $u = u_0 \ldots u_{n-1} \in D^n$, decide whether there is a tiling $t : [2^n] \times [2^n] \to D$ such that

- horizontal constraint: for all $i, j \in [2^n]$, if $t(i, j) = \mathfrak{d}$ and $t(i + 1, j) = \mathfrak{d}'$, then $(\mathfrak{d}, \mathfrak{d}') \in H$,
- vertical constraint: for all $i, j \in [2^n]$, if $t(i, j) = \mathfrak{d}$ and $t(i, j + 1) = \mathfrak{d}'$, then $(\mathfrak{d}, \mathfrak{d}') \in V$,
- initial condition: for every $i \in [n]$, $t(i, 0) = u_i$.

A tiling $t$ is equivalent to the set $X_t = \{(i, j, t(i, j)) : i, j \in [2^n]\}$ and therefore we explain below how to encode such a set by a heap. Given a heap $\mathfrak{h}$, we say that a location $\mathfrak{l}$ has a backward path of length exactly $i \geq 1$ iff there are locations $\mathfrak{l}_0, \ldots, \mathfrak{l}_i$ such that $\mathfrak{l}_i = \mathfrak{l}$, $\mathfrak{l}_0$ has no predecessor and for every $j \in [1, i]$, $\mathfrak{h}(\mathfrak{l}_{j-1}) = \mathfrak{l}_j$. A triple $(i, j, \mathfrak{d})$ in $X_t$ is encoded by a connected component $\mathcal{C}$ satisfying the following constraints.

- There is a location $\mathfrak{l}$ in $\mathcal{C}$ without successor and with at least one predecessor, which is identified by the following formula $\texttt{tile}(x) \stackrel{\text{def}}{=} (\exists y. (y \hookrightarrow x)) \wedge \forall y. \neg(x \hookrightarrow y)$.
- For every $k \in [1, s]$, $\mathfrak{d} = \mathfrak{d}_k$ iff $\mathfrak{l}$ has a backward path of length exactly $2n + k$. It is not hard to construct an SPSL2($\hookrightarrow$) formula $\mathsf{d}_k$ to describe this property.
- For every $k \in [1, n]$, the $k$th bit in the binary representation of $i$ (here the left-most bit is the first bit) is equal to 1 iff $\mathfrak{l}$ has a backward path of length exactly $k$. Similarly, this property can be described by an SPSL2($\hookrightarrow$) formula $\mathsf{h}_k$.
- For every $k \in [1, n]$, the $k$th bit in the binary representation of $j$ is equal to 1 iff $\mathfrak{l}$ has a backward path of length exactly $n + k$. Similarly, this property can be described by an SPSL2($\hookrightarrow$) formula $\mathsf{v}_k$.

Then an SPSL2($\hookrightarrow$) formula $\phi$ can be constructed so that $\phi$ is satisfiable iff the tiling problem instance has a solution.

In the following, we first define the formulae in SPSL2($\hookrightarrow$) with a unique free variable, say $\mathsf{h}_1(x), \ldots, \mathsf{h}_n(x), \mathsf{v}_1(x), \ldots, \mathsf{v}_n(x), \mathsf{d}_1(x), \ldots, \mathsf{d}_s(x)$, then $\phi$. By swapping $x$ and $y$, we also get the formulae $\mathsf{h}_1(y), \ldots, \mathsf{h}_n(y), \mathsf{v}_1(y), \ldots, \mathsf{v}_n(y)$, $\mathsf{d}_1(y), \ldots, \mathsf{d}_s(y)$.

Let us start by defining some auxiliary formulae.

1. $\psi_1(x) \stackrel{\text{def}}{=} \exists y (y \hookrightarrow x \wedge \forall x (\neg x \hookrightarrow y))$. The formula $\psi_1(x)$ simply states that $x$ has a predecessor with no predecessors (but $x$ may have other arbitrary predecessors).

2. $\psi_i(\mathbf{x}) \stackrel{\text{def}}{=} \exists\, \mathbf{y}\, (\mathbf{y} \hookrightarrow \mathbf{x} \wedge \wedge \psi_{i-1}(\mathbf{y}))$ for every $i \geq 2$. Assuming that $\mathbf{x}$ is interpreted by $\mathfrak{l}$, the formula $\psi_i(\mathbf{x})$ simply states that $\mathfrak{l}$ has a backward path of length exactly $i$.

Now let us define the formulae $\mathbf{h}_i(\mathbf{x})$, $\mathbf{v}_i(\mathbf{x})$ and $\mathbf{d}_i(\mathbf{x})$.

- For every $i \in [1, n]$, $\mathbf{h}_i(\mathbf{x}) \stackrel{\text{def}}{=} \psi_i(\mathbf{x})$.
- For every $i \in [1, n]$, $\mathbf{v}_i(\mathbf{x}) \stackrel{\text{def}}{=} \psi_{n+i}(\mathbf{x})$.
- For every $i \in [1, s]$, $\mathbf{d}_i(\mathbf{x}) \stackrel{\text{def}}{=} \psi_{2n+i}(\mathbf{x})$.

The three types of formulae are therefore only distinguished by path lengths.

Let $\phi$ be defined as the conjunction of the following formulae.

- Two locations encoding a position in the arena satisfy exactly the same formulae among $\mathbf{h}_1(\mathbf{x}), \ldots, \mathbf{h}_n(\mathbf{x}), \mathbf{v}_1(\mathbf{x}), \ldots, \mathbf{v}_n(\mathbf{x})$ are necessarily identical:

$$\phi_1 \stackrel{\text{def}}{=} \forall\, \mathbf{x}. \forall \mathbf{y}. \left( \left[ \begin{array}{c} \texttt{tile}(\mathbf{x}) \wedge \texttt{tile}(\mathbf{y}) \wedge \\ \bigwedge_{i \in [1,n]} ((\mathbf{h}_i(\mathbf{x}) \leftrightarrow \mathbf{h}_i(\mathbf{y})) \wedge (\mathbf{v}_i(\mathbf{x}) \leftrightarrow \mathbf{v}_i(\mathbf{y}))) \end{array} \right] \rightarrow \mathbf{x} = \mathbf{y} \right).$$

- There is a location that corresponds to the bottom left position:

$$\phi_2 \stackrel{\text{def}}{=} \exists\, \mathbf{x}. \left( \texttt{tile}(\mathbf{x}) \wedge \bigwedge_{i \in [1,n]} (\neg \mathbf{h}_i(\mathbf{x}) \wedge \neg \mathbf{v}_i(\mathbf{x})) \right).$$

- Each location encoding a position in the arena satisfies a unique tile:

$$\phi_3 \stackrel{\text{def}}{=} \forall\, \mathbf{x}. \left( \texttt{tile}(\mathbf{x}) \rightarrow \bigvee_{i \in [1,s]} \left( \mathbf{d}_i(\mathbf{x}) \wedge \bigwedge_{j \in [1,s] \setminus \{i\}} \neg \mathbf{d}_j(\mathbf{x}) \right) \right).$$

- Horizontal constraint for two consecutive positions within the same row:

$$\phi_4 \stackrel{\text{def}}{=} \forall\, \mathbf{x}. \left( \begin{array}{l} \left[ \texttt{tile}(\mathbf{x}) \wedge \bigwedge_{i \in [1,n]} \left( \neg \mathbf{h}_i(\mathbf{x}) \wedge \bigwedge_{i < j \leq n} \mathbf{h}_j(\mathbf{x}) \right) \right] \rightarrow \\ \exists\, \mathbf{y}. \left[ \begin{array}{l} \texttt{tile}(\mathbf{y}) \wedge \bigvee_{(\partial_l, \partial_m) \in H} (\mathbf{d}_l(\mathbf{x}) \wedge \mathbf{d}_m(\mathbf{y})) \wedge \\ \bigwedge_{j \in [1,n]} (\mathbf{v}_j(\mathbf{x}) \leftrightarrow \mathbf{v}_j(\mathbf{y})) \wedge \bigwedge_{1 \leq j < i} (\mathbf{h}_j(\mathbf{x}) \leftrightarrow \mathbf{h}_j(\mathbf{y})) \wedge \\ \mathbf{h}_i(\mathbf{y}) \wedge \bigwedge_{i < j \leq n} \neg \mathbf{h}_j(\mathbf{y}) \end{array} \right] \end{array} \right).$$

- The end of a row is immediately followed by the beginning of the next row, if any:

$$\phi_5 \stackrel{\text{def}}{=} \forall\, \mathbf{x}. \bigwedge_{i \in [1,n]} \left( \begin{array}{l} \left[ \texttt{tile}(\mathbf{x}) \wedge \left( \bigwedge_{j \in [1,n]} \mathbf{h}_j(\mathbf{x}) \right) \wedge \neg \mathbf{v}_i(\mathbf{x}) \wedge \bigwedge_{i < j \leq n} \mathbf{v}_j(\mathbf{x}) \right] \rightarrow \\ \exists\, \mathbf{y}. \left[ \begin{array}{l} \texttt{tile}(\mathbf{y}) \wedge \left( \bigwedge_{j \in [1,n]} \neg \mathbf{h}_j(\mathbf{y}) \right) \wedge \\ \left( \bigwedge_{1 \leq j < i} \mathbf{v}_j(\mathbf{x}) \leftrightarrow \mathbf{v}_j(\mathbf{y}) \right) \wedge \mathbf{v}_i(\mathbf{y}) \wedge \bigwedge_{i < j \leq n} \neg \mathbf{v}_j(\mathbf{y}) \end{array} \right] \end{array} \right).$$

Satisfaction of the formulae $\phi_2$, $\phi_3$, $\phi_4$ and $\phi_5$ guarantees that all the positions in the arena are encoded. Unicity of such an encoding is a consequence of the satisfaction of $\phi_1$.

- Vertical constraint between two consecutive vertical positions:

$$\phi_6 \overset{def}{=} \forall\, x.\ \bigwedge_{i \in [1,n]} \exists\, y. \left( \begin{array}{l} \left[ \mathtt{tile(x)} \wedge (\neg v_i(x)) \wedge \bigwedge_{i<j\leq n} v_j(x) \right] \rightarrow \\ \left[ \begin{array}{l} \mathtt{tile(y)} \wedge \bigvee_{(\partial_l, \partial_m) \in V} (d_l(x) \wedge d_m(y)) \wedge \\ \bigwedge_{j \in [1,n]} (h_j(x) \leftrightarrow h_j(y)) \wedge \bigwedge_{1 \leq j < i} (v_j(x) \leftrightarrow v_j(y)) \wedge \\ v_i(y) \wedge \bigwedge_{i<j\leq n} (\neg v_j(y)) \end{array} \right] \end{array} \right).$$

- Suppose $u_0 \ldots u_{n-1} = \partial_{j_0} \ldots \partial_{j_{n-1}}$, then the initial condition is specified as follows:

$$\phi_7 \overset{def}{=} \bigwedge_{i \in [0,n-1]} \exists\, x. \left( \begin{array}{l} \mathtt{tile(x)} \wedge \bigwedge_{j' \in [1,n]} (\neg v_{j'}(x)) \wedge d_{j_i}(x) \wedge \\ \bigwedge_{j' \in [1,n],\ j'\text{-th bit of } i \text{ is } 1} h_{j'}(x) \wedge \\ \bigwedge_{j' \in [1,n],\ j'\text{-th bit of } i \text{ is } 0} \neg h_{j'}(x) \end{array} \right).$$

Note that for readability, we choose to write the formulae $\phi_1, \ldots, \phi_7$ above not in negation normal form as required in the definition of the logic SPSL2($\hookrightarrow$) (cf. Sect. 2). Nevertheless, since the logic SPSL2($\hookrightarrow$) is closed under negations, those formulae can be easily rewritten into the required form.

**Lemma 1.** *The formula $\phi$ is satisfiable iff the tiling problem instance has a solution.*

*Proof.* First, suppose the tiling problem instance has a solution $t$. Then we construct a heap $\mathfrak{h}$ from $t$ such that

- $t$ comprises $2^n \times 2^n$ connected components, one for each $(i,j) \in [2^n] \times [2^n]$ (denoted by $\mathcal{C}_{i,j}$),
- each $\mathcal{C}_{i,j}$ is a tree, which comprises the following backward paths from the root $\mathfrak{l}$,
    - for each $k \in [1,n]$, $\mathfrak{l}$ has a backward path of length exactly $k$ (resp. $n+k$) iff the $k$th bit of the binary representation of $i$ (resp. $j$) is equal to 1,
    - let $t(i,j) = \partial_k$, then $\mathfrak{l}$ has a backward path of length exactly $2n + k$.

Since $t$ satisfies the horizontal and vertical constraint, as well as the initial condition, $\mathfrak{h} \models \phi_4 \wedge \phi_5 \wedge \phi_6 \wedge \phi_7$. Moreover, from the construction of $\mathfrak{h}$, we know that $\mathfrak{h} \models \phi_1 \wedge \phi_2 \wedge \phi_3$. Therefore, we conclude that $\mathfrak{h} \models \phi$.

Let us establish the other direction. Suppose that $\phi$ is satisfiable. Then there is a heap $\mathfrak{h}$ satisfying $\phi$. From the fact that $\mathfrak{h} \models \phi_1 \wedge \phi_2 \wedge \phi_3 \wedge \phi_4 \wedge \phi_5$, we know that for each $(i,j) \in [2^n] \times [2^n]$, there is exactly one connected component

$C_{i,j}$ such that for each $k \in [1, n]$, the root of $C_{i,j}$ has a backward path of length exactly $k$ (resp. $n + k$) iff the $k$th bit of the binary representation of $i$ (resp. $j$) is equal to 1. We construct a tiling $\mathsf{t} : [2^n] \times [2^n] \to D$ as follows: For each $(i, j) \in [2^n] \times [2^n]$, suppose $k \in [1, s]$ satisfies that the root of $C_{i,j}$ has a backward path of length exactly $2n + k$ (such a $k$ exists since $\mathfrak{h} \models \phi_3$), let $\mathsf{t}(i, j) = \mathfrak{d}_k$. Because $\mathfrak{h} \models \phi_4 \wedge \phi_6 \wedge \phi_7$, we know that $\mathsf{t}$ satisfies the horizontal and vertical constraint as well as the initial condition. Therefore, $\mathsf{t}$ is a solution of the tiling problem instance. $\qquad\square$

## 3.2   Upper Bound

The upper bound is obtained by a linear time reduction to the finite satisfiability problem of first-order logic with two-variables and counting quantifiers (denoted by $\mathcal{C}^2$), which can be decided in NEXPTIME [18, 19]. Before presenting the reduction, we first recall the definition of $\mathcal{C}^2$. A *purely relational vocabulary* $\mathcal{V}$ comprises relational symbols, but no function symbols, nor constants. The logic $\mathcal{C}^2$ over a purely relational vocabulary $\mathcal{V}$ is defined by the following rules,

$$\mathsf{v} ::= \mathsf{x} \mid \mathsf{y},$$

$$\varphi ::= \mathsf{v} = \mathsf{v} \mid R(\bar{\mathsf{v}}) \mid \neg\varphi \mid \varphi \vee \varphi \mid \exists^{\odot C}\mathsf{v}.\ \varphi,$$

where $R \in \mathcal{V}$ is of arity $k$, $\bar{\mathsf{v}} \in \{\mathsf{x}, \mathsf{y}\}^k$, $\odot \in \{<, >, \leq, \geq, =\}$, and $C \in \mathbb{N}$ is a constant. We assume that all the constants in $\mathcal{C}^2$ are encoded in binary. For a formula $\varphi \in \mathcal{C}^2$, let $|\varphi|$ denote the number of symbols occurring in $\varphi$. The formulae in $\mathcal{C}^2$ are interpreted on a triple $(A, \mathcal{I}, \mathfrak{m})$, where $A$ is a *domain*, $\mathcal{I}$ is called an *interpretation function*, which assigns each $k$-ary relation symbol $R \in \mathcal{V}$ a subset of $A^k$, and $\mathfrak{m}$ is an *assignment* that maps $\mathsf{x}$ and $\mathsf{y}$ to $A$. The semantics of $\mathcal{C}^2$ formulae are defined by a relation $(A, \mathcal{I}) \models_{\mathfrak{m}} \varphi$. The semantics of the atomic formulae, the Boolean combination, the quantifiers are standard. For the $\mathcal{C}^2$ formulae $\varphi = \exists^{=C}\mathsf{x}.\ \varphi_1$, $(A, \mathcal{I}) \models_{\mathfrak{m}} \varphi$ iff there are exactly $C$ elements of $A$, say $a_1, \ldots, a_C$, such that for each $i \in [C]$, $(A, \mathcal{I}) \models_{\mathfrak{m}[\mathsf{x} \mapsto a_i]} \varphi_1$. The semantics of the formulae $\exists^{=C}\mathsf{y}.\ \varphi_1$, $\exists^{>C}\mathsf{x}.\ \varphi_1$, $\exists^{<C}\mathsf{x}.\ \varphi_1$, etc. can be defined similarly. Let $\varphi$ be a $\mathcal{C}^2$ formula. If $(A, \mathcal{I}) \models_{\mathfrak{m}} \varphi$, then $(A, \mathcal{I}, \mathfrak{m})$ is called a *model* of $\varphi$. Then $\varphi$ is *satisfiable* if $\varphi$ has a model, and $\varphi$ is *finitely satisfiable* if $\varphi$ has a model $(A, \mathcal{I}, \mathfrak{m})$ such that $A$ is finite.

**Lemma 2** (*[18, 19]*).   *The satisfiability and finite satisfiability problem of $\mathcal{C}^2$ are* NEXPTIME-*complete.*

We are ready to present the reduction.

For each $\mathsf{f} \in \mathbb{F}$, introduce a fresh binary relation symbol $R_{\mathsf{f}}$. Let $\mathcal{V} = \mathcal{P} \cup \{R_{\mathsf{f}} \mid \mathsf{f} \in \mathbb{F}\}$. In the following, for each formula $\phi$ in $\mathrm{SPSL2}\left((\xrightarrow{\mathsf{f}})_{\mathsf{f} \in \mathbb{F}}, \mathcal{P}\right)$, we construct a $\mathcal{C}^2$ formula $\mathsf{trs}(\phi) = \phi_{\mathsf{fun}} \wedge \exists \mathsf{x}.\exists \mathsf{y}.\ (\neg\mathsf{x} = \mathsf{y} \wedge \phi_{\mathsf{rel}}(\mathsf{x}) \wedge \phi_{\mathsf{rel}}(\mathsf{y})) \wedge \phi'$ over the vocabulary $\mathcal{V}$, where

– $\phi_{\mathtt{fun}} = \bigwedge_{f \in \mathbb{F}} \forall x. \exists^{\leq 1} y.\ R_f(x, y)$ expresses that each relation symbol $R_f$ is the image of a partial function,

– $\phi_{\mathtt{rel}}(v) = \left( \bigwedge_{P \in \mathcal{P}} \neg P(v) \right) \wedge \forall v'.\ \bigwedge_{f \in \mathbb{F}} (\neg R_f(v, v') \wedge \neg R_f(v', v))$, where $v = x$ and $v' = y$, or vice versa, expresses that there is an element represented by $v$ which does not occur in any tuple from the union of the relations $P \in \mathcal{P}$ and $R_f$ for $f \in \mathbb{F}$,

– $\phi'$ is obtained from $\phi$ by replacing each atomic formula of the form $v_1 \overset{f}{\hookrightarrow} v_2$ with $R_f(v_1, v_2)$.

The formula $\exists x. \exists y.\ (\neg x = y \wedge \phi_{\mathtt{rel}}(x) \wedge \phi_{\mathtt{rel}}(y))$ expresses that there are two distinct elements satisfying the formula $\phi_{\mathtt{rel}}$.

The correctness of the reduction is guaranteed by the following result.

**Proposition 1.** *For each* $SPSL2\left( (\overset{f}{\hookrightarrow})_{f \in \mathbb{F}}, \mathcal{P} \right)$ *formula* $\phi$, $\phi$ *is satisfiable iff* $\mathtt{trs}(\phi)$ *is finitely satisfiable.*

*Proof.* Suppose that $\phi$ is an $SPSL2\left( (\overset{f}{\hookrightarrow})_{f \in \mathbb{F}}, \mathcal{P} \right)$ formula.

*"Only if" direction*: Suppose that $\phi$ is satisfiable. Then there are a labeled heap $(\mathfrak{h}, \mathfrak{I})$ and $\mathfrak{m}$ such that $(\mathfrak{h}, \mathfrak{I}) \models_{\mathfrak{m}} \phi$.

We construct a finite set $A = \mathtt{loc}(\mathfrak{h}) \cup (\bigcup_{P \in \mathcal{P}} \mathfrak{I}(P)) \cup \{\mathfrak{l}_1, \mathfrak{l}_2\}$, where

– if $\mathfrak{m}(x), \mathfrak{m}(y) \notin \mathtt{loc}(\mathfrak{h}) \cup (\bigcup_{P \in \mathcal{P}} \mathfrak{I}(P))$ such that $\mathfrak{m}(x) \neq \mathfrak{m}(y)$, then let $\mathfrak{l}_1 = \mathfrak{m}(x)$ and $\mathfrak{l}_2 = \mathfrak{m}(y)$,

– if $\mathfrak{m}(x), \mathfrak{m}(y) \notin \mathtt{loc}(\mathfrak{h}) \cup (\bigcup_{P \in \mathcal{P}} \mathfrak{I}(P))$ such that $\mathfrak{m}(x) = \mathfrak{m}(y)$, then let $\mathfrak{l}_1 = \mathfrak{m}(x)$ and $\mathfrak{l}_2$ be a location in $\mathbb{N} \setminus (\mathtt{loc}(\mathfrak{h}) \cup (\bigcup_{P \in \mathcal{P}} \mathfrak{I}(P)) \cup \{\mathfrak{m}(x)\})$,

– if $\mathfrak{m}(v) \notin \mathtt{loc}(\mathfrak{h}) \cup (\bigcup_{P \in \mathcal{P}} \mathfrak{I}(P))$ and $\mathfrak{m}(v') \in \mathtt{loc}(\mathfrak{h}) \cup (\bigcup_{P \in \mathcal{P}} \mathfrak{I}(P))$, then let $\mathfrak{l}_1 = \mathfrak{m}(v)$ and $\mathfrak{l}_2$ be a location in $\mathbb{N} \setminus (\mathtt{loc}(\mathfrak{h}) \cup (\bigcup_{P \in \mathcal{P}} \mathfrak{I}(P)) \cup \{\mathfrak{m}(v)\})$, where $v = x$ and $v' = y$, or vice versa,

– if $\mathfrak{m}(x) \in \mathtt{loc}(\mathfrak{h}) \cup (\bigcup_{P \in \mathcal{P}} \mathfrak{I}(P))$ and $\mathfrak{m}(y) \in \mathtt{loc}(\mathfrak{h}) \cup (\bigcup_{P \in \mathcal{P}} \mathfrak{I}(P))$, then let $\mathfrak{l}_1$ and $\mathfrak{l}_2$ be two distinct locations in $\mathbb{N} \setminus (\mathtt{loc}(\mathfrak{h}) \cup (\bigcup_{P \in \mathcal{P}} \mathfrak{I}(P)))$.

Consider the triple $(A, \mathcal{I}, \mathfrak{m})$, where $\mathcal{I}(P) = \mathfrak{I}(P)$ for each $P \in \mathcal{P}$, and $\mathcal{I}(R_f) = \{(\mathfrak{l}, \mathfrak{l}') \in \mathtt{loc}(\mathfrak{h}) \times \mathtt{loc}(\mathfrak{h}) \mid \mathfrak{h}_f(\mathfrak{l}) = \mathfrak{l}'\}$. We claim that $(A, \mathcal{I}) \models_{\mathfrak{m}} \mathtt{trs}(\phi)$. Since evidently $(A, \mathcal{I}) \models_{\mathfrak{m}} \psi_{\mathtt{fun}} \wedge \exists x. \exists y.\ (\neg x = y \wedge \phi_{\mathtt{rel}}(x) \wedge \phi_{\mathtt{rel}}(y))$, is sufficient to show that $(A, \mathcal{I}) \models_{\mathfrak{m}} \phi'$. In the following, we show $(A, \mathcal{I}) \models_{\mathfrak{m}} \phi'$ by proving that for each assignment $\mathfrak{m}'$ such that $\mathtt{ran}(\mathfrak{m}') \subseteq A$ and each subformula $\phi_1$ of $\phi$, $(\mathfrak{h}, \mathfrak{I}) \models_{\mathfrak{m}'} \phi_1$ iff $(A, \mathcal{I}) \models_{\mathfrak{m}'} \psi'_1$, where $\psi'_1$ is obtained from $\phi_1$ by replacing each atomic formula of the form $v_1 \overset{f}{\hookrightarrow} v_2$ with $R_f(v_1, v_2)$. We show this fact by induction on the syntax of formulae.

– The cases $\phi_1 \overset{\mathtt{def}}{=} v_1 = v_2$ and $\phi_1 \overset{\mathtt{def}}{=} \neg v_1 = v_2$ are trivial.

– Case $\phi_1 \overset{\mathtt{def}}{=} v_1 \overset{f}{\hookrightarrow} v_2$: Since $\mathtt{ran}(\mathfrak{m}') \subseteq A$, $(\mathfrak{h}, \mathfrak{I}) \models_{\mathfrak{m}'} v_1 \overset{f}{\hookrightarrow} v_2$ iff $\mathfrak{h}_f(\mathfrak{m}'(v_1)) = \mathfrak{m}'(v_2)$ iff $(\mathfrak{m}'(v_1), \mathfrak{m}'(v_2)) \in \mathcal{I}(R_f)$. Similarly for $\phi_1 \overset{\mathtt{def}}{=} \neg v_1 \overset{f}{\hookrightarrow} v_2$.

- Case $\phi_1 \overset{\text{def}}{=} P(\mathbf{v})$: $(\mathfrak{h}, \mathfrak{I}) \models_{\mathfrak{m}'} P(\mathbf{v})$ iff $\mathfrak{m}'(\mathbf{v}) \in \mathfrak{I}(P)$ iff $\mathfrak{m}'(\mathbf{v}) \in \mathcal{I}(P)$ iff $(A, \mathcal{I}) \models_{\mathfrak{m}'} P(\mathbf{v})$. Similarly for $\phi_1 \overset{\text{def}}{=} \neg P(\mathbf{v})$.

- Case $\phi_1 \overset{\text{def}}{=} \phi_2 \wedge \phi_3$ or $\phi_1 \overset{\text{def}}{=} \phi_2 \vee \phi_3$: The arguments are standard.

- Case $\phi_1 \overset{\text{def}}{=} \exists \mathbf{x}.\ \phi_2$: Our goal is to show $(\mathfrak{h}, \mathfrak{I}) \models_{\mathfrak{m}'} \phi_1$ iff $(A, \mathcal{I}) \models_{\mathfrak{m}'} \phi_1'$. Since the "if" direction is easy, we focus on the "only if" direction below. Suppose $(\mathfrak{h}, \mathfrak{I}) \models_{\mathfrak{m}'} \exists \mathbf{x}.\ \phi_2$. Then there is $\mathfrak{l}' \in \mathbb{N}$ such that $(\mathfrak{h}, \mathfrak{I}) \models_{\mathfrak{m}'[\mathbf{x} \mapsto \mathfrak{l}']} \phi_2$.
  - If $\mathfrak{l}' \in A$, then according to the induction hypothesis, $(\mathfrak{h}, \mathfrak{I}) \models_{\mathfrak{m}'[\mathbf{x} \mapsto \mathfrak{l}']} \phi_2$ iff $(A, \mathcal{I}) \models_{\mathfrak{m}'[\mathbf{x} \mapsto \mathfrak{l}']} \phi_2'$.
  - If $\mathfrak{l}' \notin A$, then $\mathfrak{l}' \neq \mathfrak{m}'(\mathbf{y})$ since $\mathfrak{m}'(\mathbf{y}) \in A$. Evidently, $\mathfrak{l}_1 \neq \mathfrak{m}'(\mathbf{y})$ or $\mathfrak{l}_2 \neq \mathfrak{m}'(\mathbf{y})$. Without loss of generality, we assume that $\mathfrak{l}_1 \neq \mathfrak{m}'(\mathbf{y})$. Because neither $\mathfrak{l}_1$ nor $\mathfrak{l}'$ belongs to $\mathtt{loc}(\mathfrak{h}) \cup \bigcup_{P \in \mathcal{P}} \mathfrak{I}(P)$, we deduce that $(\mathfrak{h}, \mathfrak{I}) \models_{\mathfrak{m}'[\mathbf{x} \mapsto \mathfrak{l}']} \phi_2$ iff $(\mathfrak{h}, \mathfrak{I}) \models_{\mathfrak{m}'[\mathbf{x} \mapsto \mathfrak{l}_1]} \phi_2$. From the induction hypothesis, $(\mathfrak{h}, \mathfrak{I}) \models_{\mathfrak{m}'[\mathbf{x} \mapsto \mathfrak{l}_1]} \phi_2$ iff $(A, \mathcal{I}) \models_{\mathfrak{m}'[\mathbf{x} \mapsto \mathfrak{l}_1]} \phi_2'$. Therefore, $(A, \mathcal{I}) \models_{\mathfrak{m}'[\mathbf{x} \mapsto \mathfrak{l}_1]} \phi_2'$ and $(A, \mathcal{I}) \models_{\mathfrak{m}'} \exists \mathbf{x}.\ \phi_2'$.

- Case $\phi_1 \overset{\text{def}}{=} \exists \mathbf{y}.\ \phi_2$: Similarly to the previous case.

- Case $\phi_1 \overset{\text{def}}{=} \forall \mathbf{x}.\ \phi_2$: Suppose that $(\mathfrak{h}, \mathfrak{I}) \models_{\mathfrak{m}'} \forall \mathbf{x}.\ \phi_2$. Then for each $\mathfrak{l}' \in \mathbb{N}$, $(\mathfrak{h}, \mathfrak{I}) \models_{\mathfrak{m}'[\mathbf{x} \mapsto \mathfrak{l}']} \phi_2$. For each $\mathfrak{l}' \in A$, according to the induction hypothesis, $(\mathfrak{h}, \mathfrak{I}) \models_{\mathfrak{m}'[\mathbf{x} \mapsto \mathfrak{l}']} \phi_2$ iff $(A, \mathcal{I}) \models_{\mathfrak{m}'[\mathbf{x} \mapsto \mathfrak{l}']} \phi_2'$. Therefore, for each $\mathfrak{l}' \in A$, we have $(A, \mathcal{I}) \models_{\mathfrak{m}'[\mathbf{x} \mapsto \mathfrak{l}']} \phi_2'$. We conclude that $(A, \mathcal{I}) \models_{\mathfrak{m}'} \forall \mathbf{x}.\ \phi_2' = \phi_1'$. On the other hand, suppose that $(A, \mathcal{I}) \models_{\mathfrak{m}'} \phi_1' = \forall \mathbf{x}.\ \phi_2'$. Then for each $\mathfrak{l}' \in A$, $(A, \mathcal{I}) \models_{\mathfrak{m}'[\mathbf{x} \mapsto \mathfrak{l}']} \phi_2'$. By the induction hypothesis, for each $\mathfrak{l}' \in A$, $(\mathfrak{h}, \mathfrak{I}) \models_{\mathfrak{m}'[\mathbf{x} \mapsto \mathfrak{l}']} \phi_2$ iff $(A, \mathcal{I}) \models_{\mathfrak{m}'[\mathbf{x} \mapsto \mathfrak{l}']} \phi_2'$. Therefore, for each $\mathfrak{l}' \in A$, $(\mathfrak{h}, \mathfrak{I}) \models_{\mathfrak{m}'[\mathbf{x} \mapsto \mathfrak{l}']} \phi_2$. Now suppose $\mathfrak{l}' \in \mathbb{N} \setminus A$. Without loss of generality, suppose that $\mathfrak{l}_1 \neq \mathfrak{m}'(\mathbf{y})$. Because neither $\mathfrak{l}_1$ nor $\mathfrak{l}'$ belongs to $\mathtt{loc}(\mathfrak{h}) \cup \bigcup_{P \in \mathcal{P}} \mathfrak{I}(P)$, $(\mathfrak{h}, \mathfrak{I}) \models_{\mathfrak{m}'[\mathbf{x} \mapsto \mathfrak{l}']} \phi_2$ iff $(\mathfrak{h}, \mathfrak{I}) \models_{\mathfrak{m}'[\mathbf{x} \mapsto \mathfrak{l}_1]} \phi_2$. From this, we deduce that for each $\mathfrak{l}' \in \mathbb{N} \setminus A$, $(\mathfrak{h}, \mathfrak{I}) \models_{\mathfrak{m}'[\mathbf{x} \mapsto \mathfrak{l}']} \phi_2$. Thus for each $\mathfrak{l}' \in \mathbb{N}$, $(\mathfrak{h}, \mathfrak{I}) \models_{\mathfrak{m}'[\mathbf{x} \mapsto \mathfrak{l}']} \phi_2$, that is, $(\mathfrak{h}, \mathfrak{I}) \models_{\mathfrak{m}'[\mathbf{x} \mapsto \mathfrak{l}']} \forall \mathbf{x}.\ \phi_2 = \phi_1$.

- Case $\phi_1 \overset{\text{def}}{=} \forall \mathbf{y}.\ \phi_2$: Similarly to the previous case.

*"If" direction*: Suppose that $\mathtt{trs}(\phi)$ is finitely satisfiable. Then there is a model $(A, \mathcal{I}, \mathfrak{m})$ of $\mathtt{trs}(\phi)$ such that $A$ is finite. Without loss of generality, we assume that $A$ is a subset of $\mathbb{N}$.

We construct $(\mathfrak{h}, \mathfrak{I})$ such that $\mathfrak{h}_\mathfrak{f}(\mathfrak{l}) = \mathfrak{l}'$ iff $(\mathfrak{l}, \mathfrak{l}') \in \mathcal{I}(R_\mathfrak{f})$ for each $\mathfrak{f} \in \mathbb{F}$, and $\mathfrak{l} \in \mathfrak{I}(P)$ iff $\mathfrak{l} \in \mathcal{I}(P)$ for each $P \in \mathcal{P}$. Since $(A, \mathcal{I}) \models_\mathfrak{m} \psi_{\mathtt{fun}} \wedge \exists \mathbf{x}.\exists \mathbf{y}.\ (\neg \mathbf{x} = \mathbf{y} \wedge \phi_{\mathtt{rel}}(\mathbf{x}) \wedge \phi_{\mathtt{rel}}(\mathbf{y}))$, we know that each $\mathfrak{h}_\mathfrak{f}$ for $\mathfrak{f} \in \mathbb{F}$ is a partial function, and there are two distinct locations $\mathfrak{l}_1, \mathfrak{l}_2 \in A \setminus \left( \mathtt{loc}(\mathfrak{h}) \cup \bigcup_{P \in \mathcal{P}} \mathfrak{I}(P) \right)$. Similarly to the arguments in the "Only if" direction, we can show that for each assignment $\mathfrak{m}'$ such that $\mathtt{ran}(\mathfrak{m}') \subseteq A$ and each subformula $\phi_1$ of $\phi$, $(\mathfrak{h}, \mathfrak{I}) \models_{\mathfrak{m}'} \phi_1$ iff $(A, \mathcal{I}) \models_{\mathfrak{m}'} \phi_1'$. From this, we deduce that $(\mathfrak{h}, \mathfrak{I}) \models_\mathfrak{m} \phi$ iff $(A, \mathcal{I}) \models_\mathfrak{m} \phi'$. We then conclude that $(\mathfrak{h}, \mathfrak{I}) \models_\mathfrak{m} \phi$. $\qquad\square$

# 4    ESPSL2 $\left( (\xrightarrow{\mathsf{f}})_{\mathsf{f}\in\mathbb{F}}, *, \xrightarrow{\neg *} \right)$

In this section, we present the main result of this paper.

**Theorem 2.** *The satisfiability problem of* $ESPSL2 \left( (\xrightarrow{\mathsf{f}})_{\mathsf{f}\in\mathbb{F}}, *, \xrightarrow{\neg *} \right)$ *is* NEXPTIME-*complete.*

The rest of this section is devoted to the proof of Theorem 2. The basic idea of the proof is to reduce in polynomial time the satisfiability of $ESPSL2\left( (\xrightarrow{\mathsf{f}})_{\mathsf{f}\in\mathbb{F}}, *, \xrightarrow{\neg *} \right)$ formulae to that of $SPSL2\left( (\xrightarrow{\mathsf{f}})_{\mathsf{f}\in\mathbb{F}}, \mathcal{P}' \right)$ formulae (for some $\mathcal{P}'$), which is NEXPTIME-complete (cf. Theorem 1). The main idea of the reduction is as follows: For a heap $\mathfrak{h}$, an assignment $\mathfrak{m}$, and an $ESPSL2\left( (\xrightarrow{\mathsf{f}})_{\mathsf{f}\in\mathbb{F}}, *, \xrightarrow{\neg *} \right)$ formula $\psi$ such that $\mathfrak{h} \models_{\mathfrak{m}} \psi$, in order to witness the fact $\mathfrak{h} \models_{\mathfrak{m}} \psi$, some other heaps should be added to $\mathfrak{h}$. Nevertheless, these additional heaps may conflict with each other. For instance, in Example 1, two heaps corresponding to $\psi'_1$ and $\psi'_3$ should be added, but these two heaps conflict with each other, since $\psi'_1$ says that $\mathsf{y}$ is the $\mathsf{f}$-successor of $\mathsf{x}$, while $\psi'_3$ says that the $\mathsf{f}$-successor of $\mathsf{x}$ but is different from $\mathsf{y}$. In the reduction from the satisfiability of $\psi$ to that of a $SPSL2\left( (\xrightarrow{\mathsf{f}})_{\mathsf{f}\in\mathbb{F}}, \mathcal{P}' \right)$ formula below, the fields in the subformulae of $\psi$ conflicting with each other are renamed, so that after the renaming, these subformulae refer to different fields. In addition, to guarantee the correctness of the reduction, some necessary constraints should be added to these fields as well as the unary predicates from $\mathcal{P}'$.

We introduce a concept of syntax trees which will be used in the reduction. Let $\psi$ be an $ESPSL2\left( (\xrightarrow{\mathsf{f}})_{\mathsf{f}\in\mathbb{F}}, *, \xrightarrow{\neg *} \right)$ formula. A *syntax tree* $\mathcal{T}_\psi = (T, E, L)$ can be constructed inductively as follows, where $T$ is a set of nodes, $E$ is the child-parent relation, and $L : T \to \mathtt{Sub}(\psi)$ is a labeling function.

- Case $\psi \overset{\text{def}}{=} \phi$, where $\phi$ is a $SPSL2\left( (\xrightarrow{\mathsf{f}})_{\mathsf{f}\in\mathbb{F}} \right)$ formula: Then $\mathcal{T}_\psi = (\{t\}, \emptyset, L)$ such that $L(t) = \psi$,
- Case $\psi \overset{\text{def}}{=} \psi_1 \odot \psi_2$ (where $\odot = \vee, \wedge, *, \xrightarrow{\neg *}$): Suppose two syntax trees $\mathcal{T}_{\psi_1} = (T_1, E_1, L_1)$ and $\mathcal{T}_{\psi_2} = (T_2, E_2, L_2)$ have been constructed for $\psi_1$ and $\psi_2$ respectively. Without loss of generality, suppose $T_1 \cap T_2 = \emptyset$ and the roots of $\mathcal{T}_{\psi_1}$ and $\mathcal{T}_{\psi_2}$ are $t_1$ and $t_2$ respectively. Then $\mathcal{T}_\psi = (T_1 \cup T_2 \cup \{t\}, E_1 \cup E_2 \cup \{(t_1, l, t), (t_2, r, t)\}, L_1 \cup L_2 \cup \{t \mapsto \psi\})$, where $t$ is a new node not in $T_1$ or $T_2$, and the label s$l, r$ denote the left and right child respectively.
- Case $\psi \overset{\text{def}}{=} \exists \mathsf{v}. \psi_1$: Suppose a syntax tree $\mathcal{T}_{\psi_1} = (T_1, E_1, L_1)$ has been constructed for $\psi_1$. Then $\mathcal{T}_\psi = (T_1 \cup \{t\}, E_1 \cup \{(t_1, l, t)\}, L_1 \cup \{t \mapsto \psi\})$, where $t$ is a new node not in $T_1$.

Let $\psi$ be an $\mathrm{ESPSL2}\left((\overset{f}{\hookrightarrow})_{f\in\mathbb{F}}, *, \overset{\rightharpoondown}{*}\right)$ formula and $\mathcal{T}_\psi = (T, E, L)$. For each $t \in T$, we introduce a fresh unary predicate $P'_t$. In addition, for each node $t \in T$ and $f \in \mathbb{F}$, introduce a fresh field $f'_t$. Let $\mathcal{P}'$ be the set of freshly introduced unary predicates, and $\mathbb{F}'$ be the set of freshly introduced fields. Our goal is to use $\mathcal{T}_\psi$ to construct an $\mathrm{SPSL2}\left((\overset{f'_t}{\hookrightarrow})_{f'_t\in\mathbb{F}'}, \mathcal{P}'\right)$ formula $\mathbf{trs}(\psi)$ so that $\psi$ is satisfiable iff $\mathbf{trs}(\psi)$ is satisfiable. Toward this purpose, for each node $t \in T$, we construct an $\mathrm{SPSL2}\left((\overset{f}{\hookrightarrow})_{f\in\mathbb{F}}, \mathcal{P}'\right)$ formula $\phi_t$. Then let $\mathbf{trs}(\psi) \overset{\mathrm{def}}{=}$

$$\bigwedge_{t\in T}\left(\forall \mathrm{x}.\ P'_t(\mathrm{x}) \leftrightarrow \bigvee_{f\in\mathbb{F}} \exists \mathrm{y}.\ \mathrm{x} \overset{f'_t}{\hookrightarrow} \mathrm{y}\right) \wedge \phi_{t_0},$$ 

where $t_0$ is the root of $\mathcal{T}_\psi$. The formulae $\phi_t$ for $t \in T$ are computed inductively as follows.

- If $L(t) = \phi$ for some $\mathrm{SPSL2}\left((\overset{f}{\hookrightarrow})_{f\in\mathbb{F}}\right)$ formula $\phi$, then $\phi_t \overset{\mathrm{def}}{=} \phi'$, where $\phi'$ is obtained from $\phi$ by replacing each occurrence of $f \in \mathbb{F}$ with $f'_t$.
- If $L(t) = \psi_1 \odot \psi_2$ (where $\odot \in \{\vee, \wedge\}$) and $t_1, t_2$ are two children of $t$ such that $L(t_1) = \psi_1$ and $L(t_2) = \psi_2$, suppose $\phi_{t_1}$ and $\phi_{t_2}$ have been computed from $t_1$ and $t_2$ respectively, then

$$\phi_t \overset{\mathrm{def}}{=} P'_t = P'_{t_1} = P'_{t_2} \wedge \bigwedge_{f\in\mathbb{F}} f'_t = f'_{t_1} = f'_{t_2} \wedge (\phi_{t_1} \odot \phi_{t_2}),$$

where $P'_t = P'_{t_1} = P'_{t_2}$ is an abbreviation of $\forall \mathrm{x}.\ (P'_t(\mathrm{x}) \leftrightarrow P'_{t_1}(\mathrm{x})) \wedge (P'_t(\mathrm{x}) \leftrightarrow P'_{t_2}(\mathrm{x}))$ and $f'_t = f'_{t_1} = f'_{t_2}$ is an abbreviation of $\forall \mathrm{x}.\forall \mathrm{y}.\ \left(\mathrm{x} \overset{f'_t}{\hookrightarrow} \mathrm{y} \leftrightarrow \mathrm{x} \overset{f'_{t_1}}{\hookrightarrow} \mathrm{y}\right) \wedge \left(\mathrm{x} \overset{f'_t}{\hookrightarrow} \mathrm{y} \leftrightarrow \mathrm{x} \overset{f'_{t_2}}{\hookrightarrow} \mathrm{y}\right)$.

- If $L(t) = \exists \mathrm{v}.\ \psi_1$ and $t_1$ is the only child of $t$, suppose $\phi_{t_1}$ has been computed, then $\phi_t \overset{\mathrm{def}}{=} P'_t = P'_{t_1} \wedge \bigwedge_{f\in\mathbb{F}} f'_t = f'_{t_1} \wedge \exists \mathrm{v}.\ \phi_{t_1}$, where $P'_t = P'_{t_1}$ and $f'_t = f'_{t_1}$ are abbreviations of formulae defined similarly to the previous case.
- If $L(t) = \psi_1 * \psi_2$ and $t_1, t_2$ are two children of $t$ such that $L(t_1) = \psi_1$ and $L(t_2) = \psi_2$, suppose $\phi_{t_1}$ and $\phi_{t_2}$ have been computed, then

$$\phi_t \overset{\mathrm{def}}{=} P'_t = P'_{t_1} \uplus P'_{t_2} \wedge \bigwedge_{f\in\mathbb{F}} f'_t = f'_{t_1} \uplus f'_{t_2} \wedge \phi_{t_1} \wedge \phi_{t_2},$$

where $P'_t = P'_{t_1} \uplus P'_{t_2}$ is an abbreviation of $\forall \mathrm{x}.\ (P'_t(\mathrm{x}) \leftrightarrow (P'_{t_1}(\mathrm{x}) \vee P'_{t_2}(\mathrm{x}))) \wedge \forall \mathrm{x}.\ (\neg P'_{t_1}(\mathrm{x}) \vee \neg P'_{t_2}(\mathrm{x}))$ and $f'_t = f'_{t_1} \uplus f'_{t_2}$ is an abbreviation of $\forall \mathrm{x}.\ \forall \mathrm{y}.\ \mathrm{x} \overset{f'_t}{\hookrightarrow} \mathrm{y} \leftrightarrow \left(\mathrm{x} \overset{f'_{t_1}}{\hookrightarrow} \mathrm{y} \vee \mathrm{x} \overset{f'_{t_2}}{\hookrightarrow} \mathrm{y}\right) \wedge \forall \mathrm{x}.\ \forall \mathrm{y}.\ \left(\neg \mathrm{x} \overset{f'_{t_1}}{\hookrightarrow} \mathrm{y} \vee \neg \mathrm{x} \overset{f'_{t_2}}{\hookrightarrow} \mathrm{y}\right)$.

– If $L(t) = \psi_1 \mathrel{-\!\!*} \psi_2$ and $t_1, t_2$ are two children of $t$ such that $L(t_1) = \psi_1$ and $L(t_2) = \psi_2$, suppose $\phi_{t_1}$ and $\phi_{t_2}$ have been computed, then

$$\phi_t \stackrel{\text{def}}{=} P'_{t_2} = P'_t \uplus P'_{t_1} \wedge \bigwedge_{\mathfrak{f} \in \mathbb{F}} \mathfrak{f}'_{t_2} = \mathfrak{f}'_t \uplus \mathfrak{f}'_{t_1} \wedge \phi_{t_1} \wedge \phi_{t_2},$$

where $P'_t = P'_{t_1} \uplus P'_{t_2}$ and $\mathfrak{f}'_{t_2} = \mathfrak{f}'_t \uplus \mathfrak{f}'_{t_1}$ are abbreviations of the formulae that can be defined similarly to the previous case.

*Example 2.* Let $\psi \stackrel{\text{def}}{=} (\neg\, \mathsf{x} = \mathsf{y}) \wedge (\psi'_1 \mathrel{-\!\!*} \psi'_2) \mathrel{-\!\!*} (\psi'_3 \mathrel{-\!\!*} \psi'_4)$ be the formula in Example 1. Then $\mathcal{T}_\psi$ is illustrated in Fig. 1. By a bottom-up computation, we get $\phi_{t_5} \stackrel{\text{def}}{=} \mathsf{x} \stackrel{\mathfrak{f}'_{t_5}}{\hookrightarrow} \mathsf{y}$,

$$\phi_{t_3} \stackrel{\text{def}}{=} P'_{t_6} = P'_{t_5} \uplus P'_{t_3} \wedge \mathfrak{f}'_{t_6} = \mathfrak{f}'_{t_5} \uplus \mathfrak{f}'_{t_3} \wedge \mathsf{x} \stackrel{\mathfrak{f}'_{t_5}}{\hookrightarrow} \mathsf{y} \wedge \exists \mathsf{y}.\ \mathsf{x} \stackrel{\mathfrak{f}'_{t_6}}{\hookrightarrow} \mathsf{y} \wedge \exists \mathsf{y}.\ \mathsf{y} \stackrel{\mathfrak{f}'_{t_6}}{\hookrightarrow} \mathsf{x},$$

$$\phi_{t_9} \stackrel{\text{def}}{=} P'_{t_9} = P'_{t_{11}} \uplus P'_{t_{12}} \wedge \mathfrak{f}'_{t_9} = \mathfrak{f}'_{t_{11}} \uplus \mathfrak{f}'_{t_{12}} \wedge \exists \mathsf{y}.\ \mathsf{y} \stackrel{\mathfrak{f}'_{t_{11}}}{\hookrightarrow} \mathsf{x} \wedge \exists \mathsf{y}.\ \mathsf{y} \stackrel{\mathfrak{f}'_{t_{12}}}{\hookrightarrow} \mathsf{x},$$

and $\phi_{t_8} \stackrel{\text{def}}{=} P'_{t_8} = P'_{t_9} = P'_{t_{10}} \wedge \bigwedge_{\mathfrak{f} \in \mathbb{F}} \mathfrak{f}'_{t_8} = \mathfrak{f}'_{t_9} = \mathfrak{f}'_{t_{10}} \wedge \phi_{t_9} \wedge \phi_{t_{10}}$, where $\phi_{t_{10}}$ is the formula corresponding to $t_{10}$, and $\phi_{t_7} \stackrel{\text{def}}{=} \neg\mathsf{x} \stackrel{\mathfrak{f}'_{t_7}}{\hookrightarrow} \mathsf{y} \wedge \exists \mathsf{y}.\ \left( \mathsf{x} \stackrel{\mathfrak{f}'_{t_7}}{\hookrightarrow} \mathsf{y} \wedge \forall \mathsf{x}.\ \neg \mathsf{y} \stackrel{\mathfrak{f}'_{t_7}}{\hookrightarrow} \mathsf{x} \right)$,

in addition, $\phi_{t_4}$ can be constructed from $\phi_{t_7}$ and $\phi_{t_8}$, similarly to the construction of $\phi_{t_3}$, and $\phi_{t_2} \stackrel{\text{def}}{=} P'_{t_4} = P'_{t_3} \uplus P'_{t_2} \wedge \mathfrak{f}'_{t_4} = \mathfrak{f}'_{t_3} \uplus \mathfrak{f}'_{t_2} \wedge \phi_{t_3} \wedge \phi_{t_4}$. The formula $\phi_{t_3}$ contains a conjunct $\phi_{t_5} = \mathsf{x} \stackrel{\mathfrak{f}'_{t_5}}{\hookrightarrow} \mathsf{y}$, while $\phi_{t_4}$ contains a conjunct $\phi_{t_7} = \neg\mathsf{x} \stackrel{\mathfrak{f}'_{t_7}}{\hookrightarrow} \mathsf{y} \wedge \exists \mathsf{y}.\ \left( \mathsf{x} \stackrel{\mathfrak{f}'_{t_7}}{\hookrightarrow} \mathsf{y} \wedge \forall \mathsf{x}.\ \neg \mathsf{y} \stackrel{\mathfrak{f}'_{t_7}}{\hookrightarrow} \mathsf{x} \right)$. Thus the conflict between $\psi'_1$ and $\psi'_3$ is resolved. □

**Proposition 2.** *For each ESPSL2* $\left( (\stackrel{\mathfrak{f}}{\hookrightarrow})_{\mathfrak{f} \in \mathbb{F}}, *, \mathrel{-\!\!*} \right)$ *formula* $\psi$, $\psi$ *is satisfiable iff* $\mathsf{trs}(\psi)$ *is satisfiable.*

*Proof.* Suppose $\psi$ is an ESPSL2$\left( (\stackrel{\mathfrak{f}}{\hookrightarrow})_{\mathfrak{f} \in \mathbb{F}}, *, \mathrel{-\!\!*} \right)$ formula.

*"Only if" direction*: Suppose that $\psi$ is satisfiable, that is, there is a pair $(\mathfrak{h}, \mathfrak{m})$ such that $\mathfrak{h} \models_\mathfrak{m} \psi$.

Let $\mathsf{Leaves}(\mathcal{T}_\psi)$ denote the set of leaves of $\mathcal{T}_\psi$. Then $\{L(t) \mid t \in \mathsf{Leaves}(\mathcal{T}_\psi)\}$ is a subset of SPSL2$\left( (\stackrel{\mathfrak{f}}{\hookrightarrow})_{\mathfrak{f} \in \mathbb{F}} \right)$ formulae. Since $\mathfrak{h} \models_\mathfrak{m} \psi$, we know that there is a subset of $\mathsf{Leaves}(\mathcal{T}_\psi)$, say $T'$, such that each $t' \in T'$ can be assigned a heap

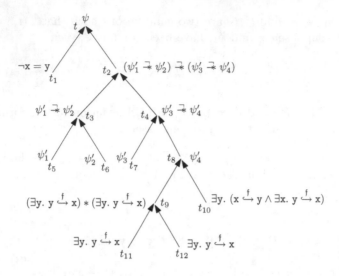

**Fig. 1.** The syntax tree $\mathcal{T}_\psi$: an example

$\mathfrak{h}_{t'}$ with a nonempty-domain, in order to witness the fact $\mathfrak{h} \models_m \psi$. Then we construct a heap $\mathfrak{h}'$ to satisfy $\mathbf{trs}(\psi)$ as follows:

1. For each $t' \in T'$, let $(\mathfrak{h}'_{t'}, \mathfrak{I}'_{t'})$ be the labeled heap such that for each $\mathfrak{f} \in \mathbb{F}$, $(\mathfrak{h}'_{t'})_{\mathfrak{f}'_{t'}}(\mathfrak{l}) = \mathfrak{l}'$ iff $(\mathfrak{h}_{t'})_{\mathfrak{f}}(\mathfrak{l}) = \mathfrak{l}'$, in addition, $\mathfrak{I}'_{t'}(P'_{t'}) = \mathrm{dom}(\mathfrak{h}'_{t'})$ and $\mathfrak{I}'_{t'}(P') = \emptyset$ for each other unary predicate $P' \in \mathcal{P}'$. Moreover, for each leaf $t' \notin T'$, let $(\mathfrak{h}_{t'}, \mathfrak{I}'_{t'})$ be the labeled heap such that $\mathfrak{h}_{t'}$ has an empty domain and $\mathfrak{I}'_{t'}(P') = \emptyset$ for each $P' \in \mathcal{P}'$.
2. By induction on the structure of $\mathcal{T}_\psi$, we can construct bottom-up a labeled heap $(\mathfrak{h}'_t, \mathfrak{I}'_t)$ for each node $t \in T$. In the construction, we need trace the relationship between unary predicates in $\mathcal{P}'$ and the relationship between the fields in $\mathbb{F}'$ which are enforced by the nodes in $\mathcal{T}_\psi$. For instance, if $t$ is a node such that $L(t) = \psi_1 * \psi_2$ and $t$ has two children $t_1$ and $t_2$, suppose $(\mathfrak{h}'_{t_1}, \mathfrak{I}'_{t_1})$ and $(\mathfrak{h}'_{t_2}, \mathfrak{I}'_{t_2})$ have been computed, then $\mathfrak{h}'_t$ is computed as the domain-disjoint union of $\mathfrak{h}'_{t_1}$ and $\mathfrak{h}'_{t_2}$, in addition, for each $\mathfrak{f} \in \mathbb{F}$ and each pair of locations $(\mathfrak{l}, \mathfrak{l}')$, $(\mathfrak{h}'_t)_{\mathfrak{f}'_t}(\mathfrak{l}) = \mathfrak{l}'$ iff $(\mathfrak{h}'_{t_1})_{\mathfrak{f}'_{t_1}}(\mathfrak{l}) = \mathfrak{l}'$ or $(\mathfrak{h}'_{t_2})_{\mathfrak{f}'_{t_2}}(\mathfrak{l}) = \mathfrak{l}'$ (here $(\mathfrak{h}'_t)_{\mathfrak{f}'_t}$ is well-defined since $(\mathfrak{h}'_{t_1})_{\mathfrak{f}'_{t_1}}$ and $(\mathfrak{h}'_{t_2})_{\mathfrak{f}'_{t_2}}$ are domain-disjoint). Moreover, $\mathfrak{I}'_t(P'_t) = \mathfrak{I}'_{t_1}(P'_{t_1}) \cup \mathfrak{I}'_{t_2}(P'_{t_2})$, and for each other unary predicate $P' \in \mathcal{P}'$, $\mathfrak{I}'_t(P') = \mathfrak{I}'_{t_1}(P') \cup \mathfrak{I}'_{t_2}(P')$.

*"If" direction*: Suppose that $\mathbf{trs}(\psi)$ is satisfiable. Then there is a labeled heap $(\mathfrak{h}, \mathfrak{I})$ and an assignment $\mathfrak{m}$ such that $(\mathfrak{h}, \mathfrak{I}) \models_m \mathbf{trs}(\psi)$. The by induction on the structure of $\mathcal{T}_\psi$, we can compute bottom-up a heap $\mathfrak{h}'$ such that $\mathfrak{h}' \models_m \psi$. The construction is essentially just a renaming of the fields.     $\square$

# 5    Conclusion

In this paper, we proposed SPSL2, semi-positive first-order logic with two variables, and investigated the complexity of the satisfiability problem of several fragments of SPSL2. Our main result is that the satisfiability of $\mathrm{ESPSL2}\Big((\overset{f}{\hookrightarrow})_{f\in\mathbb{F}}, *, \overset{\rightharpoonup}{-*}\Big)$, the fragment of SPSL2 where separating conjunction $*$ and septraction $\overset{\rightharpoonup}{-*}$ (the dual operator of magic wand $-\!*$) may occur, but none of them occurs in the scope of universal quantifiers, is NEXPTIME-complete. The proof of this result relies on the NEXPTIME-completeness result of $\mathrm{SPSL2}\Big((\overset{f}{\hookrightarrow})_{f\in\mathbb{F}}, \mathcal{P}\Big)$, the fragment of SPSL2 where separating operators do not occur, but unary predicates are available. A byproduct of this work is that the finite satisfiability of first order logic with two variables and one function symbol (without unary predicates) is NEXPTIME-complete. Although some interesting questions, e.g. the decidability of $\mathrm{SPSL2}\Big((\overset{f}{\hookrightarrow})_{f\in\mathbb{F}}, *, -\!*\Big)$, are left open in this paper, we believe that this work can be seen as a substantial step towards solving them in the future.

**Acknowledgements.** This work was partially done when I was a visiting researcher at LIAFA, Université Paris Diderot, from June 2014 to June 2015, supported by China Scholarship Council. My great thanks go to Stéphane Demri for the numerous discussions with him when I did this work. At last, I would like to thank the reviewers for their valuable comments.

# References

1. Antonopoulos, T., Gorogiannis, N., Haase, C., Kanovich, M., Ouaknine, J.: Foundations for decision problems in separation logic with general inductive predicates. In: Muscholl, A. (ed.) FoSSaCS 2014. LNCS, vol. 8412, pp. 411–425. Springer, Heidelberg (2014). doi:10.1007/978-3-642-54830-7_27
2. Benaim, S., Benedikt, M., Charatonik, W., Kieroński, E., Lenhardt, R., Mazowiecki, F., Worrell, J.: Complexity of two-variable logic on finite trees. In: Fomin, F.V., Freivalds, R., Kwiatkowska, M., Peleg, D. (eds.) ICALP 2013. LNCS, vol. 7966, pp. 74–88. Springer, Heidelberg (2013). doi:10.1007/978-3-642-39212-2_10
3. Bojańczyk, M., David, C., Muscholl, A., Schwentick, T., Segoufin, L.: Two-variable logic on data words. ACM Trans. Comput. Logic **12**(4), 27 (2011)
4. Brochenin, R., Demri, S., Lozes, E.: On the almighty wand. Inf. Comput. **211**, 106–137 (2012)
5. Calcagno, C., Yang, H., O'Hearn, P.W.: Computability and complexity results for a spatial assertion language for data structures. In: Hariharan, R., Vinay, V., Mukund, M. (eds.) FSTTCS 2001. LNCS, vol. 2245, pp. 108–119. Springer, Heidelberg (2001). doi:10.1007/3-540-45294-X_10
6. Charatonik, W., Kieroński, E., Mazowiecki, F.: Decidability of weak logics with deterministic transitive closure. In: CSL-LICS (2014)

7. Cook, B., Haase, C., Ouaknine, J., Parkinson, M., Worrell, J.: Tractable reasoning in a fragment of separation logic. In: Katoen, J.-P., König, B. (eds.) CONCUR 2011. LNCS, vol. 6901, pp. 235–249. Springer, Heidelberg (2011). doi:10.1007/978-3-642-23217-6_16

8. Demri, S., Deters, M.: Expressive completeness of separation logic with two variables and no separating conjunction. In: CSL-LICS, pp. 1–37 (2014)

9. Demri, S., Deters, M.: Two-variable separation logic and its inner circle. ACM Trans. Comput. Logic 16(2), 15 (2015)

10. Demri, S., Galmiche, D., Larchey-Wendling, D., Méry, D.: Separation logic with one quantified variable. In: Hirsch, E.A., Kuznetsov, S.O., Pin, J.É., Vereshchagin, N.K. (eds.) CSR 2014. LNCS, vol. 8476, pp. 125–138. Springer, Heidelberg (2014). doi:10.1007/978-3-319-06686-8_10

11. Etessami, K., Vardi, M.Y., Wilke, T.: First-order logic with two variables and unary temporal logic. Inf. Comput. 179(2), 279–295 (2002)

12. Grädel, E., Otto, M.: On logics with two variables. Theor. Comput. Sci. 224(1–2), 73–113 (1999)

13. Grädel, E., Otto, M., Rosen, E.: Two-variable logic with counting is decidable. In: LICS, pp. 306–317 (1997)

14. Grädel, E., Otto, M., Rosen, E.: Undecidability results on two-variable logics. Arch. Math. Log. 38(4–5), 313–354 (1999)

15. Minsky, M.L.: Computation: Finite and Infinite Machines. Prentice-Hall Inc., Upper Saddle River (1967)

16. Pacholski, L., aw Szwast, W., Tendera, L.: Complexity results for first-order two-variable logic with counting. SIAM J. Comput. 29(4), 1083–1117 (2000)

17. Piskac, R., Wies, T., Zufferey, D.: Automating separation logic using SMT. In: Sharygina, N., Veith, H. (eds.) CAV 2013. LNCS, vol. 8044, pp. 773–789. Springer, Heidelberg (2013). doi:10.1007/978-3-642-39799-8_54

18. Pratt-Hartmann, I.: Complexity of the two-variable fragment with counting quantifiers. J. Logic Lang. Inf. 14(3), 369–395 (2005)

19. Pratt-Hartmann, I.: The two-variable fragment with counting revisited. In: Dawar, A., Queiroz, R. (eds.) WoLLIC 2010. LNCS (LNAI), vol. 6188, pp. 42–54. Springer, Heidelberg (2010). doi:10.1007/978-3-642-13824-9_4

20. Reynolds, J.C.: Separation logic: a logic for shared mutable data structures. In: LICS, pp. 55–74 (2002)

21. Segoufin, L., ten Cate, B.: Unary negation. LMCS 9(3), 1–46 (2013)

22. Thakur, A., Breck, J., Reps, Th.: Satisfiability modulo abstraction for separation logic with linked lists. In: SPIN, pp. 58–67 (2014)

23. Trakhtenbrot, B.: Impossibility of an algorithm for the decision problem in finite classes. AMS Translations Ser. 2(23), 1–5 (1963)

# Distributed Computation of Fixed Points on Dependency Graphs

Andreas Engelbredt Dalsgaard, Søren Enevoldsen[✉], Kim Guldstrand Larsen, and Jiří Srba

Department of Computer Science, Aalborg University,
Selma Lagerlöfs Vej 300, 9220 Aalborg East, Denmark
{andrease,senevoldsen,kgl,srba}@cs.aau.dk

**Abstract.** Dependency graph is an abstract mathematical structure for representing complex causal dependencies among its vertices. Several equivalence and model checking questions, boolean equation systems and other problems can be reduced to fixed-point computations on dependency graphs. We develop a novel distributed algorithm for computing such fixed points, prove its correctness and provide an efficient, open-source implementation of the algorithm. The algorithm works in an on-the-fly manner, eliminating the need to generate a priori the entire dependency graph. We evaluate the applicability of our approach by a number of experiments that verify weak simulation/bisimulation equivalences between CCS processes and we compare the performance with the well-known CWB tool. Even though the fixed-point computation, being a P-complete problem, is difficult to parallelize in theory, we achieve significant speed-ups in the performance as demonstrated on a Linux cluster with several hundreds of cores.

## 1 Introduction

Formal verification techniques are increasingly applied in industrial development of software and hardware systems, both to ensure safe and reliable behaviour of the final system, and to reduce cost and time by finding bugs at early development stages. In particular industrial take-up has been boosted by the maturing of computer aided verification, where development of a variety of techniques helps in applying verification to critical parts of systems. Heuristics for SAT solving, abstraction, decomposition, symbolic execution, partial order reduction, and other techniques are used to speed up the verification of systems with various characteristics. Still, the problem of automatic verification is hard, and some difficult cases occur frequently in practical experience. For this reason, we aim in this paper at exploiting the computational power of parallel and distributed machine architectures to further enlarge the scope of automated verification.

Automated verification methods contain a large variety of model-checking and equivalence/preorder-checking algorithms. In the former, a system model is (dis-)proved correct with respect to a logical property expressed in a suitable temporal logic. In the latter, the system model is compared with an abstract

© Springer International Publishing AG 2016
M. Fränzle et al. (Eds.): SETTA 2016, LNCS 9984, pp. 197–212, 2016.
DOI: 10.1007/978-3-319-47677-3_13

model of the system with respect to a suitable behavioural equivalence or preorder, e.g. trace-equivalence, weak or strong bisimulation equivalence. Aiming at providing parallel and distributed support to (essentially) all of these problems, we design a distributed algorithm based on the notion of *dependency graphs* [1,2]. In particular, dependency graphs have proven a useful and universal formalism for representing several verification problems, offering efficient analysis through linear-time (local and global) algorithms [2] for fixed-point computation of the corresponding dependency graph. The challenge we undertake here is to provide a distributed algorithm for this fixed-point computation. The fact that dependency graphs allow for representation of bisimulation equivalences between system models suggests that we should not expect our distributed algorithm to exhibit linear speed-up in all cases as bisimulation equivalence is known to be P-complete [3]. Our experiments though still document significant speed-ups that together with the on-the-fly nature of our algorithm (where we possibly avoid the construction of the entire dependency graph in situations where it is not necessary) allow us to outperform the tool CWB [4] for equivalence/model checking of processes described in the CCS process algebra [5].

*Related Work.* Most closely related to our work are those of [6–8] offering parallel algorithms for model-checking systems with respect to the alternation-free modal $\mu$-calculus. The approach in [6] is based on games and tree decomposition but the tool prototype mentioned in the paper is not available anymore. The work in [8] reduces $\mu$-calculus formulae into alternation free Boolean equation systems. Finally [7] uses a global symbolic BDD-based distributed algorithm for modal $\mu$-calculus but does not mention any implementation. We share the on-the-fly technique with some of these works but our framework is more universal in the sense that we deal with the general dependency graphs where the problems above are reducible to. There also exist several mature tools with modern designs like FDR3 [9], CADP [10], SPIN [11] and mCRL2 [12], some of them offering also distributed and/or on-the-fly algorithms. The input language of the tools is however often strictly defined and extensions to these languages as well as the range of verification methods require nontrivial changes in the implementation. The advantage of our approach is that we first reduce a wide range of problems into dependency graphs and then use our optimized distributed implementation on these generic graphs. Finally, we have recently introduced CAAL [13] as a tool for teaching CCS and verification techniques. The tool CAAL, running in a browser and implemented in TypeScript (a typed superset of JavaScript), is also based on dependency graphs but offers only the sequential version of the local algorithm by Liu and Smolka [2]. Here we provide an optimized C++ implementation of the distributed algorithm thus laying the foundation for offering CAAL verification tasks as a cloud service.

## 2    Definitions

A *labelled transition system* (LTS) is a triple $(S, A, \rightarrow)$ where $S$ is a set of states, $A$ is a set of actions that includes the silent action $\tau$, and $\rightarrow \subseteq S \times A \times S$ is the

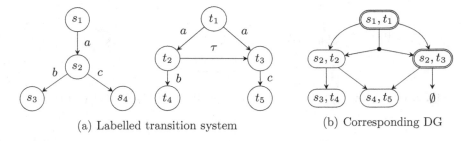

(a) Labelled transition system          (b) Corresponding DG

**Fig. 1.** Dependency graph for weak bisimulation

transition relation. Instead of $(s, a, t) \in \rightarrow$ we write $s \xrightarrow{a} t$. We also write $s \xRightarrow{a} t$ if either $a = \tau$ and $s \xrightarrow{\tau}{}^* t$, or if $a \neq \tau$ and $s \xrightarrow{\tau}{}^* s' \xrightarrow{a} t' \xrightarrow{\tau}{}^* t$ for some $s', t' \in S$.

A binary relation $R \subseteq S \times S$ over the set of states of an LTS is *weak simulation* if whenever $(s, t) \in R$ and $s \xrightarrow{a} s'$ for some $a \in A$ then also $t \xRightarrow{a} t'$ such that $(s', t') \in R$. If both $R$ and $R^{-1} = \{(t, s) \mid (s, t) \in R\}$ are weak simulations then $R$ is a *weak bisimulation*.

We say that $s$ is *weakly simulated* by $t$ and write $s \ll t$ (resp. $s$ and $t$ are *weakly bisimilar* and write $s \approx t$) if there is a weak simulation (resp. weak bisimulation) relation $R$ such that $(s, t) \in R$.

Consider the LTS in Fig. 1a (even though it consists of two disconnected parts, it can still be considered as a single LTS). It is easy to see that $s_1$ weakly simulates $t_1$ and vice versa. For example the weak simulation relation $R = \{(s_1, t_1), (s_2, t_2), (s_3, t_4), (s_4, t_5)\}$ shows that $s_1$ is weakly simulated by $t_1$. However, $s_1$ and $t_1$ are not weakly bisimilar. Indeed, if $s_1$ and $t_1$ were weakly bisimilar, the transition $t_1 \xrightarrow{a} t_3$ can only be matched by $s_1 \xrightarrow{a} s_2$ but $s_2$ has a transition under the label $b$ whereas $t_3$ does not offer such a transition.

## 2.1   Dependency Graphs

A dependency graph [2] is a general structure that expresses dependencies among the vertices of the graph and by this allows us to solve a large variety of complex computational problems by means of fixed-point computations.

**Definition 1 (Dependency Graph).** *A dependency graph is a pair* $(V, E)$ *where* $V$ *is a set of vertices and* $E \subseteq V \times 2^V$ *is a set of hyperedges. For a hyperedge* $(v, T) \in E$, *the vertex* $v \in V$ *is called the* source vertex *and* $T \subseteq V$ *is the* target set.

Let $G = (V, E)$ be a fixed dependency graph. An *assignment* on $G$ is a function $A : V \to \{0, 1\}$. Let $\mathcal{A}$ be the set of all assignments on $G$. A *fixed-point assignment* is an assignment $A$ that for all $(v, T) \in E$ satisfies the following condition: if $A(v') = 1$ for all $v' \in T$ then $A(v) = 1$.

Figure 2 shows an example of a dependency graph. The hyperedge $(a, \emptyset)$ with the empty target set is depicted by the arrow from $a$ to the symbol $\emptyset$. The figure

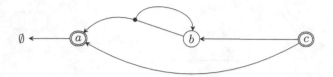

**Fig. 2.** Dependency graph $G = (\{a,b,c\}, \{(a,\emptyset), (b,\{a,b\}), (c,\{b\}), (c,\{a\})\})$

also denotes a particular assignment $A$ such that vertices with a single circle have the value 0 and vertices with a double circle have the value 1, in order words $A(a) = A(c) = 1$ and $A(b) = 0$. It can be easily verified that the assignment $A$ is a fixed-point assignment.

We are interested in the minimum fixed-point assignment. Let $A_1, A_2 \in \mathcal{A}$ be assignments. We write $A_1 \sqsubseteq A_2$ if $A_1(v) \leq A_2(v)$ for all $v \in V$, where we assume that $0 \leq 1$. Clearly $(A, \sqsubseteq)$ is a complete lattice. Let us also define a function $F : \mathcal{A} \to \mathcal{A}$ such that $F(A)(v) = 1$ if there is a hyperedge $(v, T) \in E$ such that $A(v') = 1$ for all $v' \in T$, otherwise $F(A)(v) = A(v)$. Observe that an assignment $A$ is a fixed-point assignment iff $F(A) = A$, and that the function $F$ is monotonic w.r.t. $\sqsubseteq$. By Knaster-Tarski theorem [14] there exists a unique minimum fixed-point assignment, denoted by $A_{min}$. The assignment $A_{min}$ on a finite dependency graph can be computed by a repeated application of the function $F$ on the assignment $A_0$ where $A_0(v) = 0$ for all $v \in V$, and we are guaranteed that there is a number $m$ such that $F^m(A_0) = F^{m+1}(A_0) = A_{min}$.

Consider again our example from Fig. 2 and assume that each assignment $A$ is represented by the vector $(A(a), A(b), A(c))$. We can see that $A_0 = (0,0,0)$, $F(A_0) = (1,0,0)$ and $F^2(A_0) = (1,0,1) = F^3(A_0)$. Hence the depicted assignment $(1,0,1)$ is the minimum fixed-point assignment.

## 2.2   Applications of Dependency Graphs

Many verification problems can be encoded as fixed-point computations on dependency graphs. We shall demonstrate this on the cases of weak simulation and bisimulation, however other equivalences and preorders from the linear/branching-time spectrum [15] can also be encoded as dependency graphs [16] as well as model checking problems e.g. for the CTL logic [17], reachability problems for timed games [18] and the general framework of Boolean equation systems [2], just to mention a few applications of dependency graphs.

Let $T = (S, A, \to)$ be an LTS. We define a dependency graph $G_\approx(T) = (V, E)$ such that $V = \{(s,t) \mid s, t \in S\}$ and the hyperedges are given by

$$E = \{((s,t), \{(s',t') \mid t \xRightarrow{a} t'\}) \mid s \xrightarrow{a} s'\} \cup \{((s,t), \{(s',t') \mid s \xRightarrow{a} s'\}) \mid t \xrightarrow{a} t'\}.$$

The general construction is depicted in Fig. 3 and its application to the LTS from Fig. 1a, listing only the pairs of states reachable from $(s_1, t_1)$, is shown in Fig. 1b. Observe that the size of the produced dependecy graph is polynomial with respect to the size of the input LTS.

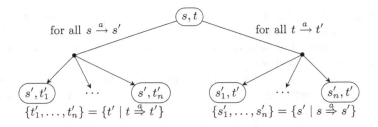

**Fig. 3.** Bisimulation reduction to dependency graph

**Proposition 1.** *Let $T = (S, A, \rightarrow)$ be an LTS and $s, t \in S$. We have $s \approx t$ if and only if $A_{min}((s, t)) = 0$ in the dependency graph $G_\approx(T)$.*

*Proof (Sketch).* "$\Rightarrow$": Let $R$ be a weak bisimulation such that $(s, t) \in R$. The assignment $A$ defined as $A((s', t')) = 0$ iff $(s', t') \in R$ can be shown to be a fixed-point assignment. Then clearly $A_{min} \sqsubseteq A$ and because $A((s, t)) = 0$ then also $A_{min}((s, t,)) = 0$. "$\Leftarrow$": Let $A_{min}((s, t)) = 0$. We construct a binary relation $R = \{(s', t') \mid A_{min}((s', t')) = 0\}$. Surely $(s, t) \in R$ and we invite the reader to verify that $R$ is a weak bisimulation. $\square$

In our example in Fig. 1b we can see that $A_{min}((s_1, t_1)) = 1$ and hence $s_1 \not\approx t_1$. The construction of the dependency graph for weak bisimulation can be adapted to work also for the weak simulation preorder by removing the hyperedges that originate by transitions performed by the right hand-side process.

We know that computing $A_{min}$ for a given dependency graph can be done in linear time [19]. By the facts that deciding bisimulation on finite LTS is P-hard [3] and the polynomial time reduction described above, we can conclude that determining the value of a vertex in the minimum fixed-point assignment of a given dependency graph is a P-complete problem.

**Proposition 2.** *The problem whether $A_{min}(v) = 1$ for a given dependency graph and a given vertex $v$ is P-complete.*

## 3   Distributed Fixed-Point Algorithm

We shall now describe our distributed algorithm for computing minimum fixed-points on dependency graphs. Let $G = (V, E)$ be a dependency graph. For the purpose of the on-the-fly computation, we represent $G$ by the function

$$\textsc{Successors}(v) = \{(v, T) \mid (v, T) \in E\}$$

that returns for each vertex $v \in V$ the set of outgoing hyperedges from $v$.

We assume a fixed number of $n$ workers. Let $i$, $1 \le i \le n$, denote a worker with id $i$. Each worker $i$ uses the following local data structures.

- A local assignment function $A^i : V \rightharpoonup \{0,1\}$, which is a partial function mapping each vertex to the values *undefined*, 0 or 1.
- A local dependency function $D^i : V \to 2^E$ returning the current set of dependent edges for each vertex.
- A local waiting set $W^i \subseteq E$ containing edges that are waiting for processing.
- A local request function $R^i : V \to 2^{\{1,...,n\}}$ where the worker $i$ remembers who requested the value for a given vertex.
- A local input message set $M^i \subseteq \{$ *"value of v needed by worker j"* $\mid v \in V, 1 \leq j \leq n\} \cup \{$ *"v has value 1"* $\mid v \in V\}$. For syntactic convenience, we assume that a worker $i$ can add a message $m$ to $M^j$ of another worker $j$ simply by executing the assignment $M^j \leftarrow M^j \cup \{m\}$.

We moreover assume some standard function TERMINATIONDETECTION, computed distributively, that returns true if there are no messages in transit and all waiting sets of all workers are empty, in other words if $\bigcup_{1 \leq i \leq n} M^i \cup W^i = \emptyset$. Finally, we assume a global partitioning function $\delta : V \to \{1, \ldots, n\}$ that partitions vertices to workers.

The distributed algorithm for computing the minimal fixed-point assignment for a given vertex $v_s$ is presented in Algorithm 1. First, all $n$ workers are initialized and the worker that owns the vertex $v_s$ updates its local assignment to 0 and adds the successor edges to its local waiting set. Then the workers start processing the edges on the waiting sets and the messages in their input message sets until they terminate either by one worker announcing that $A_{min}(v_s) = 1$ at line 18 or all waiting edges and messages have been processed and then the workers together claim that $A_{min}(v_s) = 0$ at line 13.

**Lemma 1 (Termination).** *Algorithm 1 terminates.*

*Proof.* First observe that for each vertex $v$ and each local assignment $A^i$ the value of $A^i(v)$ is first undefined. Then when $v$ is discovered either as a target vertex of some hyperedge on the waiting set (line 23) or when the value of $v$ gets requested by another worker (line 37), the value $A^i(v)$ changes to 0. Finally the value of $A^i(v)$ can be upgraded to the value 1 either by the presence of a hyperedge where all target vertices already have the value 1 (line 17) or by receiving a message from another worker (line 33). The point is that for every $v$, each of the assignments $A^i(v) \leftarrow 0$ and $A^i(v) \leftarrow 1$ is executed at most once during any execution of the algorithm. This can be easily noticed by the inspection of the conditions on the if-commands guarding these assignments.

Next we notice that new hyperedges are added to the waiting set $W^i$ only when an assignment of some vertex $v$ gets upgraded from *undefined* to 0, or from 0 to 1. As argued above, this can happen only finitely many times, hence only finitely many hyperedges can be added to each $W^i$. Similarly, new messages to the message sets can be added only at lines 19, 28 and 40. At line 19 a finite number of messages of the form *"v has value 1"* is added only when a value of $A^i(v)$ was upgraded to 1. This can happen only finitely many times. At line 28 the message *"value of v' needed by worker i"* is added only when a value of a vertex was upgraded from *undefined* to 0, hence this can happen

---

**Algorithm 1.** Distributed Algorithm for Worker $i$, $1 \leq i \leq n$

---

**Input**: A dependency graph $G = (V, E)$ represented by the function
Successors, a vertex $v_s \in V$ and a vertex partitioning function
$\delta : V \to \{1, \ldots, n\}$ where $n$ is the number of workers.

**Output**: The minimum fixed-point assignment $A_{min}(v_s)$ for the vertex $v_s$.

```
1   A^i(v) ← undefined for all v ∈ V              ▷ implemented via hashing
2   W^i ← ∅; D^i ← ∅; M^i ← ∅; R^i ← ∅
3   if δ(v_s) = i then                            ▷ initialize the computation
4   |   A^i(v_s) ← 0; W^i ← Successors(v_s)
5   repeat
6   |   while W^i ≠ ∅ or M^i ≠ ∅ do
7   |   |   Let x ∈ W^i ∪ M^i                      ▷ process message or hyperedge
8   |   |   if x ∈ W^i then
9   |   |   |   W^i ← W^i \ {x}; ProcessHyperedge(x)
10  |   |   else
11  |   |   |   M^i ← M^i \ {x}; ProcessMessage(x)
12  until TerminationDetection
13  output "A_min(v_s) = 0"
```

```
14  Procedure ProcessHyperedge((v, T)) is
15  |   if A^i(v) ≠ 1 then
16  |   |   if ∀v' ∈ T : A^i(v') = 1 then
17  |   |   |   A^i(v) ← 1
18  |   |   |   if v = v_s then output "A_min(v_s) = 1" ; terminate all workers
19  |   |   |   for all j ∈ R^i(v) do  M^j ← M^j ∪ { "v has value 1" }
20  |   |   |   R^i(v) ← ∅
21  |   |   |   W^i ← W^i ∪ D^i(v)
22  |   |   else if ∃v' ∈ T : A^i(v') is undefined then
23  |   |   |   A^i(v') ← 0
24  |   |   |   D^i(v') ← D^i(v') ∪ {(v, T)}
25  |   |   |   if δ(v') = i then                  ▷ is value of v' my responsibility?
26  |   |   |   |   W^i ← W^i ∪ Successors(v')
27  |   |   |   else                              ▷ send request for value of v'
28  |   |   |   |   M^δ(v') ← M^δ(v') ∪ { "value of v' needed by worker i" }
29  |   |   else if ∃v' ∈ T : A^i(v') = 0 then
30  |   |   |   D^i(v') ← D^i(v') ∪ {(v, T)}
```

```
31  Procedure ProcessMessage(m) is
32  |   if m = "v has value 1" and A^i(v) ≠ 1 then
33  |   |   A^i(v) ← 1
34  |   |   W^i ← W^i ∪ D^i(v)
35  |   else if m = "value of v needed by worker j" then
36  |   |   if A^i(v) is undefined then
37  |   |   |   A^i(v) ← 0
38  |   |   |   W^i ← W^i ∪ Successors(v)
39  |   |   if A^i(v) = 1 then
40  |   |   |   M^j ← M^j ∪ { "v has value 1" }   ▷ we already know it is 1
41  |   |   else
42  |   |   |   R^i(v) ← R^i(v) ∪ {j}             ▷ remember that j needs value of v
```

only finitely many times. Finally, at line 40 a message is added only when we received the message *"value of v needed by worker i"* but this message was sent only finitely many times. All together, only finitely many elements can be added to the waiting and message sets and as the main while-loop repeatedly removes elements from those sets, eventually they must become empty and the algorithm terminates at line 13 (unless it terminated earlier at line 18). □

We can now observe that if a vertex is assigned the value 1 for any worker, then the value of the vertex in the minimal fixed-point assignment is also 1.

**Lemma 2 (Soundness).** *At any moment of the execution of Algorithm 1 and for all $i$, $1 \leq i \leq n$, and all $v \in V$ it holds that*

*(a) if $A^i(v) = 1$ then $A_{min}(v) = 1$, and*
*(b) if "v has value 1" $\in M^i$ then $A_{min}(v) = 1$.*

*Proof.* The invariant holds initially as $A^i(v)$ is undefined for all $i$ and all $v$ and all input message sets are empty.

Let us assume that both condition a) and b) hold and that we assign the value 1 to $A^i(v)$ for some worker $i$ and a vertex $v$. This can only happen at lines 17 and 33. In the first assignment at line 17 we know that there is a hyperedge $(v, T)$ such that all vertices $v' \in T$ satisfy that $A^i(v') = 1$. However, this by our invariant part a) implies that $A_{min}(v') = 1$ and then necessarily also $A_{min}(v) = 1$ by the definition of fixed-point assignment. Hence the invariant for the case a) is preserved. Similarly, if $A^i(v)$ gets the value 1 at line 33, this can only happen if "v has value 1" $\in M^i$ and by the invariant part b) this implies that $A_{min}(v) = 1$ and hence the invariant for the condition a) is established.

Similarly, let us assume that conditions a) and b) hold and that a message *"v has value 1"* gets inserted into $M^j$ by some worker $i$. This can only happen at lines 19 and 40. In both situations it is guaranteed that $A^i(v) = 1$ and hence by the invariant part a) we know that $A_{min}(v) = 1$, implying that adding these messages to $M^j$ is safe. □

The next lemma establishes an important invariant of the algorithm.

**Lemma 3.** *For any vertex $v \in V$, whenever during the execution of Algorithm 1 the worker $\delta(v)$ is at line 6 then the following invariant holds: either*

*(a) $A^{\delta(v)}(v) = 1$, or*
*(b) $A^{\delta(v)}(v)$ is undefined, or*
*(c) $A^{\delta(v)}(v) = 0$ and for all $(v, T) \in E$ either*
    *(i) $(v, T) \in W^{\delta(v)}$, or*
    *(ii) there is $v' \in T$ such that $A^{\delta(v)}(v') = 0$, and $(v, T) \in D^{\delta(v)}(v')$.*

*Proof.* Initially, the invariant is satisfied as $A^{\delta(v)}(v)$ is undefined and the invariant, more specifically the subcase (i), clearly holds also when $v = v_s$ and the worker $\delta(v_s)$ performed the assignments at line 4.

Assume now that the invariant holds. Clearly, if it is by case (a) where $A^{\delta(v)}(v) = 1$ then the value of $v$ will remain 1 until the end of the execution.

If the invariant holds by case (b) then it is possible that the value of $A^{\delta(v)}(v)$ changes from *undefined* to 0. This can happen either at lines 23 or 37. If the assignment took place at line 23 (note that here $v = v'$) then clearly line 26 will be executed too and all successor edges of $v$ will be inserted into the waiting set and hence the invariant subcase (i) will hold once the execution of the procedure is finished. Similarly, if the assignment took place at line 37 then all successors of $v$ are at the next line 38 immediately added to the waiting set, hence again satisfying the invariant subcase (i).

Consider now the case (c). The invariant can be challenged by either removing the hyperedge $(v, T)$ from $W^{\delta(v)}$ hence invalidating the subcase i) or by upgrading the value of the vertex $v'$ in case (ii) such that $A^{\delta(v)}(v') = 1$. In the first case where the subcase (i) gets invalidated we can notice that this can happen only at line 9 after which the removed hyperedge $(v, T)$ is processed. There are two possible scenarios now. Either all vertices from $T$ have the value 1 and then the value of $A^{\delta(v)}(v)$ also gets upgraded to 1 at line 17 hence satisfying the invariant (a), or there is a vertex $v' \in T$ such that $A^{\delta(v)}(v') = 0$ and then the hyperedge $(v, T)$ is added at line 30 to the dependency set $D^{\delta(v)}(v')$ satisfying the subcase (ii) of the invariant. In the second subcase, we assume that the vertex $v'$ satisfying the subcase (ii) gets upgraded to the value 1. This can happen at line 17 or line 33. In both cases the dependency set $D^{\delta(v)}(v')$ (that by our invariant assumption contains the hyperedge $(v, T)$) is added to the waiting set (lines 21 and 34) implying that the invariant subpart (i) holds.                    □

The following lemmas shows that after the termination, the value 0 for a vertex $v$ in a local assignment of some worker implies the same value also in the assignment of the worker that owns the vertex $v$. This is an important fact for showing completeness of our algorithm.

**Lemma 4.** *Once all workers in Algorithm 1 terminate at line 13 then for all vertices $v \in V$ and all workers $i$ holds that if $A^i(v) = 0$ then $A^{\delta(v)}(v) = 0$.*

*Proof.* Observe that the assignment of 0 to $A^i(v)$ where $i \neq \delta(v)$ can happen only at line 23 (the assignment at line 37 is performed only if $i = \delta(v)$ as the message *"value of v is needed by worker i"* is sent only to the owner of the vertex $v$). After the assignment at line 23 done by worker $i$, the message requesting the value of the vertex is sent to its owner at line 28. Clearly, before the workers terminate, this message must be read by the owner and the value of the vertex is either set to 0 at line 37, or if the value is already known to be 1 the worker $i$ is informed about this via the message *"v has value 1"* at line 40 and this message will be necessarily read by the worker $i$ before the termination and the value $A^i(v)$ will be updated to 1. Otherwise we remember the worker's id requesting the assignment value at line 42. Should the owner upgrade the value of $v$ to 1 at some moment, all workers that requested its value will be informed about this by a message sent at line 19 and before the termination these workers must read these messages and update the local values for $v$ to 1. It is hence impossible for the algorithm to terminate while the owner of $v$ set its value to 1 and some other worker still has only the value 0 for the vertex $v$.                    □

**Lemma 5 (Completeness).** *If all workers in Algorithm 1 terminate at line 13 then for all vertices $v \in V$ the fact $A^{\delta(v)}(v) = 0$ implies that $A_{min}(v) = 0$.*

*Proof.* Note that after the termination we have $W^i = M^i = \emptyset$ for all $i$. Assume now that $A^{\delta(v)}(v) = 0$. Then by Lemma 3 and the fact that $W^{\delta(v)} = \emptyset$ we can conclude that for all $(v, T) \in E$ there exists $v' \in T$ such that $A^{\delta(v)}(v') = 0$. By Lemma 4 this means that also $A^{\delta(v')}(v') = 0$. Let us now define an assignment $A$ such that $A(v) = A^{\delta(v)}(v)$. By the arguments above, $A$ is a fixed-point assignment. As $A_{min}$ is the minimum fixed-point assignment, we have $A_{min} \sqsubseteq A$ and because $A(v) = 0$ we can conclude that $A_{min}(v) = 0$. □

**Theorem 1 (Correctness).** *Algorithm 1 terminates and outputs either*

- *"$A_{min}(v_s) = 1$" implying that $A_{min}(v_s) = 1$, or*
- *"$A_{min}(v_s) = 0$" implying that $A_{min}(v_s) = 0$.*

*Proof.* Termination is proved in Lemma 1. The algorithm can terminate either at line 18 provided that $A^i(v_s) = 1$ but then by Lemma 2 clearly $A_{min}(v_s) = 1$. Otherwise the algorithm terminates when all workers reach line 13. This can only happen when $A^{\delta(v_s)}(v_s) = 0$ and by Lemma 5 we get $A_{min}(v_s) = 0$. □

Note that the algorithm is proved correct without imposing any specific order by which messages and hyperedges are selected from the sets $W^i$ and $M^i$ or what target vertices are selected in the expressions like $\exists v' \in T$. In the next section we discuss some of the choices we have made in our implementation.

## 4  Implementation and Evaluation

The distributed algorithm described in the previous section is implemented as an MPI-program in C++, enabling the workers to cooperate not only on a single machine but also across multiple machines. The MPI-program requires a successor generator to explore the dependency graph, a partitioning function and a (de)serialisation function for the vertices (we use LZ4 compression on the generated hyperedges before they leave the successor generator). For our experiments, these functions were implemented for the case of weak bisimulation/simulation on CCS processes but they can be easily replaced with other custom implementations to support other equivalence and model checking problems, without the need of modifying the distributed engine itself.

In our implementation we use hash tables to store the assignments ($A^i$) and the dependent edges ($D^i$). The algorithm does not constrain specific structures on $W^i$ or $M^i$. For the waiting list ($W^i$) two deques are used, one for the forward propagation (outgoing hyperedges of newly discovered vertexes) and one for the backwards propagation (hyperedges that were inserted due to dependencies). Then the graph can be explored depth-first, or breadth-first, or a probabilistic combination of those, independently for both the forward and backwards propagation. Our experiments showed that it is preferable to prioritize processing of messages rather than hyperedges to free up buffers used by the senders. The distributed termination detection is determined using [20].

The implementation is open-source and available at http://code.launchpad. net/pardg/ in the branch dfpdg-paper that includes also all experimental data. The distributed engine is currently being integrated within the CAAL [13] user interface.

## 4.1   Evaluation

We evaluate the performance of our implementation on the traditional leader election protocol [21] where we scale the number of processes and on the alternating bit protocol (ABP) [22] where we scale the size of communication buffers. We ask the question whether the specification and implementation (both described as CCS processes) are weakly bisimilar. For both cases we consider a variant where the weak bisimulation holds and where it does not hold (by injecting an error). Finally, we also ask about the schedulability of 180 different task graphs from the well known benchmark database [23] on two processors within a given deadline. Whenever applicable, the performance is compared with the tool Concurrency WorkBench (CWB) [4] version 7.1 using 1 core (there is no parallel/distributed version of CWB). CWB implements the best performing global algorithms for bisimulation checking on CCS processes.

All experiments are performed on a Linux cluster, composed of compute nodes with 1 TB of DDR3 1600 mhz memory, four AMD Opteron 6376 processors (in total 64 cores@2,3 Ghz with speedstep disabled) and interconnected using Intel True Scale InfiniBand (40 Gb/s) for low latency communication. All nodes run an identical image of Ubuntu 14.04 and MPICH 3.2 was used for MPI communication. We use the depth first search order for the forward search strategy and the breadth first search order for the backwards search strategy.

The results for the leader election and ABP are presented in Tables 1 and 2, respectively. For each entry in the tables, four runs were performed and the mean run time and the relative sample standard deviation are reported. We also report on how many microseconds were used (in parallel) per explored vertex of the dependency graph. This measure gives an idea of the speedup achieved when more cores are available. We note that for small instances this time can be very high due to the initialization of the distributed algorithm and memory allocations for dynamic data structures.

We observe that in the positive cases where the entire dependency graph must be explored, we achieve (with 256 cores) speedups 32 and 52 for leader election with 9 and 10 processes, respectively. For ABP with buffer sizes 3 and 4 the speedups are 102 and 98, respectively. However we do see a relative high standard deviation for 8–32 cores if the run time is short. This is because the scheduler is not configured to ensure locality among NUMA nodes. Compared to the performance of CWB, we observe that on the smallest instances we need up to 64 cores in leader election and 16 cores in ABP to match the run time of CWB. However, on the next instance the run time of CWB is matched already

**Table 1.** Time is reported in seconds, RSD is the relative sample standard deviation in percentage and μs/tv is the time spend per vertex in micro seconds. The star in RSD column means that only one run finished within the given timeout.

| Leader election where implementation and specification are weakly bisimilar | | | | | | | | | | | | |
|---|---|---|---|---|---|---|---|---|---|---|---|---|
| | 9 processes | | | 10 processes | | | 11 processes | | | 12 processes | | |
| cores | time | RSD | μs/tv | time | RSD | μs/tv | time | RSD | μs/tv | time | RSD | μs/tv |
| CWB | 8.21 | 0.2 | N/A | 328 | 0.5 | N/A | - | - | N/A | - | - | N/A |
| 1 | 187 | 0.6 | 6399 | 1957 | 1.0 | 17921 | - | - | - | - | - | - |
| 2 | 102 | 0.7 | 482 | 1020 | 0.6 | 9338 | - | - | - | - | - | - |
| 4 | 55.7 | 1.0 | 907 | 553 | 1.1 | 5065 | - | - | - | - | - | - |
| 8 | 38.6 | 31.0 | 322 | 304 | 6.3 | 2783 | 2885.7 | 1.1 | 7013 | - | - | - |
| 16 | 28.5 | 17.6 | 975 | 208 | 5.9 | 1903 | 2098.6 | 1.1 | 5100 | - | - | - |
| 32 | 16.8 | 14.3 | 574 | 120 | 6.9 | 1099 | 1172.6 | 0.5 | 2850 | - | - | - |
| 64 | 9.7 | 3.0 | 332 | 81 | 3.5 | 738 | 723.9 | 1.7 | 1759 | - | - | - |
| 128 | 7.0 | 1.7 | 241 | 53 | 6.3 | 489 | 407.4 | 2.9 | 990 | 3464 | 1.3 | 2221 |
| 256 | 5.8 | 1.9 | 200 | 38 | 2.8 | 345 | 276.8 | 1.4 | 673 | 2115 | 1.0 | 1356 |
| Leader election where implementation and specification are not weakly bisimilar | | | | | | | | | | | | |
| | 8 processes | | | 9 processes | | | 10 processes | | | 11 processes | | |
| CWB | 4.1 | 0.4 | N/A | 33.7 | 1.3 | N/A | 3765.0 | 0.9 | N/A | - | - | N/A |
| 1 | 1.5 | 5.5 | 349.8 | 13.1 | 7.9 | 521.6 | 122.3 | 7.0 | 736.0 | 1110 | 0.1 | 920 |
| 2 | 1.1 | 12.7 | 258.2 | 5.0 | 10.0 | 908.6 | 7.8 | 39.8 | 178011 | 236 | 58.8 | 959 |
| 4 | 2.1 | 79.1 | 157.1 | 8.5 | 24.8 | 74.5 | 303.4 | 47.7 | 97.4 | 2148 | * | 82 |
| 8 | 4.5 | 46.0 | 25.9 | 37.6 | 151.9 | 37.0 | 516.6 | 164.1 | 52.7 | 2764 | 8.2 | 104 |
| 16 | 3.6 | 97.1 | 21.1 | 31.8 | 103.2 | 55.1 | 83.3 | 31.7 | 69.7 | 1078 | 7.5 | 342 |
| 32 | 1.7 | 30.9 | 4.7 | 10.7 | 67.7 | 19.0 | 49.4 | 12.7 | 28.5 | 1072 | 15.4 | 107 |
| 64 | 0.9 | 2.2 | 3.6 | 5.2 | 5.8 | 7.9 | 75.0 | 5.0 | 9.9 | 1231 | 26.1 | 19 |
| 128 | 0.8 | 13.0 | 3.5 | 6.4 | 10.3 | 2.7 | 28.5 | 13.0 | 8.3 | 812 | 32.7 | 7 |
| 256 | 1.2 | 13.4 | 9.4 | 5.6 | 6.7 | 1.5 | 22.6 | 6.9 | 1.5 | 243 | 23.8 | 6 |

by 8 and 2 cores, respectively. This demonstrates that the performance of our distributed algorithm considerably improves with the increasing problem size.

In the negative cases, it is often enough to explore only a smaller portion of the dependency graph in order to provide a conclusive answer and here the on-the-fly nature of our distributed algorithm shows a real advantage compared to the global algorithms implemented in CWB. For on-the-fly exploration the search order is very important and we can note that increasing the number of cores does not necessarily imply that we can compute the fixed-point value for the root faster. Even though the algorithm scales still very well and with more cores explores a substantially larger part of the dependency graph, it may (by the combined search strategy of the workers) explore large parts of the graph that are not needed for finding the answer. For example in leader election for 10 processes, two cores produced a very successful search strategy that needed only 7.8 s to find the answer, however, increasing the number of cores led the search in a wrong direction.

**Table 2.** Time is reported in seconds, RSD is the relative sample standard deviation in percentage and μs/tv is the time spend per vertex in micro seconds.

ABP where implementation and specification are weakly bisimilar

| cores | buffer size 3 | | | buffer size 4 | | | buffer size 5 | | |
|---|---|---|---|---|---|---|---|---|---|
| | time | RSD | μs/tv | time | RSD | μs/tv | time | RSD | μs/tv |
| CWB | 9.7 | 0.6 | N/A | 1610.3 | 1.3 | N/A | - | - | N/A |
| 1 | 81.3 | 0.5 | 113.6 | 2409.5 | 0.3 | 161.4 | - | - | - |
| 2 | 42.0 | 0.7 | 58.7 | 1268.5 | 3.8 | 85.0 | - | - | - |
| 4 | 22.4 | 2.1 | 31.3 | 650.3 | 1.2 | 43.6 | - | - | - |
| 8 | 13.8 | 11.6 | 19.3 | 332.0 | 1.9 | 22.2 | - | - | - |
| 16 | 10.2 | 13.6 | 14.3 | 239.1 | 6.2 | 16.0 | - | - | - |
| 32 | 5.9 | 14.4 | 8.2 | 127.0 | 3.9 | 8.5 | 3314.7 | 1.0 | 10.8 |
| 64 | 3.4 | 1.2 | 4.7 | 78.8 | 2.5 | 5.3 | 1970.5 | 0.4 | 6.4 |
| 128 | 2.1 | 3.7 | 3.0 | 42.4 | 0.8 | 2.8 | 1020.3 | 1.2 | 3.3 |
| 256 | 1.8 | 23.1 | 2.5 | 24.7 | 2.7 | 1.7 | 551.2 | 0.6 | 1.8 |

ABP where implementation and specification are not weakly bisimilar

| cores | buffer size 4 | | | buffer size 5 | | | buffer size 6 | | |
|---|---|---|---|---|---|---|---|---|---|
| CWB | 8.3 | 0.9 | N/A | 170.2 | 0.5 | N/A | - | - | N/A |
| 1 | 5.0 | 0.4 | 15365.9 | 3.4 | 0.3 | 109113 | 4.1 | 0.4 | 584643 |
| 2 | 15.0 | 1.2 | 56.9 | 1.3 | 14.8 | 179286 | 4.1 | 2.8 | 590714 |
| 4 | 7.8 | 4.4 | 37.8 | 168.3 | 0.5 | 95.9 | 3125.1 | 0.8 | 202.2 |
| 8 | 6.4 | 25.6 | 65.0 | 98.1 | 17.0 | 297.3 | 1602.2 | 1.0 | 669.4 |
| 16 | 4.4 | 20.0 | 45.5 | 66.1 | 13.2 | 108.7 | 1128.2 | 1.1 | 15391.5 |
| 32 | 2.2 | 3.5 | 694.9 | 35.8 | 1.6 | 1792.8 | 649.6 | 9.9 | 7481.1 |
| 64 | 1.3 | 7.3 | 367.4 | 21.8 | 1.5 | 1006.6 | 370.9 | 0.4 | 3869.9 |
| 128 | 0.8 | 3.7 | 289.2 | 14.4 | 1.4 | 755.7 | 197.5 | 1.2 | 2482.5 |
| 256 | 0.5 | 3.9 | 127.6 | 7.9 | 2.1 | 436.1 | 107.7 | 1.1 | 1305.1 |

Finally, results for checking the simulation preorder on the task graph benchmark can be seen in Table 3. As this is a large number of experiments requiring nontrivial time to run, we tested the scaling only up to 64 cores. We queried whether all the tasks in the task graph (or rather their initial prefixes) can be completed within 25 time units. Out of the 180 task graphs, 61 of them are solvable in one hour (and 34 of them are schedulable while 27 are not schedulable). As CWB does not support simulation preorder, the weaker trace inclusion property is used but CWB cannot solve any of the task graphs within one hour. We achieve an average 25 times speedup using 64 cores, both for the positive and negative cases, showing a very satisfactory performance on this large collection of experiments.

**Table 3.** Number of solved task graphs within 1 h for all, positive and negative instances. The accumulated average time (AAT) is projected on 9 task graphs that 1 core is able to solve between 20 min and 1 h.

| | Weak simulation preorder on task graphs | | | | | |
|---|---|---|---|---|---|---|
| | Total | | Positive | | Negative | |
| cores | solved | AAT | solved | AAT | solved | AAT |
| 1 | 35 | 19660 | 16 | 7818 | 19 | 11841 |
| 2 | 39 | 10278 | 18 | 4085 | 21 | 6192 |
| 4 | 43 | 5301 | 21 | 2095 | 22 | 3205 |
| 8 | 49 | 2996 | 26 | 1201 | 23 | 1794 |
| 16 | 51 | 2240 | 28 | 858 | 23 | 1381 |
| 32 | 57 | 1271 | 33 | 493 | 24 | 777 |
| 64 | 61 | 798 | 34 | 310 | 27 | 487 |

## 5    Conclusion

We presented a distributed algorithm for computing fixed points on dependency graphs and showed on weak bisimulation/simulation checking between CCS processes that, even though the problem is in general P-hard, we can in many cases obtain reasonable speed-ups as we increase the number of cores. Our algorithm works on-the-fly and hence for the cases where only a small portion of the dependency graphs needs to be explored to provide the answer, we perform significantly better than the global algorithms implemented in the CWB tool. Compared to CWB we also scale better with the increasing instance sizes, even for the cases where the whole dependency graph must be explored. The advantage of our approach based on dependency graphs is that we provide a general distributed algorithm and its efficient implementation that can be directly applied also to other problems like e.g. model checking—most importantly without the need of designing and coding specific single-purpose distributed algorithms for the different applications. In our future work we plan to look into finding better parallel search strategies that will allow for early termination in the cases where the fixed-point value of the root is 1 and also terminating the parallel search of the graph once we know that the exploration is not needed any more.

**Acknowledgments.** The present work was supported by the Danish e-Infrastructure Cooperation by co-funding acquisitions of the MCC Linux cluster at Aalborg University and received funding from the Sino-Danish Basic Research Center IDEA4CPS funded by the Danish National Research Foundation and the National Science Foundation, China, the Innovation Fund Denmark center DiCyPS, as well as the ERC Advanced Grant LASSO. The fourth author is partially affiliated with FI MU in Brno.

# References

1. Liu, X., Ramakrishnan, C.R., Smolka, S.A.: Fully local and efficient evaluation of alternating fixed points. In: Steffen, B. (ed.) TACAS 1998. LNCS, vol. 1384, pp. 5–19. Springer, Heidelberg (1998). doi:10.1007/BFb0054161
2. Liu, X., Smolka, S.A.: Simple linear-time algorithms for minimal fixed points. In: Larsen, K.G., Skyum, S., Winskel, G. (eds.) ICALP 1998. LNCS, vol. 1443, pp. 53–66. Springer, Heidelberg (1998). doi:10.1007/BFb0055040
3. Balcázar, L.J., Gabarró, J., Santha, M.: Deciding bisimilarity is p-complete. Formal Asp. Comput. **4**(6A), 638–648 (1992)
4. Cleaveland, R., Parrow, J., Steffen, B.: The concurrency workbench: a semantics-based tool for the verification of concurrent systems. ACM Trans. Program. Lang. Syst. **15**(1), 36–72 (1993)
5. Milner, R.: A Calculus of Communicating Systems. LNCS, vol. 92. Springer, Berlin (1980)
6. Bollig, B., Leucker, M., Weber, M.: Local parallel model checking for the alternation-free $\mu$-calculus. In: Bošnački, D., Leue, S. (eds.) SPIN 2002. LNCS, vol. 2318, pp. 128–147. Springer, Heidelberg (2002). doi:10.1007/3-540-46017-9_11
7. Grumberg, O., Heyman, T., Schuster, A.: Distributed symbolic model checking for $\mu$-calculus. Formal Meth. Syst. Des. **26**(2), 197–219 (2005)
8. Joubert, C., Mateescu, R.: Distributed on-the-fly model checking and test case generation. In: Valmari, A. (ed.) SPIN 2006. LNCS, vol. 3925, pp. 126–145. Springer, Heidelberg (2006). doi:10.1007/11691617_8
9. Gibson-Robinson, T., Armstrong, P., Boulgakov, A., Roscoe, A.W.: FDR3 — a modern refinement checker for CSP. In: Ábrahám, E., Havelund, K. (eds.) TACAS 2014. LNCS, vol. 8413, pp. 187–201. Springer, Heidelberg (2014). doi:10.1007/978-3-642-54862-8_13
10. Garavel, H., Lang, F., Mateescu, R., Serwe, W.: CADP 2011: a toolbox for the construction and analysis of distributed processes. Int. J. Softw. Tools Technol. Transf. **15**(2), 89–107 (2013)
11. Holzmann, G.: Spin Model Checker, the: Primer and Reference Manual, 1st edn. Addison-Wesley Professional, Boston (2003)
12. Groote, J.F., Mousavi, M.R.: Modeling and Analysis of Communicating Systems. The MIT Press, Cambridge (2014)
13. Andersen, J.R., Andersen, N., Enevoldsen, S., Hansen, M.M., Larsen, K.G., Olesen, S.R., Srba, J., Wortmann, J.K.: CAAL: concurrency workbench, aalborg edition. In: Leucker, M., Rueda, C., Valencia, F.D. (eds.) ICTAC 2015. LNCS, vol. 9399, pp. 573–582. Springer, Heidelberg (2015). doi:10.1007/978-3-319-25150-9_33
14. Tarski, A.: A lattice-theoretical fixpoint theorem and its applications. Pacific J. Math **5**(2), 285–309 (1955)
15. Glabbeek, R.J.: The linear time - branching time spectrum. In: Baeten, J.C.M., Klop, J.W. (eds.) CONCUR 1990. LNCS, vol. 458, pp. 278–297 Springer, Heidelberg (1990). doi:10.1007/BFb0039066
16. Andersen, J.R., Hansen, M.M., Andersen, N.: CAAL 2.0: equivalences, preorders and games for CCS and TCCS. Master's thesis, Aalborg University (2015)
17. Jensen, J.F., Larsen, K.G., Srba, J., Oestergaard, L.K.: Local model checking of weighted CTL with upper-bound constraints. In: Bartocci, E., Ramakrishnan, C.R. (eds.) SPIN 2013. LNCS, vol. 7976, pp. 178–195. Springer, Heidelberg (2013). doi:10.1007/978-3-642-39176-7_12

18. Cassez, F., David, A., Fleury, E., Larsen, K.G., Lime, D.: Efficient on-the-fly algorithms for the analysis of timed games. In: Abadi, M., Alfaro, L. (eds.) CONCUR 2005. LNCS, vol. 3653, pp. 66–80. Springer, Heidelberg (2005). doi:10.1007/11539452_9

19. Liu, X., Smolka, S.A.: Simple linear-time algorithms for minimal fixed points. In: Larsen, K.G., Skyum, S., Winskel, G. (eds.) ICALP 1998. LNCS, vol. 1443, pp. 53–66. Springer, Heidelberg (1998). doi:10.1007/BFb0055040

20. Dijkstra, E.W.: Shmuel Safra's version of termination detection. EWD Manuscript **998**, (1987)

21. Chang, E., Roberts, R.: An improved algorithm for decentralized extrema-finding in circular configurations of processes. Commun. ACM **22**(5), 281–283 (1979)

22. Bartlett, K.A., Scantlebury, R.A., Wilkinson, P.T.: A note on reliable full-duplex transmission over half-duplex links. Commun. ACM **12**(5), 260–261 (1969)

23. Kwok, Y.-K., Ahmad, I.: Benchmarking and comparison of the task graph scheduling algorithms. J. Parallel Distrib. Comput. **59**(3), 381–422 (1999)

# A Complete Approximation Theory
# for Weighted Transition Systems

Mikkel Hansen$^{(\boxtimes)}$, Kim Guldstrand Larsen, Radu Mardare,
Mathias Ruggaard Pedersen, and Bingtian Xue

Department of Computer Science, Aalborg University,
Selma Largerlöfsvej 300, Aalborg, Denmark
{mhan,kgl,mardare,mrp,bingt}@cs.aau.dk

**Abstract.** We propose a way of reasoning about minimal and maximal values of the weights of transitions in a weighted transition system (WTS). This perspective induces a notion of bisimulation that is coarser than the classic bisimulation: it relates states that exhibit transitions to bisimulation classes with the weights within the same boundaries. We propose a customized modal logic that expresses these numeric boundaries for transition weights by means of particular modalities. We prove that our logic is invariant under the proposed notion of bisimulation. We show that the logic enjoys the finite model property which allows us to prove the decidability of satisfiability and provide an algorithm for satisfiability checking. Last but not least, we identify a complete axiomatization for this logic, thus solving a long-standing open problem in this field. All our results are proven for a class of WTSs without the image-finiteness restriction, a fact that makes this development general and robust.

## 1 Introduction

Weighted transition systems (WTSs) are used to model concurrent and distributed systems in the case where some resources are involved, such as time, bandwidth, fuel, or energy consumption. Recently, the concept of a cyber-physical system (CPS), which considers the integration of computation and the physical world has become relevant in modeling various real-life situations. In these models, sensor feedback affects computation, and through machinery, computation can further affect physical processes. The quantitative nature of weighted transition systems is well-suited for the quantifiable inputs and sensor measurements of CPSs, but their rigidity makes them less well suited for the uncertainty inherent in CPSs. In practice, there is often some uncertainty attached to the resource cost, whereas weights in a WTS are precise. Thus, the model may be too restrictive and unable to capture the uncertainties inherent in the domain that is being modeled.

In this paper, we attempt to remedy this shortcoming by introducing a modal logic for WTSs that allows for approximate reasoning by speaking about upper and lower bounds for the weights of the transitions. The logic has two types

© Springer International Publishing AG 2016
M. Fränzle et al. (Eds.): SETTA 2016, LNCS 9984, pp. 213–228, 2016.
DOI: 10.1007/978-3-319-47677-3_14

of modal operators that reason about the minimal and maximal weights on transitions, respectively. This allows reasoning about models where the quantitative information may be imprecise (e.g. due to imprecisions introduced when gathering real data), but where we can establish a lower and upper bound for transitions.

In order to provide the semantics for this logic, we use the set of possible transition weights from one state to a set of states as an abstraction of the actual transition weights. The logic is expressive enough to characterize WTSs up to a relaxed notion of weighted bisimilarity, where the classical conditions are replaced with conditions requiring that the minimal and maximal weights on transitions are matched. This logical characterization works for a class of WTSs that is strictly larger than the class of image-finite WTSs.

Our main contribution is a complete axiomatization of our logic, showing that any validity in this logic can be proved as a theorem from the axiomatic system. This solves a long-standing open problem in the field of weighted systems. Completeness allows us to transform any validity checking problem into a theorem proving one that can be solved automatically by modern theorem provers, thus bridging the gap to the theorem proving community. The completeness proof adapts the classical filtration method, which allows one to construct a (canonical) model using maximal consistent sets of formulae. The main difficulty of adapting this method to our setting is that we must establish both lower and upper bounds for the transitions in this model.

To achieve this result, we firstly demonstrate that our logic enjoys the finite model property. This property allows us not only to achieve the completeness proof, but also to address the problem of decidability of satisfiability. This is our second significant contribution in this paper: we propose a decision procedure for determining the satisfiability of formulae in our logic. This decision procedure makes use of the finite model property to automatically generate a finite model for any satisfiable formula.

**Related Work.** Several logics have been proposed in the past to express properties of quantified (weighted, probabilistic or stochastic) systems [5,6,12,15,17]. They typically use modalities indexed with real numbers to express properties such as "$\varphi$ holds with at least probability $b$", "we can reach a state satisfying $\varphi$ with a cost at least $r$", etc. While our logical syntax resemble these, our semantics is different in the sense that we argue not about one value (a probability or a cost), but about a compact interval of possible costs. For instance, in the aforementioned logics we have a validity of type $\vdash \neg L_r \phi \rightarrow M_r \phi$ saying that the value of the transition from the current state to $\phi$ is either at least $r$ or at most $r$; on the other hand, in our logic the formula $\neg L_r \phi \wedge \neg M_r \phi$ might have a model since $L_r \phi$ and $M_r \phi$ express the fact that the lower cost of a transition to $\phi$ is at least $r$ and the highest cost is at most $r$ respectively.

However, our completeness proof uses a technique similar to the one used for weighted modal logic [13] and Markovian logic [12,16]. It is however different from these related constructions since our axiomatization is finitary, while the aforementioned ones require infinitary proof rules. Our axiomatic systems are related to the ones mentioned above and the mathematical structures revealed by this work are also similar to the related ones. This suggest a natural extension towards a Stone duality type of result on the line of [11], which we will consider in a future work.

Satisfiability results have been given for some related logics too, such as weighted modal logics [14] and probabilistic versions of CTL and the $\mu$-calculus [4]. However, the satisfiability problem is known to be undecidable for other related logics, in particular timed logics such as TCTL [1] and timed modal logic [8]. This fact suggests our logic as an interesting one which, despite its expressivity, remains decidable.

Our approach of considering upper and lower bounds is related to interval-based formalisms such as interval Markov chains (IMCs) [9] and interval weighted modal transition systems (WMTSs) [10]. Much like our approach, IMCs consider upper and lower bounds on transitions in the probabilistic case. WMTSs add intervals of weights to individual transitions of modal transition systems, in which there can be both may- and must-transitions. A main focus of the work both on IMCs and WMTSs have been a process of refinement, making the intervals progressively smaller until an implementation is obtained. However, none of these works have explored the logical perspective up to the level of axiomatization or satisfiability results, which is the focus of our paper.

## 2  Model

The models addressed in this paper are weighted transition systems, in which transitions are labeled with numbers to specify the cost of the corresponding transition. In order to specify and reason about properties regarding imprecision, such as "the maximum cost of going to a safe state is 10" and "the minimum cost of going to a halting state is 5", we will abstract away the individual transitions and only consider the minimum and maximum costs from a state to another. We will do this by constructing for any two states the set of weights that are allowed from one to the other.

First we recap the definition of a weighted transition system. A WTS is formally defined as follows:

**Definition 1.** *A weighted transition system (WTS) is a tuple* $\mathcal{M} = (S, \rightarrow, \ell)$, *where*

- $S$ *is a non-empty set of* states,
- $\rightarrow \subseteq S \times \mathbb{R}_{>0} \times S$ *is the* transition relation, *and*
- $\ell : S \rightarrow 2^{AP}$ *is a* labeling function *mapping to each state a set of atomic propositions.*

Note that we impose no restrictions on the state space $S$; it can be uncountable.

Consider now a WTS as in Fig. 1a. If this is a CPS, then the weights may have been obtained by measurements, simulations, or educated guesses, which may be imprecise. However, it may be that we can establish 1 as a lower bound and 10

(a) We may not know the   (b) We add transitions from   (c) We add infinitely many
precise weights from $s$ to $t$,   $s$ to $t$ with weights that are   transitions from $s$ to $t$ with
but we can establish 1 as a   between 1 and 10.   each real weight between 1
lower bound, and 10 as an   and 10.
upper bound.

**Fig. 1.** Possible ways to address the problem of not knowing the precise weight for
each transition.

as an upper bound on the actual weight. We could then address this problem by
making more measurements and adding the results as weights on transitions, as
in Fig. 1b but as long as we only introduce finitely many new transitions, there
will still be some imprecision. Instead, we could add infinitely many transitions,
for example one for each real or rational weight that lies between 1 and 10, as in
Fig. 1c. However, then our WTS is no longer image-finite, so it no longer satisfies
the Hennessy-Milner theorem [7].

In this paper, we will address this problem by abstracting away the individual
transitions, and instead consider the set of weights between a state and a set of
states.

**Definition 2.** *For arbitrary WTS* $\mathcal{M} = (S, \rightarrow, \ell)$ *the function* $\theta_{\mathcal{M}} : S \rightarrow \left(2^S \rightarrow 2^{\mathbb{R}_{\geq 0}}\right)$ *is defined for any state* $s \in S$ *and set of states* $T \subseteq S$ *as*

$$\theta_{\mathcal{M}}(s)(T) = \{r \in \mathbb{R}_{\geq 0} \mid \exists t \in T \: such \: that \: s \xrightarrow{r} t\}.$$

Thus $\theta_{\mathcal{M}}(s)(T)$ is the set of all possible weights of going from $s$ to a state in $T$.
We will sometimes refer to $\theta(s)(T)$ as the *image from* $s$ *to* $T$ or simply as an
*image set*.

Next, we introduce the notion of an image-compact WTS, which imposes
a requirement on the image sets. This notion is very closely related to that of
compactly branching introduced by van Breugel [3].

**Definition 3.** *Let* $\mathcal{M} = (S, \rightarrow, \ell)$ *be a WTS. We say that* $\mathcal{M}$ *is* image-compact
*if for any* $s \in S$ *and* $T \subseteq S$, $\theta_{\mathcal{M}}(s)(T)$ *is a compact set, i.e. a closed and
bounded set.*

Intuitively, one can think of a WTS being image-compact if each state can not
take transitions with arbitrarily large weights and whenever a state can take
transitions with weights arbitrarily close to some real number $x$ it can also take
a transition with exactly the weight $x$. We will drop the subscript $\mathcal{M}$ from $\theta$
unless we wish to differentiate between the image sets of two different WTSs. For
the bisimulation invariance theorem that we will discuss later, it will be necessary
to restrict ourselves to only considering image-compact WTSs. However, this will
be the only place in the paper where this restriction is needed.

Consider a state $s$ that can take a transition with weight $\frac{1}{2^i}$ for any $i \in \mathbb{N}$ to some state in a set $T$. We then have $\theta(s)(T) = \{\frac{1}{2^i} \mid i \in \mathbb{N}\}$ which is clearly not a closed set, since $\frac{1}{2^i} \xrightarrow{i \to \infty} 0$ and $0 \notin \theta(s)(T)$, hence it is non-compact. Consider now a state $s'$ that has the same outgoing transitions as $s$ except that also $s' \xrightarrow{0} t$ for some $t \in T$. We then have $\theta(s')(T) = \{\frac{1}{2^i} \mid i \in \mathbb{N}\} \cup \{0\}$ which is a closed and bounded set, hence it is compact.

Note that any image-finite WTS is also image-compact, since any finite set is compact. However, an image-compact WTS is not always image-finite. In the rest of the paper, we will use the notation $\theta^-(s)(T) = \inf \theta(s)(T)$ and $\theta^+(s)(T) = \sup \theta(s)(T)$ with the convention that $\inf \emptyset = -\infty$ and $\sup \emptyset = \infty$. Note that this convention is the opposite of the one usually adopted.

*Example 4.* Figure 2 shows a simple model of a robot vacuum cleaner that can be in a waiting state, a cleaning state, or a charging state. This is an example of a cyber-physical system where the costs of transitions are necessarily imprecise. The time it takes to recharge the batteries depends on the condition of the batteries as well as that of the charger; the time it takes to clean the room depends on how dirty the room is, and how free the floor is from obstacles; and the time it takes to reach the charger depends on where in the room the robot is when it needs to be recharged. By constructing the image sets, we can abstract away from the individual transitions. For example, we have $\theta(s_2)(\{s_1\}) = \{5, 10, 15\}$, so $\theta^-(s_2)(\{s_1\}) = 5$ and $\theta^+(s_2)(\{s_1\}) = 15$.

We will now establish some useful properties of image sets. We first show that the transition function is monotonic with respect to set inclusion, meaning that if $T_1$ is a subset of $T_2$ then, the image from any state $s$ to $T_1$ is also a subset of the image from $s$ to $T_2$.

**Lemma 5 (Monotonicity of $\theta$).** *Let $\mathcal{M} = (S, \to, \ell)$ be a WTS and let $T_1$ and $T_2$ be subsets of $S$. If $T_1 \subseteq T_2$, then $\theta(s)(T_1) \subseteq \theta(s)(T_2)$.*

Next, we show that union and intersection over image sets distribute as usual.

**Lemma 6.** *Let $\mathcal{M} = (S, \to, \ell)$ be a WTS. For any $s \in S$ and $T_1, T_2 \subseteq S$, it holds that*

*1. $\theta(s)(T_1 \cup T_2) = \theta(s)(T_1) \cup \theta(s)(T_2)$ and*
*2. $\theta(s)(T_1 \cap T_2) = \theta(s)(T_1) \cap \theta(s)(T_2)$.*

As usual we would like some way of relating model states with equivalent behavior. To this end we define the notion of a bisimulation relation. The classical notion of a bisimulation relation for weighted transition systems [2], which we term weighted bisimulation, is defined as follows.

**Definition 7.** *Given a WTS $\mathcal{M} = (S, \rightarrow, \ell)$, an equivalence relation $\mathcal{R} \subseteq S \times S$ on $S$ is called a* weighted bisimulation relation *iff for all $s, t \in S$, $s\mathcal{R}t$ implies*

- *(Atomic harmony) $\ell(s) = \ell(t)$,*
- *(Zig) if $s \xrightarrow{r} s'$ then there exists $t' \in S$ such that $t \xrightarrow{r} t'$ and $s'\mathcal{R}t'$, and*
- *(Zag) if $t \xrightarrow{r} t'$ then there exists $s' \in S$ such that $s \xrightarrow{r} s'$ and $s'\mathcal{R}t'$.*

We say that $s, t \in S$ are weighted bisimilar, written $s \sim_W t$, iff there exists a weighted bisimulation relation $\mathcal{R}$ such that $s\mathcal{R}t$. Weighted bisimilarity, $\sim_W$, is the largest weighted bisimulation relation. Note that we could replace the zig-zag conditions by the condition that $\theta(s)(T) = \theta(t)(T)$ for all $\mathcal{R}$-equivalence classes $T \subseteq S$.

Since it is our goal to abstract away from the exact weights on the transitions, the bisimulation that we will now introduce does not impose the classical zig-zag conditions [2] of a bisimulation relation, but instead require that bounds be matched for any bisimulation class.

**Definition 8.** *Given a WTS $\mathcal{M} = (S, \rightarrow, \ell)$, an equivalence relation $\mathcal{R} \subseteq S \times S$ on $S$ is called a* generalized weighted bisimulation relation *iff for all $s, t \in S$, $s\mathcal{R}t$ implies*

- *(Atomic harmony) $\ell(s) = \ell(t)$,*
- *(Lower bound) $\theta^-(s)(T) = \theta^-(t)(T)$, and*
- *(Upper bound) $\theta^+(s)(T) = \theta^+(t)(T)$*

*for any $\mathcal{R}$-equivalence class $T \subseteq S$.*

Given $s, t \in S$ we say that $s$ and $t$ are generalized weighted bisimilar, written $s \sim t$, iff there exists a generalized weighted bisimulation relation $\mathcal{R}$ such that $s\mathcal{R}t$. Generalized weighted bisimilarity, $\sim$, is the largest generalized weighted bisimulation relation.

In what follows, we will use bisimulation to mean generalized weighted bisimulation and bisimilarity to mean generalized weighted bisimilarity. We now show the relationship between $\sim$ and $\sim_W$.

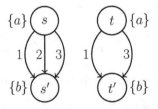

**Fig. 3.** $s \sim t$ but $s \not\sim_W t$.

*Example 9.* Consider the WTS depicted in Fig. 3. It is easy to see that $\{s', t'\}$ is a $\sim$-equivalence class, and in fact it is the only $\sim$-equivalence class with ingoing transitions. Since $\theta^-(s)(\{s', t'\}) = \theta^-(t)(\{s', t'\}) = 1$ and $\theta^+(s)(\{s', t'\}) = \theta^+(t)(\{s', t'\}) = 3$ we must have $s \sim t$, but because $s \xrightarrow{2} s'$ and $t \xrightarrow{2}\!\!\!\!\!/ \,$ it cannot be the case that $s \sim_W t$.

**Theorem 10.** *Generalized weighted bisimilarity is coarser than weighted bisimilarity, i.e.*

$$\sim_W \subsetneq \sim$$

This result is not surprising, as our bisimulation relation only looks at the extremes of the transition weights, whereas weighted bisimulation looks at all of the transition weights.

## 3   Logic

In this section we introduce a modal logic. Our aim is that our logic should be able to capture the notion of bisimilar states as presented in the previous section, and as such it must be able to reason about the lower and upper bounds on transition weights.

**Definition 11.** *The formulae of the logic $\mathcal{L}$ are induced by the abstract syntax*

$$\mathcal{L}: \quad \varphi, \psi ::= p \mid \neg\varphi \mid \varphi \wedge \psi \mid L_r\varphi \mid M_r\varphi$$

*where $r \in \mathbb{Q}_{\geq 0}$ is a non-negative rational number and $p \in \mathcal{AP}$ is an atomic proposition.*

$L_r$ and $M_r$ are modal operators. An illustration of how $L$ and $M$ are interpreted can be seen in Fig. 4. Intuitively, $L_r\varphi$ means that the cost of transitions to where $\varphi$ holds is *at least* $r$ (see Fig. 4a), and $M_r\varphi$ means that the the cost of transitions to where $\varphi$ holds is *at most* $r$ (see Fig. 4b).

We now give the precise semantics interpreted on WTSs.

**Definition 12.** *Given a WTS $\mathcal{M} = (S, \rightarrow, \ell)$, a state $s \in S$ and a formula $\varphi \in \mathcal{L}$, the satisfiability relation $\models$ is defined inductively as:*

$$
\begin{aligned}
\mathcal{M}, s &\models p && iff\ p \in \ell(s), \\
\mathcal{M}, s &\models \neg\varphi && iff\ \mathcal{M}, s \not\models \varphi, \\
\mathcal{M}, s &\models \varphi \wedge \psi && iff\ \mathcal{M}, s \models \varphi\ and\ \mathcal{M}, s \models \psi, \\
\mathcal{M}, s &\models L_r\varphi && iff\ \theta^-(s)(\llbracket\varphi\rrbracket_\mathcal{M}) \geq r, \\
\mathcal{M}, s &\models M_r\varphi && iff\ \theta^+(s)(\llbracket\varphi\rrbracket_\mathcal{M}) \leq r,
\end{aligned}
$$

*where $\llbracket\varphi\rrbracket_\mathcal{M} = \{s \in S \mid \mathcal{M}, s \models \varphi\}$.*

We will omit the subscript $\mathcal{M}$ from $\llbracket\varphi\rrbracket_\mathcal{M}$ whenever the model is clear from the context. If $\mathcal{M}, s \models \varphi$ we say that $\mathcal{M}$ is a model of $\varphi$. A formula is said to be *satisfiable* if it has at least one model. We say that $\varphi$ is a *validity* and write $\models \varphi$ if $\neg\varphi$ is not satisfiable. In addition to the operators defined by the syntax of $\mathcal{L}$, we also have the derived operators such as $\bot, \rightarrow$, etc. defined in the usual way.

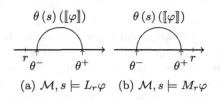

(a) $\mathcal{M}, s \models L_r\varphi$   (b) $\mathcal{M}, s \models M_r\varphi$

**Fig. 4.** $L_r$ and $M_r$ semantics.

The formula $L_0\varphi$ has special significance in our logic, as this formula means that there exists some transition to where $\varphi$ holds. In fact, it follows in a straightforward manner from the semantics that $\mathcal{M}, s \models L_0\varphi$ if and only if $\theta(s)(\llbracket\varphi\rrbracket) \neq \emptyset$.

*Example 13.* Consider again our model of a robot vacuum cleaner depicted in Fig. 2. Perhaps we want a guarantee that it takes no more than one time unit to go from a waiting state to a charging state. This can be expressed by the formula waiting $\rightarrow M_1$charging, but since we know the only waiting state in our model is $s_1$ this can be simplified to simply checking whether $\mathcal{M}, s_1 \models M_1$charging. We thus have to check that $\theta^+(s_1)(\llbracket$charging$\rrbracket) \leq 1$. We do this by constructing the image set $\theta(s_1)(\llbracket$charging$\rrbracket)$. Since $\llbracket$charging$\rrbracket = \{s_3\}$, we have $\theta(s_1)(\{s_3\}) = \{1, 2\}$. Hence $\theta^+(s_1)(\llbracket$charging$\rrbracket) = 2 \nleq 1$, so $\mathcal{M}, s_1 \nvDash M_1$charging.

Next we show that our logic $\mathcal{L}$ is invariant under bisimulation, which is also known as the Hennessy-Milner property.

**Theorem 14 (Bisimulation invariance).** *For any image-compact WTS $\mathcal{M} = (S, \rightarrow, \ell)$ and states $s, t \in S$ it holds that*

$$s \sim t \quad iff \quad [\forall\varphi \in \mathcal{L}. \mathcal{M}, s \models \varphi \; iff \; \mathcal{M}, t \models \varphi].$$

The proof strategy follows a classical pattern: The left to right direction is shown by induction on $\varphi$ for $\varphi \in \mathcal{L}$. The right to left direction is shown by constructing a relation $\mathcal{R}$ relating those states that satisfy the same formulae and showing that this relation is a bisimulation relation.

# 4   Metatheory

In this section we propose an axiomatization for our logic that we prove not only sound, but also complete with respect to the proposed semantics.

## 4.1   Axiomatic System

Let $r, s \in \mathbb{Q}_{\geq 0}$. Then the deducibility relation $\vdash \subseteq 2^{\mathcal{L}} \times \mathcal{L}$ is a classical conjunctive deducibility relation, and is defined as the smallest relation which satisfies the axioms of propositional logic in addition to the axioms given in Table 1. We will write $\vdash \varphi$ to mean $\emptyset \vdash \varphi$, and we say that a formula or a set of formulae is *consistent* if it can not derive $\bot$.

Axiom A1 captures the notion that since $\bot$ is never satisfied, we can never take a transition to where $\bot$ holds. Axiom A2 says that if we know some value is the lower bound for going to where $\varphi$ holds, then any lower value is also a lower bound for going to where $\varphi$ holds. Axiom A2$'$ is the analogue for upper bounds. Axioms A3–A4 show how $L_r$ and $M_r$ distribute over conjunction and disjunction. The version of axiom A4 where $L_r$ is replaced with $M_r$ is also sound,

**Table 1.** The axioms for our axiomatic system, where $\varphi, \psi \in \mathcal{L}$ and $q, r \in \mathbb{Q}$.

| | |
|---|---|
| (A1): $\vdash \neg L_0 \bot$ | |
| (A2): $\vdash L_{r+q}\varphi \rightarrow L_r\varphi$ | if $q > 0$ |
| (A2'): $\vdash M_r\varphi \rightarrow M_{r+q}\varphi$ | if $q > 0$ |
| (A3): $\vdash L_r\varphi \wedge L_q\psi \rightarrow L_{\min\{r,q\}}(\varphi \vee \psi)$ | |
| (A3'): $\vdash M_r\varphi \wedge M_q\psi \rightarrow M_{\max\{r,q\}}(\varphi \vee \psi)$ | |
| (A4): $\vdash L_r(\varphi \vee \psi) \rightarrow L_r\varphi \vee L_r\psi$ | |
| (A5): $\vdash \neg L_0\psi \rightarrow (L_r\varphi \rightarrow L_r(\varphi \vee \psi))$ | |
| (A5'): $\vdash \neg L_0\psi \rightarrow (M_r\varphi \rightarrow M_r(\varphi \vee \psi))$ | |
| (A6): $\vdash L_{r+q}\varphi \rightarrow \neg M_r\varphi$ | if $q > 0$ |
| (A7): $\vdash M_r\varphi \rightarrow L_0\varphi$ | |
| (R1): $\vdash \varphi \rightarrow \psi \implies \vdash ((L_r\psi) \wedge (L_0\varphi)) \rightarrow L_r\varphi$ | |
| (R1'): $\vdash \varphi \rightarrow \psi \implies \vdash ((M_r\psi) \wedge (L_0\varphi)) \rightarrow M_r\varphi$ | |
| (R2): $\vdash \varphi \rightarrow \psi \implies \vdash L_0\varphi \rightarrow L_0\psi$ | |

but it can be proven from the other axioms. Axioms A5 and A5' say that if it is not possible to take a transition to where $\psi$ holds, then requiring that $\psi$ also holds does not change the bounds. Axioms A6 and A7 show the relationship between $L_r$ and $M_r$. In particular, A6 ensures that all bounds are well-formed. Notice also that the contrapositive of axiom A2 and A7 together gives us that $\neg L_0\varphi$ implies $\neg L_r\varphi$ and $\neg M_r\varphi$ for any $r \in \mathbb{Q}_{\geq 0}$. The axioms R1 and R1' give a sort of monotonicity for $L_r$ and $M_r$, and axiom R2 says that if $\psi$ follows from $\varphi$, then if it is possible to take a transition to where $\varphi$ holds, it is also possible to take a transition to where $\psi$ holds.

**Theorem 15 (Soundness)**

$$\vdash \varphi \quad implies \quad \models \varphi.$$

## 4.2    Finite Model Property and Completeness

With our axiomatization proven sound we are now ready to present our main results, namely that our logic has the finite model property and that our axiomatization is complete.

To show the finite model property we will adapt the classical filtration method to our setting. Starting from an arbitrary formula $\rho$, we define a finite fragment of our logic, $\mathcal{L}[\rho]$, which we then use to construct a finite model for $\rho$. The main difference from the classical filtration method is that we must find an upper and a lower bound for the transitions in the model. For an arbitrary formula $\rho \in \mathcal{L}$ we define the following based on $\rho$:

- Let $Q_\rho \subseteq \mathbb{Q}_{\geq 0}$ be the set of all rational numbers $r \in \mathbb{Q}_{\geq 0}$ such that $L_r$ or $M_r$ appears in the syntax of $\rho$.
- Let $\Sigma_\rho$ be the set of all atomic propositions $p \in \mathcal{AP}$ such that $p$ appears in the syntax of $\rho$.
- The *granularity* of $\rho$, denoted as $gr(\rho)$, is the least common denominator of all the elements in $Q_\rho$.
- The *range* of $\rho$, denoted as $R_\rho$, is defined as

$$R_\rho = \begin{cases} \emptyset & \text{if } Q_\rho = \emptyset \\ I_\rho \cup \{0\} & \text{otherwise ,} \end{cases}$$

where $I_\rho = \left\{ q \in \mathbb{Q}_{\geq 0} \mid \exists j \in \mathbb{N}.\ q = \frac{j}{gr(\rho)} \text{ and } \min Q_\rho \leq q \leq \max Q_\rho \right\}$. Note that we need to add 0 to $R_\rho$ whether or not $\rho$ actually contains 0 in any of its modalities. This is because, as we have pointed out before, formulae involving $L_0$ have special significance in our logic.
- The *modal depth* of $\rho$, denoted as $md(\rho)$, is defined inductively as:

$$md(\rho) = \begin{cases} 0 & \text{if } \rho = p \in \mathcal{AP} \\ md(\varphi) & \text{if } \rho = \neg\varphi \\ \max\{md(\varphi_1), md(\varphi_2)\} & \text{if } \rho = \varphi_1 \wedge \varphi_2 \\ 1 + md(\varphi) & \text{if } \rho = L_r\varphi \text{ or } \rho = M_r\varphi. \end{cases}$$

Since all formulae are finite, the modal depth is always a non-negative integer.

The *language* of $\rho$, denoted by $\mathcal{L}[\rho]$, is defined as

$$\mathcal{L}[\rho] = \{\varphi \in \mathcal{L} \mid R_\varphi \subseteq R_\rho, md(\varphi) \leq md(\rho) \text{ and } \Sigma_\varphi \subseteq \Sigma_\rho\}.$$

Because all formulae are finite, $\mathcal{L}[\rho]$ must also be finite (modulo logical equivalence), and as we shall see, it contains all the formulae that are necessary to construct a model for $\rho$.

In order to define the model, we need the notion of filters and ultrafilters.

**Definition 16.** *A non-empty subset $F$ of $\mathcal{L}[\rho]$ is called a* filter *on $\mathcal{L}[\rho]$ iff*

- $\perp \notin F$,
- $\varphi \in F$ and $\vdash \varphi \rightarrow \psi$ implies $\psi \in F$, and
- $\varphi \in F$ and $\psi \in F$ implies $\varphi \wedge \psi \in F$.

Intuitively, one can think of a filter as a consistent set of formulae closed under conjunction and deduction.

**Definition 17.** *A filter $F \in \mathcal{F}$ is called an* ultrafilter *iff for all formulae $\varphi \in \mathcal{L}$ either $\varphi \in F$ or $\neg\varphi \in F$.*

The ultrafilters on $\mathcal{L}[\rho]$ correspond to the maximal consistent sets of $\mathcal{L}[\rho]$. We let $\mathcal{U}[\rho]$ denote the set of all ultrafilters on $\mathcal{L}[\rho]$. Since $\mathcal{L}[\rho]$ is finite $\mathcal{U}[\rho]$ is also finite and consequently, any ultrafilter $u \in \mathcal{U}[\rho]$ must be a finite set of formulae.

Hence the formula obtained by taking the conjunction over all the formulae of $u$ tells us exactly what formulae $u$ contains.

For any set of formulae $\Phi \subseteq \mathcal{L}[\rho]$, the characteristic formula of $\Phi$, denoted $(\!|\Phi|\!)$, is defined as $(\!|\Phi|\!) = \bigwedge_{\varphi \in \Phi} \varphi$. Note that $(\!|\Phi|\!) \in \mathcal{L}[\rho]$ is a finite formula, and that if $u \in \mathcal{U}[\rho]$, then $(\!|u|\!) \in u$.

We will now construct a (finite) model, $\mathcal{M}_\rho$, for $\rho$. In order to define the transition relation $\rightarrow_\rho \subseteq \mathcal{U}[\rho] \times \mathbb{R}_{\geq 0} \times \mathcal{U}[\rho]$, we consider any two ultrafilters $u, v \in \mathcal{U}[\rho]$ and define two functions $L, M : \mathcal{U}[\rho] \times \mathcal{U}[\rho] \rightarrow 2^{R_\rho}$ as

$$L(u,v) = \{r \mid L_r(\!|v|\!) \in u\} \quad \text{and} \quad M(u,v) = \{s \mid M_s(\!|v|\!) \in u\}.$$

The following lemma establishes a relationship between $L$ and $M$, that we will need to define the transition relation. The lemma is a straightforward consequence of axiom $A7$.

**Lemma 18.** *Given any ultrafilters $u, v \in \mathcal{U}[\rho]$, it can not be the case that $L(u,v) = \emptyset$ and $M(u,v) \neq \emptyset$.*

We can now define the transition relation in terms of $L(u,v)$ and $M(u,v)$. In Fig. 5, we have illustrated the different cases that we must consider. For any of the arches in the figure, we have the following correspondence with $L_r$ and $M_r$.

- If a number $r$ on the real line is contained within the arch, then we have $\neg L_r(\!|v|\!) \in u$ and $M_r(\!|v|\!) \in u$.
- If a number $r$ on the real line is to the left of the arch, then we have $L_r(\!|v|\!) \in u$ and $\neg M_r(\!|v|\!) \in u$.
- If a number $r$ on the real line is to the right of the arch, then we have $M_r(\!|v|\!) \in u$ and $\neg L_r(\!|v|\!) \in u$.

In case (a), we therefore have $L(u,v) \neq \emptyset$ and $M(u,v) \neq \emptyset$, so we have all the information we need to define the transition. In case (b) and (f), we have $L(u,v) \neq \emptyset$ and $M(u,v) = \emptyset$, so we have enough information to define the minimum transition, but we do not know what the maximum transition is. Note that we can not simply say that the maximum transition is $\max Q_\rho$, because that would imply $M_{\max Q_\rho}(\!|v|\!) \in u$, but we know that $M(u,v) = \emptyset$. Hence we need to pick a number that is to the right of $\max Q_\rho$ as the maximum. In case (d), we have both $L(u,v) = \emptyset$ and $M(u,v) = \emptyset$. This implies that $\neg L_0(\!|v|\!) \in u$, which means that there should be no transition from $u$ to $v$. In case (c) and (e), we have $L(u,v) = \emptyset$ and $M(u,v) \neq \emptyset$, but according to Lemma 18 these cases can never occur.

We therefore distinguish the following three cases in order to define the transition relation:

1. If $L(u,v) \neq \emptyset$ and $M(u,v) \neq \emptyset$, then we add the two transitions $u \xrightarrow{r_1} v$ and $u \xrightarrow{r_2} v$ where $r_1 = \max L(u,v)$ and $r_2 = \min M(u,v)$.
2. If $L(u,v) \neq \emptyset$ and $M(u,v) = \emptyset$, then we add the two transitions $u \xrightarrow{r_1} v$ and $u \xrightarrow{r_2} v$ where $r_1 = \max L(u,v)$ and $r_2 = \max Q_\rho + \frac{1}{gr(\rho)}$.
3. If $L(u,v) = \emptyset$ and $M(u,v) = \emptyset$, then there is no transition from $u$ to $v$.

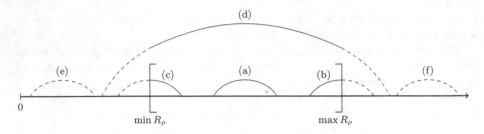

**Fig. 5.** When constructing a transition from $u$ to $v$, we will only have information about what happens in the region $R_\rho$ (which always includes 0). The line represents the non-negative real line and the arches represent the transitions that would be possible in a full model (i.e. one not restricted to $\mathcal{L}[\rho]$). The dashed part of the arches represent the part of the transition that we do not have information about.

Finally we define the labeling function $\ell_\rho : \mathcal{U}[\rho] \to 2^{\mathcal{AP}}$ for any $u \in \mathcal{U}[\rho]$ as $\ell_\rho(u) = \{p \in \mathcal{AP} \mid p \in u\}$. We then have a model $\mathcal{M}_\rho = (\mathcal{U}[\rho], \to_\rho, \ell_\rho)$, and it is not difficult to prove that $\mathcal{M}_\rho$ is a WTS. The following lemma shows that any formula $\varphi$ in the language of $\rho$ that is contained in some ultrafilter $u$ must be satisfied by the state $u$ in the finite model $\mathcal{M}_\rho$.

**Lemma 19 (Truth lemma).** *If $\rho \in \mathcal{L}$ is a consistent formula, then for all $\varphi \in \mathcal{L}[\rho]$ and $u \in \mathcal{U}[\rho]$ we have $\mathcal{M}_\rho, u \models \varphi$ iff $\varphi \in u$.*

To prove the truth lemma, we first establish the following two theorems.

$$\vdash \varphi \leftrightarrow \psi \implies \vdash L_r \varphi \leftrightarrow L_r \psi \qquad \vdash \varphi \leftrightarrow \psi \implies \vdash M_r \varphi \leftrightarrow M_r \psi$$

The proof then proceeds by induction on the structure of $\varphi$. For the only-if-case of $\varphi = L_r \psi$, it is easy to see that $[\![\psi]\!] \neq \emptyset$. We then partition the ultrafilters $v \in [\![\psi]\!]$ by $[\![\psi]\!] = E \cup N$ where $E = \{v \in [\![\psi]\!] \mid L(u,v) = \emptyset\}$ and $N = \{v \in [\![\psi]\!] \mid L(u,v) \neq \emptyset\}$. Because $u$ is an ultrafilter, we have $\bigwedge_{v \in E} \neg L_0(v) \wedge \bigwedge_{v \in N} L_r(v) \in u$, which we prove implies $L_r \psi \in u$. For the if-case, it is straightforward to show by contradiction that $\theta^-(u)([\![\psi]\!]) \geq r$, if we know that $\theta(u)([\![\psi]\!]) \neq \emptyset$. To show this, assume towards a contradiction that $\theta(u)([\![\psi]\!]) = \emptyset$. Then $\neg L_r(v) \in u$ for all $v \in [\![\psi]\!]$, which we can enumerate as $\neg L_r(v_1) \wedge \cdots \wedge \neg L_r(v_n) \in u$. This can then be shown to imply $\neg L_r \psi \in u$, which is a contradiction.

Having established the truth lemma, we can now show that any consistent formula is satisfied by some finite model.

**Theorem 20 (Finite model property).** *For any consistent formula $\varphi \in \mathcal{L}$, there exists a finite WTS $\mathcal{M} = (S, \to, \ell)$ and a state $s \in S$ such that $\mathcal{M}, s \models \varphi$.*

We are now able to state our main result, namely that our axiomatization is complete.

**Theorem 21 (Completeness).** *For any formula $\varphi \in \mathcal{L}$, it holds that*

$$\models \varphi \quad implies \quad \vdash \varphi.$$

We have thus established completeness for our logic. There is also a stronger notion of completeness, often called strong completeness, which asserts that $\Phi \models \varphi$ implies $\Phi \vdash \varphi$ for any set of formulae $\Phi \subseteq \mathcal{L}$. Completeness is a special case of strong completeness where $\Phi = \emptyset$. In the case of compact logics, strong completeness follows directly from completeness. However, our logic is non-compact.

**Theorem 22.** *Our logic is non-compact, meaning that there exists an infinite set $\Phi \subseteq \mathcal{L}$ such that each finite subset of $\Phi$ admits a model, but $\Phi$ does not.*

*Proof.* Consider the set $\Phi = \{L_q\varphi \mid q < r\} \cup \{\neg L_r\varphi\}$. For any finite subset of $\Phi$, it is easy to construct a model. However, if $\mathcal{M}, s \models L_q\varphi$ for all $q < r$ where $q, r \in \mathbb{Q}_{\geq 0}$, then by the Archimedean property of the rationals, we also have $\mathcal{M}, s \models L_r\varphi$. Hence there can be no model for $\Phi$. $\qquad\square$

## 5   Satisfiability

The finite model property gives us a way of deciding in general whether there exists a WTS and a state in that WTS that satisfies a given formula. We do so by constructing a model $\mathcal{M}_\rho$ such that if $\rho$ is satisfiable there exists a state $\Gamma$ in $\mathcal{M}_\rho$ such that $\mathcal{M}_\rho, \Gamma \models \rho$. The model construction closely mimics the finite model construction in Sect. 4.2. We will not go into the details of the construction here, but instead point out where the construction differs from that in Sect. 4.2.

Given an arbitrary formula $\rho \in \mathcal{L}$, we construct the language of $\rho$, $\mathcal{L}[\rho]$, in the same way as we did in Sect. 4.2. In this section we will not use ultrafilters as states in our model, but rather their semantic counterpart which we term *maximal* sets of formulae.

**Definition 23.** *We say that a set $\Gamma \subseteq \mathcal{L}[\rho]$ of formulae is* propositionally maximal *if it satisfies the following where $\varphi, \psi \in \mathcal{L}[\rho]$:*

(P1): $\forall \varphi \in \mathcal{L}[\rho].\ \varphi \in \Gamma$ iff $\neg\varphi \notin \Gamma$
(P2): $\varphi \wedge \psi \in \Gamma$ implies $\varphi \in \Gamma$ and $\psi \in \Gamma$
(P3): $\varphi \vee \psi \in \Gamma$ implies $\varphi \in \Gamma$ or $\psi \in \Gamma$.

In addition to the conditions for propositional maximality listed in Definition 23, we also have another notion of maximality that we term *quantitative maximality*.

**Definition 24.** *We say that a set $\Gamma \subseteq \mathcal{L}[\rho]$ of formulae is* quantitatively maximal *if it satisfies the following:*

(Q1): $\neg L_0\bot \in \Gamma$
(Q2): $L_{r+q}\varphi \in \Gamma$ implies $L_r\varphi \in \Gamma$
(Q2'): $M_r\varphi \in \Gamma$ implies $M_{r+q}\varphi \in \Gamma$
(Q3): $L_r\varphi \wedge L_q\psi \in \Gamma$ implies $L_{\min\{r,q\}}(\varphi \vee \psi) \in \Gamma$
(Q3'): $M_r\varphi \wedge M_q\psi \in \Gamma$ implies $M_{\max\{r,q\}}(\varphi \vee \psi) \in \Gamma$
(Q4): $L_r(\varphi \vee \psi) \in \Gamma$ implies $L_r\varphi \vee L_r\psi \in \Gamma$

(Q4'): $M_r(\varphi \vee \psi) \in \Gamma$ implies $M_r\varphi \vee M_r\psi \in \Gamma$
(Q5): $\neg L_0\psi \in \Gamma$ and $L_r\varphi \in \Gamma$ implies $L_r(\varphi \vee \psi) \in \Gamma$
(Q5'): $\neg L_0\psi \in \Gamma$ and $M_r\varphi \in \Gamma$ implies $M_r(\varphi \vee \psi) \in \Gamma$
(Q6): $L_{r+q}\varphi \in \Gamma$ implies $\neg M_r\varphi \in \Gamma$
(Q7): $M_r\varphi \in \Gamma$ implies $L_0\varphi \in \Gamma$
(Q8): $\varphi \rightarrow \psi \in \Gamma$ and $((L_r\psi) \wedge (L_0\varphi)) \in \Gamma$ implies $L_r\varphi \in \Gamma$
(Q8'): $\varphi \rightarrow \psi \in \Gamma$ and $((M_r\psi) \wedge (L_0\varphi)) \in \Gamma$ implies $M_r\varphi \in \Gamma$
(Q9): $\varphi \rightarrow \psi \in \Gamma$ and $L_0\varphi \in \Gamma$ implies $L_0\psi \in \Gamma$

where $\varphi, \psi \in \mathcal{L}[\rho]$ and $r, q \in R_\rho$.

The conditions for quantitative maximality are semantic analogues of the axioms listed in Table 1. We will say that a set $\Gamma \subseteq \mathcal{L}[\rho]$ of formulae is *maximal* if it is both propositionally maximal and quantitatively maximal.

The transitions between states and their associated weights are derived in the same was as in Sect. 4.2. We can now formally define the WTS $\mathcal{M}_\rho$.

**Definition 25.** *Given a formula $\rho \in \mathcal{L}$, we define the WTS $\mathcal{M}_\rho = (S_\rho, \rightarrow_\rho, \ell_\rho)$ as follows.*

- $S_\rho = \{\Gamma \in 2^{\mathcal{L}[\rho]} \mid \Gamma \text{ is maximal}\}$.
- $\rightarrow_\rho \subseteq S_\rho \times \mathbb{R}_{\geq 0} \times S_\rho$ *is defined as: for any $\Gamma, \Gamma' \in S_\rho$, $\Gamma \xrightarrow{x}_\rho \Gamma'$ if $L_0(\!(\Gamma')\!) \in \Gamma$ and either*
  1. $M(\Gamma, \Gamma') = \emptyset$ *and* $x \in \left\{\max L(\Gamma, \Gamma'), \max Q_\rho + \frac{1}{gr(\rho)}\right\}$, *or*
  2. $M(\Gamma, \Gamma') \neq \emptyset$ *and* $x \in \{\max L(\Gamma, \Gamma'), \min M(\Gamma, \Gamma')\}$.
- $\ell_\rho : S_\rho \rightarrow 2^{\mathcal{AP}}$ *is defined for any $\Gamma \in S_\varphi$ as $\ell_\rho(\Gamma) = \{p \in \mathcal{AP} \mid p \in \Gamma\}$.*

The following lemma shows that any formula contained in a maximal set in the language of $\rho$ has at least one model, namely the model $\mathcal{M}_\rho$.

**Lemma 26.** *For an arbitrary formula $\varphi \in \mathcal{L}[\rho]$ and maximal set of formulae $\Gamma \in 2^{\mathcal{L}[\rho]}$ it holds that $\varphi \in \Gamma$     iff     $\mathcal{M}_\rho, \Gamma \models \varphi$.*

With the preceding result, we are now able to show that any formula in the language of $\rho$ which has a model, must also be contained in a maximal set and vice versa.

**Theorem 27.** *For any formula $\rho \in \mathcal{L}$, the following two statements are equivalent:*

1. *There exists a maximal set $\Gamma \in 2^{\mathcal{L}[\rho]}$ such that $\rho \in \Gamma$.*
2. *There exists a model $\mathcal{M} = (S, \rightarrow, \ell)$ and a state $s \in S$ such that $\mathcal{M}, s \models \rho$.*

A consequence of Theorem 27 is that if we can find a maximal set $\Gamma \in 2^{\mathcal{L}[\rho]}$ such that $\rho \in \Gamma$, then $\rho$ is satisfiable, and in particular it is satisfied by $\Gamma$ in the WTS $\mathcal{M}_\rho$. Also, if we can find no such maximal set, then $\rho$ is not satisfiable. This gives a way of deciding satisfiability of a given formula. For any formula $\varphi \in \mathcal{L}$, the following algorithm decides whether $\varphi$ is satisfiable, and constructs a model if it is satisfiable.

**Algorithm 28**

1. Construct the finite language $\mathcal{L}[\varphi]$.
2. Construct the finite set $2^{\mathcal{L}[\varphi]}$ of all subsets of $\mathcal{L}[\varphi]$.
3. Go through all elements $\Gamma \in 2^{\mathcal{L}[\varphi]}$ and check whether they satisfy the conditions for maximality. If they do not, remove them.
4. Go through all the remaining maximal sets $\Gamma$ and check whether $\varphi \in \Gamma$. If there is no such $\Gamma$, then $\varphi$ is not satisfiable. If there is one such $\Gamma$, then $\varphi$ is satisfiable, and the finite model $\mathcal{M}_\varphi$ is a model for $\varphi$.

*Example 29.* Applying Algorithm 28 on the formula $M_1 \texttt{charging}$ yields a model $\mathcal{M}_{M_1 \texttt{charging}}$ with a state $\Gamma$ such that $\mathcal{M}_{M_1 \texttt{charging}}, \Gamma \models M_1 \texttt{charging}$, thus showing the satisfiability of the formula $M_1 \texttt{charging}$. We will not go through the construction here, but consider the WTS depicted in Fig. 6. It is easy to verify that $\mathcal{M}, s_1 \models M_1 \texttt{charging}$.

# 6 Concluding Remarks

Our contributions in this paper have been to define a new bisimulation relation for weighted transition systems (WTSs), which relates those states that have similar behavior with respect to their minimum and maximum weights on transitions, as well as an accompanying modal logic to reason about the upper and lower bounds of weights on transitions. We have shown that this logic characterizes exactly those states that are bisimilar. This characterization holds for WTSs that we call image-compact, which is a weaker requirement than image-finiteness. Furthermore, we have provided a complete axiomatization of our logic, and we have shown that it enjoys the finite model property. Based on this finite model property, we have developed an algorithm which decides the satisfiability of a formula in our logic and constructs a finite model for the formula if it is satisfiable.

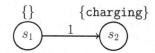

**Fig. 6.** A model for $M_1 \texttt{charging}$.

This work could be extended in different ways. Since our logic is non-compact, strong completeness does not follow directly from weak completeness, and hence it would be interesting to explore a strong-complete axiomatization of the proposed logic. Such an axiomatization would need additional, infinitary axioms. An example of such axioms would be $\{L_q\varphi \mid q < r\} \vdash L_r\varphi$ and $\{M_q\varphi \mid q < r\} \vdash M_r\varphi$, which are easily proven sound and describe the Archimedean property discussed in Theorem 22.

Although we have shown that our logic is expressive enough to capture bisimulation, it would also be of interest to extend our logic with a kind of fixed point operator or standard temporal logic operators such as until in order to increase its expressivity, and hence its practical use. We envisage two ways in which such a logic could be given semantics: either by accumulating weights or by taking the maximum or minimum of weights. In the accumulating case in particular, one could also allow negative weights to model that the system gains resources.

**Acknowledgements.** The authors would like to thank the anonymous reviewers for their useful comments and suggestions. This research was supported by the Danish FTP project ASAP: "Approximate Stochastic Analysis of Processes", the ERC Advanced Grant LASSO: "Learning, Analysis, Synthesis and Optimization of Cyber Physical Systems" as well as the Sino-Danish Basic Research Center IDEA4CPS.

# References

1. Alur, R., Courcoubetis, C., Dill, D.: Model-checking in dense real-time. Inf. Comput. **104**(1), 2–34 (1993)
2. Blackburn, P., van Benthem, J.F.A.K., Wolter, F.: Handbook of Modal Logic. Studies in Logic and Practical Reasoning. Elsevier Science, Amsterdam (2006)
3. Breugel, F.: Generalizing finiteness conditions of labelled transition systems. In: Abiteboul, S., Shamir, E. (eds.) ICALP 1994. LNCS, vol. 820, pp. 376–387. Springer, Heidelberg (1994). doi:10.1007/3-540-58201-0_83
4. Chakraborty, S., Katoen, J.P.: On the satisfiability of some simple probabilistic logics. In: LICS (2016 to appear)
5. Fagin, R., Halpern, J.Y.: Reasoning about knowledge and probability. J. ACM **41**, 340 (1994)
6. Heifetz, A., Mongin, P.: Probability logic for type spaces. Games Econ. Behav. **35**, 31 (2001)
7. Hennessy, M., Milner, R.: Algebraic laws for nondeterminism and concurrency. J. ACM **32**, 137 (1985)
8. Jaziri, S., Larsen, K.G., Mardare, R., Xue, B.: Adequacy and complete axiomatization for timed modal logic. Electr. Notes Theor. Comput. Sci. **308**, 183–210 (2014)
9. Jonsson, B., Larsen, K.G.: Specification and refinement of probabilistic processes. In: Proceedings of the Sixth Annual Symposium on Logic in Computer Science (LICS 1991) (1991)
10. Juhl, L., Larsen, K.G., Srba, J.: Modal transition systems with weight intervals. J. Log. Algebr. Program. **81**, 408 (2012)
11. Kozen, D., Larsen, K.G., Mardare, R., Panangaden, P.: Stone duality for Markov processes. In: 2013 28th Annual IEEE/ACM Symposium on Logic in Computer Science (LICS), pp. 321–330, June 2013
12. Kozen, D., Mardare, R., Panangaden, P.: Strong completeness for Markovian logics. In: Chatterjee, K., Sgall, J. (eds.) MFCS 2013. LNCS, vol. 8087, pp. 655–666. Springer, Heidelberg (2013). doi:10.1007/978-3-642-40313-2_58
13. Larsen, K.G., Mardare, R.: Complete proof systems for weighted modal logic. Theor. Comput. Sci. **546**, 164 (2014)
14. Larsen, K.G., Mardare, R., Xue, B.: On decidability of recursive weighted logics. Soft Comput. **20**, 1–18 (2016)
15. Larsen, K.G., Mardare, R., Panangaden, P.: Taking it to the limit: approximate reasoning for Markov processes. In: Rovan, B., Sassone, V., Widmayer, P. (eds.) MFCS 2012. LNCS, vol. 7464, pp. 681–692. Springer, Heidelberg (2012). doi:10.1007/978-3-642-32589-2_59
16. Mardare, R., Cardelli, L., Larsen, K.G.: Continuous Markovian logics-axiomatization and quantified metatheory. Log. Meth. Comput. Sci. **8**(4) (2012)
17. Zhou, C.: A complete deductive system for probability logic. J. Log. Comput. **19**, 1427 (2009)

# Zephyrus2: On the Fly Deployment Optimization Using SMT and CP Technologies

Erika Ábrahám[1], Florian Corzilius[1], Einar Broch Johnsen[2], Gereon Kremer[1], and Jacopo Mauro[2(✉)]

[1] RWTH Aachen University, Aachen, Germany
[2] Department of Informatics, University of Oslo, Oslo, Norway
jacopom@ifi.uio.no

**Abstract.** Modern cloud applications consist of software components deployed on multiple virtual machines. Deploying such applications is error prone and requires detailed system expertise. The deployment optimization problem is about how to configure and deploy applications correctly while at the same time minimizing resource cost on the cloud. This problem is addressed by tools such as Zephyrus, which take a declarative specification of the components and their configuration requirements as input and propose an optimal deployment. This paper presents Zephyrus2, a new tool which addresses deployment optimization by exploiting modern SMT and CP technologies to handle larger and more complex deployment scenarios. Compared to Zephyrus, Zephyrus2 can solve problems involving hundreds of components to be deployed on hundreds of virtual machines *in a matter of seconds instead of minutes*. This significant speed-up, combined with an improved specification format, enables Zephyrus2 to interactively support on the fly decision making.

## 1 Introduction

Modern software systems are often developed to be highly configurable both in the functionality they offer and in their deployment architecture. Applications targeting the cloud need to adapt their deployment to the virtual machines (VMs) that the cloud makes available. Cloud applications typically consist of a large number of interconnected software components (such as packages or services) that must be deployed on VMs that can be created on-the-fly by means of cloud computing technologies. The correct deployment and configuration of cloud applications is a challenging task and a major source of errors. In fact, inappropriate deployment and configuration are the second cause of errors in Google data centers, only after software bugs [6]. The deployment flexibility of cloud applications is further restricted by the availability and price of resources offered by the cloud. The *deployment optimization problem* is the problem of how

Supported by the EU project FP7-644298 *HyVar: Scalable Hybrid Variability for Distributed, Evolving Software Systems*, by the DAAD project SMT4ABS and by the CDZ project *CAP*.

M. Fränzle et al. (Eds.): SETTA 2016, LNCS 9984, pp. 229–245, 2016.
DOI: 10.1007/978-3-319-47677-3_15

to correctly deploy all the software components needed by a cloud application on suitable VMs on the cloud at minimal cost.

The deployment and on-the-fly configuration of systems on the cloud are handled by so-called DevOps teams, which address efficient system delivery and frequent infrastructure changes by combining development and operations experts. Different tools and technologies have been developed to support the work of DevOps teams. The mainstream approach restricts solutions to a fixed set of pre-configured VM images, which offers all the needed software packages and services, and which can be launched directly on the VMs of the targeted cloud system (e.g., Bento Boxes [24], Cloud Blueprints [12], or AWS CloudFormation [4]). The main drawback of this approach is that, as the deployment may use only the pre-configured VM images, it might use more resources than necessary or force the software to only run on specific cloud providers (resulting in vendor lock-in). More advanced techniques allow application architects to design their own software architectures using *high-level description languages* such as the graphical drag-and-drop approach of Juju [31] or the TOSCA (Topology and Orchestration Specification for Cloud Applications) standard [41]. Unfortunately, the use of these languages is knowledge-intensive since they require the architect to design the entire architecture and have a deep understanding of all the components to deploy. Furthermore, these languages do not address deployment optimization.

To overcome these limitations and address the deployment optimization problem, *declarative approaches* have recently been proposed which enable the DevOps teams to automatically generate optimal VM configurations from high-level specifications [23,30]. In particular, the automatic configuration generator tool *Zephyrus* [17] has been applied in a number of industrial settings [15,19,26]. Starting from a description of the available VMs and the components that need to be deployed, the architect can specify requirements in the form of constraints and use the tool to generate optimal machine configurations and deployment at minimal cost. The application architect can exploit the expressiveness of the constraints to focus on the most important aspects of the application, leaving to the tool the task of deducing other components that are needed to obtain a correct configuration and where to deploy them.

The contribution of this paper is to present *Zephyrus2* , a new tool to tackle the deployment optimization problem, inspired by Zephyrus. Zephyrus2 overcomes some limitations of Zephyrus by using different solving approaches, which enables us to solve problems involving hundreds of components to be deployed on hundreds of VMs *in a matter of seconds instead of minutes*. We report on the obtained performance gains with different solving approaches. Based on industrial experiences with declarative deployment optimization [15,19,26], Zephyrus2 allows a more direct and concise specification of deployment scenarios and user requirements than Zephyrus. The simplified input format combined with a significant speed-up (i.e., seconds instead of minutes) allows Zephyrus2 to be used by DevOps teams in a more interactive way for on the fly decision making [20].

*Paper Structure.* Section 2 gives a brief overview of the declarative deployment optimization problem. Section 3 introduces Zephyrus2 and Sect. 4 evaluates its performance on a set of industry inspired instances. Section 5 discusses the main differences between Zephyrus2 and Zephyrus and Sect. 6 discusses other related work. Section 7 concludes the paper, indicating directions for future research.

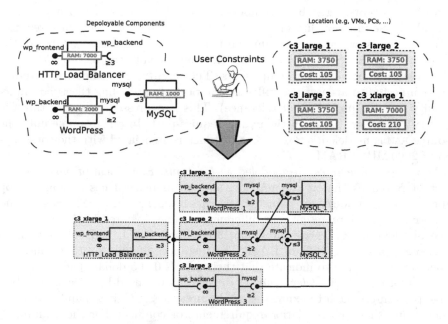

**Fig. 1.** The deployment problem.

## 2    Preliminaries

We give an overview of the declarative deployment optimization problem [11,17]. The basic deployment problem is illustrated by Fig. 1; we assume three different inputs:

   (i)  a description of the *components* that can be deployed,
  (ii)  a description of the *virtual machines* where the components can run, and
 (iii)  the *constraints* that capture the specific requirements of the DevOps teams.

   We specify *components* in the Aeolus [18] modeling language as black-boxes that expose *require-* and *provide-ports* to capture required and provided functionalities respectively. *Connections (bindings)* from require- to provide-ports model the usage of services. *Capacity constraints* associated to the ports might constrain those connections: (i) for provide-ports they can specify how many require-ports might be connected (maximal number of served components), or impose that no other component can provide the same functionality (used to model the notion of conflicts among components), (ii) for require-ports multiple providers offering the given functionality can be required (used to model replication requirements). Every component instance consumes resources such as memory or processing power when deployed, which is also captured in the model. Some examples for component descriptions are graphically illustrated in Fig. 1 (top left). For instance, the open-source content management system WordPress is represented by a component named WordPress which provides, if installed, the functionality wp_backend via a provide-port and which requires the functionality mysql via a require-port. By associating the $\infty$ symbol to the provide-port wp_backend we model that the functionality can be provided to an unbounded number of other components. The WordPress component requires to be connected to at least two different components providing the mysql functionality (e.g., to provide fault tolerance). This is expressed by associating the capacity constraint "$\geq 2$" to the mysql require-port. In our example only the resource RAM is modeled: a WordPress instance is associated with the consumption of 2000 MB of RAM.

The *virtual machines* are modeled as *locations*. Each location has a name, a list of resources that it can provide, and an associated cost. Figure 1 (top right) shows four locations named c3_large_1, c3_large_2, c3_large_3, and c3_xlarge_1. They represent four different machines inspired by Amazon EC2, three of them are c3_large, providing 3.75 GB of RAM, and one of them is c3_xlarge, providing 7 GB of RAM. The instances of type c3_large have an associated cost of 105 to indicate that their cost is 0.105 dollars per hour.

The user can specify (*deployment*) *constraints* in an ad-hoc declarative language powerful enough to express, e.g., the presence of a given number of components and their co-installation requirements or conflicts. For the example in Fig. 1, the user might require the presence of at least one HTTP_Load_Balancer and impose that, for fault tolerance reasons, no two WordPress or MySQL instances should be installed on the same virtual machine.

The goal of the *declarative deployment optimization problem* is to find a configuration distributing components on a set of locations such that:

 (i) the constraints reflecting the user requirements are satisfied,
 (ii) every functionality required by a deployed component is provided,
(iii) in each location, the available resources are sufficient to cover the resource needs of all components deployed on it, and
(iv) the values of some user-defined (prioritized) objective functions are minimized.

## 3   Zephyrus2

Zephyrus2 is a tool to solve the declarative deployment optimization problem. It offers a concise language to specify deployment optimization problems, and it can use different technologies to solve them. Zephyrus2 is written in Python ($\sim 3k$ lines of code) and is open source and freely available [34]. In order to increase portability, Zephyrus2 can be installed using Docker containers [22].

## 3.1   Problem Specification Language

The use of Zephyrus in an industrial environment [15, 19, 26] has emphasized the need to have (i) a simpler way to define components and locations, and (ii) a more concise specification language to describe deployment constraints.

To tackle the first concern, Zephyrus2 supports for component and location specifications the JavaScript Object Notation (JSON) format. Due to the lack of space, here we only show some examples; the formal JSON Schema of the input is available at [34]. The following JSON snippet defines the WordPress component in Fig. 1:

```
"WordPress ": {
  "resources": { "RAM": 2000 },
  "requires": { "mysql": 2 },
  "provides": [ { "ports": [ "wp_backend" ], "num": -1 } ]
}
```

In the second line, with the keyword `resources`, it is declared that WordPress consumes 2000 MB of RAM. The keyword `requires` defines that the component has a require-port requiring the service `mysql` with a capacity constraint "$\geq 2$". Similarly, the `provides` keyword declares that WordPress provides `wp_backend` to a possibly unbounded number of components (represented by $-1$).

The definition of locations is also done in JSON. For instance, the JSON input to define 10 `c3_large` Amazon virtual machines is the following:

```
"c3_large": {
  "num": 10,
  "resources": { "RAM": 3750 },
  "cost": 105
}
```

To tackle the second concern, Zephyrus2 introduces a new specification language for deployment constraints. This language is a key factor for the usability of the tool: while users who want to deploy their applications on a cloud usually need rather simple deployment constraints (requiring, e.g., that one instance of the main application component should be deployed), the language allows DevOps teams to express also more complex cloud- and application-specific constraints. In the following we describe some main features of the language by means of simple examples, referring the interested reader to [34] for the formal grammar of the language and more examples.

A deployment constraint is a logical combination of comparisons between arithmetic expressions. Besides integers, expressions may refer to component names representing the total number of deployed instances of a component. Location instances are identified by a location name followed by the instance index (starting at zero) in square brackets. A component name prefixed by a location instance stays for the number of component instances deployed on the given location instance. For example, the following formula requires the presence of at least one HTTP Load Balancer instance, and exactly one WordPress server instance on the second `c3_large` location instance:

```
HTTP_Load_Balancer > 0 and c3_large[1].WordPress  = 1
```

For quantification and for building sum expressions, we use identifiers pre-fixed with a question mark as variables. Quantification and sum building can range over components, locations, or over components/locations whose names match a given regular expression. Using such constraints, it is possible to express more elaborate properties such as the co-location or distribution of components, or limit the amount of components deployed on a given location. For example, the constraint

```
forall ?x in locations: ( ?x.WordPress  > 0  impl ?x.MySQL > 0)
```

states that the presence of an instance of `WordPress` deployed on any location x implies the presence of an instance of `MySQL` deployed on the same location x. As another example, requiring the HTTP Load Balancer to be installed alone on a virtual machine can be done by requiring that if a Load Balancer is installed on a given location then the sum of the components installed on that location should be exactly 1.

```
forall ?x in locations: ( ?x.HTTP_Load_Balancer > 0 impl
  (sum ?y in components: ?x.?y) = 1 )
```

For optimization, Zephyrus2 allows the user to express her preferences over valid configurations in the form of a list of arithmetic expressions whose values should be minimized in the given priority order. The keyword `cost` can be used to require the minimization of the total cost of the application. The following list specifies the metric to minimize first the total cost of the application and then the total number of components:

```
cost; ( sum ?x in components: ?x )
```

This is also the default metric used if the user does not specify her own prefer-ences.

## 3.2   Solving Technologies

Zephyrus2 solves deployment optimization problems, specified in the above-described languages, by translating them into *Constraint Optimization Problems* (*COP*) encoded in MiniZinc [39].[1] By default, Zephyrus2 solves the resulting multi-objective optimization problems by optimizing the first objective function value and then optimizing the other objective function values sequentially fol-lowing their order after substituting the previously determined optimal values. We believe that this solution is particularly effective since usually minimizing the first objective (e.g., the cost) has a significant impact on the performance when reducing the second objective (e.g., the number of components). However, this solution has the drawback that we need to restart the solver. In order to exploit the capabilities of further multi-objective optimization techniques, Zephyrus2 also supports MiniSearch [46], a meta-search language for MiniZinc that allows

---

[1] For the interested reader, an example of the encoding in MiniZinc is reported in [1].

to solve MiniZinc models with (heuristic) meta-searches, such as large neighborhood search (LNS), lexicographic branch-and-bound, and And/Or search. Currently, Zephyrus2 uses MiniSearch to execute a lexicographic branch-and-bound search procedure. Unfortunately, for the time being, there is a limited amount of solvers supporting the programmatic APIs of the version 2.0 of MiniZinc that eliminates the need to communicate through text files and enables the addition of constraints at runtime without restarting the solvers. However, from an engineering point of view, we believe that the support of MiniSearch is important since it allows to explore and try different search procedures and improve the current performance as soon as more constraint programming solvers will adopt the MiniZinc 2.0 APIs.

Zephyrus2 also supports the use of *satisfiability-modulo-theories* (*SMT*) *solvers*. SMT solving extends and improves upon SAT solving by introducing the possibility of stating constraints in some expressive theories, e.g., arithmetic or bit-vector expressions. For our application, we need a solver that supports integer arithmetic and also features optimization. One that is capable of doing that is Z3 [7,21], one of the most prominent SMT solvers. The last version of Z3 (4.4.2) has introduced some optimization features as an extension of the SMT-LIBv2 input language [5], i.e., the standard format to define SMT instances. This is very suitable for our purpose since Z3 can solve the multi-objective optimization problems directly, and we do not need to develop search strategies on top of it. To use Z3, the optimization problems were translated into the SMT-LIB format using *fzn2smt* [8] and further processed to simplify equations and reduce the number of variables. For more details we refer the reader to [34]. Note that optimization of SMT formulas is a very recent feature and still subject to a lot of research. Though it makes use of optimization techniques known from linear programming, significant progress can be expected in the future from which Zephyrus2 will directly benefit.

## 4   Experimental Results

In this section we describe the performance of Zephyrus2 while using different settings and solving engines.

To the best of our knowledge, due to the novelty of these approaches, there are no established benchmarks for application deployment. Moreover, the first industrial problems solved by Zephyrus in [15,19,26] were not challenging, taking only a few seconds to be solved. For this reason, to compare Zephyrus2 with Zephyrus, in this work we rely on the synthetic benchmark proposed in [17,50]. The instances of the benchmark are derived from a parametrized variant of the WordPress deployment scenario presented in [15] and partially depicted in Fig. 1. This scenario was parametrized in three dimensions to allow the analysis of scalability issues: (1) the parameter mysql_req encodes the number of MySQL instances a WordPress requires, (2) the parameter wp_req encodes the number of WordPress instances the HTTP Load Balancer requires and (3) the parameter vm_amount represents the amount of the four different types of Amazon

EC2 virtual machines that can be used to deploy the application. The benchmark instances are obtained by varying the parameters mysql_req and wp_req within $\{6, ..., 12\}$, and vm_amount within $\{6, ..., 25\}$ (this corresponds to considering up to one hundred virtual machines). The goal for all the instances is to deploy an HTTP Load Balancer with the additional requirement that it is not possible to install two MySQL instances or two WordPress instances on the same machine.

We compare several different configurations of Zephyrus2 against the original approach of Zephyrus denoted as zephyrus. As solver backends for Zephyrus2 we use the lexicographic minimization approach with restarts using the solvers Chuffed [13] (lex-chuffed), Gecode [27] (lex-gecode), and Or-tools [28] (lex-ortools), as well as the SMT solver Z3 [21] (smt). We run all approaches using AMD Opteron 6172 processors and a timeout of 300 s (per problem instance), which is usually the time it takes to require and obtain a virtual machine from a cloud provider.

We remark that the times of zephyrus should just be used as an indicative measure of the original performance of the Zephyrus approach. Indeed, these times include also the time taken to generate the connections (bindings) between the components that, as better explained in Sect. 5, in Zephyrus2 is a task that by design is deferred to an external utility. However, the times of zephyrus are still significant for a comparison since the generation of the bindings in Zephyrus takes just few milliseconds and the runtimes of zephyrus, especially for big problem instances, are greatly dominated by the time needed to prove that a found configuration, if any, is optimal.

We also performed experiments using MiniSearch with the aforementioned CP solvers as backends. However, due to a bug in MiniSearch,[2] it often returns a suboptimal solution. As Zephyrus2 and Zephyrus are supposed to provide optimal solutions only, we have excluded the MiniSearch approach in the following comparison.

**Table 1.** Experimental results for all the approaches.

| Solver | Solved | | Timeout | Seconds |
|--------|--------|------|---------|---------|
| zephyrus | 261 | 27 % | 719 | 67.81 |
| lex-chuffed | 980 | 100 % | 0 | 4.45 |
| **lex-gecode** | **980** | **100 %** | **0** | **2.25** |
| lex-ortools | 975 | 99 % | 5 | 7.13 |
| smt | 960 | 98 % | 20 | 50.23 |

A summary of the results is presented in Table 1. The columns *Solved* and *Timeout* denote the number of instances that were solved correctly and the number of instances where the solver was terminated due to a timeout, respectively.

---

[2] MiniSearch is a very recent framework, only available in a beta version.

The last column gives the average time needed to solve the instances that could be solved within the timeout.

As can be seen, all approaches based on Zephyrus2 can solve almost all the benchmarks. While `lex-chuffed` and `lex-gecode` solve all, `lex-ortools` and smt lack only five and twenty, respectively, of the 980 benchmarks. As for the comparison with the original Zephyrus approach `zephyrus`, it is immediately visible that Zephyrus2 is faster and able to solve more instances, whatever solver is used as backend. While `zephyrus` is able to solve only 27 % of the benchmarks in less than 5 min, `lex-gecode` which resulted as the best solver in average, is able to solve all benchmarks in at most 25 s (with only 10 benchmarks taking more than 10 s to be solved). Surprisingly, Gecode is more efficient than both Chuffed and Or-tools which, based on the results of the last MiniZinc Challenge [40] – an annual competition of constraint programming solvers – are among the best CP solvers available today. We believe that this is probably due to the nature of the deployment problem that favors the pure propagation and search approach used by Gecode. As we will see later, although smt does not manage to solve all the benchmarks and it is much slower than the other approaches of Zephyrus2, it is not dominated by `lex-chuffed` or `lex-gecode` and for a few hard instances it is faster than both of them.

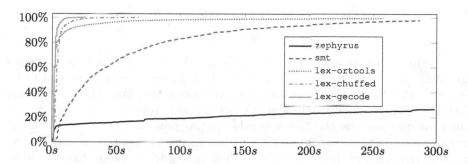

**Fig. 2.** Percentage of the instances that could be solved within a given timeout.

Figure 2 shows for varying timeout values ($x$ axis) the percentage of the problem instances that could be solved within that timeout ($y$ axis) by the different approaches.

In Fig. 3 we compare the individual results of the best approach `lex-gecode` with the other approaches. Each plot relates the benchmark results of two approaches on a logarithmic scale where every cross represents a single problem instance. Firstly we can see that `lex-chuffed` is almost dominated by `lex-gecode` as there are no examples that took more than 2 s in which `lex-chuffed` performs significantly better. The same conclusion can be drawn also for `lex-ortools` that is sometimes faster than `lex-gecode`, but only on comparably easy instances that are quickly solved by both solvers. In contrast to that, we believe that smt can be a valuable complement to `lex-gecode`,

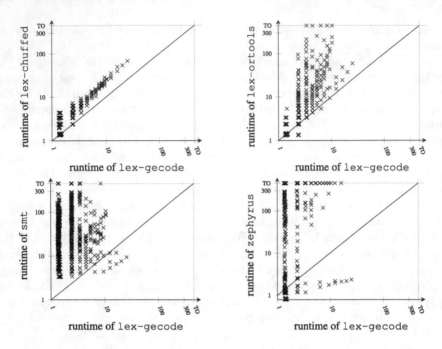

**Fig. 3.** Comparison of `lex-gecode` to the other approaches.

as it performs better on exactly those benchmarks that `lex-gecode` struggles to solve. For every benchmark that takes `lex-gecode` more than 10 s to be solved, `smt` is faster than `lex-gecode`. We conjecture that this is due to the fact that the dynamic search heuristics of the `smt` approach are more robust than the ones used by the `lex-gecode` for this problem type. A deeper comparison between these two approaches on harder instances is left as future work. Surprisingly, some of the instances that took `lex-gecode` more than 10 s to be solved are instead solved by Zephyrus in a shorter time. We conjecture that for these instances the search heuristics used by Zephyrus lead to a good solution faster, thus allowing to prove the optimality in shorter time.

## 5    Discussion

Zephyrus was the first tool to tackle the deployment optimization problem as defined in Sect. 2. The development of Zephyrus2 was triggered by experiences in applying Zephyrus in industrial case studies [15,19,26]. In this section, we discuss in detail the technical differences between Zephyrus2 and Zephyrus [17].

*Modeling.* Whereas Zephyrus directly uses the Aeolus [18] component model for the cloud, this component model has been extended for Zephyrus2 in order to model locations and components more naturally. A concern that arose in industrial case studies was that components may expose different interfaces at the same time. In Aeolus this is modeled by one provide-port for every interface. However, this encoding leads to an exponential blow-up in the number of components when capacity constraints are associated to the provide-port. To avoid this exponential blow-up, Zephyrus2 allows provide-ports to have one or more provided functionalities. For instance, assuming that a new version of a server is able to provide its functionalities to at most two clients via a protocol $a$ or $b$, in Zephyrus2 we can specify this by adding only one provide-port offering concurrently the possibility to use both protocols to at most 2 clients. Conversely, Zephyrus requires the introduction of different components: a component offering to two clients the functionality using protocol $a$, a component offering to two clients the functionality with protocol $b$, and a component offering to one client the functionality with protocol $a$ and to another client the functionality with protocol $b$.

In the deployment constraint language, support for quantifiers and sum terms (cf. Sect. 3) allows Zephyrus2 to express properties in a more concise way. In addition, Zephyrus2 supports user-defined metric functions to better customize the optimization, whereas users of Zephyrus are restricted to a predefined set of metrics.

*Constraint solving.* Zephyrus solves deployment optimization problems by translating them into a COP encoded in MiniZinc [39] and then uses standard CP solvers (such as the Gecode [27] and G12 solvers [48]) to iteratively minimize a number of cost functions through sequential invocations of an optimizing solver. Zephyrus2 offers more possibilities for the choice of solving technologies, as discussed in Sect. 4.

Another major difference between the tools is that Zephyrus2 relies on a completely new MiniZinc model[3], which enables Zephyrus2 to better exploit the symmetries of the deployment problem. Zephyrus initially has to consider a large number of locations to ensure that there are enough inexpensive virtual machines available. Every additional location extends the search space considerably. To cope with this problem, Zephyrus relies on ad-hoc location trimming heuristics [50] which tries to reduce the number of locations. Instead of relying on ad-hoc heuristics, it is also possible to use symmetry breaking constraints [47] to reduce the search space by removing some symmetric solutions (unless a user has machine-specific constraints on where to deploy components such as, for example, that WordPress needs to be installed on c3_large_2 but not on the similar VM c3_large_1). For this purpose, Zephyrus2 uses well-known symmetry breaking constraints for the bin packing problem [45]. In particular, Zephyrus2 establishes an order between locations with the same resources and enforces the deployment of the components in the cheapest location available, following the pre-determined order on these locations. This removes some of the symmetries between the locations.

Zephyrus2 supports similar constraints to break symmetries between the components, which establish an order between components and then enforce

---

[3] The MiniZinc model, available from [34], was submitted to the MiniZinc Challenge 2016.

the deployment in the location following the lexicographic order. Compared to Zephyrus2, Zephyrus only uses symmetry breaking constraints for locations, but not for the components. The effectiveness of the symmetry breaking constraints in Zephyrus2 allows the tool to reach better performance than Zephyrus, even without the use of location trimming heuristics that are effective only in a limited number of deploying scenarios.

Zephyrus2 simplifies the non-linear constraints used in Zephyrus into a conjunction of linear implications, by means of encoding techniques [16]. These techniques have proven effective to increase the performance of SMT solvers and allows the use of Chuffed, which does not support the non-linear constraints of the original formalization.

We would like to emphasize that the performance of the different solvers heavily depends on the encoding of the constraints, and the addition of redundant or symmetry breaking constraints. For instance, we noticed that without the so-called Ralf's redundant constraints [50], the performance of the CP solvers degrades considerably while for SMT solving they do not have any strong impact. Conversely, the simplification of the non-linear constraints allows the use of Chuffed and improves the performance of the SMT approach, but it has no impact on the performance of Gecode or Or-tools. Symmetry breaking constraints have a huge impact on the performances of both the CP and SMT approaches. More details on these are presented in the technical report [1].

Based on the use of Zephyrus in [15,19,26], we noticed that users often have preferences over the bindings between the components. For instance, it is often better to have bindings between co-located components and avoid configurations in which, e.g., a WordPress uses the functionalities of a MySQL deployed on another location while it could have used the functionalities of a MySQL deployed in its own location. In Zephyrus, the resolution of the deployment problem is tied to the generation of the bindings performed by means of an ad-hoc polynomial algorithm. Unfortunately, Zephyrus does not take into account preferences between bindings. For this reason, as a design choice, Zephyrus2 separates the task of computing the distribution of the components in the various locations from the task of connecting the components. Contrary to Zephyrus, the generation of the connections between the components is therefore not part of the core of Zephyrus2 and is instead deferred to an external utility. In particular, Zephyrus2 comes with a default simple bindings generation utility that maximizes the number of local bindings in few seconds (for further details, see [34]).

## 6   Related Work

Whereas in Sect. 5 we discussed the differences between Zephyrus and Zephyrus2, this section considers the deployment optimization problem addressed by these tools in a broader context.

With the increasing popularity of cloud computing, the problem of automating application deployment has recently attracted a lot of attention. Many system management tools have been developed for this purpose, including popular tools such as CFEngine [9], Puppet [32], MCollective [43], and Chef [42].

Despite their differences, such tools allow to declare the components that should be installed on each machine, together with their configuration parameters. In order to use such tools, the DevOps architect needs to know how components should be distributed and configured.

Some tools aim to compute an (optimal) configuration of a distributed system without computing the deployment steps needed to reach it. CP appears to be one of the best methods today for solving different configuration problems [47]. The structure of a configured system depends on the application domain and this knowledge is exploited to speed up the search for valid configurations. CP techniques have already been applied with success to the problem of deciding the allocation of resources in data centers and clouds [10,29,33,36,37]. Zephyrus2 relies on solver techniques similar to those adopted by these tools. Indeed, the COP problems solved by Zephyrus can be seen as an extension of the well-known bin packing problem [14] where some items, corresponding to components, have to be included into bins, corresponding to locations. However, in contrast to these approaches, Zephyrus2 not only computes the optimal allocation but also identifies the additional components that are needed to form a valid configuration. Formally, even without considering the allocation of components, the problem of deciding if there is a correct configuration is already NP-hard [16]. To the best of our knowledge, no trivial encoding exists that allows the reuse of [10,29,33,36,37] to solve the deployment optimization problem tackled by Zephyrus2.

Perhaps the most similar to our approach is ConfSolve [30], which uses a formulation based on constraints to propose an optimal allocation of virtual machines to servers and of applications to virtual machines. Similar to Aeolus [18], a declarative language is used to describe the entities (e.g., machines and services), and the deployment problem is then solved by translating the declarative specification into MiniZinc. However, in contrast to Zephyrus2, ConfSolve does not handle capacity and replication constraints, so there is no obvious representation of our benchmarks for ConfSolve. Another interesting related work is Saloon [44], where a deployment problem is described by means of a feature model extended with feature cardinalities. Saloon applies CP technologies to determine a deployment. While Saloon is able to automatically detect inconsistencies between components, it does not address the optimization problem; i.e., the solutions proposed by Saloon do not minimize the number of resources and virtual machines to be used.

Other approaches rely on a range of techniques to compute optimal component allocation. For instance, SAT solvers are used to solve network configuration problems [38], a prediction-based online approach [35] is proposed to find optimal reconfiguration policies, a genetic-based algorithm [25] is used to support the migration and deployment of enterprise software with their reconfiguration policies, and integer linear programming has been used to find energy-efficient VM configurations [49]. However, none of the tools that we are aware of allows capacity and replication constraints to be expressed, which are essential non-functional constraints for any non-trivial, scalable application. Furthermore, most of them give no optimality guarantee on the solution.

## 7    Conclusions

In this paper we presented Zephyrus2, a tool that is advancing the state-of-the-art by computing the *cheapest* way to deploy complex cloud applications based on declarative specifications. Optimal deployments involving up to a hundred components and virtual machines can be generated within seconds. This allows Zephyrus2 to be used in a more interactive way by the DevOps architect, who does not need to wait for minutes to inspect the proposed optimal solution and restart the computation in case she has forgotten to elicit one constraint or preference.

Zephyrus2 has already been tested in an industrial environment to check the cost optimality of currently deployed solutions and to devise the optimal allocation for deploying new components. The feedback obtained so far is positive both for the tool's usability and for its running time. As witnessed in [20], the support of a more concise language to specify user constraints and the improved performance makes Zephyrus2 a better alternative to the original version of Zephyrus.

To further improve the performance of Zephyrus2 in future work, we plan to study whether the SMT encoding can be improved and especially to consider whether SMT solvers can be extended with modules that perform propagation similar to those implemented by CP solvers. Moreover, exploiting the flexibility of MiniSearch, we plan to study local search procedures, which could be extremely useful in scenarios where the user is not interested in the optimal solution but just in finding a sufficiently good solution very quickly (e.g., in a second or less).

Based on the good results obtained by the Sunny portfolio approach [2] on the last MiniZinc Challenge, we also plan to study how the different search procedures can be combined to obtain a globally better solver. In particular, we are interested in combining the strengths of the different solvers by using, e.g., the bound sharing with restarts approach of the parallel version of the Sunny solver [3].

Finally, we are also interested in enriching the formal model behind Zephyrus2 to allow constraints on the bindings between components. This will allow configurations with more complex properties to be generated, such as the non-transferability of data between borders or the load balancing of traffic between parts of the system deployed on different data centers.

**Acknowledgements.** We would like to thank Andreas Schutt from NICTA (National ICT of Australia) for proposing a search annotation for the deployment problem when submitting Zephyrus2 instances to the MiniZinc Challenge 2016.

# References

1. Ábrahám, E., Corzilius, F., Johnsen, E.B., Kremer, G., Mauro, J.: Zephyrus2: On the Fly deployment optimization using SMT and CP technologies. Technical report, University of Oslo (2016)
2. Amadini, R., Gabbrielli, M., Mauro, J.: SUNNY: a lazy portfolio approach for constraint solving. TPLP **14**(4–5), 509–524 (2014)
3. Amadini, R., Gabbrielli, M., Mauro, J.: A multicore tool for constraint solving. In IJCAI, pp. 232–238, 2015
4. Amazon: AWS CloudFormation. http://aws.amazon.com/cloudformation/
5. Barrett, C., Fontaine, P., Tinelli, C.: The SMT-LIB Standard Version 2.6 (2015)
6. Barroso, L.A., Clidaras, J., Hölzle, U.: The Datacenter as a Computer: An Introduction to the Design of Warehouse-Scale Machines. Morgan and Claypool Publishers, San Rafael (2013)
7. Bjørner, N., Phan, A.-D., Fleckenstein, L.: $\nu Z$ - an optimizing SMT solver. In TACAS, pp. 194–199, 2015
8. Bofill, M., Palahí, M., Suy, J., Villaret, M.: Solving constraint satisfaction problems with SAT modulo theories. Constraints **17**(3), 273–303 (2012)
9. Burgess, M.: A site configuration engine. Comput. Syst. **8**(2), 309–337 (1995)
10. Cambazard, H., Mehta, D., O'Sullivan, B., Simonis, H.: Bin packing with linear usage costs - an application to energy management in data centres. In: CP, pp. 47–62 (2013)
11. Catan, M., et al.: Aeolus: mastering the complexity of cloud application deployment. In: Lau, K.-K., Lamersdorf, W., Pimentel, E. (eds.) ESOCC 2013. LNCS, vol. 8135, pp. 1–3. Springer, Heidelberg (2013). doi:10.1007/978-3-642-40651-5_1
12. CenturyLink: Cloud Blueprints. https://www.centurylinkcloud.com/blueprints/
13. Chuffed: The CP solver. https://github.com/geoffchu/chuffed
14. Coffman Jr., E.G., Garey, M.R., Johnson, D.S.: Approximation algorithms for bin packing: a survey. In: Approximation Algorithms for NP-hard Problems, pp. 46–93 (1997)
15. Cosmo, R., Eiche, A., Mauro, J., Zacchiroli, S., Zavattaro, G., Zwolakowski, J.: Automatic deployment of services in the cloud with aeolus blender. In: Barros, A., Grigori, D., Narendra, N.C., Dam, H.K. (eds.) ICSOC 2015. LNCS, vol. 9435, pp. 397–411. Springer, Heidelberg (2015). doi:10.1007/978-3-662-48616-0_28
16. Cosmo, R.D., Lienhardt, M., Mauro, J., Zacchiroli, S., Zavattaro, G., Zwolakowski, J.: Automatic application deployment in the cloud: from practice to theory and back. In: CONCUR, pp. 1–16 (2015)
17. Cosmo, R.D., Lienhardt, M., Treinen, R., Zacchiroli, S., Zwolakowski, J., Eiche, A., Agahi, A.: Automated synthesis and deployment of cloud applications. In: ASE, pp. 211–222 (2014)
18. Cosmo, R.D., Mauro, J., Zacchiroli, S., Zavattaro, G.: Aeolus: a component model for the cloud. Inf. Comput. **239**, 100–121 (2014)
19. Gouw, S., Lienhardt, M., Mauro, J., Nobakht, B., Zavattaro, G.: On the integration of automatic deployment into the ABS modeling language. In: Dustdar, S., Leymann, F., Villari, M. (eds.) ESOCC 2015. LNCS, vol. 9306, pp. 49–64. Springer, Heidelberg (2015). doi:10.1007/978-3-319-24072-5_4
20. Gouw, S., Mauro, J., Nobakht, B., Zavattaro, G.: Declarative elasticity in ABS. In: Aiello, M., Johnsen, E.B., Dustdar, S., Georgievski, I. (eds.) ESOCC 2016. LNCS, vol. 9846, pp. 118–134. Springer, Heidelberg (2016). doi:10.1007/978-3-319-44482-6_8

21. Moura, L., Bjørner, N.: Z3: an efficient SMT solver. In: Ramakrishnan, C.R., Rehof, J. (eds.) TACAS 2008. LNCS, vol. 4963, pp. 337–340. Springer, Heidelberg (2008). doi:10.1007/978-3-540-78800-3_24

22. Docker Inc.: Docker. https://www.docker.com/

23. Fischer, J., Majumdar, R., Esmaeilsabzali, S.: Engage: a deployment management system. In: PLDI, pp. 263–274 (2012)

24. Flexiant:        Bento        Boxes.        http://www.flexiant.com/2012/12/03/application-provisioning/

25. Frey, S., Fittkau, F., Hasselbring, W.: Search-based genetic optimization for deployment and reconfiguration of software in the cloud. In: ICSE, pp. 512–521 (2013)

26. Gabbrielli, M., Giallorenzo, S., Guidi, C., Mauro, J., Montesi, F.: Self-reconfiguring microservices. In: Ábrahám, E., Bonsangue, M., Johnsen, E.B. (eds.) Theory and Practice of Formal Methods. LNCS, vol. 9660, pp. 194–210. Springer, Heidelberg (2016). doi:10.1007/978-3-319-30734-3_14

27. GECODE: An open, free, efficient constraint solving toolkit. http://www.gecode.org

28. Google: Optimization tools. https://developers.google.com/optimization/

29. Hermenier, F., Demassey, S., Lorca, X.: Bin repacking scheduling in virtualized datacenters. In: Lee, J. (ed.) CP 2011. LNCS, vol. 6876, pp. 27–41. Springer, Heidelberg (2011). doi:10.1007/978-3-642-23786-7_5

30. Hewson, J.A., Anderson, P., Gordon, A.D.: A declarative approach to automated configuration. In: LISA, pp. 51–66 (2012)

31. Juju: DevOps Distilled. https://jujucharms.com/

32. Kanies, L.: Puppet: next-generation configuration management. Login: the USENIX Mag. 31(1), 19–25 (2006)

33. Malapert, A., Régin, J., Parpaillon, J.: The package server location problem. In: ICORES, pp. 193–204 (2013)

34. Mauro, J.: Zephyrus2. https://bitbucket.org/jacopomauro/zephyrus2/src

35. Mi, H., Wang, H., Yin, G., Zhou, Y., xi Shi, D., Yuan, L.: Online self-reconfiguration with performance guarantee for energy-efficient large-scale cloud computing data centers. In: SCC, pp. 514–521 (2010)

36. Michel, L., Shvartsman, A.A., Sonderegger, E.L., Hentenryck, P.V.: Load balancing and almost symmetries for RAMBO quorum hosting. In: CP, pp. 598–612 (2010)

37. Michel, L., Hentenryck, P., Sonderegger, E., Shvartsman, A., Moraal, M.: Bandwidth-limited optimal deployment of eventually-serializable data services. In: Hoeve, W.-J., Hooker, J.N. (eds.) CPAIOR 2009. LNCS, vol. 5547, pp. 193–207. Springer, Heidelberg (2009). doi:10.1007/978-3-642-01929-6_15

38. Narain, S., Levin, G., Malik, S., Kaul, V.: Declarative infrastructure configuration synthesis and debugging. J. Netw. Syst. Manage. 16(3), 235–258 (2008)

39. Nethercote, N., Stuckey, P.J., Becket, R., Brand, S., Duck, G.J., Tack, G.: MiniZinc: towards a standard CP modelling language. In: Bessière, C. (ed.) CP 2007. LNCS, vol. 4741, pp. 529–543. Springer, Heidelberg (2007). doi:10.1007/978-3-540-74970-7_38

40. NICTA: MiniZinc Challenge 2015. http://www.minizinc.org/challenge2015/results2015.html

41. OASIS: Topology and orchestration specification for cloud applications (TOSCA) version       1.0.http://docs.oasis-open.org/tosca/TOSCA/v1.0/cs01/TOSCA-v1.0-cs01.html

42. Opscode: Chef. https://www.chef.io/chef/

43. Puppet Labs: Marionette collective. http://docs.puppetlabs.com/mcollective/

44. Quinton, C., Pleuss, A., Berre, D.L., Duchien, L., Botterweck, G.: Consistency checking for the evolution of cardinality-based feature models. In: SPLC, pp. 122–131 (2014)
45. Régin, J., Rezgui, M.: Discussion about constraint programming bin packing models. In: AI for Data Center Management and Cloud, Computing, pp. 21–23 (2011)
46. Rendl, A., Guns, T., Stuckey, P.J., Tack, G.: MiniSearch: a solver-independent meta-search language for minizinc. In: Pesant, G. (ed.) CP 2015. LNCS, vol. 9255, pp. 376–392. Springer, Heidelberg (2015). doi:10.1007/978-3-319-23219-5_27
47. Rossi, F., van Beek, P., Walsh, T. (eds.): Handbook of Constraint Programming. Elsevier, Amsterdam (2006)
48. Stuckey, P.J., Banda, M.G., Maher, M., Marriott, K., Slaney, J., Somogyi, Z., Wallace, M., Walsh, T.: The G12 project: mapping solver independent models to efficient solutions. In: Gabbrielli, M., Gupta, G. (eds.) ICLP 2005. LNCS, vol. 3668, pp. 9–13. Springer, Heidelberg (2005). doi:10.1007/11562931_3
49. Tran, P.N., Casucci, L., Timm-Giel, A.: Optimal mapping of virtual networks considering reactive reconfiguration. In: CLOUDNET, pp. 35–40 (2012)
50. Zwolakowski, J.: A Formal Approach to Distributed Application Synthesis and deployment automation. Ph.D. thesis, Université Paris Diderot Paris 7 (2015)

# Exploiting Symmetry for Efficient Verification of Infinite-State Component-Based Systems

Qiang Wang[(⊠)]

École Polytechnique Fédérale de Lausanne, Lausanne, Switzerland
qiang.wang@epfl.ch

**Abstract.** In this paper we present an efficient verification algorithm for infinite-state component-based systems modeled in the behavior-interaction-priority (BIP) framework. Our algorithm extends the persistent set partial order reduction by taking into account system symmetries, and further combines it with lazy predicate abstraction. We have implemented the new verification algorithm in our model checker for BIP. The experimental evaluation shows that for systems exhibiting certain symmetries, our new algorithm outperforms the existing algorithms significantly.

## 1 Introduction

Recently rigorous system design [33] has been proposed as a formal, accountable and coherent process for building large complex systems in the correctness-by-construction manner. In this design methodology, trustworthy system implementations are derived from the high-level system models by applying a series of property preserving source-to-source model transformations, and refined with details specific to the target platforms. Correctness of the system is guaranteed at the earliest possible design phase by applying algorithmic verification to high-level system models. In practice, BIP framework [3] has been developed to support the rigorous system design methodology and actively used in many industrial applications [1,4].

BIP provides a general and expressive component-based framework for modeling complex concurrent systems. One of the key underlying principles is the separation of concerns (i.e. computation and coordination). System models are constructed by superposing three layers of modeling: Behaviour, Interaction and Priority. Behaviour models the computation and is characterized by a set of components. A component in BIP is specified as an automaton extended with linear arithmetic. Interaction models the communication and coordination of components. Intuitively, an interaction represents a guarded multi-party synchronisation of components, among which data transfer may take place. Priority is used to schedule the set of interactions to be executed, or resolve conflicts when several interactions are enabled simultaneously.

Various techniques have been developed to verify BIP designs automatically in our recent work [7,29,38]. In this work, we go one step further and present

© Springer International Publishing AG 2016
M. Fränzle et al. (Eds.): SETTA 2016, LNCS 9984, pp. 246–263, 2016.
DOI: 10.1007/978-3-319-47677-3_16

an efficient verification technique for infinite-state system models, which have certain symmetric structure features, e.g. component symmetries. Such symmetries are common in component-based designs. For instance, a system model consisting of one server and several identical users is usually symmetric with respect to the users. Moreover, permutating the users would not affect the satisfaction of certain safety properties, e.g. deadlocks, mutual exclusion. Based on this observation, we investigate in this paper how to exploit such symmetries to verify infinite-state concurrent systems more efficiently. To this end, we made the following contributions:

1. We extend the notion of interaction independence by taking into account the system symmetries, i.e. two interactions are independent if they commute under some symmetries. The original definition of independence is then a special case of this one with identical symmetry. we also extend persistent set based partial order reduction technique [22] by relying on this new independence relation and show how to compute the persistent set by adapting the Stubborn set approach [34].
2. We propose an integration of our new partial order reduction with lazy predicate abstraction [26], and prove that though on abstraction structures, commutativity of independent transition in general does not hold, our technique still computes a sound over-approximation such that no real counterexamples would be missed. We have also implemented the proposed verification algorithm and performed a set of experiments. The results show that this new algorithm outperforms the others significantly, for systems with certain symmetries.

The rest of this paper is organized as follows. In Sect. 2, we provide the preliminaries. In Sect. 3, we review the BIP models without priority and illustrate our verification idea with an example. In Sect. 4, we present the new persistent set based partial order reduction technique extended with system symmetries. In Sect. 5, we present the integration of the new partial order reduction with lazy abstraction and an algorithm to compute the persistent set for BIP models. In Sect. 6, we present the experimental evaluation. In Sects. 7 and 8, we review the most related work and draw the conclusions respectively.

## 2    Formal Preliminaries

We denote by $\mathbb{V}$ a set of variables of integer domain $\mathbb{Z}$, and by $\mathbf{V}$ their valuations, i.e., a mapping from $\mathbb{V}$ to $\mathbb{Z}$. We also denote by $\mathcal{E}_\mathbb{V}$ the set of expressions, and $\mathcal{F}_\mathbb{V}$ the set of formulae in the theory of linear arithmetic over $\mathbb{V}$. We denote by $\mathbf{V} \models g$ when a valuation $\mathbf{V}$ satisfies a formula $g \in \mathcal{F}_\mathbb{V}$. A labeled transition system $\mathcal{T} = \langle \mathcal{C}, \Sigma, \mathcal{R}, \mathcal{C}_0 \rangle$ consists of 1. a set of states $\mathcal{C}$; 2. a set of labels $\Sigma$; 3. a set of transitions $\mathcal{R} \subseteq \mathcal{C} \times \Sigma \times \mathcal{C}$, and 4. a set of initial states $\mathcal{C}_0 \subseteq \mathcal{C}$. A trace is a sequence of connected transitions, denoted by $\sigma^* \in \Sigma^*$. A state $c$ is reachable if there is a trace from an initial state $c_0$ to $c$, denoted by $c_0 \xrightarrow{\sigma^*} c$ in the sequel.

## 2.1  Symmetry Reduction

In order to avoid exploring the entire state space of a transition system, symmetry reduction [12,17,28] exploits the symmetries of a transition system. Intuitively, a transition system has symmetry if the transition relations remain invariant when states are rearranged by certain permutations.

**Definition 1.** *A symmetry of a labeled transition system $\mathcal{T} = \langle \mathcal{C}, \Sigma, \mathcal{R}, \mathcal{C}_0 \rangle$ is a permutation $\pi$ over $\mathcal{C} \cup \Sigma$, such that 1. $\pi(\mathcal{C}) = \mathcal{C}$ and $\pi(\Sigma) = \Sigma$, and 2. $\langle c_1, \gamma, c_2 \rangle \in \mathcal{R}$ iff $\langle \pi(c_1), \pi(\gamma), \pi(c_2) \rangle \in \mathcal{R}$, 3. $\pi(\mathcal{C}_0) = \mathcal{C}_0$.*

The set of all symmetries of $\mathcal{T}$ forms a group under the function composition, denoted by $Aut(\mathcal{T})$. However, obtaining $Aut(\mathcal{T})$ is computationally expensive, since one has to explore the whole state space. In practice, subgroups of $Aut(\mathcal{T})$, which can be obtained from the high-level system structure, are used. Example subgroups include rotation group, full component symmetry group and the Cartesian product of such groups. A subgroup $\mathcal{G} \subseteq Aut(\mathcal{T})$ induces an equivalence relation $\equiv_\mathcal{G}$ on $\mathcal{T}$ as follows: $s \equiv_\mathcal{G} t \Leftrightarrow \exists \pi \in \mathcal{G}.s = \pi(t)$. The equivalence relation $\equiv_\mathcal{G}$ is also called the orbit relation, and it induces a quotient model $\mathcal{T}_\mathcal{G}$, which is bisimilar to $\mathcal{T}$ [12,17]. Model checking of a symmetric property, i.e. a property remains invariant under permutations in $\mathcal{G}$, can be performed on the quotient model. We remark that deadlock states are trivially invariant under symmetry permutations.

As noticed in [11], under arbitrary symmetries, detecting state equivalence is as hard as the graph isomorphism problem. In order to bypass the orbit relation, for some specific symmetry subgroups, e.g. full component symmetry, rotation symmetry, one can select some representatives from the orbit relation and define a mapping function that computes these representatives [12,18,19]. Then during the state space exploration, states are dynamically mapped to their respective representatives.

## 2.2  Partial Order Reduction

For efficient verification of concurrent systems, partial order reduction [13,22,32, 34] exploits commutativity of two independent transitions, i.e. two independent transitions will lead to the same state when they are executed in different orders. Sequences of transitions that can be obtained by successively permuting two adjacent independent transitions, are equivalent in the sense that they all result in the same final state. Thus, it is sufficient to explore only one sequence out of all the equivalent ones, if the property of interest is irrelevant to the intermediate states of interleavings.

**Definition 2.** *Two transitions $\gamma_1$ and $\gamma_2$ are independent iff the following two conditions hold on every state $c$: 1. if $\gamma_1$ is enabled on $c$, then $\gamma_2$ is enabled on $c$ iff $\gamma_2$ is enabled on $c'$, where $c \xrightarrow{\gamma_1} c'$; 2. if $\gamma_2$ is enabled on $c$, then $\gamma_1$ is enabled on $c$ iff $\gamma_1$ is enabled on $c'$, where $c \xrightarrow{\gamma_2} c'$; 3. if $\gamma_1$ and $\gamma_2$ are both enabled on $c$, then $c'_1 = c'_2$, where $c \xrightarrow{\gamma_1 \gamma_2} c'_1$, and $c \xrightarrow{\gamma_2 \gamma_1} c'_2$.*

Relying on the transition independence, partial order reduction performs a selective search to reduce the number of visited states in the state space exploration, i.e., on every state reached during the exploration, only a subset of the enabled transitions on the state is explored. A widely used technique to perform selective search is the persistent set approach [20,22,37].

**Definition 3.** *A set $\Sigma'$ of transitions on a state $c$ is persistent iff, for all traces $c \xrightarrow{\gamma_1} c_1 \xrightarrow{\gamma_2} \dots c_{n-1} \xrightarrow{\gamma_n} c_n$ with $\gamma_i \notin \Sigma', i \in [1,n]$, $\gamma_n$ is independent of all transitions in $\Sigma'$.*

Intuitively, a subset of enabled transitions on a state is called persistent if whatever one reaches through the transitions outside of this subset, remains independent with transitions in the subset. Correctness of persistent set reduction with respect to safety property verification has been proved in [2,22].

## 2.3 Symbolic Structures and Abstraction

For infinite-state systems verification, symbolic algorithms manipulate a symbolic region structure [26], which is formally defined as follows.

**Definition 4.** *A symbolic region structure for a labeled transition system $T$ is a tuple $\mathcal{S} = \langle \mathcal{Q}, \bot, \sqcup, \sqcap, post, \beta \rangle$, where 1. $\mathcal{Q}$ is a set of regions, and $\bot \in \mathcal{Q}$; 2. $\sqcup$ and $\sqcap$ are two total functions: $\mathcal{Q} \times \mathcal{Q} \to \mathcal{Q}$, computing the greatest lower bound and least upper bound respectively; 3. $post : \mathcal{Q} \times \Sigma \to \mathcal{Q}$ is the strongest post operator over regions; 4. $\beta$ is a total function: $\mathcal{Q} \to 2^{\mathcal{C}}$, mapping a region to the set of representing states.*

A region can be understood as an abstract representation of a set of states of $T$. A region structure carries a natural preorder $\sqsubseteq$, which is defined by $r \sqsubseteq r'$ iff $\beta(r) \subseteq \beta(r')$, for $r, r' \in \mathcal{Q}$. A region structure is computable if the functions $\sqcup, \sqcap, post$ and $\sqsubseteq$ are computable.

**Definition 5.** *A symbolic abstraction structure for a transition system $T$ is a tuple $\mathcal{A} = (\mathcal{S}, \widehat{post})$, where 1. $\mathcal{S}$ is a computable region structure for $T$; 2. $\widehat{post} : \mathcal{Q} \times \Sigma \to \mathcal{Q}$ is an approximation of $post$, such that for every region $r \in \mathcal{Q}$ and every label $\gamma \in \Sigma$, $post(r, \gamma) \sqsubseteq \widehat{post}(r, \gamma)$ and $\widehat{post}$ is monotonic with respect to the preorder $\sqsubseteq$.*

Given a labeled transition system $T$, a predicate language $\mathcal{L}$ is a set of predicates that are interpreted over the state space of $T$. Further, we assume the following two properties: 1. the boolean closure of $\mathcal{L}$ is a decidable theory; 2. the boolean closure of $\mathcal{L}$ is closed under the *post* function. Thus, for every formula $\phi$ in the boolean closure of $\mathcal{L}$, and for every label $\gamma \in \Sigma$, one can compute a formula $\phi'$, such that $\beta(\phi') = post(\beta(\phi), \gamma)$. A particular symbolic abstraction structure is the predicate abstraction structure [14,21,23] defined as follows.

**Definition 6.** *Given a predicate language $\mathcal{L}$ for a labelled transition system $\mathcal{T}$, the predicate abstraction structure of $\mathcal{T}$ is a tuple $\mathcal{A}_\mathcal{L} = \langle \mathcal{S}_\mathcal{L}, \widehat{post} \rangle$, where 1. $\mathcal{S}_\mathcal{L} = \langle \mathcal{Q}, \bot, \sqcup, \sqcap, post, \beta \rangle$ is a symbolic region structure, where (a) $\mathcal{Q}$ is the set of formulae in boolean closure of $\mathcal{L}$; (b) $\bot = false$; (c) $\phi_1 \sqcup \phi_2 = \phi_1 \vee \phi_2$; (d) $\phi_1 \sqcap \phi_2 = \phi_1 \wedge \phi_2$; (e) $post(\phi, \gamma) = \phi'$, such that $\beta(\phi') = post(\beta(\phi), \gamma)$; (f) $\beta(\phi)$ is the set of states satisfying $\phi$. 2. for each region $\phi$, and label $\gamma \in \Sigma$, $\widehat{post}(\phi, \gamma) = \widehat{\phi}$, and $\widehat{\phi}$ the strongest formula such that $post(\phi, \gamma) \implies \widehat{\phi}$.*

Predicate abstraction structure over-approximates the state space of the transition system $\mathcal{T}$. A widely used algorithm to construct a predicate abstraction structure is the lazy abstraction with interpolant-based refinement [25,26,30].

# 3   BIP Model

A BIP model [7,38] is composed of a finite set of components, each of which is formally defined as follows.

**Definition 7.** *A BIP component is a tuple $\mathbb{B} = \langle \mathbb{V}, \mathbb{L}, \mathbb{P}, \mathbb{E}, \ell \rangle$, where 1. $\mathbb{V}$ is a finite set of variables; 2. $\mathbb{L}$ is a finite set of control locations; 3. $\mathbb{P}$ is a finite set of communication ports; 4. $\mathbb{E} \subseteq \mathbb{L} \times \mathbb{P} \times \mathcal{F}_\mathbb{V} \times \mathcal{E}_\mathbb{V} \times \mathbb{L}$ is a finite set of transition edges extended with guards in $\mathcal{F}_\mathbb{V}$ and operations in $\mathcal{E}_\mathbb{V}$; 5. $\ell \in \mathbb{L}$ is an initial control location.*

Transition edges in a component are labelled by ports, which form the interface of the component. We assume that, from each control location, every pair of outgoing transitions have different ports, and the ports of different components are disjoint. Given a component violating such assumptions, one can easily transform it into the required form by renaming the ports, while retaining the BIP expressiveness power.

We denote by $\mathcal{B} = \{\mathbb{B}_i \mid i \in [1, n]\}$ a set of components. In BIP, coordinations of components are specified by using interactions.

**Definition 8.** *An interaction for $\mathcal{B}$ is a tuple $\gamma = \langle g, \mathcal{P}, f \rangle$, where $g \in \mathcal{F}_\mathbb{V}$, $f \in \mathcal{E}_\mathbb{V}$ and $\mathcal{P} \subseteq \bigcup_{i=1}^{n} \mathbb{P}_i$, $\mathcal{P} \neq \emptyset$, and for all $i \in [1, n]$, $|\mathcal{P} \cap \mathbb{P}_i| \leq 1$.*

Intuitively, an interaction defines a guarded multi-party synchronization with data transfer: when the guard $g$ of an interaction $\mathcal{P}$ is enabled, then the data transfer specified by $f$ can be executed, and after that the transitions labelled by the ports in $\gamma$ can be taken simultaneously. We denote by $\Gamma$ a finite set of interactions. A BIP model is constructed by composing a number of components with interactions.

**Definition 9.** *A BIP model $\mathcal{M}_{\text{BIP}}$ is a tuple $\langle \mathcal{B}, \Gamma \rangle$, where $\mathcal{B}$ is a finite set of components, and $\Gamma$ is a finite set of interactions for $\mathcal{B}$.*

A tuple $c = \langle \langle l_1, \mathbf{V}_1 \rangle, \dots, \langle l_n, \mathbf{V}_n \rangle \rangle$ is a state of a BIP model $\mathcal{M}_{\text{BIP}}$, if for all $i \in [1, n]$, $l_i \in \mathbb{L}_i$ and $\mathbf{V}_i$ is a valuation of $\mathbb{V}_i$. A state $c_0$ is an initial state if, for all $i \in [1, n]$, $l_i = \ell_i$ and $\mathbf{V}_i$ is the initial valuation of $\mathbb{V}_i$. A state $c$ is an error state if, for some $i \in [1, n]$, $l_i$ is an error location. An interaction $\gamma = \langle g, \mathcal{P}, f \rangle \in \Gamma$ is enabled in a state $c$ if, $\wedge_{i=1}^{n} \mathbf{V}_i \models g$ and for every component $\mathbb{B}_i \in \mathcal{B}$ such that $\mathcal{P} \cap \mathbb{P}_i \neq \emptyset$, there is an edge $\langle l_i, \gamma \cap \mathbb{P}_i, g_i, f_i, l_i' \rangle \in \mathbb{E}_i$ and $\mathbf{V}_i \models g_i$.

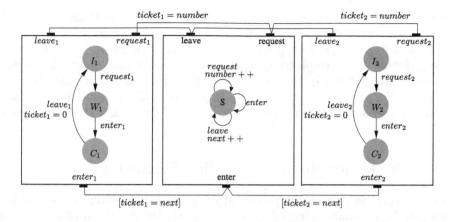

**Fig. 1.** Ticket mutual exclusion protocol

**Definition 10.** *Given a BIP model $\mathcal{M}_{\mathrm{BIP}} = \langle \mathcal{B}, \Gamma \rangle$, its operational semantics is defined by a transition system $\mathcal{T}_{\mathrm{BIP}} = \langle \mathcal{C}, \Sigma, \mathcal{R}, \mathcal{C}_0 \rangle$, where 1. $\mathcal{C}$ is the set of states, 2. $\Sigma = \Gamma$, 3. $\mathcal{R}$ is the set of transitions, and we say there is a transition from a state $c$ to another $c'$, if there is an interaction $\gamma$ such that, (a) $\gamma$ is enabled in $c$; (b) for all $\mathbb{B}_i \in \mathbb{B}$ such that $\gamma \cap \mathbb{P}_i \neq \emptyset$, there is an edge $\langle l_i, \gamma \cap \mathbb{P}_i, g_i, f_i, l'_i \rangle \in \mathbb{E}_i$, then $\mathbf{V}'_i = f_i(f(\cup_{i=1}^{n} \mathbf{V}_i))$; (c) for all $\mathbb{B}_i \in \mathbb{B}$ such that $\gamma \cap \mathbb{P}_i = \emptyset$, $l'_i = l_i$ and $\mathbf{V}'_i = \mathbf{V}_i$. (4.) $\mathcal{C}_0$ is the set of initial states.*

Instead of specifying safety properties by using logics, we recognize a set of states in a BIP model as error states. A BIP model is safe if no error states are reachable. Notice that any safety property can be encoded as an error state reachability problem by necessarily adding additional components.

We illustrate the BIP model as well as the basic idea of our verification approach by using the following mutual exclusion protocol.

*Example 1.* Figure 1 depicts a BIP model of the ticket mutual exclusion protocol with two processes. Upon entering the critical section $C_i, i = 1, 2$, each process requests a fresh ticket from the controller, then the process waits until its ticket equals to the number to be served next. When leaving the critical section, the process resets the ticket and the controller increases the number to be served by one.

Six interactions are defined to coordinate the transitions of controller and processes. Each interaction is depicted as a wire in Fig. 1. The ports connected by a wire are synchronized, and an interaction may also have a guard, e.g. $[ticket_1 = next]$, and an operation, e.g. $ticket_1 = number$.

Assume we start the state space exploration from the initial state $\langle \langle I_1, ticket_1 = 0 \rangle, \langle S, number = 1, next = 1 \rangle, \langle I_2, ticket_2 = 0 \rangle \rangle$, then the following two interactions $\gamma_1, \gamma_2$ are expanded, where $\gamma_1 = \langle true, \{request, request_1\}, ticket_1 = number \rangle$, and $\gamma_2 = \langle true, \{request, request_2\}, ticket_2 = number \rangle$.

Apparently, $\gamma_1, \gamma_2$ are not independent, since the interleavings $\gamma_1; \gamma_2$ and $\gamma_2; \gamma_1$ lead to two different states $\langle\langle W_1, ticket_1 = 1\rangle, \langle S, number = 3, next = 1\rangle, \langle W_2, ticket_2 = 2\rangle\rangle$ and $\langle\langle W_1, ticket_1 = 2\rangle, \langle S, number = 3, next = 1\rangle, \langle W_2, ticket_2 = 1\rangle\rangle$ respectively. However, we notice that under the following permutation of processes $\pi = \{1 \mapsto 2, 2 \mapsto 1\}$, the above states become the same, thus the two interactions $\gamma_1, \gamma_2$ commute and become independent. In this case, we may consider exploring only one interleaving and achieve better reductions.

## 4    Partial Order Reduction Under Symmetry

In this section, we extend the persistent set partial order reduction by taking symmetry into account. First of all, we generalize the definition of interaction independence.

**Definition 11.** *Given a symmetry group $\mathcal{G}$, two interactions $\gamma_1$ and $\gamma_2$ are independent under symmetry, if and only if for every state $c$ in the global system, there is a symmetry $\pi \in \mathcal{G}$ such that the following two conditions hold: 1. if $\gamma_1$ is enabled in $c$, then $\gamma_2$ is enabled in $c$ iff $\gamma_2$ is enabled in $c'$, where $c \xrightarrow{\gamma_1} c'$. 2. if $\gamma_1$ is enabled in $c$, then $\gamma_2$ is enabled in $c$ iff $\gamma_2$ is enabled in $c'$, where $c \xrightarrow{\gamma_1} c'$. 3. if $\gamma_1$ and $\gamma_2$ are both enabled in $c$, then $c'_1 = \pi(c'_2)$, where $c \xrightarrow{\gamma_1\gamma_2} c'_1$, and $c \xrightarrow{\gamma_2\gamma_1} c'_2$.*

This new definition differs the one in Definition 2 in that two interactions are viewed as independent if their executions commute under some symmetry permutation. We denote by $\mathcal{D}_\gamma$ the set of interactions that are not independent under symmetry with $\gamma$.

In previous work [7,38], we obtain an under-approximation of independence relation statically from system specification: two interactions are *independent* if they do not share a common component. Though being easy to obtain, this approximation is too coarse, and many independent transitions are ignored. For instance, the following two interactions in Example 1, $\gamma_1 = \langle[ticket_1 = next], \{enter, enter_1\}, skip\rangle$, and $\gamma_2 = \langle[ticket_2 = next], \{enter, enter_2\}, skip\rangle$, are independent, but cannot be obtained using the previous analysis.

In this work, we apply a finer static analysis to check if two interactions are independent or not. Given two interactions $\gamma_1 = \langle g_1, \mathcal{P}_1, f_1\rangle$, $\gamma_2 = \langle g_2, \mathcal{P}_2, f_2\rangle$, we check if they are independent by checking the validity of the following three formulae:

1. $\forall c. \exists c'. c \models g_1 \wedge c \xrightarrow{\gamma_1} c' \implies (c \models g_2 \equiv c' \models g_2)$
2. $\forall c. \exists c'. c \models g_2 \wedge c \xrightarrow{\gamma_2} c' \implies (c \models g_1 \equiv c' \models g_1)$
3. there is a permutation $\pi \in \mathcal{G}$, such that the formula $\forall c. \exists c_1, c_2. c \models g_1 \wedge c \models g_2 \wedge c \xrightarrow{\gamma_1\gamma_2} c_1 \wedge c \xrightarrow{\gamma_2\gamma_1} c_2 \implies c_1 = \pi(c_2)$ is valid.

Considering the complexity, the number of interactions is linear to the size of the system model. Thus, the number of validity checks is also linear to the size of system model. In order to detect the state equivalence under symmetry, one intuitive approach is to traverse all permutations in the symmetry group $\mathcal{G}$. However, this would blow up the analysis, even for full component symmetry group, whose complexity is factorial in the number of components. As in [19], we use a sorting function that maps a state to a representative in the orbit relation, then two states are equivalent if they can be mapped to the same representative. The sorting function requires a total order on the symbolic states of each component. We say a symbolic state $c_1$ is greater than another $c_2$ if $c_1 > c_2$ is valid.

However, since we focus on infinite state systems, our partial order reduction should apply to a symbolic abstraction structure instead of the labelled transition system. We remark that on symbolic abstraction structures, the independent transitions do not commute anymore. For instance, consider two interactions $\gamma_1 = \langle true, \{p_1\}, x_1 + + \rangle$ and $\gamma_2 = \langle true, \{p_2\}, x_2 + + \rangle$, it is obvious they are independent in the concrete state space. Suppose the predicate language of the abstraction structure is given by predicates $b_1 = (x_1 > x_2)$ and $b_2 = (x_1 = x_2)$, and then starting from the initial state $\neg b_1 \wedge b_2$, transition sequence $x_1 + +; x_2 + +$ leads to the state $b_1 \vee b_2$, while another transition sequence $x_2 + +; x_1 + +$ leads to a different state $\neg b_1 \vee b_2$. Then the question is whether it is still safe if we only explore one transition sequence.

The following lemma shows that independent transitions still commute under symmetry on the concrete states represented by abstraction structures. Thus, exploiting independence on the symbolic abstraction structure is still sound.

**Lemma 1.** *Let $\gamma_1$ and $\gamma_2$ be two independent transitions under symmetry $\pi$, and let $r$ be an abstract region, then for all $c \in \beta(\widehat{post}(\widehat{post}(r, \gamma_1)), \gamma_2)$, we have that if there is a state $c_1 \in \beta(r)$, such that $c = post(post(c_1, \gamma_1), \gamma_2)$, then $\pi(c) \subset \beta(\widehat{post}(\widehat{post}(r, \gamma_2)), \gamma_1)$.*

*Proof.* Assume we have $c = post(post(c_1, \gamma_1), \gamma_2)$, since $\gamma_1$ and $\gamma_2$ are independent under symmetry $\pi$, then we also have $\pi(c) = post(post(c_1, \gamma_2), \gamma_1)$. According to the semantics of $\widehat{post}$, it holds that $\pi(c) \in \beta(\widehat{post}(\widehat{post}(r, \gamma_2), \gamma_1))$.

We then extend the persistent set in Definition 3 by relying on the notion of independence under symmetry and by generalising to the symbolic abstraction structure.

**Definition 12.** *A set of interactions $\Gamma$ on an abstract region $r$ is persistent iff, for some state $c$, such that $c \models r$ and for all traces $c \xrightarrow{\gamma_1} c_1 \xrightarrow{\gamma_2} ... \xrightarrow{\gamma_n} c_n$ in the original state space with $\gamma_i \notin \Gamma, i \in [1, n]$, $\gamma_n$ is independent of all interactions in $\Gamma$ under symmetry.*

The following theorem states the correctness of selective search over symbolic abstraction structure by using persistent sets on abstract regions.

**Theorem 1.** *Let $\mathcal{T}$ be a labeled transition system and $\mathcal{A} = \langle \mathcal{S}, \widehat{post} \rangle$ be a symbolic abstraction struction for $\mathcal{T}$, where $\mathcal{S} = \langle \mathcal{Q}, \bot, \sqcup, \sqcap, post, \beta \rangle$, selective search over $\mathcal{A}$ with persistent set under symmetry preserves all deadlock states of $\mathcal{T}$.*

*Proof.* We prove by induction on the length of the trace to deadlock state. The basis case trivially holds. Suppose deadlock states are preserved by traces of length $n$, then we prove that they are also preserved by traces of length $n + 1$.

Let $\gamma_1, ..., \gamma_n, \gamma_{n+1}$ be a trace from some state $c$ in the concrete state space that leads to a deadlock, we first prove that there is another trace from $c$, which leads to the same deadlock state and starts with an interaction from the persistent set of $c$. Consider the set of traces of the form $\gamma'_1, ..., \gamma'_n, \gamma'_{n+1}$, which are obtained by permutating two adjacent transtions independent under symmetry, then according to Definition 11, we know that any of the above traces also leads to the same deadlock since permuting adjacent independent interaction under symmetry would lead to states, which are equivalent under symmetry.

Then we need to prove that there is at least one trace from the above set, whose first interaction $\gamma'_1$ is in the persistent set of $c$. Suppose that none of interactions $\gamma_i, i \in [1, n]$ is in the persistent set of $c$, then by Definition 12, the interactions in persistent set of $c$ is still enabled, which contradicts the deadlock state assumption. Thus, there is at least one interaction $\gamma_i, i \in [1, n]$ in the persistent set of $c$, assume the first such interaction is $\gamma_j, j \in [1, n]$, then for all interactions $\gamma_k, k < j$, we have $\gamma_j$ is independent with $\gamma_k$. Thus, $\gamma_k$ can be moved to the beginning of the trace, which proves the existence of $\gamma'_1$.

Then according to the hypothesis that deadlocks are preserved by traces of length $n$ and the fact that at least one $\gamma'_1$ is in the persistent set of $c$, we know that deadlock states are also reachable in the reduced concrete state space by traces of length $n + 1$.

Finally, according to the definition of symbolic abstraction struction, we know that any trace on concrete state $c$ is also possible on the abstract region $r$, which over-approximates $c$, i.e. $c \models r$. Then by Lemma 1, we can conclude that both traces $\gamma_1, ..., \gamma_{n+1}$ and $\gamma'_1, ..., \gamma'_{n+1}$ lead to the same deadlock state on the symbolic abstraction structure.

# 5    Combining with Lazy Abstraction

In this section, we present how to combine the new partial order reduction with lazy abstraction of BIP [38]. The integrated algorithm is shown in Algorithm 1. As in lazy abstraction [26], the algorithm computes an over-approximation of the reachable states by constructing a reduced abstract reachability tree (ART), whose nodes consist of both control locations and abstract data regions of all components.

**Definition 13.** *An ART node is a tuple* $\eta = \langle \langle l_1, \phi_1 \rangle, ..., \langle l_n, \phi_n \rangle, \phi \rangle$, *where each* $\langle l_i, \phi_i \rangle$ *is an abstract component region consisting of the control location* $l_i$ *and the abstract data region* $\phi_i$ *of component* $B_i$, *and* $\phi$ *is the global data region.*

An abstract component region $\langle l_i, \phi_i \rangle$ over-approximates the set of concrete states with control location $l_i$. A global data region $\phi$ keeps track of the variables in data transfer. An ART node is an error node if at least one of the control location $l_i$ is an error location and the data regions are consistent, i.e. $\phi \wedge \bigwedge_{i=1}^{n} \phi_i$ is satisfiable. A state $c = \langle \langle l_1, \mathbf{V}_1 \rangle, \ldots, \langle l_n, \mathbf{V}_n \rangle \rangle$ is covered by a node $\eta = \langle \langle l'_1, \phi_1 \rangle, \ldots, \langle l'_n, \phi_n \rangle, \phi \rangle$, denoted by $c \models \eta$, if for all $i \in [1, n]$, $l_i = l'_i$ and $\mathbf{V}_i \models \phi_i$ and $\wedge_{i=1}^{n} \mathbf{V}_i \models \phi$. A node can also be covered by another node. The covering relation is defined as follows.

**Definition 14.** *An ART node $\eta = \langle \langle l_1, \phi_1 \rangle, \ldots, \langle l_n, \phi_n \rangle, \phi \rangle$ is covered by another node $\eta' = \langle \langle l'_1, \phi'_1 \rangle, \ldots, \langle l'_n, \phi'_n \rangle, \phi' \rangle$ if $l_i = l'_i$ and the implication $\phi_i \Rightarrow \phi'_i$ is valid for all $i \in [1, n]$, and $\phi \Rightarrow \phi'$ is valid.*

A covered node will not be explored in the future exploration. However, a covered node may be uncovered, if the covering relation is no longer valid. An ART is complete if all the nodes are either fully expanded or covered. An ART is safe if it is complete, and contains no error nodes.

The algorithm constructs the ART by expanding the ART nodes progressively, starting from the initial one. Upon expanding a node, it first checks if an error node is encountered. If an error node is detected, it generates a counterexample and checks if the counterexample is real. If the counterexample is real, it reports the counterexample and stops the analysis. Otherwise, it refines the abstraction and continues the exploration. Then it checks if a cycle occurs. We say a cycle occurs, if the control locations of the node have been visited before in the trace to this node. If a cycle is detected, the predecessor of this node will be fully expanded in order to avoid the 'ignoring problem' [13]. Function FullyExpand expands the set of interactions, which have been removed by partial order reduction. Then it checks if the node can be covered by another one. If not, it will expand the node, where the partial order reduction is incorporated to reduce the number of successors.

To expand a node, the set of enabled interactions is computed via function EnabledInteraction. We say that an interaction $\gamma = \langle g, \mathcal{P}, f \rangle$ is enabled on an ART node $\langle \langle l_1, \phi_1 \rangle, \ldots, \langle l_n, \phi_n \rangle, \phi \rangle$ if for each component $B_i$ such that $\mathcal{P} \cap \mathbb{P}_i = \{p_i\}$, there is an outgoing transition $\langle l_i, g_i, p_i, f_i, l'_i \rangle \in \mathbb{E}_i$. Notice that we do not check the satisfiability of the guards on the ART node, since in lazy abstraction if an interaction is disabled on the ART node, the successor node will be inconsistent, i.e. the conjunction $\wedge_{i=1}^{n} \phi'_i \wedge \phi'$ is unsatisfiable. Thus, the successor node will be discarded. Then the persistent set is obtained via function PersistentSet, which will be elaborated in the next section. Finally, the node is expanded according the persistent set and the successors are added into the worklist. We refer to [38] for more details about node expansion and abstraction refinement.

The following theorem states the correctness of Algorithm 1.

**Theorem 2.** *Given a BIP model $\mathcal{M}_{\text{BIP}}$, for every terminating execution of Algorithm 1, the following properties hold:*

1. *If a counterexample is returned, then there is concrete counterexample in $\mathcal{M}_{\text{BIP}}$;*
2. *If an ART is returned, then it is safe with respect to safety properties, which are invariant under symmetry.*

---

**Algorithm 1.** Lazy abstraction with partial order reudction for BIP

---

**Input:** a BIP model $\mathcal{M}_{\text{BIP}}$

    create an ART *art* with initial node $\eta_0$

    create a worklist *wl* and push $\eta_0$ into *wl*

    **while** $wl \neq \emptyset$ **do**

        $\eta \leftarrow \text{pop}(wl)$

        **if** $\text{IsError}(\eta)$ **then**

            $cex \leftarrow \text{BuildCEX}(\eta)$

            **if** *cex* is real **then**

                **return** *cex*

            **else**

                $\text{Refine}(art, cex)$

        **else if** $\text{Cycle}(\eta)$ **then**

            $\eta' \leftarrow \text{Predecessor}(\eta)$

            $\Gamma_E \leftarrow \text{EnabledInteractions}(\eta')$

            $\text{FullyExpand}(\eta', \Gamma_E)$

            push all successors of $\eta'$ into *wl*

            mark $\eta$ as covered

        **else if** $\text{Covering}(\eta)$ **then**

            mark $\eta$ as covered

        **else**

            $\Gamma_E \leftarrow \text{EnabledInteractions}(\eta)$

            $\Gamma_P \leftarrow \text{PersistentSet}(\Gamma_E, \eta)$

            $\text{Expand}(\eta, \Gamma_P)$

            push all successors of $\eta$ into *wl*

    **return** *art*

---

*Proof.* The correctness relies on the result reported in [38] that lazy abstraction of BIP without partial order reduction constructs a symbolic abstraction structure that over-approximates all the reachable states. Then according to Theorem 1, applying our partial order reduction technique to a symbolic abstraction structure preserves deadlock-freedom. The preservation of general safety properties, which are invariant under symmetry, follows from the full expansion strategy we use to avoid the 'ignoring problem' in [13] and in Chap. 8.2 in [2].

### 5.1   Computing Persistent Set

In this section, we present an algorithm to compute a persistent set on an ART node, i.e. the implementation of function PersistentSet in Algorithm 1. We adopt the Stubborn set approach [22,34] and also make use of the independence relation under symmetry.

First, we introduce the definition of an enabling set for a disabled interaction on an ART node.

**Definition 15.** *Let $\gamma \in \Gamma$ be a disabled interaction on an ART node $\eta$, an enabling set for $\gamma$ on $\eta$ is a set of interactions $\mathcal{N}_\gamma$ such that for all sequences of* *interactions of form* $\eta \xrightarrow{\gamma_1 \ldots \gamma_n} \eta' \xrightarrow{\gamma}$, *there is at least one interaction* $\gamma_i \in \mathcal{N}_\gamma$, *for some* $i \in [1, n]$.

This definition of enabling set is on an abstract region, which is different from the necessary enabling set on a concrete state in [22,34]. An enabling set of a disabled interaction on an abstract region characterizes the interactions that may interfere with the disabled interaction in the symbolic exploration of lazy abstraction. To obtain an enabling set for a disabled transition, fine-grained static analysis is used.

Given a disabled interaction $\gamma = \langle g, \mathcal{P}, f \rangle$ on an ART node $\eta = \langle \langle l_1, \phi_1 \rangle, \ldots, \langle l_n, \phi_n \rangle, \phi \rangle$, for any component $B_i$ such that $\mathcal{P} \cap \mathbb{P}_i = \{p_i\}$, but there is no such an outgoing transition $\langle l_i, g_i, p'_i, f_i, l'_i \rangle \in \mathbb{E}_i$ that $p'_i = p_i$, we say another interaction $\gamma' = \langle g', \mathcal{P}', f' \rangle$ is in the enabling set $\mathcal{N}_\gamma$, if $\mathcal{P}' \cap \mathbb{P}_i = \{p'_i\}$, and there is a path in $B_i$ from $l_i$ to a control location, where $p_i$ is an outgoing transition.

**Definition 16.** *A set of interactions $\Gamma$ is a Stubborn set in an ART node $\eta$ if the following conditions hold: 1. $\Gamma$ contains at least one enabled interaction if the set of enabled interactions on $\eta$ is non-empty; 2. for each disabled interaction $\gamma \in \Gamma$, there is an enabling set $\mathcal{N}_\gamma$ s.t. $\mathcal{N}_\gamma \subseteq \Gamma$; 3. for each enabled interaction $\gamma \in \Gamma$, then $\mathcal{D}_\gamma \subseteq \Gamma$.*

This definition suggests a method (shown in Algorithm 2) to compute a Stubborn set: it constructs the set incrementally by making sure that each new interaction added to the set fulfills the Stubborn set conditions. We remark that in Stubborn set computation, a weaker notion of independence and its complement can be used: two coenabled interactions are independent if they commute under symmetry and do not disable each other. These notions correspond to the accord and do-not-accord relation in [22,34].

The following theorem states that the set of enabled interactions in a Stubborn set is indeed a persistent set. Thus, Algorithm 2 can be easily incorporated in Algorithm 1, replacing the function PersistentSet.

---

**Algorithm 2.** Stubborn set computation

---

**procedure** STUBBORN($\eta$, $\mathcal{M}_{\text{BIP}}$)
    $\Gamma_{work} = \{\gamma\}$ such that $\gamma$ is enabled on $\eta$
    $\Gamma_{stubborn} = \emptyset$
    **while** $\Gamma_{work} \neq \emptyset$ **do**
        pick some $\gamma \in \Gamma_{work}$
        $\Gamma_{work} = \Gamma_{work} - \gamma$, $\Gamma_{stubborn} = \Gamma_{stubborn} \cup \{\gamma\}$
        **if** $\gamma$ is enabled **then**
            $\Gamma_{work} = \Gamma_{work} \cup \mathcal{D}_\gamma \backslash \Gamma_{stubborn}$
        **else**
            $\mathcal{N}_\gamma = NES(\gamma, \eta, \mathcal{M}_{\text{BIP}})$
            $\Gamma_{work} = \Gamma_{work} \cup \mathcal{N}_\gamma \backslash \Gamma_{stubborn}$
    return $\Gamma_{stubborn}$

---

**Theorem 3.** *Let $\Gamma$ be a Stubborn set returned by Algorithm 2, and let $\Gamma'$ be the set of enabled interactions in $\Gamma$, then $\Gamma'$ is a persistent set under symmetry on an ART node $\eta$.*

*Proof.* Suppose $\Gamma'$ is not a persistent set, then for all states $c$, such that $c \models \eta$, there is a trace $c \xrightarrow{\gamma_1} c_1 \xrightarrow{\gamma_2} c_2... \xrightarrow{\gamma_n} c_n$ with $\gamma_i \notin \Gamma'$, $i \in [1, n]$, $\gamma_n$ depends on some interaction $\gamma'$ in $\Gamma'$.

Assume $\gamma_n$ is enabled on $c$, then $\gamma_n$ is also enabled on $\eta$ and should be included in $\Gamma$, and thus $\Gamma'$ since it depends on $\gamma'$, which contradicts the assumption. Assume the guard of $\gamma_n$ is diabled on $c$, however, since we perform lazy abstraction, $\gamma_n$ is still enabled on the ART node $\eta$, thus $\gamma_n$ is in $\Gamma'$, also contradicting the assumption.

Assume $\gamma_n$ is disabled on $\eta$, however, since it is enabled on $c_n$, there must be a nonempty enabling set for $\gamma_n$ on node $\eta$, and there is at least one interaction $\gamma_j, 1 \leq j \leq n$ in this enabling set, and according to the assumption, $\gamma_j$ is disabled on $c$, otherwise $\gamma_j$ is in $\Gamma'$. Then by repeating the same reasoning, there is an interaction $\gamma_{j'}, 1 \leq j' < j$ in the enabling set for $\gamma_j$ and $\gamma_{j'}$ is diabled on $\eta$. In the end, we can conclude that $\gamma_1$ is in some enabling set and is disabled in $\eta$, which contradicts the assumption. Thus, $\Gamma'$ is indeed a persistent set.

## 6    Experimental Evaluation

We have implemented the proposed verification technique in our prototype model checker for BIP. In the experimental evaluation, we took a set of benchmarks from the literature and our previous work, including the ticket mutual exclusion protocol in star topology, a leader election protocol in ring topology, and a consensus protocol in star topology. All these benchmarks are scalable in terms of the number of components, and all are infinite-state, and they all use data transfer on interactions. We model them in BIP and for each benchmark, we create a safe and an unsafe version, and for each version, we have 10 instances. All the experiments are performed on a 64-bit Linux PC with a 2.8 GHz Intel i7-2640M CPU, with a memory limit of 4 GB and a time limit of 300 seconds per benchmark. We refer to our website[1] for all the benchmarks and the tool.

We run the following configurations and compare the running time of solving the benchmarks: plain lazy abstraction of BIP [38] (represented as 'plain' in the figures), lazy abstraction with persistent set reduction [7] (represented as 'pset' in the figures), lazy abstraction with simultaneous set reduction [38] (represented as 'simset' in the figures) and our new algorithm[2] (represented as 'sympor' in the figures). We also compare with the state-of-the-art invariant verification algorithm IC3 [8,9]. We do not compare with DFinder [5], or VCS [24], since they do not handle data transfer or infinite-state models respectively.

---

[1] http://risd.epfl.ch/bipchecker.

[2] We include the independence detection time in the running time of our new algorithm.

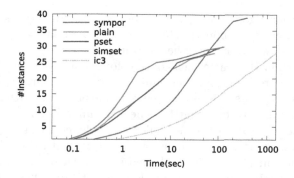

**Fig. 2.** Cumulative plot of time for solving all benchmarks

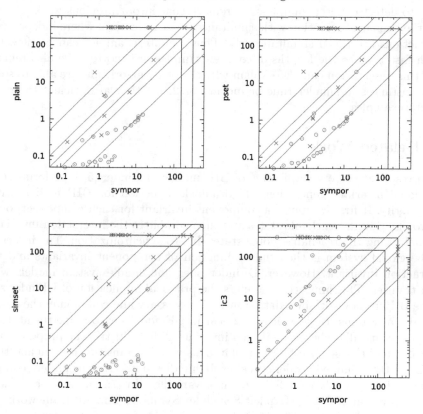

**Fig. 3.** Scatter plot of analysis time

In Fig. 2, we plot the cumulative time (x-axis) of solving a number of benchmarks (y-axis). A point $(x, y)$ in the plot tells us the number $y$ of benchmarks, which can be verified in the given time bound $x$. We remark that time $x$ is not the accumulation of the analysis time of all $y$ benchmarks. We see that our new algorithm can always solve more instances in a given time bound than IC3, while comparing to other algorithms, it is not always faster, but can still solve more instances in a larger time bound. This tells that the analysis time of our new algorithm grows slower than the other algorithms.

In Fig. 3, we show the scatter plots of time for solving each benchmark.[3] In all plots, symbol × represents a safe benchmark, and ○ represents an unsafe benchmark. A point in the plots tells us the analysis time of the algorithms represented by x-axis and y-axis. We find that for safe benchmarks, our new algorithm is always more efficient that the others, while for unsafe models, though being more efficient than IC3, it does not demonstrate clear strength over the other algorithms. This phenomenon can be explained since with partial order reduction, independent interactions are chosen non-deterministically, thus, the interactions leading to an error may be delayed, which would increase the length of execution steps to detect counterexamples. We remark that in our previous work [38], we had the similar observation with simultaneous set reduction, where independent interactions are taken simultaneously. Thus, counterexamples can be detected much faster, since the lengths of counterexamples are shorter. We also remark that this observation can also explain why the other algorithms are able to solve more instances in smaller time bound in Fig. 2, i.e. they are efficient to detect counterexamples.

## 7 Related Work

With respect to safety verification of BIP models, DFinder [5,6] performs compositional invariant generation and deadlock detection for BIP models without priority. It first computes a component invariant for each component, over-approximating its reachable states and then computes the interaction invariant, characterising the reachable global states due to synchronization. The invariant of the global system is then the conjunction of component invariants and the interaction invariant. However, DFinder does not handle system models with data transfer. This limitation hampers the practical application of DFinder and of the BIP framework, since data transfer is necessary and common in the design of real-life systems. Besides, it is not clear in DFinder how to refine the abstraction automatically when the inferred invariant fails to justify the property. The VCS [24] tool translates a BIP model into a symbolic transition system and then performs the bounded model checking. It handles data transfer among components, but only deals with finite domain variables. In [7], the authors present an instantiation of ESST (Explicit Scheduler Symbolic Thread) framework [10] and several optimisations for BIP. In [38], the authors present a lazy abstraction algorithm for BIP and also propose a simultaneous set reduction technique, which is further combined with lazy abstraction to boost the analysis. However, neither takes into account symmetry features of the BIP model.

Both state space symmetries and partial order reductions have been extensively investigated in model checking community over the decades, leading to a

---

[3] Red diagonal guides provide a reference for comparison, each indicating shift of one order of magnitude.

variety of symmetry reduction techniques [12,17,19,28,31,36], and partial order reduction techniques [13,20,32,34,37]. In [15], the authors propose a symmetry aware counter-example guarded abstraction refinement technique for replicated non-recursive C programs. Their abstraction technique is eager in the sense that an abstraction model is constructed first, which differs from our lazy abstraction technique. The work in [35] combines lazy abstraction with interpolant and partial order reduction under conditional independence [37] for the verification of generic multi-threaded programs. We remark that though in [16,27], the authors have studied how to combine symmetry reduction with ample set based partial order reduction. However, both work focus on finite state models, and no abstraction techniques are used.

# 8   Conclusion

In this paper we extend our previous work on safety property verification of BIP models, and present a new lazy reachability analysis algorithm that combines persistent set partial order reduction and symmetry reduction. The experimental evaluation demonstrates that for system models that have inherent symmetries, exploiting symmetries can boost the abstract reachability analysis significantly. As future work, we will investigate more efficient reduction techniques for component-based systems, that can be combined with abstract reachability analysis, such as directed model checking.

**Acknowledgements.** I wish to express my sincere gratitude to Prof. Joseph Sifakis for his inspirations on BIP, and to Dr. Alessandro Cimatti, Dr. Marco Roveri, Dr. Sergio Mover for their instructive guidance and help with Kratos model checker, and to Dr. Rongjie Yan, Dr. Simon Bliudze for the discussion and proofreading this manuscript, and all the anonymous reviewers for their careful reading of the paper.

# References

1. Abdellatif, T., Bensalem, S., Combaz, J., de Silva, L., Ingrand, F.: Rigorous design of robot software: a formal component-based approach. Robot. Auton. Syst. **60**, 1563–1578 (2012)
2. Baier, C., Katoen, J.P., Larsen, K.G.: Principles of Model Checking. MIT Press, Cambridge (2008)
3. Basu, A., Bensalem, S., Bozga, M., Combaz, J., Jaber, M., Nguyen, T.H., Sifakis, J.: Rigorous component-based system design using the BIP framework. IEEE Softw. **28**, 41–48 (2011)
4. Basu, A., Bensalem, S., Bozga, M., Delahaye, B., Legay, A.: Statistical abstraction and model-checking of large heterogeneous systems. STTT **14**, 53–72 (2012)
5. Bensalem, S., Bozga, M., Nguyen, T.-H., Sifakis, J.: D-Finder: a tool for compositional deadlock detection and verification. In: Bouajjani, A., Maler, O. (eds.) CAV 2009. LNCS, vol. 5643, pp. 614–619. Springer, Heidelberg (2009). doi:10.1007/978-3-642-02658-4_45

6. Bensalem, S., Griesmayer, A., Legay, A., Nguyen, T.-H., Sifakis, J., Yan, R.: D-Finder 2: towards efficient correctness of incremental design. In: Bobaru, M., Havelund, K., Holzmann, G.J., Joshi, R. (eds.) NFM 2011. LNCS, vol. 6617, pp. 453–458. Springer, Heidelberg (2011). doi:10.1007/978-3-642-20398-5_32

7. Bliudze, S., Cimatti, A., Jaber, M., Mover, S., Roveri, M., Saab, W., Wang, Q.: Formal verification of infinite-state BIP models. In: Finkbeiner, B., Pu, G., Zhang, L. (eds.) ATVA 2015. LNCS, vol. 9364, pp. 326–343. Springer, Heidelberg (2015). doi:10.1007/978-3-319-24953-7_25

8. Bradley, A.R.: SAT-based model checking without unrolling. In: Jhala, R., Schmidt, D. (eds.) VMCAI 2011. LNCS, vol. 6538, pp. 70–87. Springer, Heidelberg (2011). doi:10.1007/978-3-642-18275-4_7

9. Cimatti, A., Griggio, A., Mover, S., Tonetta, S.: IC3 modulo theories via implicit predicate abstraction. In: Ábrahám, E., Havelund, K. (eds.) TACAS 2014. LNCS, vol. 8413, pp. 46–61. Springer, Heidelberg (2014). doi:10.1007/978-3-642-54862-8_4

10. Cimatti, A., Narasamdya, I., Roveri, M.: Software model checking with explicit scheduler and symbolic threads. Log. Meth. Comput. Sci. 8(2) (2012). doi:10.2168/LMCS-8(2:18)2012

11. Clarke, E.M., Emerson, E.A., Jha, S., Sistla, A.P.: Symmetry reductions in model checking. In: Hu, A.J., Vardi, M.Y. (eds.) CAV 1998. LNCS, vol. 1427, pp. 147–158. Springer, Heidelberg (1998). doi:10.1007/BFb0028741

12. Clarke, E.M., Enders, R., Filkorn, T., Jha, S.: Exploiting symmetry in temporal logic model checking. Formal Methods Syst. Des. 9, 77–104 (1996)

13. Clarke, E.M., Grumberg, O., Minea, M., Peled, D.: State space reduction using partial order techniques. STTT 2, 279–287 (1999)

14. Das, S., Dill, D.L., Park, S.: Experience with predicate abstraction. In: Halbwachs, N., Peled, D. (eds.) CAV 1999. LNCS, vol. 1633, pp. 160–171. Springer, Heidelberg (1999). doi:10.1007/3-540-48683-6_16

15. Donaldson, A.F., Kaiser, A., Kroening, D., Tautschnig, M., Wahl, T.: Counterexample-guided abstraction refinement for symmetric concurrent programs. Formal Methods Syst. Des. 41, 25–44 (2012)

16. Emerson, E.A., Jha, S., Peled, D.: Combining partial order and symmetry reductions. In: Brinksma, E. (ed.) TACAS 1997. LNCS, vol. 1217, pp. 19–34. Springer, Heidelberg (1997). doi:10.1007/BFb0035378

17. Emerson, E.A., Sistla, A.P.: Symmetry and model checking. Formal Methods Syst. Des. 9, 105–131 (1996)

18. Emerson, E.A., Trefler, R.J.: From asymmetry to full symmetry: new techniques for symmetry reduction in model checking. In: Pierre, L., Kropf, T. (eds.) CHARME 1999. LNCS, vol. 1703, pp. 142–157. Springer, Heidelberg (1999). doi:10.1007/3-540-48153-2_12

19. Emerson, E.A., Wahl, T.: Dynamic symmetry reduction. In: Halbwachs, N., Zuck, L.D. (eds.) TACAS 2005. LNCS, vol. 3440, pp. 382–396. Springer, Heidelberg (2005). doi:10.1007/978-3-540-31980-1_25

20. Flanagan, C., Godefroid, P.: Dynamic partial-order reduction for model checking software. In: POPL (2005)

21. Flanagan, C., Qadeer, S.: Predicate abstraction for software verification. ACM SIGPLAN Not. 37, 191–202 (2002). ACM

22. Godefroid, P.: Partial-Order Methods for the Verification of Concurrent Systems: An Approach to the State-Explosion Problem. Springer, New York (1996)

23. Graf, S., Saidi, H.: Construction of abstract state graphs with PVS. In: Grumberg, O. (ed.) CAV 1997. LNCS, vol. 1254, pp. 72–83. Springer, Heidelberg (1997). doi:10.1007/3-540-63166-6_10

24. He, F., Yin, L., Wang, B.-Y., Zhang, L., Mu, G., Meng, W.: VCS: a verifier for component-based systems. In: Hung, D., Ogawa, M. (eds.) ATVA 2013. LNCS, vol. 8172, pp. 478–481. Springer, Heidelberg (2013). doi:10.1007/978-3-319-02444-8_39

25. Henzinger, T.A., Jhala, R., Majumdar, R., McMillan, K.L.: Abstractions from proofs. ACM SIGPLAN Not. **39**, 232–244 (2004). ACM

26. Henzinger, T.A., Jhala, R., Majumdar, R., Sutre, G.: Lazy abstraction. ACM SIGPLAN Not. **37**, 58–70 (2002)

27. Iosif, R.: Symmetry reductions for model checking of concurrent dynamic software. STTT **6**, 302–319 (2004)

28. Ip, C.N., Dill, D.L.: Better verification through symmetry. Formal Methods Syst. Des. **9**, 41–75 (1996)

29. Konnov, I., Kotek, T., Wang, Q., Veith, H., Bliudze, S., Sifakis, J.: Parameterized systems in BIP: design and model checking. In: CONCUR (2016, to appear)

30. McMillan, K.L.: Lazy abstraction with interpolants. In: Ball, T., Jones, R.B. (eds.) CAV 2006. LNCS, vol. 4144, pp. 123–136. Springer, Heidelberg (2006). doi:10.1007/11817963_14

31. Miller, A., Donaldson, A., Calder, M.: Symmetry in temporal logic model checking. ACM Comput. Surv. (2006)

32. Peled, D.: Combining partial order reductions with on-the-fly model-checking. In: Dill, D.L. (ed.) CAV 1994. LNCS, vol. 818, pp. 377–390. Springer, Heidelberg (1994). doi:10.1007/3-540-58179-0_69

33. Sifakis, J.: Rigorous system design. Found. Trends Electron. Des. Autom. **6**(4), 293–362 (2013)

34. Valmari, A.: A stubborn attack on state explosion. In: Clarke, E.M., Kurshan, R.P. (eds.) CAV 1990. LNCS, vol. 531, pp. 156–165. Springer, Heidelberg (1991). doi:10.1007/BFb0023729

35. Wachter, B., Kroening, D., Ouaknine, J.: Verifying multi-threaded software with Impact. In: FMCAD (2013)

36. Wahl, T., Donaldson, A.: Replication and abstraction: symmetry in automated formal verification. Symmetry **2**, 799–847 (2010)

37. Wang, C., Yang, Z., Kahlon, V., Gupta, A.: Peephole partial order reduction. In: Ramakrishnan, C.R., Rehof, J. (eds.) TACAS 2008. LNCS, vol. 4963, pp. 382–396. Springer, Heidelberg (2008). doi:10.1007/978-3-540-78800-3_29

38. Qiang, W., Bliudze, S.: Verification of component-based systems via predicate abstraction and simultaneous set reduction. In: Ganty, P., Loreti, M. (eds.) TGC 2015. LNCS, vol. 9533, pp. 147–162. Springer, Heidelberg (2016). doi:10.1007/978-3-319-28766-9_10

# Formalization of Fault Trees in Higher-Order Logic: A Deep Embedding Approach

Waqar Ahmad[✉] and Osman Hasan

School of Electrical Engineering and Computer Science,
National University of Sciences and Technology, Islamabad, Pakistan
{waqar.ahmad,osman.hasan}@seecs.nust.edu.pk

**Abstract.** Fault Tree (FT) is a standard failure modeling technique that has been extensively used to predict reliability, availability and safety of many complex engineering systems. In order to facilitate the formal analysis of FT based analyses, a higher-order-logic formalization of FTs has been recently proposed. However, this formalization is quite limited in terms of handling large systems and transformation of FT models into their corresponding Reliability Block Diagram (RBD) structures, i.e., a frequently used transformation in reliability and availability analyses. In order to overcome these limitations, we present a deep embedding based formalization of FTs. In particular, the paper presents a formalization of AND, OR and NOT FT gates, which are in turn used to formalize other commonly used FT gates, i.e., NAND, NOR, XOR, Inhibit, Comparator and majority Voting, and the formal verification of their failure probability expressions. For illustration purposes, we present a formal failure analysis of a communication gateway software for the next generation air traffic management system.

**Keywords:** Higher-order logic · Fault Tree · Theorem proving

## 1 Introduction

Fault Tree (FT) is used as a standard failure modeling technique in various safety-critical domains, including nuclear power industry, civil aerospace and military systems. It mainly provides a graphical model for analyzing the conditions and factors causing an undesired top event, i.e., a critical event, which can cause the complete system failure upon its occurrence. The preceding nodes of the FT are represented by gates, like OR, AND and XOR, which are used to link two or more cause events of a fault in a prescribed manner. Using these FT gates, a FT model of a given system is constructed either on paper or by utilizing graphical editors provided by FT-based computer simulation tools, such as Relia-Soft [1] and ASENT [2]. In the paper-and-pencil proof methods, this

---

The original version of this chapter was revised. The spelling of the author Waqar Ahmad has been corrected. The erratum to this chapter is available at DOI: 10.1007/978-3-319-47677-3_21

© Springer International Publishing AG 2016
M. Fränzle et al. (Eds.): SETTA 2016, LNCS 9984, pp. 264–279, 2016.
DOI: 10.1007/978-3-319-47677-3_17

obtained FT model is then used for the identification of the Minimal Cut Set (MCS) of failure events that are associated with the components of the given system. This is followed by associating the failure random variables, i.e., exponential or Weibull, to these MCS failure events. The Probabilistic Inclusion-Exclusion (PIE) principle [3] is then used to evaluate the exact probability of failure of the overall system. On the other hand, the FT-based computer tools can be utilized to build a FT model by associating appropriate random variables with each component of the system. The reliability and the failure probability analysis of the complete system is then carried out by using computer arithmetic and numerical techniques on the generated samples from these random variables. However, both these methods cannot ascertain absolute correctness due to their inherent inaccuracy limitations. For instance, paper-and-pencil methods are prone to human errors, especially for large and complex systems, where a FT may consist of 50–130 levels of logic gates [4]. Manually manipulating such a large data makes it quite probable that some of MCS failure events may be overlooked, which would in turn lead to an erroneous design [4]. On the other hand, software tools can efficiently handle the analysis of large FTs but the computational requirements drastically increase as the size of the FT increases.

To overcome the above-mentioned limitations, a higher-order-logic formalization of some basic FT gates and their corresponding failure probability expressions [5] has been recently proposed. However, a major drawback of this formalization is the increase in complexity when analyzing FT of large and complex system. This formalization was primarily based on a shallow embedding approach, where the notion of each FT gate was explicitly defined on an event list and then its corresponding failure probability relationship was verified on the given failure event list. This approach makes the FT gate formalization noncompositional in nature, i.e., the basic FT gates, such as AND, OR and NOT, cannot be used to formalize other FT gates that are usually composed from these basic FT gates. Also, this work [5] utilizes the PIE principle to formally compute the exact failure probability of the given system, which limits its usability for complex system due to the involvement of large number of PIE terms. In the literature, several methods have been used to deal with this inherent complexity issue of the PIE principle. A tractable solution is to transform the given system FT to its equivalent Reliability Block Diagram (RBD) [6], which is also a well-known reliability modeling technique. This transformation considerably reduces the analysis complexity due to the fact that RBD offers closed form expressions compared to a FT, which requires unfolding of all the PIE terms.

In order to overcome the above-mentioned scalability issues of the existing formalization of FT gates [5] and thus broaden the scope of formal FT analysis, we propose a deep embedding approach to formalize the commonly used FT gates, such as AND, OR and NOT. This proposed formalization approach is compositional in nature and can be easily extended to formalize other FT gates, such as NAND, NOR, XOR, Inhibit, Comparator and majority Voting. It also enables us to transform the given system FT model to its equivalent RBD model, without any loss of valuable information. The RBD model can then be formally analyzed using our recently proposed formal reasoning support for RBDs [7].

To illustrate the practical effectiveness of our proposed approach, we present a formal failure analysis of a Next Generation (NextGen) Air Traffic Management (ATM) gateway system, which is primarily used to enhance the safety and reliability of air transportation, to improve efficiency in the air transportation and to reduce aviation impact on the environment. The FT of the NextGen ATM gateway, which consists of more than 40 basic failure events including software, hardware, database update and transmission system is divided into four levels. The formally verified failure probability expressions of individual levels are then used to reason about the failure probability of the overall NextGen system. In addition, we also provide some automated reasoning support for the FT based failure analysis. This automation allows us to automatically simplify the failure expression of the NextGen system from the given values of the failure rates.

## 2   Related Work

The COMPASS tool-set [8] supports the dynamic FT analysis specifically for aerospace systems using the NuSMV and MRMC model checkers. The Interval Temporal Logic (ITS), i.e., a temporal logic that supports first-order logic, has been used, along with the Karlsruhe Interactive Verifier (KIV), for formal FT analysis of a rail-road crossing [9]. A deductive method for FT construction, in contrast to the intuitive approach followed in [9], by using the Observational Transition Systems (OTS), is presented in [10]. The formal analysis of this FT is then carried out using CafeOBJ [11], which is a formal specification language with interactive verification support. However, the scope of these tools is somewhat limited in terms of handling larger systems, due to the inherent state-space explosion problem of model checking. Moreover, either some of these approaches [9,10] do not cater for probabilities or if they do cater for them then the computation of probabilities in these methods [8] involves numerical techniques, which compromises the accuracy of the results.

Leveraging upon the high expressiveness of higher-order logic and the inherent soundness of theorem proving, Mhamdi's formalized probability theory [12] has been recently used for the formalization of RBDs [7], including series [13], parallel [14], parallel-series [14] and series-parallel [15]. These formalizations have been used for the reliability analysis of many applications including simple oil and gas pipelines with serial components [13], wireless sensor network protocols [14] and logistic supply chains [14]. Similarly, Mhamdi's probability theory have also been used for the formalization of commonly used FT gates, such as AND, OR, NAND, NOR, XOR and NOT, and the PIE principle [5]. In addition, the above-mentioned RBD and FT formalizations have been recently utilized for availability analysis [16]. In this paper, we have formalized the FT gates using a deep embedding approach to facilitate the analysis of larger FTs. Besides the existing formalization of FT gates [5], this paper also provides the formalization of inhibit, 2-bit comparator and Majority voting FT gates. Moreover, we have combined our existing formalizations of RBDs [13–15] to make the formal FT based analysis more scalable.

# 3    Probability Theory and Fault Trees in HOL

Mathematically, a measure space is defined as a triple $(\Omega, \Sigma, \mu)$, where $\Omega$ is a set, called the sample space, $\Sigma$ represents a $\sigma$-algebra of subsets of $\Omega$, where the subsets are usually referred to as measurable sets, and $\mu$ is a measure with domain $\Sigma$. A probability space is a measure space $(\Omega, \Sigma, Pr)$, such that the measure, referred to as the probability and denoted by $Pr$, of the sample space is 1. In the HOL4 formalization of probability theory [12], given a probability space $p$, the functions space, subsets and prob return the corresponding $\Omega$, $\Sigma$ and $Pr$, respectively. This formalization also includes the formal verification of some of the most widely used probability axioms, which play a pivotal role in formal reasoning about reliability properties.

A random variable is a measurable function between a probability space and a measurable space. The measurable functions belong to a special class of functions, which preserves the property that the inverse image of each measurable set is also measurable. A measurable space refers to a pair $(S, \mathcal{A})$, where $S$ denotes a set and $\mathcal{A}$ represents a nonempty collection of sub-sets of $S$. Now, if $S$ is a set with finite elements, then the corresponding random variable is termed as a discrete random variable otherwise it is called a continuous one.

The cumulative distribution function (CDF) is defined as the probability of the event where a random variable $X$ has a value less than or equal to some value $t$, i.e., $Pr(X \leq t)$. This definition characterizes the distribution of both discrete and continuous random variables and has been formalized [13] as follows:

$\vdash \forall$ p X t. CDF p X t = distribution p X $\{$y $\mid$ y $\leq$ Normal t$\}$

The function Normal takes a *real* number as its input and converts it to its corresponding value in the *extended-real* data-type, i.e., it is the *real* data-type with the inclusion of positive and negative infinity. The function distribution takes three parameters: a probability space $p : (\alpha \rightarrow bool)\#((\alpha \rightarrow bool) \rightarrow bool)\#((\alpha \rightarrow bool) \rightarrow real)$, a random variable $X : (\alpha \rightarrow extreal)$ and a set of *extended-real* numbers and returns the probability of the given random variable $X$ acquiring all the values of the given set in probability space $p$.

The unreliability or the probability of failure $F(t)$ is defined as the probability that a system or component will fail by the time $t$. It can be described in terms of CDF, known as the failure distribution function, if the random variable $X$ represent a time-to-failure of the component. This time-to-failure random variable $X$ usually exhibits the exponential or Weibull distribution.

The notion of mutual independence of $n$ random variables is a major requirement for reasoning about the failure analysis of most of the FT gates. According to this notion, a list of $n$ events are mutual independent if and only if for each set of $k$ events, such that $(1 \leq k \leq n)$, we have:

$$Pr(\bigcap_{i=1}^{k} A_i) = \prod_{i=1}^{k} Pr(A_i) \tag{1}$$

It is important to note that mutual independence is a much stronger property compared to pairwise independence [3], which ensures independence between two events only. On the other hand, mutual independence makes sure that any subset of events are independent with each other. Also, we can verify many interesting properties of independence using the mutual independence property. For instance, given a list of mutually independent events, say $L$, we can verify that an element $h \in L$ is independent with the list $L - [h]$ representing the list $L$ without element $h$.

The mutual independence concept is formalized in HOL4 as follows [13]:

```
⊢ ∀ p (L:α → bool). mutual_indep p L = ∀ L1 (n:num). PERM L L1 ∧
   1 ≤ n ∧ n ≤ LENGTH L ⇒
   prob p (inter_list p (TAKE n L1)) = list_prod (list_prob p (TAKE n L1))
```

The function `mutual_indep` accepts a list of events $L$ and probability space $p$ and returns $True$ if the events in the given list are mutually independent in the probability space $p$. The predicate `PERM` ensures that its two lists as its arguments form a permutation of one another. The function `LENGTH` returns the length of the given list. The function `TAKE` returns the first $n$ elements of its argument list as a list. The function `inter_list` performs the intersection of all the sets in its argument list of sets and returns the probability space if the given list of sets is empty. The function `list_prob` takes a list of events and returns a list of probabilities associated with the events in the given list of events in the given probability space. Finally, the function `list_prod` recursively multiplies all the elements in the given list of real numbers. Using these functions, the function `mutual_indep` models the mutual independence condition such that for $n$ events taken from any permutation of the given list $L$, Eq. (1) holds.

## 3.1 Formalization of Fault Tree Gates

The proposed formalization is primarily based on defining a new polymorphic datatype *gate* that encodes the notion of AND, OR and NOT FT gates. Then a semantic function is defined on that *gate* datatype yielding an event for the corresponding FT gate. This semantic function allows us to verify the generic failure probability expressions of the FT gates by utilizing the underlying probability theory within the sound core of the HOL4 theorem prover. Such a deep embedding considerably simplifies the FT gate modeling approach, compared to our previous work [5] (shallow embedding), and also enables us to develop a framework that can deal with arbitrary levels of FTs, which can be used to cater for a wide variety of real-world failure analysis problems.

We start the formalization process by type abbreviating the notion of event, which is essentially a set of observations with type `'a->bool` as follows:

```
type_abbrev ("event", ''':'a ->bool'')
```

We then define a recursive datatype *gate* in the HOL4 system as follows:

```
Hol_datatype 'gate = AND of gate list | OR of gate list | NOT of gate |
                     atomic of 'a event'
```

The type constructors AND and OR recursively function on *gate*-typed lists and the type constructor NOT operates on *gate*-type variable. The type constructor atomic is basically a typecasting operator between *event* and *gate*-typed variables. These type constructors allow us to encode the notion of all the basic FT gates.

We define a semantic function $FTree : \alpha \ event \ \# \ \alpha \ event \ event \ \# \ (\alpha \ event \rightarrow real) \rightarrow \alpha \ gate \rightarrow \alpha \ event$ over the above-defined *gate* datatype that can yield the corresponding event from the given FT gate as follows:

**Definition 1.** ⊢ (∀ p. FTree p (AND [])) = p_space p) ∧
(∀ xs x p. FTree p (AND (x::xs)) = FTree p x ∩ FTree p (AND xs)) ∧
(∀ p. FTree p (OR []) = {}) ∧
(∀ xs x p. FTree p (OR (x::xs)) = FTree p x ∪ FTree p (OR xs)) ∧
(∀ p a. FTree p (NOT a) = p_space p DIFF FTree p a) ∧
(∀ p a. FTree p (atomic a) = a)

The above function decodes the semantic embedding of a FT by yielding a corresponding failure event, which can then be used to determine the failure probability of a given FT. The function FTree takes a list of type *gate*, identified by a type constructor AND, and returns the whole probability space if the given list is empty and otherwise returns the intersection of the events that are obtained after applying the function FTree on each element of the given list in order to model the AND FT gate behaviour. Similarly, to model the behaviour of the OR FT gate, the function FTree operates on a list of datatype *gate*, encoded by a type constructor OR. It then returns the union of the events after applying the function FTree on each element of the given list or an empty set if the given list is empty. The function FTree takes a type constructor NOT and returns the complement of the failure event obtained from the function FTree. The function FTree returns the failure event using the type constructor atomic.

If the occurrence of the failure event at the output is caused by the occurrence of all the input failure events then this kind of behavior can be modeled by using the AND FT gate. The failure probability expression of the AND FT gate can be expressed mathematically as follows:

$$F_{AND\_gate}(t) = Pr(\bigcap_{i=2}^{N} A_i(t)) = \prod_{i=2}^{N} F_i(t) \tag{2}$$

Using Definition 1, we can verify the above equation in HOL4 as follows:

**Theorem 1.** ⊢ ∀ p L. prob_space p ∧
(∀x'. MEM x' L ⇒ x' ∈ events p) ∧ 2 ≤ LENGTH L ∧
mutual_indep p L ⇒
  (prob p (FTree p (AND (gate_list L))) = list_prod (list_prob p L))

The first two assumptions, in Theorem 1, ensures that *p* is a valid probability space and each element of a given event list *L* must be in event space *p* based on the probability theory in HOL4 [12]. The function MEM finds an element in a given list and returns false, if a match does not occur. The next two assumptions guarantee that the list of events $L$, representing the failure probability of individual components, must have at least two events and the failure events are mutually independent. The conclusion of the theorem represents Eq. (2). The function gate_list generates a list of type *gate* by mapping the function atomic to each element of the given event list $L$ to make it consistent with the assumptions of Theorem 1. It can be formalized in HOL4 as: $\forall$ L. gate_list L = MAP ($\lambda$a. atomic a) L

The proof of Theorem 1 is primarily based on a mutual independence property and some fundamental axioms of probability theory.

In the OR FT gate, the occurrence of the output failure event depends upon the occurrence of any one of its input failure event. Mathematically, the failure probability of an OR FT gate can be expressed as:

$$F_{OR\_gate}(t) = Pr(\bigcup_{i=2}^{N} A_i(t)) = 1 - \prod_{i=2}^{N}(1 - F_i(t)) \tag{3}$$

By following the approach, used in Theorem 1, we can formally verify the failure probability expression OR FT gate, given in Eq. (3), in HOL4:

**Theorem 2.** $\vdash \forall$ p L. prob_space p $\wedge$ 2 $\leq$ LENGTH L $\wedge$
($\forall$x'. MEM x' L $\Rightarrow$ x' $\in$ events p) $\wedge$ mutual_indep p L $\Rightarrow$
(prob p (FTree p (OR (gate_list L))) =
1 - list_prod (one_minus_list (list_prob p L)))

The above theorem is verified under the same assumptions as Theorem 1. The conclusion of the theorem represents Eq. (3) where, the function one_minus_list accepts a list of *real* numbers $[x1, x2, x3, \cdots, xn]$ and returns the list of *real* numbers such that each element of this list is 1 minus the corresponding element of the given list, i.e., $[1 - x1, 1 - x2, 1 - x3, \cdots, 1 - xn]$.

The NOT FT gate can be used in conjunction with the AND and OR FT gates to formalize other FT gates. The formalization of these gates is given in Table 1. The NAND FT gate, represented by the function NAND_FT_gate in Table 1, models the behavior of the occurrence of an output failure event when at least one of the failure events at its input does not occur. This type of gate is used in FTs when the non-occurrence of the failure event in conjunction with the other failure events causes the top failure event to occur. This behavior can be expressed as the intersection of complementary and normal events, where the complementary events model the non-occurring failure events and the normal events model the occurring failure events. The output failure event occurs in the 2-input XOR FT gate if only one, and not both, of its input failure events occur. The inhibit FT gate produces an output failure event only if the conditional event occurs at the same time when the input failure event occurs. The HOL4 function inhibit_FT_gate, given in Table 1, models the behavior of a 2-input inhibit FT gate by composing the type constructors *AND*, *OR* and *NOT*. In the comparator FT gate, the output failure event occurs if all the failure events at its input occur or if all of the them do not occur. In the majority voting gate, the output failure event occurs if at least $m$ out of $n$ input failure events occurs. This behaviour can be modeled by utilizing the concept of binomial trials, which are used to find the chances of at least $m$ success in $n$ trials. The function major_voting_FT_gate accepts a probability space $p$, a binomial random variable $X$ and two variables, $m$ and $n$, which represent the number of successes and total number of trials, respectively. It then returns the union of the corresponding events that are associated with the binomial random variable $X$, which takes values from the set $\{x \mid k \leq x \wedge x < \text{SUC } n\}$. The

**Table 1.** HOL4 Formalization of fault tree gates

| FT Gates | Formalization |
|---|---|
| NAND | $\vdash \forall$ p L1 L2. NAND_FT_gate p L1 L2 =<br>FTree p (AND (gate_list (compl_list p L1 ++ L2))) |
| NOR | $\vdash \forall$ p L. NOR_FT_gate p L = FTree p (NOT (OR (gate_list L))) |
| XOR | $\vdash \forall$ p A B. XOR_FT_gate p A B =<br>FTree p (OR [AND [NOT A; B]; AND [A; NOT B]]) |
| Inhibit | $\vdash \forall$ p A B C. inhibit_FT_gate p A B C =<br>FTree p (AND [OR [A; B]; NOT C]]) |
| Comp | $\vdash \forall$ p A B. comp_FT_gate p A B =<br>FTree p (OR [AND [A; B]; NOR_FT_gate p [A; B]]) |
| m | $\vdash \forall$ p X m n. major_voting_FT_gate p X m n =<br>BIGUNION (IMAGE ($\lambda$x. PREIMAGE X {Normal (&x)} $\cap$ p_space p)<br>{x \| k $\leq$ x $\wedge$ x < SUC n}) |

function IMAGE takes a function $f$ and an arbitrary domain set and returns a range set by applying the function $f$ to all the elements of the given domain set. The function BIGUNION returns the union of all the element of given set of sets.

The verification of the corresponding failure probability expressions, of the above-mentioned FT gates, is presented in Table 2. These expressions are verified under the same assumptions as the ones used for Theorems 1 and 2. However, some additional provisos are required for the verification of majority voting gate as follows: (i) prob_space ensures that $p$ is a valid probability space; (ii) m $\leq$ n makes sure that the number of successes of trails $m$ must be less than or equal to the total number of trials $n$; (iii) ($\lambda$x. PREIMAGE X Normal(&x) $\cap$ p_space p) $\in$ ((count (SUC n)) $\rightarrow$ events p) ensures that all the corresponding events that are associated with the binomial random variable $X$ are drawn from the events space $p$; and (iv) ($\forall$x. distribution p X {Normal (&x)} = (&binomial n x)*(F pow x)*(1 - F) pow (n-x)) guarantees that the random variable $X$ is exhibiting the binomial distribution.

## 3.2 Formalization of Probabilistic Inclusion-Exclusion Principle

In FT analysis, firstly all the basic failure events are identified that can cause the occurrence of the system top failure event. These failure events are then combined to model the overall fault behavior of the given system by using the fault gates. These combinations of basic failure events, called cut sets, are then reduced to minimal cut sets (MCS) by using some set-theory rules, such as idempotent, associative and commutative. Then, the Probabilistic Inclusion Exclusion (PIE) principle is used to evaluate the overall failure probability of the given system based on the MCS events. According to the PIE principle, if $A_i$ represents the $i^{th}$ basic failure event or a combination of failure events then the overall failure probability of the given system can be expressed as follows:

$$\mathbb{P}(\bigcup_{i=1}^{n} A_i) = \sum_{t\neq\{\},t\subseteq\{1,2,\dots,n\}} (-1)^{|t|+1}\mathbb{P}(\bigcap_{j\in t} A_j) \tag{4}$$

The above equation has been formally verified in HOL as follows [5]:

**Table 2.** Probability of failures of fault tree gates

| Mathmatical Expressions | Theorem's Conclusion |
|---|---|
| $F_{NAND}(t) = Pr(\bigcap_{i=2}^{k} \overline{A}_i(t) \cap \bigcap_{j=k}^{N} A_i(t))$ $= \prod_{i=2}^{k}(1 - F_i(t)) * \prod_{j=k}^{N}(F_j(t))$ | $\vdash \forall$ p L1 L2. (prob p (NAND_FT_gate p L1 L2) = list_prod ((list_prob p (compl_list p L1))) * list_prod (list_prob p L2)) |
| $F_{NOR}(t) = 1 - F_{OR}(t) = \prod_{i=2}^{N}(1 - F_i(t))$ | $\vdash \forall$ p L. (prob p (NOR_FT_gate p L) = list_prod (one_minus_list (list_prob p L))) |
| $F_{XOR}(t) = Pr(\overline{A}(t)B(t) \cup A(t)\overline{B}(t))$ $= (1 - F_A(t))F_B(t)+$ $F_A(t)(1 - F_B(t))$ | $\vdash \forall$p A B. prob_space p $\wedge$ A $\in$ events p $\wedge$ B $\in$ events p (prob p (XOR_FT_gate p (atomic A) (atomic B) = (1- prob p A)*prob p B + prob p A*(1 - prob p B) |
| $F_{inhibit}(t) = Pr((A(t) \cup B(t)) \cap \overline{C(t)})$ $= (1 - (1 - F_A(t))*$ $(1 - F_B(t))) * (1 - F_C(t))$ | $\vdash \forall$p A B C. (prob p (inhibit_FT_gate p (atomic A) (atomic B) (atomic C) = (1 - (1 - prob p A) * (1 - prob p B))*(1 - prob p C) |
| $F_{comp}(t) = Pr((A(t) \cap B(t)) \cup \overline{(A(t) \cup B(t))})$ $= (1 - (1 - F_A(t)F_B(t))*$ $(1 - (1 - F_A(t)) * (1 - F_B(t)))$ | $\vdash \forall$p A B C. (prob p (comp_FT_gate p (atomic A) (atomic B) = (1 - (1 - prob p A * prob p B)* (1 - (1 - prob p A)*(1- prob p B)) |
| $F_{m\|n}(t) = Pr(\bigcup_{i=k}^{n}\{exactly\ i\ components\ are$ $functioning\ properly\})$ $= \sum_{i=m}^{n}(\binom{n}{m}F^i(1 - F)^{n-1})$ | $\vdash \forall$p n k X F (prob p (major_voting_FT_gate p X m n) = sum (m, SUC n - m) ($\lambda$x. (&binomial n x)*(F pow x)* (1- F) pow (n-x))) |

**Theorem 3.** $\vdash \forall$ p L. prob_space p $\wedge$ ($\forall$ x. MEM x L $\Rightarrow$ x $\in$ events p) $\Rightarrow$
(prob p (union_list L) =
    sum_set {t | t $\subseteq$ set L $\wedge$ t $\neq$ {} }
        ($\lambda$t. -1 pow (CARD t + 1) * prob p (BIGINTER t)))

The assumptions of the above theorem are the same as the ones used in Theorem 1. The function sum_set takes an arbitrary set $s$ with element of type $\alpha$ and a real-valued function $f$ and recursively sums the return values of the function $f$, when applied on each element of the given set $s$. In the above theorem, the set $s$ is represented by the term $\{x | C(x)\}$ that contains all the values of $x$, which satisfy condition $C$. Whereas, the $\lambda$ abstraction function ($\lambda$t. -1 pow (CARD t + 1) * prob p (BIGINTER t)) models $(-1)^{|t|+1}\mathbb{P}(\bigcap_{j\in t} A_j)$, such that the functions CARD and BIGINTER return the number of elements and the intersection of all the elements of the given set, respectively.

## 3.3   Formalization of Reliability Block Diagrams

Transformation of a system FT to its equivalent reliability block diagram (RBD) has been proposed as a viable solution to reduce the complexity associated with finding the failure probability of large systems [17]. The proposed deep embedding based formalization of FT gates allows the establishment of this link and thus we have used the existing formalization of RBDs [7] to make the formal analysis of FTs more scalable. In this paper, we only describe the formalization of the parallel-series RBD configuration

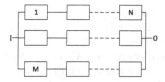

**Fig. 1.** Parallel-series reliability block diagrams

because it is required to conduct the formal failure analysis of ASN gateway system, described in the next section.

In a parallel-series RBD configuration, as shown in Fig. 1, the reserved *subsystems* are connected serially and it can be considered as the nested form of series RBD in a parallel RBD configuration. If $A_{ij}(t)$ is the event corresponding to the reliability of the $j^{th}$ component connected in a $i^{th}$ subsystem at time $t$, then parallel-series RBD configuration can be expressed as:

$$R_{parallel-series}(t) = Pr(\bigcup_{i=1}^{M} \bigcap_{j=1}^{N} A_{ij}(t)) = 1 - \prod_{i=1}^{M}(1 - \prod_{j=1}^{N}(R_{ij}(t))) \tag{5}$$

The HOL4 formalization of the above equation is as follows [7]:

**Theorem 4.** ⊢ ∀ p L. prob_space p ∧ (∀z. MEM z L ⇒ ⌐NULL z) ∧
(∀x'. MEM x' (FLAT L) ⇒ x' ∈ events p) ∧
mutual_indep p (FLAT L) ⇒
(prob p (rbd_struct p ((parallel of (λa. series (rbd_list a))) L)) =
(1 - list_prod (one_minus_list) of (λa. list_prod (list_prob p a))) L)

where the function **rbd_struct** is defined on a recursive datatype *rbd* and can take any combination of type constructors **series** and **parallel**. It then yields the corresponding event of the given RBD configuration constituted by these type-constructors. The function **rbd_list** serves similar functionality as that of the function **gate_list**. The assumptions are quite similar to the ones used for Theorems 1 and 2. The conclusion models Eq. (5) and the infixr function **of** connects two *rbd* type-constructors by using the HOL4 **MAP** function.

# 4    Formalization of the NextGen ASN Gateway System

NextGen is supported by the nation-wide Aviation Simulation Network (ASN), which is an environment including simulated and human-in-the-loop (HIL) real-life components, e.g., pilots and air traffic controllers. The Real Time Distributed Simulation (RTDS) application suite [18] is used to facilitate the ASN by providing low and medium fidelity en-route simulation capabilities. An ASN gateway software system acts as an intermediary between RTDS and ASN by providing logic for data translation, two-way communication and transfer messages among them. The overall NextGen ASN gateway FT can be viewed as a four level FT [19]. The first or top level of the ASN gateway FT models an aviation accident caused by the lack of appropriate control, equipment, internal and external malfunctions. The internal failure event opens up to a second level of the ASN gateway FT, which comprises of failures related to the flight function mishap and transmissions. The flight mishap failure is caused by the failure of the Auto Pilot (AP) or Flight Director (FD) along with the failure not mitigated in time (FF1). The Transmission failure event captures the failure events due to data/message not correctly transmitted (A), failure to display (NotShown), and not performing transmission in a timely manner (RT). The third level of the ASN gateway FT is composed of several sub-FTs, given in Table 3, representing the RT and failure event A. The RT failure event occurs if the delay is too long for the transmission to meet its deadline (Time) and a latency problem occurs related to either the application (AL), serialization (SL), propagation delay (PD) or any other relevant sources. Similarly, the failure event A represents a failure to correctly transmit a message and consists of two events. i.e., B1: failure to transfer a message from ASN to RTDS and B2: failure to transfer a message from RTDS to ASN of the communication link. The FT of the events B1 and B2 are given at the fourth level of the ASN gateway FT [19]. The overall ASN gateway FT consists of 47 basic failure events that are related to messages transmission failures, propagation delays, software and hardware equipment failures, database update failures and human mistakes.

### 4.1 Formal Fault Tree Models for ASN Gateway System

The formal definitions of FT gates [5] along with Definition 1 can be utilized to formally represent the FT of the ASN gateway in terms of its failure events. We systematically present the formalization of the ASN gateway FT by starting from the fourth level, i.e., the formalization of B1 sub-FT:

**Definition 2.** ∀p t D1 D4 E1 E2 E3 E4 E5 E6 E7 E8 E9 E10 E21.
B1_FT p t D1 D4 E1 E2 E3 E4 E5 E6 E7 E8 E9 E10 E21 =
(OR [OR [atomic (fail_event p D1 t);
         AND [OR (gate_list (fail_event_list p [E1; E2] t));
         atomic (fail_event p E21 t)];
         OR (gate_list (fail_event_list p [E3; E4; E5] t))];
    OR [atomic (fail_event p D4 t);
         AND [OR (gate_list (fail_event_list p [E6; E7] t));
         atomic (fail_event p E21 t)];
         OR (gate_list (fail_event_list p [E8; E9; E10] t))]])

Where the random variables $D1$, $D4$, $E1 - E10$ and $E21$ model the time-to-failure of the communication process ASN to RTDS. The diagram of B1 FT is similar to B2 FT, which can be seen in Table 3. Additionally, the cut-set failure events in the above definition is already minimal, i.e., there are no combination of redundant failure events to be removed [19]. Therefore, the cut-sets and MCS for B1 sub-FT, in this case, are equivalent.

Similarly, other sub-FTs, such as B2-FT, A-FT, RT-FT and Internal-FT, which are at the fourth, third and second level of the ASN gateway FT can be formalized in HOL4 as shown in Table 3. It is important to note that the formal definition of the top level or first level FT, in Table 3, builds upon the formal definitions of all the other sub-FTs and models the complete ASN gateway FT.

We consider that the random variables, associated with the failure events of the ASN gateway FT, exhibit the exponential distribution:

**Definition 3.** ⊢ ∀ p X l. exp_dist p X l =
  ∀ x. (CDF p X x = if 0 ≤ x then 1 - exp (-l * x) else 0)

**Table 3.** ASN Gateway FT levels with their HOL formalizations

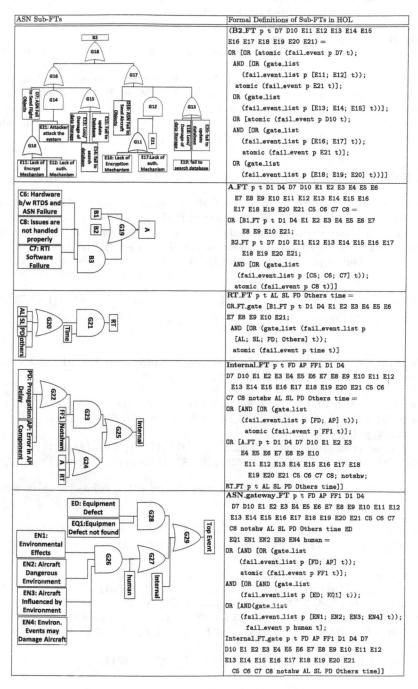

| ASN Sub-FTs | Formal Definitions of Sub-FTs in HOL |
|---|---|
| | (B2_FT p t D7 D10 E11 E12 E13 E14 E15<br>E16 E17 E18 E19 E20 E21) =<br>OR [OR [atomic (fail_event p D7 t);<br>  AND [OR (gate_list<br>    (fail_event_list p [E11; E12] t));<br>  atomic (fail_event p E21 t)];<br>  OR (gate_list<br>    (fail_event_list p [E13; E14; E15] t))];<br>OR [atomic (fail_event p D10 t);<br>  AND [OR (gate_list<br>    (fail_event_list p [E16; E17] t));<br>  atomic (fail_event p E21 t)];<br>  OR (gate_list<br>    (fail_event_list p [E18; E19; E20] t))]] |
| | A_FT p t D1 D4 D7 D10 E1 E2 E3 E4 E5 E6<br>E7 E8 E9 E10 E11 E12 E13 E14 E15 E16<br>E17 E18 E19 E20 E21 C5 C6 C7 C8 =<br>OR [B1_FT p t D1 D4 E1 E2 E3 E4 E5 E6 E7<br>  E8 E9 E10 E21;<br>B2_FT p t D7 D10 E11 E12 E13 E14 E15 E16 E17<br>  E18 E19 E20 E21;<br>AND [OR (gate_list<br>  (fail_event_list p [C5; C6; C7] t));<br>  atomic (fail_event p C8 t)]] |
| | RT_FT p t AL SL PD Others time =<br>OR_FT_gate [B1_FT p t D1 D4 E1 E2 E3 E4 E5 E6<br>E7 E8 E9 E10 E21;<br>AND [OR (gate_list (fail_event_list p<br>[AL; SL; PD; Others] t));<br>atomic (fail_event p time t)]] |
| | Internal_FT p t FD AP FF1 D1 D4<br>D7 D10 E1 E2 E3 E4 E5 E6 E7 E8 E9 E10 E11 E12<br>E13 E14 E15 E16 E17 E18 E19 E20 E21 C5 C6<br>C7 C8 notshw AL SL PD Others time =<br>OR [AND [OR (gate_list<br>    (fail_event_list p [FD; AP] t));<br>    atomic (fail_event p FF1 t)];<br>OR [A_FT p t D1 D4 D7 D10 E1 E2 E3<br>    E4 E5 E6 E7 E8 E9 E10<br>    E11 E12 E13 E14 E15 E16 E17 E18<br>    E19 E20 E21 C5 C6 C7 C8; notshw;<br>RT_FT p t AL SL PD Others time]] |
| | ASN_gateway_FT p t FD AP FF1 D1 D4<br>  D7 D10 E1 E2 E3 E4 E5 E6 E7 E8 E9 E10 E11 E12<br>  E13 E14 E15 E16 E17 E18 E19 E20 E21 C5 C6 C7<br>  C8 notshw AL SL PD Others time ED<br>  EQ1 EN1 EN2 EN3 EN4 human =<br>OR [AND [OR (gate_list<br>    (fail_event_list p [FD; AP] t));<br>    atomic (fail_event p FF1 t)];<br>AND [OR [AND (gate_list<br>    (fail_event_list p [ED; EQ1] t));<br>OR [AND(gate_list<br>    (fail_event_list p [EN1; EN2; EN3; EN4] t));<br>    fail_event p human t];<br>Internal_FT_gate p t FD AP FF1 D1 D4 D7<br>D10 E1 E2 E3 E4 E5 E6 E7 E8 E9 E10 E11 E12<br>E13 E14 E15 E16 E17 E18 E19 E20 E21<br>  C5 C6 C7 C8 notshw AL SL PD Others time]] |

The function exp_dist guarantees that the CDF of the random variable $X$ is that of an exponential random variable with a failure rate $l$ in a probability space $p$. We classify a list of exponentially distributed random variables as follows:

**Definition 4.** ⊢ ∀p L. list_exp p [] L = T ∧
∀p h t L. list_exp p (h::t) L = exp_dist p (HD L) h ∧ list_exp p t (TL L)

The function list_exp accepts a list of failure rates, a list of random variables $L$ and a probability space $p$. It guarantees that all elements of the list $L$ are exponentially distributed with the corresponding failure rates, given in the other list, within the probability space $p$. For this purpose, it utilizes the list functions HD and TL, which return the *head* and *tail* of a list, respectively.

## 4.2   Failure Assessment of NextGen ASN Gateway System

We now present the formal verification of all the sub-FTs, such as B1-FT, B2-FT, A-FT, RT-FT and Internal-FT. The formally verified results of these sub-FTs are then used to reason about the failure probability of overall ASN gateway communication system. Using the closed form expression of parallel-series RBD configuration, given in Eq. (5), the failure probability of the B1-FT can be expressed mathematically as follows:

$$F_{B1}(t) = (1 - e^{-(c_1+c_2+c_3+c_4)t}) * (1 - (1 - e^{-C_{E1}t})(1 - e^{-C_{E21}t}))(1 - (1 - e^{-C_{E2}t})$$
$$(1 - e^{-C_{E21}t}))(1 - (1 - e^{-C_{E6}t})(1 - e^{-C_{E21}t}))(1 - (1 - e^{-C_{E7}t})(1 - e^{-C_{E21}t}))$$

$$(6)$$

To verify Eq. (6), we first verify a lemma that transforms the B1 sub-FT to its equivalent parallel-series RBD model as follow:

**Lemma 1.** ⊢ ∀ p t D1 D4 E1 E2 E3 E4 E5 E6 E7 E8 E9 E10 E21.
FTree p (B1_FT p t D1 D4 E1 E2 E3 E4 E5 E6 E7 E8 E9 E10 E21) =
(rbd_struct p ((parallel of
(λa. series (rbd_list (fail_event_list a)))) [[D1];[D4];[E1;E21];
 [E2;E21]; [E3];[E4];[E5];[E6;E21];[E7;E21];[E8];[E9];[E10]]))

Now, using the formal definition of B1-FT and Lemma 1, the failure probability of B1 sub-FT can be verified in HOL4 as follows:

**Theorem 5.** ⊢ ∀p t D1 D4 E1 E2 E3 E4 E5 E6 E7 E8 E9 E10 E21 C_E1 C_E2
C_E6 C_E7 C_D1 C_D4 C_E3 C_E4 C_E5 C_E8 C_E9 C_E10 C_21.
time_positive t ∧ prob_space p ∧
in_events p (fail_event_list p [D1;D4;E1;···;E10;E21] t) ∧
mutual_indep p (fail_event_list p [D1;D4;E1;···;E10;E21] t) ∧
list_exp p [C_D1;C_D4;C_E1;···;C_E10;C_E21] [D1;D4;E1;···;E10;E21] ⇒
(prob p (B1_FT p t D1 D4 E1 E2 E3 E4 E5 E6 E7 E8 E9 E10 E21) =
  1 - exp(-(t * list_sum [C_D1;C_D4;C_E3;C_E4;C_E5;C_E8;C_E9;C_E10])) *
  list_prod(one_minus_exp_prod t
   [[C_E1;C_E21];[C_E2;C_E21];[C_E6;C_E21];[C_E7;C_E21]]))

The function **exp** represents the exponential function. The function **list_sum** is used to sum all the elements of the given list of failure rates, the function **one_minus_exp** accepts a list of failure rates and returns a one minus list of exponentials and the function **one_minus_exp_prod** accepts a two dimensional list of failure rates and returns a list with one minus product of one minus exponentials of every sub-list. For example, **one_minus_exp_prod**$[[c1; c2; c3]; [c4; c5]; [c6; c7; c8]]$ $x = [1 - ((1 - e^{-(c1)x}) * (1 - e^{-(c2)x}) * (1 - e^{-(c3)x})); (1 - (1 - e^{-(c4)x}) * (1 - e^{-(c5)x})); (1 - (1 - e^{-(c6)x}) * (1 - e^{-(c7)x}) * (1 - e^{-(c8)x}))]$. The first assumption ensures that the variable t models time $t$ as it can acquire positive integer values only. The next assumption ensures that p is a valid probability space based on the probability theory in HOL [12]. The next two assumptions ensure that the events corresponding to the failures modeled by the random variables D1, D2, E1 to E10 and E21 are valid events from the probability space p and they are mutually independent. Finally, the last assumption characterizes the random variables D1, D2, E1 to E10 and E21, as exponential random variables with failure rates C_D1, C_D2, C_E1 to C_E10 and C_E21, respectively. The conclusion of Theorem 5 represents the failure probability of the communication process between ASN to RTDS in terms of the failure rates of the components involved during the communication process. The proof of Theorem 5 is primarily based on Theorem 3 and some fundamental facts and axioms of probability.

Similarly, the failure probabilities of other sub-FTs, i.e., B1-FT, B2-FT, A-FT, RT-FT and Internal-FT, are verified in HOL4 [20]. These theorems are verified under the same assumptions as the one used in Theorem 5.

Now, using the formal definitions of ASN gateway sub-FTs, given in Table 3, and their verified failure probability results [20], we formally verified the failure probability of the complete ASN gateway system as follows:

**Theorem 6.** ⊢ (prob p (ASN_gateway_FT p t FD AP FF1 D1 D4 D7 D10 E1 ⋯ E21 C5 C6 C7 C8 notshw AL SL PD Others time ED EQ. 1 EN1 ⋯ EN4 human) =
1 - (list_prod(one_minus_exp_prod t [[C_ED;C_EQ1];
      [C_EN1;C_EN2;C_EN3;C_EN4];[C_E6;C_E21]])) *
exp (-(t*C_human)) * exp -(t*C_notshw) *
1 - (list_prod(one_minus_exp_prod t [[C_FD;C_FF1];[C_AP;C_FF1]]) *
1 - (1 - exp(-(t*list_sum  [C_D1;C_D4;C_E3;C_E4;C_E5;C_E8;C_E9;C_E10])) *
list_prod(one_minus_exp_prod t [[C_E1;C_E21];[C_E2;C_E21];
    [C_E6;C_E21];[C_E7;C_E21]])))*
1 - exp(-(t*list_sum[C_D7;C_D10; C_E13;C_E14;C_E15;C_E18;C_E19;C_E20])) *
list_prod(one_minus_exp_prod t
    [[C_E11;C_E21];[C_E12;C_E21];[C_E16;C_E21];[C_E17;C_E21]])) *
list_prod(one_minus_exp_prod t [[C_C5;C_C8];
    [C_C6;C_C8];[C_C7;C_C8]])))))*
list_prod(one_minus_exp_prod t [[C_AL;C_time];
    [C_SL;C_time];[C_PD;C_time]; [C_other;C_time]]))))

The assumptions of the above theorem are similar to the ones used in Theorem 5 and its proof is based on Theorem 3 and some basic arithmetic lemmas and probability theory axioms. The proof of Theorems 5 and 6 and the formalization of sub-FTs, presented in Table 3, with their corresponding probability of failure took more than 2500 lines of HOL codes [20] and about 125 man-hours.

In order to facilitate the use of our formally verified results by industrial design engineers for their failure analysis, we have also developed a set of SML scripts to automate the simplification step of these theorems for any given failure rate list corresponding to the NextGen ATM system components. For instance, the output of the auto_ASN_gateway_FT script [20] for the automatic simplification of Theorem 6 is as follows:

⊢ (prob p (ASN_gateway_FT p t FD AP FF1 D1 D4 D7 D10 E1 ⋯ E21 C5 C6 C7 C8 notshw AL SL PD Others time ED EQ1 EN1 ⋯ EN4 human) =

$1 - (1 - (1 - e^{(-5/2)}) * (1 - e^{(-3/2)})) * ((1 - (1 - e^{(-1/2)}) * ((1 - e^{(-2)}) *$
$((1 - e^{(-3/2)}) * (1 - e^{(-4)})))) * e^{(-9/2)}) * ((1 - (1 - e^{(-7/2)}) * (1 - e^{(-3)})) *$
$(1 - (1 - e^{(-4)}) * (1 - e^{(-3)})) * (e^{(-4)} * ((1 - (1 - e^{(-1/2)}) * (1 - e^{(-3)})) *$
$((1 - (1 - e^{(-1/2)}) * (1 - e^{(-3)})) * ((1 - (1 - e^{(-1/2)}) * (1 - e^{(-3)})) *$
$(1 - (1 - e^{(-1/2)}) * (1 - e^{(-3)})))) * (e^{(-321/20)} * ((1 - (1 - e^{(-1/2)}) * (1 - e^{(-3)})) *$
$((1 - (1 - e^{(-1/2)}) * (1 - e^{(-3)})) * ((1 - (1 - e^{(-1/2)}) * (1 - e^{(-3)})) *$
$(1 - (1 - e^{(-1/2)}) * (1 - e^{(-3)})))))) * ((1 - (1 - e^{(-3/2)}) * (1 - e^{(-2)})) *$
$((1 - (1 - e^{(-1/2)}) * (1 - e^{(-2)})) * (1 - (1 - e^{(-1/2)}) * (1 - e^{(-2)}))))) * e^{(-1)} *$
$((1 - (1 - e^{(-7/2)}) * (1 - e^{(-3)})) * ((1 - (1 - e^{(-3/2)}) * (1 - e^{(-3)})) *$
$((1 - (1 - e^{(-1/2)}) * (1 - e^{(-3)})) * (1 - (1 - e^{(-5/2)}) * (1 - e^{(-3)})))))))$

With a very little modification, these kind of automation scripts can facilitate industrial design engineers to accurately determine the failure probability of many other safety-critical systems.

## 5   Conclusion

The accuracy of failure analysis is a dire need for safety and mission-critical applications, like the avionic ASN gateway communication system, where a slight error in the failure analysis may lead to disastrous situations including the death of innocent human lives or heavy financial setbacks. In this paper, we presented a deep embedding based formalization of commonly used FT gates, which facilitates the transformation of a FT model to its equivalent RBD model. The transformation considerably reduces the complexity of the FT analysis compared to our earlier FT formalization [5]. For illustration, the paper presents the formalization of each level of ASN gateway FT and then building upon this formalization the failure probability of overall ASN gateways communication system is verified.

## References

1. ReliaSoft (2016). http://www.reliasoft.com/
2. ASENT (2016). https://www.raytheoneagle.com/asent/rbd.htm
3. Trivedi, K.S.: Probability and Statistics with Reliability, Queuing and Computer Science Applications. Wiley, New York (2002)
4. Epstein, S., Rauzy, A.: Can we trust PRA? Reliab. Eng. Syst. Saf. 88(3), 195–205 (2005)
5. Ahmed, W., Hasan, O.: Towards formal fault tree analysis using theorem proving. In: Kerber, M., Carette, J., Kaliszyk, C., Rabe, F., Sorge, V. (eds.) CICM 2015. LNCS (LNAI), vol. 9150, pp. 39–54. Springer, Heidelberg (2015). doi:10.1007/978-3-319-20615-8_3
6. Bilintion, R., Allan, R.: Reliability Evaluation of Engineering Systems. Springer, New York (1992)

7. Ahmed, W., Hasan, O., Tahar, S.: Formalization of reliability block diagrams in higher-order logic. J. Appl. Logic **18**, 19–41 (2016)
8. Bozzano, M., Cimatti, A., Katoen, J.-P., Nguyen, V.Y., Noll, T., Roveri, M.: The COMPASS approach: correctness, modelling and performability of aerospace systems. In: Buth, B., Rabe, G., Seyfarth, T. (eds.) SAFECOMP 2009. LNCS, vol. 5775, pp. 173–186. Springer, Heidelberg (2009). doi:10.1007/978-3-642-04468-7_15
9. Ortmeier, F., Schellhorn, G.: Formal fault tree analysis-practical experiences. Electron. Notes Theoret. Comput. Sci. **185**, 139–151 (2007). Elsevier
10. Xiang, J., Futatsugi, K., He, Y.: Fault tree and formal methods in system safety analysis. In: IEEE Computer and Information Technology, pp. 1108–1115 (2004)
11. Futatsugi, K., Nakagawa, A.T., Tamai, T.: CAFE: An Industrial-Strength Algebraic Formal Method. Elsevier, Elsevier (2000)
12. Mhamdi, T., Hasan, O., Tahar, S.: On the formalization of the Lebesgue integration theory in HOL. In: Kaufmann, M., Paulson, L.C. (eds.) ITP 2010. LNCS, vol. 6172, pp. 387–402. Springer, Heidelberg (2010). doi:10.1007/978-3-642-14052-5_27
13. Ahmed, W., Hasan, O., Tahar, S., Hamdi, M.S.: Towards the formal reliability analysis of oil and gas pipelines. In: Watt, S.M., Davenport, J.H., Sexton, A.P., Sojka, P., Urban, J. (eds.) CICM 2014. LNCS (LNAI), vol. 8543, pp. 30–44. Springer, Heidelberg (2014). doi:10.1007/978-3-319-08434-3_4
14. Ahmed, W., Hasan, O., Tahar, S.: Formal reliability analysis of wireless sensor network data transport protocols using HOL. In: IEEE Wireless and Mobile Computing, Networking and Communications, pp. 217–224 (2015)
15. Ahmad, W., Hasan, O., Tahar, S., Hamdi, M.: Towards formal reliability analysis of logistics service supply chains using theorem proving. In: Implementation of Logics, pp. 111–121 (2015)
16. Ahmed, W., Hasan, O.: Formal availability analysis using theorem proving. In: International Conference on Formal Engineering Methods. LNCS, pp. 1–16. Springer, Switzerland (2016, to appear). arXiv:1608.01755
17. Kuykendall, T.A.: Section 3.9, fault tree to RBD transformation. In: Systems Engineering "Toolbox" for Design-Oriented Engineers, pp. 52–52. NASA (1994)
18. Törngren, M.: Fundamentals of implementing real-time control applications in distributed computer systems. Real-Time Syst. **14**(3), 219–250 (1998)
19. Kornecki, A.J., Liu, M.: Fault tree analysis for safety/security verification in aviation software. Electronics **2**(1), 41–56 (2013)
20. Ahmad, W.: Formalization of fault trees in higher-order logic: a deep embedding approach (2016). http://save.seecs.nust.edu.pk/fault-tree/

# An Efficient Synthesis Algorithm for Parametric Markov Chains Against Linear Time Properties

Yong Li[1,2], Wanwei Liu[3], Andrea Turrini[1(✉)], Ernst Moritz Hahn[1],
and Lijun Zhang[1,2]

[1] State Key Laboratory of Computer Science, Institute of Software,
CAS, Beijing, China
turrini@ios.ac.cn
[2] University of Chinese Academy of Sciences, Beijing, China
[3] College of Computer Science, National University of Defense Technology,
Changsha, China

**Abstract.** In this paper, we propose an efficient algorithm for the parameter synthesis of PLTL formulas with respect to parametric Markov chains. The PLTL formula is translated to an almost fully partitioned Büchi automaton which is then composed with the parametric Markov chain. We then reduce the problem to solving an optimisation problem, allowing to decide the satisfaction of the formula using an SMT solver. The algorithm works also for interval Markov chains. The complexity is linear in the size of the Markov chain, and exponential in the size of the formula. We provide a prototype and show the efficiency of our approach on a number of benchmarks.

## 1 Introduction

Model checking, an automatic verification technique, has attracted much attention as it can be used to verify the correctness of software and hardware systems [1,10,12]. In classical model checking, temporal formulas are often used to express properties that one wants to check.

Probabilistic verification problems have been studied extensively in recent years. Markov chains (MCs) are a prominent probabilistic model used for modelling probabilistic systems. Properties are specified using probabilistic extensions of temporal logics such as probabilistic CTL (PCTL) [22] and probabilistic LTL (PLTL) [4] and their combination PCTL*. In the probabilistic setting, most of the observations about CTL carry over to their probabilistic counterpart. An exception is the complexity for verifying PLTL: here one could have a double exponential blowup. This is the case, because in general nondeterministic Büchi automata cannot be used directly to verify LTL properties, as they will cause imprecise probabilities in the product. In turn, it is often necessary to construct their deterministic counterparts in terms of other types of automata, for instance Rabin or Parity automata, which adds another exponential blowup. As a result, most of the work in literature focuses on branching time verification problems.

M. Fränzle et al. (Eds.): SETTA 2016, LNCS 9984, pp. 280–296, 2016.
DOI: 10.1007/978-3-319-47677-3_18

Moreover, state-of-the-art tools such as PRISM [30] and MRMC [26] can handle large systems with PCTL specifications, but rather small systems –if at all– for PLTL specifications.

In the seminal paper by Courcoubetis and Yannakakis [13], it is shown that for MCs the PLTL model checking problem is in **PSPACE**. They perform transformations of the Markov chain model recursively according to the LTL formula. At each step, the algorithm replaces a subformula rooted at a temporal operator with a newly introduced proposition; meanwhile, it refines the Markov chain with that proposition, and such refinement preserves the distribution. Then, it is finally boiled down to the probabilistic model checking upon a propositional formula. At the refinement step the state space is doubled, thus resulting in a **PSPACE** algorithm. Even if it is theoretically a single exponential algorithm for analysing MCs with respect to PLTL, it has not been exploited in the state-of-the-art probabilistic model checkers.

In automata-based approaches, one first translates the LTL formula into a Büchi automaton and then analyses the product of the MC and the Büchi automaton. This is sufficient for non-probabilistic model checking. For the probabilistic setting, the Büchi automaton is usually further transformed into a deterministic variant. Such a determinisation step usually exploits Safra's determinisation construction [35]. Several improvements have been made in recent years, see for instance [32,33,36]. Model checkers such as PRISM [30] and LiQuor [9] handle PLTL formulas by using off-the-shelf tools (e.g. (J)Ltl2Dstar [29]) to perform this determinisation step. To avoid the full complexity of the deterministic construction, Chatterjee et al. [7] have proposed an improved algorithm for translating the formulas of the FG-fragment of LTL to an extension of Rabin automata. Recently [18], this algorithm has been extended to the complete LTL.

Despite the above improvements, the size of the resulting deterministic automaton is still the bottleneck of the approach for linear temporal properties. In [14], it is first observed that the second blowup can be circumvented by using *unambiguous Büchi automata* (UBAs) [6]. The resulting algorithm has the same complexity as the one in [13]. Despite the importance of probabilistic model checking, unfortunately, the algorithm in [14] is less recognised. To the best of the authors knowledge, it is not applied in any of the existing model checkers. Later in [5], the problem of verifying $\omega$-regular properties over MCs by using alternating Büchi infinite-word automata has been considered. Recently, in [28], the authors construct the so called *limit deterministic* Büchi automata that are exponential in the size of LTL\GU formula $\phi$, which is another fragment of LTL. The approach is only applied to the analysis of qualitative PLTL of the form $\mathbb{P}_{>0}[\phi]$.

In this paper, we present a further improvement of the solution proposed in [14], adapted directly to solving the parameter synthesis problem for parametric Markov chains. We exploit a simple construction translating the given LTL formula to a reverse deterministic UBA, and then build the product of the parametric Markov chains. We then extract an equation system from the product, then the synthesis problem reduces to the existence of a solution of the equation

system. Further, we remark that the related interval Markov chains can be han-
dled by our approach as well. We integrate our approach in the model checker
IscasMC [21], and employ SMT solver to solving the obtained equation system.
We present detailed experimental results, and observe that our implementation
can deal with some real-world probabilistic systems modelled by parametric
Markov chains.

*Related Work.* In [20], they first use state elimination to compute the reachability probability for parametric Markov models. This has be improved by Dehnert
*et al.* [17]. Another related model is interval Markov chains, which can be interpreted as a family of Markov chains [8,25,37] whose transition probabilities lie
within the interval ranges. The **PSPACE** complexity of model checking PCTL
has been established in [37] while the same complexity has been shown in [8]
for $\omega$-regular properties such as $\omega$-PCTL; in [25] interval Markov chains have
been used as abstraction models by using three-valued abstraction for Markov
chains. To our best knowledge, it is the first time that one can easily integrate
parameter synthesis algorithm that is exponential in the size of LTL formulas
over parametric Markov chains.

A long version of this paper containing further examples and proofs can be
found in [31]. We remark that a single exponential algorithm is presented for
verifying PLTL properties on Markov chains in [3], which, however, contains a
flaw. In [2], the authors have corrected the flaw of [3]. In this paper we have
developed a similar approach which fixes the flaw; we remark that our approach
has been developed in parallel.

## 2    Preliminaries

Given a function $\mathbf{f}\colon X_1 \times \cdots \times X_n \to 2^Y$, we may alternatively consider it as the
set $\{(x_1,\ldots,x_n,y) \in X_1 \times \cdots \times X_n \times Y \mid y \in \mathbf{f}(x_1,\ldots,x_n)\}$.

Given a set $W$, we say that an infinite sequence $\varpi = w_0 w_1 \ldots$ is an $\omega$-*word*
over $W$, if $\varpi \in W^\omega$.

Given a finite word $\nu = v_0 \ldots v_k$ and a finite or infinite word $\varpi = w_0 w_1 \ldots$,
we denote by $\nu \cdot \varpi$ the *concatenation* of $\nu$ and $\varpi$, i.e., the finite or infinite word
$\nu \cdot \varpi = v_0 \ldots v_k w_0 w_1 \ldots$, respectively. We may just write $\nu\varpi$ instead of $\nu \cdot \varpi$.
We denote by $[1..n]$ the set of natural numbers $\{1, \cdots, n\}$.

Given a set $X$, we denote by $\mathrm{Disc}(X)$ the set of discrete probability measures
over $X$, and by $\mathrm{SubDisc}(X)$ the set of discrete sub-probability measures over $X$.
Given a discrete probability measure $\mu$, we denote by $\mathrm{Supp}(\mu)$ the *support* of $\mu$,
i.e., $\mathrm{Supp}(\mu) = \{x \in X \mid \mu(x) > 0\}$. Moreover, we denote by $\delta_x$, for $x \in X$, the
*Dirac* measure such that for each $y \in X$, $\delta_x(y) = 1$ if $y = x$ and $\delta_x(y) = 0$ if
$y \neq x$.

A *directed graph* $G$ is a pair $G = (V, E)$ where $V$ is a finite non-empty set of
*vertices*, also called *nodes*, and $E \subseteq V \times V$ is the set of *edges* or *arcs*. Given an
arc $e = (u, v)$, we call the vertex $u$ the *head* of $e$, denoted as $u = \mathrm{head}(e)$, and
the vertex $v$ the *tail* of $e$, denoted as $v = \mathrm{tail}(e)$. In the remainder of the paper
we consider only directed graphs and we refer to them just as graphs.

A *path* $\pi$ is a sequence of edges $\pi = e_1e_2\ldots e_n$ such that for each $1 \leq i < n$, tail$(e_i) = $ head$(e_{i+1})$. We say that $v$ is *reachable* from $u$ if there exists a path $\pi = e_1\ldots e_n$ such that head$(e_1) = u$ and tail$(e_n) = v$.

A *strongly connected component* (SCC) is a set of vertices $C \subseteq V$ such that for each pair of vertices $u, v \in C$, $u$ is reachable from $v$ and $v$ is reachable from $u$. We say that a graph $G = (V, E)$ is *strongly connected* if $V$ is an SCC. An SCC $C$ is *non-extensible* if for each SCC $C'$ of $G$ we have $C \subseteq C'$ implies $C' = C$. Without loss of generality, in the remainder of this paper we consider only non-extensible SCCs.

We define the partial order $\preceq$ over the SCCs of the graph $G$ as follows: given two SCCs $C_1$ and $C_2$, $C_1 \preceq C_2$ if there exist $v_1 \in C_1$ and $v_2 \in C_2$ such that $v_2$ is reachable from $v_1$. We say that an SCC $C$ is *maximal* with respect to $\preceq$ if for each SCC $C'$ of $G$, $C \preceq C'$ implies $C' = C$. We may call a maximal SCC as *bottom SCC* (BSCC).

A graph can be enriched with labels as follows: a *labelled graph* $G$ is a triple $G = (V, \Sigma, E)$ where $V$ is a finite non-empty set of *vertices*, $\Sigma$ is a finite set of *labels*, and $E \subseteq V \times \Sigma \times V$ is the set of *labelled edges*. The notations and concepts on graphs trivially extend to labelled graphs.

A *generalized Büchi automaton* (GBA) $\mathcal{A}$ is a tuple $(\Sigma, Q, \mathbf{T}, Q_0, ACC)$ where $\Sigma$ is a finite *alphabet*, $Q$ is a finite set of *states*, $\mathbf{T} : Q \times \Sigma \to 2^Q$ is the *transition function*, $Q_0 \subseteq Q$ is the set of *initial states*, and $ACC = \{F_i \subseteq \mathbf{T} \mid i \in [1..k]\}$ is the set of *accepting sets*.

A *run* $\sigma$ of $\mathcal{A}$ over an infinite word $w = a_0a_1\ldots \in \Sigma^\omega$ is an infinite sequence $\sigma = q_0a_0q_1a_1q_2 \ldots \in (Q \cdot \Sigma)^\omega$ such that $q_0 \in Q_0$ and for each $i \in \mathbb{N}$ it is $q_{i+1} \in \mathbf{T}(q_i, a_i)$. Similarly, a *run* of $\mathcal{A}$ over a finite word $w = a_0a_1\ldots a_k \in \Sigma^*$ is a finite sequence $\sigma = q_0a_0q_1a_1q_2\ldots a_kq_{k+1} \in Q \cdot (\Sigma \cdot Q)^*$ such that $q_0 \in Q_0$ and for each $i \in \{0, \ldots, k\}$ it is $q_{i+1} \in \mathbf{T}(q_i, a_i)$. Let Inf$(\sigma) = \{(q, a, q') \in \mathbf{T} \mid \forall i \in \mathbb{N}.\exists j \geq i.(q_j, a_j, q_{j+1}) = (q, a, q')\}$ be the set of tuples $(q, a, q')$ occurring infinitely often in $\sigma$. The run $\sigma$ is *accepting* if Inf$(\sigma) \cap F_i \neq \emptyset$ for each $i \in [1..k]$. The word $w$ is accepted by $\mathcal{A}$ if there is an accepting run of $\mathcal{A}$ over $w$; we denote by $\mathcal{L}(\mathcal{A})$ the *language* of $\mathcal{A}$, i.e., the set of infinite words accepted by $\mathcal{A}$.

Given a GBA $\mathcal{A} = (\Sigma, Q, \mathbf{T}, Q_0, ACC)$, for the sake of convenience, we denote by $\mathcal{A}^q$ the GBA $(\Sigma, Q, \mathbf{T}, \{q\}, ACC)$ with initial state $q$ and accordingly for $U \subseteq Q$ we let $\mathcal{A}^U \stackrel{\text{def}}{=} (\Sigma, Q, \mathbf{T}, U, ACC)$.

The graph $G = (V, \Sigma, E)$ underlying a GBA $\mathcal{A}$ is the graph whose set of vertices (nodes) $V$ is the set of states $S$ of $\mathcal{A}$ and there is an edge $e \in E$ labelled with $a \in \Sigma$ from $q$ to $q'$ if $q' \in \mathbf{T}(q, a)$. In this case, we say that $q$ is an $a$-*predecessor* of $q'$ and $q'$ is an $a$-*successor* of $q$.

For the GBA $\mathcal{A}$, we say that

- $\mathcal{A}$ is *deterministic*, if $|Q_0| = 1$ and $|\mathbf{T}(q, a)| = 1$ for each $q \in Q$ and $a \in \Sigma$;
- $\mathcal{A}$ is *reverse deterministic* if each state has exactly one $a$-predecessor for each $a \in \Sigma$;
- $\mathcal{A}$ is *unambiguous* if for each $q \in Q$, $a \in \Sigma$, and $q', q'' \in \mathbf{T}(q, a)$ such that $q' \neq q''$, we have $\mathcal{L}(\mathcal{A}^{q'}) \cap \mathcal{L}(\mathcal{A}^{q''}) = \emptyset$; and
- $\mathcal{A}$ is *separated* if $\mathcal{L}(\mathcal{A}^q) \cap \mathcal{L}(\mathcal{A}^{q'}) = \emptyset$ for each pair of states $q, q' \in Q$, $q \neq q'$.

We say that a state $q \in Q$ is *reenterable* if $q$ has some predecessor in $\mathcal{A}$. Let $Q'$ be the set of all reenterable states of $\mathcal{A}$ and consider the GBA $\mathcal{A}' = (\Sigma, Q', \mathbf{T}', Q', ACC')$ where $\mathbf{T}' = \mathbf{T}|_{Q' \times \Sigma}$ and $ACC' = \{F'_i = F_i \cap \mathbf{T}' \mid F_i \in ACC, i \in [1..k]\}$. Then, we say that $\mathcal{A}$ is *almost unambiguous* (respectively *almost separated, almost reverse deterministic*) if $\mathcal{A}'$ is unambiguous (respectively separated, reverse deterministic).

For an (almost) separated GBA $\mathcal{A}$, if for each $\alpha \in \Sigma^\omega$ there exists some state $q$ of $\mathcal{A}$ such that $\alpha \in \mathcal{L}(\mathcal{A}^q)$, then we say that $\mathcal{A}$ is (almost) *fully partitioned*. Clearly, if an automaton is (almost) fully partitioned, then it is also (almost) separated, (almost) unambiguous and (almost) reverse deterministic.

As an example of GBA that is reverse-deterministic and separated but not fully partitioned, consider the generalised Büchi automaton $\mathcal{A}$ depicted

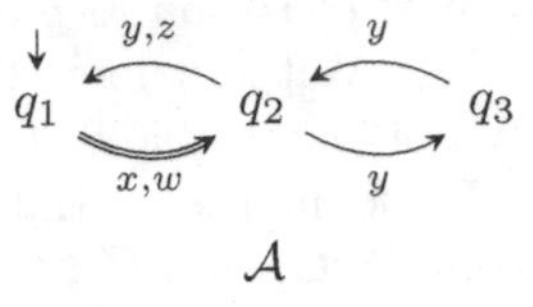

**Fig. 1.** An example of generalised Büchi automaton

in Fig. 1, where $ACC = \{\{(q_1, x, q_2), (q_1, w, q_2)\}\}$, represented by double arrows. The fact that $\mathcal{A}$ is not fully partitioned is clear since no word starting with $xw$ is accepted by any of the states $q_1$, $q_2$, or $q_3$. One can easily check that $\mathcal{A}$ is indeed reverse-deterministic but checking the separated property can be more involved. The checks involving $q_1$ are trivial, as it is the only state enabling a transition with label $x$ or $w$; for the states $q_2$ and $q_3$, the separated property implies that given any $w_1 \in \mathcal{L}(\mathcal{A}^{q_2})$, it is not possible to find some $w_2 \in \mathcal{L}(\mathcal{A}^{q_3})$ such that $w_1 = w_2$. For instance, suppose the number of the most front $y$'s in $w_1$ is odd, it must be the case that the most front $y$'s are directly followed by $x$ or $w$. In order to match $w_1$, we must choose $y$ instead of $z$ on transition $(q_2, q_1)$. It follows that the number of the most front $y$'s in $w_2$ is even. We can get similar result when the number of the most front $y$'s in $w_1$ is even. Thus $w_1$ and $w_2$ can never be the same.

# 3    Parametric Markov Chains and Probabilistic LTL

In this section we recall the definitions of parametric Markov chains as proposed in [20], interval Markov chain considered in [3,8,25,37] and of the logic PLTL.

In addition, we consider the translation of LTL formulas to GBAs which is used later for analysing PLTL properties.

## 3.1    Parametric Markov Chains

Before introducing the parametric Markov chain model, we briefly present some general notation. Given a finite set $V = \{x_1, \ldots, x_n\}$ with domain in $\mathbb{R}$, an *evaluation* is a partial function $\upsilon \colon V \to \mathbb{R}$. Let $\mathrm{Dom}(\upsilon)$ denote the domain of $\upsilon$; we say that $\upsilon$ is total if $\mathrm{Dom}(\upsilon) = V$. A *polynomial* $p$ over $V$ is a sum of monomials $p(x_1, \ldots, x_n) = \sum_{i_1, \ldots, i_n} a_{i_1, \ldots, i_n} \cdot x_1^{i_1} \cdots x_n^{i_n}$ where each $i_j \in \mathbb{N}$ and each $a_{i_1, \ldots, i_n} \in \mathbb{R}$. A *rational function* over $V$ is a fraction $f(x_1, \ldots, x_n) = \frac{p_1(x_1, \ldots, x_n)}{p_2(x_1, \ldots, x_n)}$ of two polynomials $p_1$ and $p_2$ over $V$; we denote by $\mathcal{F}_V$ the set of all rational functions over $V$. Given $f \in \mathcal{F}_V$, $V' \subseteq V$, and an evaluation $v$, we let $f[V'/v]$ denote the rational function obtained from $f$ by replacing each occurrence of $v \in V' \cap \mathrm{Dom}(v)$ with $v(v)$.

**Definition 1.** *A parametric Markov chain (PMC) is a tuple $\mathcal{M} = (S, L, \bar{s}, V, \mathbf{P})$ where: $S$ is a finite set of states; $L \colon S \to \Sigma$ is a labelling function, in which $\Sigma$ is a set of labels; $\bar{s} \in S$ is the initial state; $V$ is a finite set of parameters; and $\mathbf{P} \colon S \times S \to \mathcal{F}_V$ is a transition matrix.*

We now define the PMC induced with respect to a given evaluation:

**Definition 2.** *Given a PMC $\mathcal{M} = (S, L, \bar{s}, V, \mathbf{P})$ and an evaluation $v$, the PMC $\mathcal{M}_v$ induced by $v$ is the tuple $(S, L, \bar{s}, V \setminus \mathrm{Dom}(v), \mathbf{P}_v)$ where the transition matrix $\mathbf{P}_v \colon S \times S \to \mathcal{F}_{V \setminus \mathrm{Dom}(v)}$ is given by $\mathbf{P}_v(s, t) = \mathbf{P}(s, t)[\mathrm{Dom}(v)/v]$.*

We say that a *total* evaluation is *well-defined* for a PMC $\mathcal{M}$ if $\mathbf{P}_v(s, s') \in [0, 1]$ and $\sum_{t \in S} \mathbf{P}_v(s, t) = 1$ for each $s, s' \in S$. In the remainder of the paper we consider only well-defined evaluations, and we require that, for a given PMC $\mathcal{M}$ and two states $s, t \in S$, if $\mathbf{P}_v(s, t) > 0$ for some evaluation $v$, then $\mathbf{P}_{v'}(s, t) > 0$ for all evaluations $v'$. We may omit the actual evaluation $v$ when we are not interested in the actual value for $\mathbf{P}_v(s, t)$, such as for the case $\mathbf{P}_v(s, t) > 0$.

The *underlying graph* of a PMC $\mathcal{M}$ for a given evaluation $v$ is the graph $G = (V, E)$ where $V = S$ and $E = \{(s, s') \in S \times S \mid \mathbf{P}_v(s, s') > 0\}$. Note that the requirement that "if $\mathbf{P}_v(s, t) > 0$ for some evaluation $v$, then $\mathbf{P}_{v'}(s, t) > 0$ for all evaluations $v'$" ensures that the underlying graph is the same for all evaluations $v'$.

We use $|S|$ to denote the number of states, and $|\mathcal{M}|$ for the number of non-zero probabilistic transitions, i.e., $|\mathcal{M}| = |\{(s, s') \in S \times S \mid \mathbf{P}(s, s') > 0\}|$.

A *path* is a sequence of states $\pi = s_0 s_1 \ldots$ satisfying $\mathbf{P}(s_i, s_{i+1}) > 0$ for all $i \geq 0$. We call a path $\pi$ *finite* or *infinite* if the sequence $\pi$ is finite or infinite, respectively. We use $\pi(i)$ to denote the suffix $s_i s_{i+1} \ldots$ and we denote by $Paths^{\mathcal{M}}$ and $Paths_{fin}^{\mathcal{M}}$ the set of all infinite and finite paths of $\mathcal{M}$, respectively. An infinite path $\pi = s_0 s_1 \ldots$ defines the $\omega$-word $w_0 w_1 \ldots \in \Sigma^\omega$ such that $w_i = L(s_i)$ for $i \in \mathbb{N}$.

For a finite path $s_0 s_1 \ldots s_k$, we denote by $Cyl(s_0 s_1 \ldots s_k)$ the *cylinder set* of $s_0 s_1 \ldots s_k$, i.e., the set of infinite paths starting with prefix $s_0 s_1 \ldots s_k$. Given an evaluation $v$, we define the measure of the cylinder set by

$$\mathbb{P}^{\mathcal{M}_v}\big(Cyl(s_0 s_1 \ldots s_k)\big) \stackrel{\text{def}}{=} \delta_{\bar{s}}(s_0) \cdot \prod_{i=0}^{k-1} \mathbf{P}_v(s_i, s_{i+1}).$$

For a given PMC $\mathcal{M}$ and an evaluation $v$, we can extend $\mathbb{P}^{\mathcal{M}_v}$ uniquely to a probability measure over the $\sigma$-field generated by cylinder sets [27].

We call the BSCCs of the underlying graph $G$ *ergodic sets* and for each ergodic set $C$, we call each state $s \in C$ *ergodic*. A nice property of a BSCC $C$ is the so-called *ergodicity property*: for each $s \in C$, $s$ will be reached again in the future with probability 1 from any state $s' \in C$, including $s$ itself. Moreover, for each finite path $\pi$ within $C$, $\pi$ will be performed again in the future with probability 1.

In this paper we are particularly interested in $\omega$-regular properties $\mathcal{L} \subseteq \Sigma^\omega$ and the probability $\mathbb{P}^{\mathcal{M}}(\mathcal{L})$ for some measurable set $\mathcal{L}$. Such properties are known to be measurable in the $\sigma$-field generated by cylinders [38]. We write $\mathbb{P}_s^{\mathcal{M}}$ to denote the probability function when assuming that $s$ is the initial state of the PMC $\mathcal{M}$. To simplify the notation, we omit the superscript $\mathcal{M}$ whenever $\mathcal{M}$ is clear from the context and we use $\Pi$ as a synonym for *Paths*.

## 3.2   Interval Markov Chain

In this section we recall the definition of interval Markov chain [8,24,25] and show how it can be converted to a parametric Markov chain.

**Definition 3.** *An* interval Markov chain *(IMC) is a tuple* $\mathcal{M} = (S, L, \bar{s}, \mathbf{P}_l, \mathbf{P}_u)$ *where* $S$, $L$ *and* $\bar{s}$ *are as for PMCs while* $\mathbf{P}_l, \mathbf{P}_u \colon S \times S \to [0,1]$ *are the transition matrices such that for each* $s, s' \in S$, $\mathbf{P}_l(s, s') \leq \mathbf{P}_u(s, s')$.

We show how to convert an IMC to a PMC in the following. Given an IMC $\mathcal{M} = (S, L, \bar{s}, \mathbf{P}_l, \mathbf{P}_u)$, we define the corresponding PMC $\mathcal{M}' = (S, L, \bar{s}, \mathbf{P})$ as follows. For every pair of states, say $(s, t)$, we add a new parameter $p_{st}$ to $V$ such that $V = \{p_{st} \mid \mathbf{P}_l(s, t) \leq p_{st} \leq \mathbf{P}_u(s, t)\}$; then, we define $\mathbf{P}$ as $\mathbf{P}(s, t) = p_{st}$. For instance, suppose in an IMC, there is a state $s$ with two successors, namely $t$ and $w$, with $\mathbf{P}_l(s, t) = 0.2$, $\mathbf{P}_l(s, w) = 0.3$, $\mathbf{P}_u(s, t) = 0.7$ and $\mathbf{P}_u(s, w) = 0.5$. We add two parameters $p_{st}$ and $p_{sw}$ for the pairs $(s, t)$ and $(s, w)$ whose ranges are $[0.2, 0.7]$ and $[0.3, 0.5]$ respectively. Moreover, in order to get an instance of Markov chain from the resulting PMC, we must make sure that $p_{st} + p_{sw} = 1$.

## 3.3   Probabilistic Linear Time Temporal

Throughout the whole paper, we will assume that the state space $S$ of any PMC is always equipped with labels that identify distinguishing state properties. For this, we let $AP$ denote a set of atomic propositions. We assume $\Sigma = 2^{AP}$ as state labels, so that $L(s)$ specifies the subset of atomic propositions holding in state $s$.

We first recall the linear time temporal logic (LTL). The syntax of LTL is given by:

$$\phi \stackrel{\text{def}}{=} p \mid \neg\phi \mid \phi \wedge \phi \mid \mathsf{X}\phi \mid \phi\mathsf{U}\phi$$

where $p \in AP$. We use standard derived operators, such as: $\phi_1 \vee \phi_2 \stackrel{\text{def}}{=} \neg(\neg\phi_1 \wedge \neg\phi_2)$, *true* $\stackrel{\text{def}}{=} a \vee \neg a$, $\phi_1 \to \phi_2 \stackrel{\text{def}}{=} \neg\phi_1 \vee \phi_2$, $\mathsf{F}\phi \stackrel{\text{def}}{=} true\mathsf{U}\phi$, and $\mathsf{G}\phi \stackrel{\text{def}}{=} \neg(\mathsf{F}\neg\phi)$. Semantics is standard and is omitted here.

A PLTL formula has the form $\mathbb{P}_\mathsf{J}(\phi)$ where $\mathsf{J} \subseteq [0,1]$ is a non-empty interval with rational bounds and $\phi$ is an LTL formula. In a PMC $\mathcal{M}$ with evaluation $v$, for a state $s \in S$ and a formula $\mathbb{P}_\mathsf{J}(\phi)$, we have:

$$s \models \mathbb{P}_\mathsf{J}(\phi) \text{ if and only if } \mathbb{P}_s^{\mathcal{M}_v}(\{\pi \in \Pi \mid \pi \models \phi\}) \in \mathsf{J} \tag{1}$$

From the measurability of $\omega$-regular properties, we can easily show that for any LTL formula $\phi$, the set $\{\pi \in \Pi \mid \pi \models \phi\}$ is measurable in the $\sigma$-field generated by the cylinder sets.

## 3.4    From LTL to Büchi Automaton

The following section describes how we can transform a given LTL formula into a GBA which has the required properties for the subsequent parameter synthesis procedure.

The set of *elementary formulas* $el(\phi)$ for a given LTL formula $\phi$ is defined recursively as follows: $el(p) = \emptyset$ if $p \in AP$; $el(\neg\psi) = el(\psi)$; $el(\phi_1 \wedge \phi_2) = el(\phi_1) \cup el(\phi_2)$; $el(\mathsf{X}\psi) = \{\mathsf{X}\psi\} \cup el(\psi)$; and $el(\phi_1 \mathsf{U}\phi_2) = \{\mathsf{X}(\phi_1 \mathsf{U}\phi_2)\} \cup el(\phi_1) \cup el(\phi_2)$.

Given a set $V \subseteq el(\phi)$ and $a \in \Sigma = 2^{AP}$, we inductively define the *satisfaction relation* $\Vdash$ for each subformula of $\phi$ as follows:

$(V, a) \Vdash p$         if $p \in a$ in the case of $p \in AP$,

$(V, a) \Vdash \neg\psi$        if it is not the case that $(V, a) \Vdash \psi$,

$(V, a) \Vdash \phi_1 \wedge \phi_2$    if $(V, a) \Vdash \phi_1$ and $(V, a) \Vdash \phi_2$,

$(V, a) \Vdash \mathsf{X}\psi$        if $\mathsf{X}\psi \in V$, and

$(V, a) \Vdash \phi_1 \mathsf{U}\phi_2$    if $(V, a) \Vdash \phi_2$ or, $(V, a) \Vdash \phi_1$ and $(V, a) \Vdash \mathsf{X}(\phi_1 \mathsf{U}\phi_2)$.

Finally, $\mathcal{A}_\phi = (\Sigma = 2^{AP}, Q_\phi, \mathbf{T}_\phi, \{\phi\}, ACC_\phi)$ is the Büchi automaton where:

- $Q_\phi = \{\phi\} \cup 2^{el(\phi)}$;
- $\mathbf{T}_\phi(\{\phi\}, a) = \{V \subseteq el(\phi) \mid (V, a) \Vdash \phi\}$ and for each $V \subseteq el(\phi)$, we have: $\mathbf{T}_\phi(V, a) = \{U \subseteq el(\phi) \mid \forall \mathsf{X}\psi \in el(\phi). \mathsf{X}\psi \in V \Longleftrightarrow (U, a) \Vdash \psi\}$; and
- $ACC_\phi = \{F_\psi\}$ where for each subformula $\psi = \phi_1 \mathsf{U}\phi_2$ of $\phi$, we have $F_\psi = \{(U, a, V) \in \mathbf{T}_\phi(V, a) \Vdash \phi_2$ or $(V, a) \Vdash \neg\psi\}$.

According to the definition, each formula in $el(\phi)$ is guaranteed to be of the form $\mathsf{X}\phi'$; the size of $el(\phi)$ is precisely the number of temporal operators (i.e., $\mathsf{X}$ and $\mathsf{U}$) occurring in $\phi$.

**Theorem 1 (cf. [11,14]).** *For the automaton $\mathcal{A}_\phi$, the following holds:*

1. *For each infinite word $\pi \in \Sigma^\omega$, we have $\pi \models \phi$ if and only if $\pi \in \mathcal{L}(\mathcal{A}_\phi)$.*
2. *More generally, for each $U \subseteq el(\phi)$ and $\mathsf{X}\psi \in el(\phi)$ we have: $\pi \models \psi$ if and only if $\mathsf{X}\psi \in U$, for every $\pi \in \mathcal{L}(\mathcal{A}_\phi^U)$.*

It follows directly that

**Corollary 1.** *For each $U, V \subseteq el(\phi)$, if $U \neq V$ then $\mathcal{L}(\mathcal{A}_\phi^U) \cap \mathcal{L}(\mathcal{A}_\phi^V) = \emptyset$. Moreover, $\mathcal{A}_\phi$ is both almost unambiguous and almost separated.*

We observe that for each subset $U \subseteq el(\phi)$ and each $a \in \Sigma$, there is exactly one $a$-predecessor of $U$, namely the set $\{\mathsf{X}\psi \in el(\phi) \mid (U, a) \Vdash \psi\}$. Hence, we also have the following conclusion.

**Corollary 2.** *The automaton $\mathcal{A}_\phi$ is almost reverse deterministic.*

## 4   Parameter Synthesis Algorithm

We consider a parametric Markov chain $\mathcal{M}$ and an (almost) unambiguous automaton $\mathcal{A} = (\Sigma, Q, \mathbf{T}, Q_0, ACC)$ obtained from the LTL specification, where $\mathcal{L}(\mathcal{A}^{q_1}) \cap \mathcal{L}(\mathcal{A}^{q_2}) = \emptyset$ if $q_1, q_2 \in Q_0$ and $q_1 \neq q_2$. To simplify the notation, in the following we assume that for a given PMC $\mathcal{M}$ we have $S = \Sigma$ and $L(s) = s$ for each $s \in S$; this modification does not change the complexity of probabilistic verification [13].

In the following we shall introduce the product graph composed by the PMC and Büchi automaton as well as the corresponding synthesis algorithm.

Given the automaton $\mathcal{A} = (\Sigma, Q, \mathbf{T}, Q_0, ACC)$ the PMC $\mathcal{M} = (S, L, \bar{s}, V, \mathbf{P})$, the *product graph* of $\mathcal{A}$ and $\mathcal{M}$, denoted $\mathcal{G} = \mathcal{A} \times \mathcal{M}$, is the graph $(\Gamma, \boldsymbol{\Delta})$ where $\Gamma = \{(q, s) \mid q \in Q, s \in S\}$ and $((q, s), (q', s')) \in \boldsymbol{\Delta}$ (also written $(q, s)\boldsymbol{\Delta}(q', s')$) if and only if $\mathbf{P}(s, s') > 0$ and $q' \in \mathbf{T}(q, L(s))$.

Suppose that $ACC = \{F_1, \ldots, F_k\}$. We say that an SCC $C$ of $\mathcal{G}$ is *accepting* if for each $F_i \in ACC$, there exist $(q, s), (q', s') \in C$ such that $(q, s)\boldsymbol{\Delta}(q', s')$ and $(q, a, q') \in F_i$ for some $a \in \Sigma$.

Given an SCC $C$ of $\mathcal{G}$, we denote by $\mathscr{H}(C)$ the *corresponding SCC* of $C$, where $\mathscr{H}(C) = \{s \in S \mid (q, s) \in C\}$. We denote by Proj a function to get the corresponding path of $\mathcal{M}$ from the path of $\mathcal{G}$, i.e., $\text{Proj}((q_0, s_0)(q_1, s_1)\cdots) = s_0 s_1 \cdots$ and we usually call the path $s_0 s_1 \cdots$ the projection of $(q_0, s_0)(q_1, s_1)\cdots$. For convenience, we also write $\alpha \trianglelefteq \beta$ if the (finite) path $\alpha$ is a fragment of the (finite) path $\beta$.

Below we give the definition of *complete SCC* [14]. For an SCC $C$ of $\mathcal{G}$ and $K = \mathscr{H}(C)$ the corresponding SCC of $C$, we say that $C$ is *complete* if for each finite path $\sigma_K$ in $K$, we can find a finite path $\sigma_C$ in $C$ such that $\sigma_K = \text{Proj}(\sigma_C)$.

Consider the product $\mathcal{A} \times \mathcal{M}$ shown in Fig. 2. It has two non-trivial SCCs, namely $C_1$ and $C_2$. Clearly, $K_1$ and $K_2$ are the corresponding SCCs of $C_1$ and $C_2$, respectively. We observe that $C_1$ is a complete SCC while $C_2$ is not complete since the finite path $zz$ of $K_2$ is not a projection of any finite path in $C_2$. The key observation is that some transitions in the SCCs of $\mathcal{M}$ may be lost in building

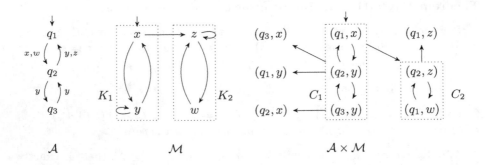

**Fig. 2.** The GBA $\mathcal{A}$ from Fig. 1, a PMC $\mathcal{M}$, and their product $\mathcal{A} \times \mathcal{M}$

the product, so we only consider the complete SCCs in the product to make sure
that no transitions of the projection SCCs are missing.

The following lemma characterises the property of a complete SCC of $\mathcal{G}$.

**Lemma 1** ([5]). *Consider a complete SCC $C$ of $\mathcal{G}$ with $\mathcal{H}(C) = K$ and an
arbitrary finite path $\rho_C$ in $C$. Then, there exists some finite path $\sigma_K$ in $K$ with
the following property: for each finite path $\sigma_C$ in $C$ with $\sigma_K = \mathrm{Proj}(\sigma_C)$, $\sigma_C$
contains $\rho_C$ as a fragment, i.e., $\rho_C \unlhd \sigma_C$.*

Based on Lemma 1, the following corollary relates the paths in the product
and the projected Markov chains:

**Corollary 3** ([5]). *Let $C$ be a complete SCC of $\mathcal{G}$ and $K = \mathcal{H}(C)$; consider
two infinite paths $\sigma_C$ in $C$ and $\sigma_K$ in $K$ such that $\sigma_K = \mathrm{Proj}(\sigma_C)$; let $P_C$ and
$P_K$ be the following properties:*

- *$P_C$: $\sigma_C$ visits each finite path in $C$ infinitely often;*
- *$P_K$: $\sigma_K$ visits each finite path of $K$ infinitely often.*

*Then $P_C$ holds if and only if $P_K$ holds.*

We say that an SCC $C$ of $\mathcal{G}$ is *locally positive* if:

1. $C$ is accepting and complete.
2. $\mathcal{H}(C)$ is a BSCC.

Consider again the example from Fig. 2. Assume the acceptance condition of $\mathcal{A}$ is
$ACC = \{\{(q_2, y, q_1)\}\}$; we observe that the SCC $C_1$ is both accepting and complete but not locally positive since $\mathcal{H}(C_1) = \{x, y\}$ is not a bottom SCC in $\mathcal{M}$.

According to Corollary 3, the ergodicity property of Markov chains, and the
definition of Büchi acceptance, we have the following result.

**Theorem 2.** *$\mathbb{P}^{\mathcal{M}}(\mathcal{L}(A)) \neq 0$ if and only if there exists some locally positive
SCC that is reachable from some initial state in $\mathcal{G}$.*

For a given SCC, in order to decide whether it is locally positive, we have to
judge whether it is complete. In general, doing so is a nontrivial task; however,
thanks to [13, Lemma 5.10], completeness can be checked efficiently:

**Lemma 2.** *If $\mathcal{A}$ is (almost) reverse deterministic, then the following two conditions are equivalent:*

(i) *$C$ is complete, i.e., each finite path of $\mathcal{H}(C)$ is a projection of some finite
path in $C$.*
(ii) *There is no other SCC $C'$ of $\mathcal{G}$ with $\mathcal{H}(C') = \mathcal{H}(C)$ such that $C' \preceq C$.*

We now turn to the problem of computing the exact probability.

**Theorem 3.** *Given a PMC $\mathcal{M}$ and an (almost) separated Büchi automaton $\mathcal{A}$, let $\mathcal{G} = \mathcal{A} \times \mathcal{M}$ be their product. Let $\text{pos}(\mathcal{G})$ be the set of all locally positive SCCs of $\mathcal{G}$ and $\text{npos}(\mathcal{G})$ be the set of all BSCCs of $\mathcal{G}$ which are not locally positive. Further, for an SCC $C$ let $C_{\mathcal{M}} = \{s \in S \mid \exists q \in Q. \; (q, s) \in C\}$ denote the set of states of $\mathcal{M}$ occurring in $C$. We define the following equation system:*

$$\mu(q, s) = \sum_{s' \in S} \left( \mathbf{P}(s, s') \cdot \sum_{(q,s)\boldsymbol{\Delta}(q',s')} \mu(q', s') \right) \qquad \forall q \in Q, s \in S \quad (2)$$

$$\sum_{\substack{q \in Q \\ (q,s) \in C}} \mu(q, s) = 1 \qquad \qquad \forall C \in \text{pos}(\mathcal{G}), s \in C_{\mathcal{M}} \quad (3)$$

$$\mu(q, s) = 0 \qquad \qquad \forall C \in \text{npos}(\mathcal{G}) \text{ and } (q, s) \in C \quad (4)$$

*Then, it holds that $\mathbb{P}^{\mathcal{M}_v}(\mathcal{L}(\mathcal{A})) = \sum_{q_0 \in Q_0} \mu(q_0, \bar{s})$ for any well-defined evaluation $v$.*

In general, all locally positive SCCs can be projected to the BSCCs in the induced MC $\mathcal{M}_v$. In the original MC $\mathcal{M}_v$, the reachability probability of every state in the accepting BSCC should be 1. Thus in a locally positive SCC of $\mathcal{G}$, the probability mass 1 distributes on the states in which they share the same second component, i.e., $s$ from state $(q, s)$.

Consider the PMC $\mathcal{M}$ and the automaton $\mathcal{A}$ as depicted in Fig. 3, together with their product graph. For clarity, we have omitted the isolated vertices like $(q_1, w)$ and $(q_5, y)$, i.e., the vertices with no incoming or outgoing edges. One may check that $\mathcal{A}$ is indeed separated, unambiguous, and reverse deterministic. The product of $\mathcal{M}$ and $\mathcal{A}$ consists of a single locally positive SCC $C = \{(q_1, x), (q_2, x), (q_3, y), (q_4, z), (q_5, w)\}$. We state the relevant part of the equation system resulting from Eqs. (2) and (3) of Theorem 3:

$$\mu(q_3, z) = 0$$
$$\mu(q_4, y) = 0$$
$$\mu(q_1, x) = (0.5 + \epsilon) \cdot \mu(q_3, y) + (0.5 - \epsilon) \cdot \mu(q_3, z) \qquad \qquad \mu(q_1, x) + \mu(q_2, x) = 1$$
$$\mu(q_2, x) = (0.5 - \epsilon) \cdot \mu(q_4, z) + (0.5 + \epsilon) \cdot \mu(q_4, y) \qquad \qquad \mu(q_3, y) = 1$$
$$\mu(q_3, y) = 1 \cdot \mu(q_5, w) \qquad \qquad \mu(q_4, z) = 1$$
$$\mu(q_4, z) = 1 \cdot \mu(q_5, w) \qquad \qquad \mu(q_5, w) = 1$$
$$\mu(q_5, w) = 1 \cdot (\mu(q_1, x) + \mu(q_2, x))$$

$$(2) \qquad \qquad \qquad \qquad \qquad (3)$$

We remark that the values for the nodes $(q_3, z)$ and $(q_4, y)$ as well as all isolated nodes like $(q_1, w)$ or $(q_5, y)$ are 0, because for them the inner summation in (2) is over the empty set. The family of solutions of this equation system has as non-zero values $\mu(q_1, x) = 0.5 + \epsilon$, $\mu(q_2, x) = 0.5 - \epsilon$, $\mu(q_3, y) = 1$, $\mu(q_4, z) = 1$,

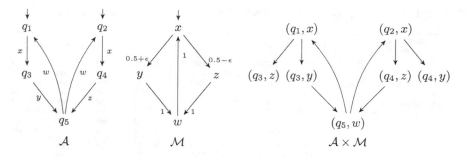

**Fig. 3.** An example of a GBA $\mathcal{A}$, a PMC $\mathcal{M}$, and their product $\mathcal{A} \times \mathcal{M}$

and $\mu(q_5, w) = 1$. From this, we have $\mathbb{P}^{\mathcal{M}}(\mathcal{L}(\mathcal{A})) = \mu(q_1, x) + \mu(q_2, x) = 1$ for any well-defined evaluation $v$, i.e., an evaluation such that $v(\epsilon) \in (-0.5, 0.5)$.

Let us summarise this section with the following results: given a parametric Markov chain $\mathcal{M}$ and an LTL formula $\phi$, both the emptiness checking and the qualitative-correctness computation could be done within time $\mathcal{O}(|\mathcal{M}| \cdot 2^{|el(\phi)|})$. When the specification is given as an almost separated automaton $\mathcal{A}$, the time complexity is $\mathcal{O}(|\mathcal{M}| \cdot |\mathcal{A}|)$. For quantitative computation, after we get all reachable locally positive SCCs, we reduce the problem to solving the equation system listed in Theorem 3.

## 5   Experiment Results

We have implemented our synthesis algorithm for LTL properties of parametric Markov chains in our tool IscasMC using an explicit state-space representation. The machine we used for the experiments is a 3.6 GHz Intel Core i7-4790 with 16 GB 1600 MHz DDR3 RAM of which 12 GB assigned to the tool; the time-out has been set to 30 min. We considered three models, namely the Bounded Retransmission Protocol (BRP) [23] in the version of [15], Randomized Protocol for Signing Contracts (RPSC) [19] and the Crowds protocol for anonymity [34]. We have replaced the probabilistic choices by parametric choices to obtain the admissible failure probabilities.

In our implementation, we use Z3 [16] to solve the equation system of Theorem 3 because by using parametric transition probabilities the equation system from Theorem 3 becomes nonlinear. Performance results of the experiments are shown in Tables 1 to 3, where the columns have the following meaning: In column "Model", the information about the model, the name of the constants influencing the model size, and the analysed property is provided; "Constants" contains the values for the constants defining the model; "$|S_{\mathcal{M}}|$" and "$|V_{\mathcal{G}}|$" denote the number of states and vertices in $\mathcal{M}$ and $\mathcal{G}$, respectively; "$SCC_{\mathcal{G}}$" reports the number of non-trivial SCCs checked in $\mathcal{G}$ out of which "$SCC_{pos}$" are the positive ones; "$T_{\mathcal{G}}$" is the time spent by constructing and checking the product graph; "$|Vars_{Z3}|$", "$|Cons_{Z3}|$" and "$T_{Z3}$" record the number of variables and

**Table 1.** Experimental Results for Parametric Markov chain models

| Model (Constants)<br>Property | Constants | $\|S_\mathcal{M}\|$ | $\|V_{\mathcal{G}}\|$ | $SCC_\mathcal{G}$ | $SCC_{pos}$ | $T_\mathcal{G}$ | $\|Vars_{Z3}\|$ | $\|Cons_{Z3}\|$ | $T_{Z3}$ | $T_{mc}$ |
|---|---|---|---|---|---|---|---|---|---|---|
| | 2,50 | 20534 | 20535 | 1372 | 1 | <1 | 314 | 320 | 13 | 13 |
| | 2,60 | 28446 | 28447 | 1938 | 1 | <1 | 374 | 380 | 42 | 43 |
| Crowds (TotalRuns, CrowdSize) | 2,70 | 39131 | 39132 | 2613 | 1 | <1 | 434 | 440 | 80 | 82 |
| | 2,80 | 51516 | 51517 | 3388 | 1 | 1 | 494 | 500 | 92 | 94 |
| $\mathbb{P} \geq 0.9[GF(newInstance \wedge$ | 2,90 | 65601 | 65602 | 4263 | 1 | 1 | 554 | 560 | 162 | 164 |
| $runCount = 0 \wedge observe0 \geq 1)]$ | 2,100 | 81386 | 81387 | 5238 | 1 | 2 | 614 | 620 | 198 | 202 |
| | 2,110 | 98871 | 98872 | 6313 | 1 | 2 | 674 | 680 | 404 | 407 |
| | 2,120 | 118056 | 118057 | 7488 | 1 | 2 | 734 | 740 | 436 | 439 |
| | 512,80 | 1084476 | 1085505 | 1111 | 0 | 15 | NE | NE | NE | 16 |
| | 512,100 | 1351476 | 1352505 | 1131 | 0 | 22 | NE | NE | NE | 22 |
| | 512,120 | 1618476 | 1619505 | 1151 | 0 | 41 | NE | NE | NE | 42 |
| BRP (N, MAX) | 512,140 | 1885476 | 1886505 | 1171 | 0 | 54 | NE | NE | NE | 54 |
| | 512,160 | 2152476 | 2153505 | 1191 | 0 | 75 | NE | NE | NE | 76 |
| $\mathbb{P} \geq 0.9[GF(s = 5 \wedge T)]$ | 512,180 | 2419476 | 2420505 | 1211 | 0 | 87 | NE | NE | NE | 87 |
| | 512,200 | 2686476 | 2687505 | 1231 | 0 | 110 | NE | NE | NE | 111 |
| | 512,220 | 2953476 | 2954505 | 1251 | 0 | 136 | NE | NE | NE | 137 |
| | 5,15 | 300030 | 305492 | 1 | 0 | 5 | NE | NE | NE | 5 |
| | 5,20 | 402430 | 409442 | 1 | 0 | 9 | NE | NE | NE | 9 |
| | 5,25 | 504830 | 513392 | 1 | 0 | 12 | NE | NE | NE | 12 |
| RPSC (N, L) | 5,30 | 607230 | 617342 | 1 | 0 | 19 | NE | NE | NE | 19 |
| | 5,35 | 709630 | 721292 | 1 | 0 | 25 | NE | NE | NE | 25 |
| $\mathbb{P} \geq 0.9[GF(\neg knowA \wedge knowB)]$ | 5,40 | 812030 | 825242 | 1 | 0 | 32 | NE | NE | NE | 32 |
| | 5,45 | 914430 | 929192 | 1 | 0 | 39 | NE | NE | NE | 39 |
| | 5,50 | 1016830 | 1033142 | 1 | 0 | 46 | NE | NE | NE | 47 |

constraints of the equation system we input into Z3 and its solution time; and "$T_{mc}$" gives the total time spent for constructing and analysing the product $\mathcal{G}$ and solving the equation system. In the tables, entries marked by "NE" mean that Z3 has not been executed as there were no locally positive SCCs, thus the construction and evaluation of the equation system can be avoided; entries marked by "–" mean that the operation has not been performed since the analysis has been interrupted in a previous stage; the marks "TO" and "MO" stand for a termination by timeout or memory out.

As we can see from Tables 1 and 2 relative to parametric Markov chains, the implementation of the analysis method presented in this paper is able to check models in the order of millions of states. The parameter synthesis time is mainly depending on the behavior of Z3 and on the number and size of the SCCs of the product, since each SCC has to be classified as positive or non-positive. Regarding Z3, we can see that the time it requires for solving the provided system is loosely related to the size of the system: the Crowds cases in Table 1 take hundreds of seconds for a system of size less than one thousand while the RPSC cases in Table 2 are completed in less than one hundred seconds even if the system size is more than one million. In Table 3, we list some experimental results for the Crowds protocol modelled as an interval Markov chain. As for the previous cases, it is the solution of the equation system to limit the size of the models we can analyse, so a more performing solver would improve considerably

**Table 2.** Experimental Results for Parametric Markov chain models

| Model (Constants) Property | Constants | $|S_\mathcal{M}|$ | $|V_\mathcal{G}|$ | $SCC_\mathcal{G}$ | $SCC_{pos}$ | $T_\mathcal{G}$ | $|Vars_{Z3}|$ | $|Cons_{Z3}|$ | $T_{Z3}$ | $T_{mc}$ |
|---|---|---|---|---|---|---|---|---|---|---|
| Crowds (TotalRuns, CrowdSize)<br><br>$\mathbb{P} \geq 0.9[(Fobserve0 > 1 \vee Gobserve1 > 1)$<br>$\wedge(Fobserve2 > 1 \vee Gobserve3 > 1)]$ | 2,50 | 20534 | 82149 | 1374 | 0 | 4 | NE | NE | NE | 5 |
| | 2,60 | 28446 | 113797 | 1940 | 0 | 6 | NE | NE | NE | 7 |
| | 2,70 | 39131 | 156537 | 2615 | 0 | 10 | NE | NE | NE | 11 |
| | 2,80 | 51516 | 206077 | 3390 | 0 | 21 | NE | NE | NE | 21 |
| | 2,90 | 65601 | 262417 | 4265 | 0 | 26 | NE | NE | NE | 27 |
| | 2,100 | 81386 | 325557 | 5240 | 0 | 108 | NE | NE | NE | 108 |
| | 2,110 | 98871 | 395497 | 6315 | 0 | 109 | NE | NE | NE | 110 |
| | 2,120 | 118056 | 472237 | 7490 | 0 | 118 | NE | NE | NE | 118 |
| BRP (N, MAX)<br><br>$\mathbb{P} \geq 0.9[F(s = 5) \wedge FG(rrep = 2)]$ | 512,10 | 149976 | 893661 | 1055 | 0 | 32 | NE | NE | NE | 33 |
| | 512,20 | 283476 | 1692861 | 1075 | 0 | 114 | NE | NE | NE | 114 |
| | 512,30 | 416976 | 2492061 | 1095 | 0 | 241 | NE | NE | NE | 241 |
| | 512,40 | 550476 | 3291261 | 1115 | 0 | 476 | NE | NE | NE | 477 |
| | 512,50 | 683976 | 4090461 | 1135 | 0 | 622 | NE | NE | NE | 623 |
| | 512,60 | 817476 | 4889661 | 1155 | 0 | 962 | NE | NE | NE | 962 |
| | 512,70 | 950976 | 5688861 | 1175 | 0 | 1219 | NE | NE | NE | 1220 |
| | 512,80 | 1084476 | MO | – | – | – | – | – | – | – |
| RPSC (N, L)<br><br>$\mathbb{P} \geq 0.9[F(\neg knowA) \vee G(knowB)]$ | 5,15 | 300030 | 1200117 | 4 | 1 | 68 | 300031 | 300036 | 32 | 106 |
| | 5,20 | 402430 | 1609717 | 4 | 1 | 122 | 402431 | 402436 | 42 | 169 |
| | 5,25 | 504830 | 2019317 | 4 | 1 | 178 | 504831 | 504836 | 49 | 230 |
| | 5,30 | 607230 | 2428917 | 4 | 1 | 278 | 607231 | 607236 | 75 | 373 |
| | 5,35 | 709630 | 2838517 | 4 | 1 | 384 | 709631 | 709636 | 86 | 488 |
| | 5,40 | 812030 | 3248117 | 4 | 1 | 490 | 812031 | 812036 | 95 | 600 |
| | 5,45 | 914430 | 3657717 | 4 | 1 | 606 | 914431 | 914436 | 104 | 719 |
| | 5,50 | 1016830 | 4067317 | 4 | 1 | 726 | 1016831 | 1016836 | 108 | 839 |

**Table 3.** Experimental results for crowds interval Markov chain model

| Model (Constants) Property | Constants | $|S_\mathcal{M}|$ | $|V_\mathcal{G}|$ | $SCC_\mathcal{G}$ | $SCC_{pos}$ | $T_\mathcal{G}$ | $|Vars_{Z3}|$ | $|Cons_{Z3}|$ | $T_{Z3}$ | $T_{mc}$ |
|---|---|---|---|---|---|---|---|---|---|---|
| Crowds (TotalRuns, CrowdSize)<br><br>$\mathbb{P} \geq 0.9[GF(newInstance\wedge$<br>$runCount = 0 \wedge observe0 \geq 1)]$ | 2,6 | 423 | 424 | 36 | 1 | <1 | 240 | 473 | <1 | <1 |
| | 2,8 | 698 | 699 | 55 | 1 | <1 | 380 | 760 | <1 | <1 |
| | 2,10 | 1041 | 1042 | 78 | 1 | <1 | 552 | 1115 | 1 | 1 |
| | 2,12 | 1452 | 1453 | 105 | 1 | <1 | 756 | 1538 | 15 | 15 |
| | 2,14 | 1931 | 1932 | 136 | 1 | <1 | 992 | 2029 | 15 | 15 |
| | 2,16 | 3093 | 3094 | 210 | 1 | <1 | 1260 | 2588 | TO | TO |

the applicability of our approach, in particular when we can not exclude that the formula is satisfiable, as happens when there are no positive SCCs.

# 6  Conclusion

In this paper we have surveyed the parameter synthesis of PLTL formulas with respect to parametric Markov chains. The algorithm first transforms the LTL specification to an almost separated automaton and then builds the product graph of the model under consideration and this automaton. Afterwards, we reduce the parameter synthesis problem to solving an (nonlinear) equation system, which allows us employ an SMT solver to obtain feasible parameter values. We have conducted experiments to demonstrate that our techniques indeed work for models of realistic size. To the best of our knowledge, our method is the first approach for the PLTL synthesis problem for parametric Markov chains which is single exponential in the size of the property.

**Acknowledgement.** This work is supported by the CDZ project CAP (GZ 1023), by the Chinese Academy of Sciences Fellowship for International Young Scientists, by the National Natural Science Foundation of China (Grants No. 61532019, 61472473, 61550110249, 61550110506, 61103012, 61379054, and 61272335), and by the CAS/SAFEA International Partnership Program for Creative Research Teams.

# References

1. Baier, C., Katoen, J.-P.: Principles of Model Checking. MIT Press, Cambridge (2008)
2. Baier, C., Kiefer, S., Klein, J., Klüppelholz, S., Müller, D., Worrell, J.: Markov chains and unambiguous Büchi automata. In: Chaudhuri, S., Farzan, A. (eds.) CAV 2016. LNCS, vol. 9779, pp. 23–42. Springer, Heidelberg (2016). doi:10.1007/978-3-319-41528-4_2
3. Benedikt, M., Lenhardt, R., Worrell, J.: LTL model checking of interval Markov chains. In: Piterman, N., Smolka, S.A. (eds.) TACAS 2013. LNCS, vol. 7795, pp. 32–46. Springer, Heidelberg (2013). doi:10.1007/978-3-642-36742-7_3
4. Bianco, A., Alfaro, L.: Model checking of probabilistic and nondeterministic systems. In: Thiagarajan, P.S. (ed.) FSTTCS 1995. LNCS, vol. 1026, pp. 499–513. Springer, Heidelberg (1995). doi:10.1007/3-540-60692-0_70
5. Bustan, D., Rubin, S., Vardi, M.Y.: Verifying $\omega$-regular properties of Markov chains. In: Alur, R., Peled, D.A. (eds.) CAV 2004. LNCS, vol. 3114, pp. 189–201. Springer, Heidelberg (2004). doi:10.1007/978-3-540-27813-9_15
6. Carton, O., Michel, M.: Unambiguous Büchi automata. TCS **297**(1–3), 37–81 (2003)
7. Chatterjee, K., Gaiser, A., Křetínský, J.: Automata with generalized Rabin pairs for probabilistic model checking and LTL synthesis. In: Sharygina, N., Veith, H. (eds.) CAV 2013. LNCS, vol. 8044, pp. 559–575. Springer, Heidelberg (2013). doi:10.1007/978-3-642-39799-8_37
8. Chatterjee, K., Sen, K., Henzinger, T.A.: Model-checking $\omega$-regular properties of interval Markov chains. In: Amadio, R. (ed.) FoSSaCS 2008. LNCS, vol. 4962, pp. 302–317. Springer, Heidelberg (2008). doi:10.1007/978-3-540-78499-9_22
9. Ciesinski, F., Baier, C.: LiQuor: a tool for qualitative and quantitative linear time analysis of reactive systems. In: QEST, pp. 131–132 (2006)
10. Clarke, E.M.: The birth of model checking. In: Grumberg, O., Veith, H. (eds.) 25 Years of Model Checking. LNCS, vol. 5000, pp. 1–26. Springer, Heidelberg (2008). doi:10.1007/978-3-540-69850-0_1
11. Clarke, E., Grumberg, O., Hamaguchi, K.: Another look at LTL model checking. In: Dill, D.L. (ed.) CAV 1994. LNCS, vol. 818, pp. 415–427. Springer, Heidelberg (1994). doi:10.1007/3-540-58179-0_72
12. Clarke, E.M., Grumberg, O., Peled, D.: Model Checking. MIT Press, Cambridge (2001)
13. Courcoubetis, C., Yannakakis, M.: The complexity of probabilistic verification. J. ACM **42**(4), 857–907 (1995)
14. Couvreur, J.-M., Saheb, N., Sutre, G.: An optimal automata approach to LTL model checking of probabilistic systems. In: Vardi, M.Y., Voronkov, A. (eds.) LPAR 2003. LNCS (LNAI), vol. 2850, pp. 361–375. Springer, Heidelberg (2003). doi:10.1007/978-3-540-39813-4_26

15. D'Argenio, P.R., Jeannet, B., Jensen, H.E., Larsen, K.G.: Reachability analysis of probabilistic systems by successive refinements. In: Alfaro, L., Gilmore, S. (eds.) PAPM-PROBMIV 2001. LNCS, vol. 2165, pp. 39–56. Springer, Heidelberg (2001). doi:10.1007/3-540-44804-7_3

16. Moura, L., Bjørner, N.: Z3: an efficient SMT solver. In: Ramakrishnan, C.R., Rehof, J. (eds.) TACAS 2008. LNCS, vol. 4963, pp. 337–340. Springer, Heidelberg (2008). doi:10.1007/978-3-540-78800-3_24

17. Dehnert, C., Junges, S., Jansen, N., Corzilius, F., Volk, M., Bruintjes, H., Katoen, J.-P., Ábrahám, E.: PROPhESY: a PRObabilistic ParamEter SYnthesis tool. In: Kroening, D., Păsăreanu, C.S. (eds.) CAV 2015. LNCS, vol. 9206, pp. 214–231. Springer, Heidelberg (2015). doi:10.1007/978-3-319-21690-4_13

18. Esparza, J., Křetínský, J.: From LTL to deterministic automata: a safraless compositional approach. In: Biere, A., Bloem, R. (eds.) CAV 2014. LNCS, vol. 8559, pp. 192–208. Springer, Heidelberg (2014). doi:10.1007/978-3-319-08867-9_13

19. Even, S., Goldreich, O., Lempel, A.: A randomized protocol for signing contracts. Commun. ACM **28**(6), 637–647 (1985)

20. Hahn, E.M., Hermanns, H., Zhang, L.: Probabilistic reachability for parametric Markov models. STTT **13**(1), 3–19 (2011)

21. Hahn, E.M., Li, Y., Schewe, S., Turrini, A., Zhang, L.: IscasMC: a web-based probabilistic model checker. In: Jones, C., Pihlajasaari, P., Sun, J. (eds.) FM 2014. LNCS, vol. 8442, pp. 312–317. Springer, Heidelberg (2014). doi:10.1007/978-3-319-06410-9_22

22. Hansson, H., Jonsson, B.: A logic for reasoning about time and reliability. FAC **6**(5), 512–535 (1994)

23. Helmink, L., Sellink, M.P.A., Vaandrager, F.W.: Proof-checking a data link protocol. In: Barendregt, H., Nipkow, T. (eds.) TYPES 1993. LNCS, vol. 806, pp. 127–165. Springer, Heidelberg (1994). doi:10.1007/3-540-58085-9_75

24. Jonsson, B., Larsen, K.G.: Specification and refinement of probabilistic processes. In: LICS, pp. 266–277 (1991)

25. Katoen, J.-P., Klink, D., Leucker, M., Wolf, V.: Three-valued abstraction for probabilistic systems. J. Log. Algebr. Program. **81**(4), 356–389 (2012)

26. Katoen, J.-P., Zapreev, I.S., Hahn, E.M., Hermanns, H., Jansen, D.N.: The ins and outs of the probabilistic model checker MRMC. Perform. Eval. **68**(2), 90–104 (2011)

27. Kemeny, J.G., Snell, J.L., Knapp, A.W.: Denumerable Markov Chains. D. Van Nostrand Company, New York (1966)

28. Kini, D., Viswanathan, M.: Limit deterministic and probabilistic automata for LTL\GU. In: Baier, C., Tinelli, C. (eds.) TACAS 2015. LNCS, vol. 9035, pp. 628–642. Springer, Heidelberg (2015). doi:10.1007/978-3-662-46681-0_57

29. Klein, J., Baier, C.: On-the-Fly Stuttering in the Construction of Deterministic ω-Automata. In: Holub, J., Žd'árek, J. (eds.) CIAA 2007. LNCS, vol. 4783, pp. 51–61. Springer, Heidelberg (2007). doi:10.1007/978-3-540-76336-9_7

30. Kwiatkowska, M., Norman, G., Parker, D.: PRISM 4.0: verification of probabilistic real-time systems. In: Gopalakrishnan, G., Qadeer, S. (eds.) CAV 2011. LNCS, vol. 6806, pp. 585–591. Springer, Heidelberg (2011). doi:10.1007/978-3-642-22110-1_47

31. Li, Y., Liu, W., Turrini, A., Hahn, E.M., Zhang, L.: An efficient synthesis algorithm for parametric Markov chains against linear time properties (2016). CoRR. http://arxiv.org/abs/1605.04400

32. Liu, W., Wang, J.: A tighter analysis of Piterman's Büchi determinization. Inf. Process. Lett. **109**(16), 941–945 (2009)

33. Piterman, N.: From nondeterministic Büchi and Streett automata to deterministic parity automata. LMCS **3**(3:5), 1–21 (2007)
34. Reiter, M., Rubin, A.: Crowds: anonymity for web transactions. ACM TISSEC **1**(1), 66–92 (1998)
35. Safra, S.: On the complexity of $\omega$-automata. In: FOCS, pp. 319–327 (1988)
36. Schewe, S.: Tighter bounds for the determinisation of Büchi automata. In: Alfaro, L. (ed.) FoSSaCS 2009. LNCS, vol. 5504, pp. 167–181. Springer, Heidelberg (2009). doi:10.1007/978-3-642-00596-1_13
37. Sen, K., Viswanathan, M., Agha, G.: Model-checking Markov chains in the presence of uncertainties. In: Hermanns, H., Palsberg, J. (eds.) TACAS 2006. LNCS, vol. 3920, pp. 394–410. Springer, Heidelberg (2006). doi:10.1007/11691372_26
38. Vardi, M.Y.: Automatic verification of probabilistic concurrent finite-state programs. In: FOCS, pp. 327–338 (1985)

# Time-Bounded Statistical Analysis
# of Resource-Constrained Business Processes
# with Distributed Probabilistic Systems

Ratul Saha[1($\boxtimes$)], Madhavan Mukund[2], and R.P. Jagadeesh Chandra Bose[3]

[1] National University of Singapore, Singapore, Singapore
ratul@comp.nus.edu.sg
[2] Chennai Mathematical Institute, Chennai, India
madhavan@cmi.ac.in
[3] Xerox Research Center India, Bengaluru, India
jcbose@gmail.com

**Abstract.** Business processes often incorporate stochastic decision points, either due to uncontrollable actions or because the control flow is not fully specified. Formal modeling of such business processes with resource constraints and multiple instances is hard because of the interplay among stochastic behavior, concurrency, real-time and resource contention. In this setting, statistical techniques are easier to use and more scalable than numerical methods to verify temporal properties. However, existing approaches towards simulation techniques of business processes typically rest on shaky theoretical foundations. In this paper, we propose a modular approach towards modeling stochastic resource-constrained business processes. We analyze such processes in presence of commonly used resource-allocation strategies. Our model, Distributed Probabilistic Systems (DPS), incorporates a set of probabilistic agents communicating among each other in fixed-duration real-time. Our methodology admits statistical analysis of business processes with provable error bounds. We also illustrate a number of real-life scenarios that can be modeled and verified using this approach.

## 1 Introduction

In recent years, a plethora of models have been proposed in the area of Business Process Management (BPM) [3,10]. These models have been used to analyze large processes from diverse industry sectors such as Internet companies, health care, and finance services. The tasks in these processes are typically mapped to a finite set of shared resources whose allocation depends on a variety of practical constraints. In addition, each process is often replicated as a large number of instances. To optimize performance, one needs to be able to analyze resource-constrained business processes with well-defined confidence in the result.

BPM systems often do not realize deterministic behavior and incorporate stochastic decision points. This is due to both the increasing complexity of such systems, which makes the exact control flow difficult to capture, as well as the

© Springer International Publishing AG 2016
M. Fränzle et al. (Eds.): SETTA 2016, LNCS 9984, pp. 297–314, 2016.
DOI: 10.1007/978-3-319-47677-3_19

inclusion of uncontrollable components in business processes. Even when the probability distributions can be measured or approximated from domain knowledge or historical data [2], model-based analysis [3] of such systems is hard due to the interplay between stochastic behavior, concurrency, time taken to perform tasks, and resource contention.

We propose a novel modular approach towards modelling resource-constrained BPM (rcBPM) systems. Such systems have a finite set of resources allocated across replicated instances of a stochastic business process. A business process is a set of tasks with logical and temporal dependencies. Each task is mapped to one of the available resources that can perform the task. Resources are assigned following a predetermined allocation strategy. Each task has an execution time, ideally drawn from a probability distribution. For simplicity, we assume the time taken by a task to be fixed—say the mean value of the distribution.

A *case* is an instance of the process model. Multiple cases run in parallel, sharing the same set of resources. Cases need not start simultaneously. We study systems with a fixed number of cases arriving within a given period of time. The cases may follow an arrival process such as a Poisson process.

An example of an rcBPM system is the loan/overdraft process in a financial institution, where cases correspond to applications from different customers. Tasks in the process may include 'submitting', 'reviewing', 'accepting' or 'declining' the application.

We observe that tasks and resources can be considered as individual agents that behave independently and communicate among each other. There are two types of communications among them: (i) task-task interaction, where a completed task passes the thread of control to some other tasks, and (ii) task-resource interaction, which describes the allocation of a task to a resource.

This observation leads us to introduce *Distributed Probabilistic Systems (DPS)* to model rcBPM systems. A DPS consists of a set of communicating probabilistic agents. Each agent has a finite local state space. Periodically, agents synchronize with each other on common *actions* to perform joint probabilistic *events*. Each event has a fixed duration and cost. A scheduler is used to resolve non-determinism: if an agent can take part in more than one synchronization at a global state, the scheduler determines which one of them is to be performed.

In general, after the scheduler resolves non-determinism, multiple independent actions are enabled at a global state of a DPS, which can be executed concurrently. Each synchronization action among a set of agents leads to an event being chosen according to a probability distribution. Each event has a fixed duration, after which the local states of the participating agents change. We show that the dynamics of a DPS with a fixed scheduler can be captured as a discrete-time Markov chain. The DPS model can be viewed as a Markov Decision Process (MDP) variant of Distributed Markov Chains (DMC) [20].

We model an rcBPM system as a DPS. Each task is an individual agent incorporating the states of a task, such as 'ready to perform', 'waiting for a resource', 'busy executing', 'finished' etc. When a task finishes, it triggers other task agents in the control flow that are ready to perform. Each resource is also a simple agent, looping between being 'available' and 'busy'. The scheduler maps each task waiting for a resource to an available resource. This results in a synchronization that generates an event whose duration captures the time taken to perform the task. We also model the start and end states of a process as agents, to model the arrival and completion of a case.

To verify temporal properties of rcBPM systems, we use Statistical Model Checking (SMC) [14,24]. A typical property is of the form "when $C$ cases arrive with constant arrival density $\lambda$, at least $x$ % of the cases will complete within time $t$, with probability at least $p$". Since the DPS model of an rcBPM system can be viewed as a discrete-time Markov chain, we can simulate an rcBPM system and use hypothesis testing to verify properties with provable error bounds.

We illustrate our approach using a business process [7] depicting loan/overdraft applications of a large financial institution. The process has been mined from real-life event logs from BPM Challenge 2012 [22]. The process has 46 resources for performing 15 tasks in the process. We scaled up to 500 cases arriving at the rate of 1 case every 10 s. We show (i) the relationship between the number of cases arriving within a fixed time bound and the fraction of cases that complete, and (ii) the relationship between the minimum time of completion and the fraction of cases completed for a fixed number of total cases.

To summarize, our contribution is as follows: (i) we propose a model of distributed probabilistic systems where events are assigned time and cost values, and demonstrate that the model acts as a Markov chain for a well-behaved class of schedulers, (ii) we provide a strong theoretical foundation for resource-constrained business processes modelled as distributed probabilistic systems, (iii) we demonstrate a statistical model checking based technique for inferring time-bounded properties of business processes with multiple cases and a finite set of shared resources. To the best of our knowledge, this is the first attempt to provide a simulation based technique for analyzing resource-constrained business processes with provable error bounds and sound sample size analysis.

The paper is structured as follows. Section 2 introduces resource-constrained BPM systems and different properties of interest. Section 3 defines the Distributed Probabilistic System (DPS) model and a statistical model checking technique for DPS. Sections 4 and 5 provide a proof-of-concept demonstration of our approach. Section 4 describes how rcBPM systems are modelled using DPS. Section 5 discusses experimental results for a simple example. Finally, Sect. 6 provides a summary and future directions.

**Related Work.** The need for an underlying formal model for business processes has been long felt. Workflow nets (WF nets, a class of Petri nets), equipped with clean graphical notation and abundance of analysis techniques, have served as a solid framework for BPM systems [1,4,6].

For this discussion, we restrict to related work that involves modelling or analyzing stochastic business processes. The most prominent modelling formalism for stochastic analysis for business processes is (generalized) stochastic WF nets [9,18]. Such a system is modelled using exponentially distributed firing delay with timed transitions. A few recent papers [8,15,21] also demonstrate a generic Markovian analysis that is mostly applicable to rigidly structured WF systems. The work of [12] focuses on modelling BPM systems as Markov decision processes in the language of PRISM [13]. These works are either very simplistic in nature, where block-like patterns are chained together, or hard-to-tackle models involving arbitrary nondeterminism that cannot be readily adopted for sound simulation techniques. Most importantly, extending these approaches to model business processes with shared resources across multiple cases is not obvious.

Analyzing resource-constrained BPM systems with simulation-based techniques is not new [16,17], but rigorous statistical analysis has often been limited to computing analysis of variance and confidence intervals. The deductions are often difficult to justify and can be arbitrarily far from truth—van der Aalst correctly points out that "simulation does not provide any proof" [5]. Our work is the first to (i) provably bound the error of analysis of business processes with finite resources shared across multiple cases arriving at different time points and (ii) provide a sound analysis of the sample size of simulation.

## 2    Resource-Constrained BPM (rcBPM) Systems

A resource-constrained BPM (rcBPM) system consists of two main components: (i) the process, instantiated as a fixed number of cases, and (ii) a finite set of resources. An rcBPM system is then accompanied by a resource allocation strategy. To explain these, we use a process model that has been mined [7] from a real-life event log of loan/overdraft applications of a large financial institution. Along with the process, the average time for each resource to perform a particular task has been mined. The event log is obtained from BPI Challenge 2012 [22].

**The Process.** A process in an rcBPM system consists of a start state, a finite set of tasks and an end state. The tasks in the process are combined in sequence or parallel. We assume that the control flow is probabilistic: a discrete probability distribution is attached to each set of outgoing choices and sequential tasks have outgoing probability 1. Each case is an instantiated copy of the process.

**The Example.** Figure 1 demonstrates, in Petri net notation, a process for loan/overdraft applications of a large financial institution. We consider two sub-processes—namely the application and offer sub-processes—of the overall loan/overdraft application process. The tasks of the application and offer sub-processes are prefixed with "A_" and "O_" respectively. From historical data, we estimate the probability values of outgoing edges from places in the Petri net. Unmarked edges have probability 1. The process starts with the submission of an application (A_Submitted/A_PartlySubmitted). An application can be declined (A_Declined) if it does not pass any checks. The probability of an application being declined outright is estimated to be 0.84. Applications that pass the checks are pre-accepted (A_PreAccepted) with probability 0.16. Often additional information is obtained by contacting the customer by phone. Based on this information, an application can be cancelled (A_Cancelled) with probability 0.63 or accepted (A_Accepted) with probability 0.37. Once an application

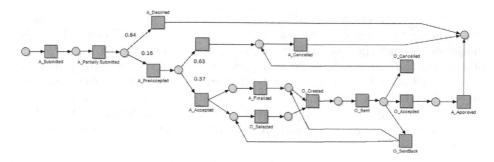

**Fig. 1.** An example process of an rcBPM system (in Petri net notation).

is accepted, it is finalized (A_Finalized) and, in parallel, an offer is selected for the customer (O_Selected). An offer is then created (O_Created) and sent to eligible applicants and their responses are assessed (O_Sent). A customer then may (i) accept the offer (O_Accepted) with probability 0.01, (ii) cancel the offer (O_Cancelled) with probability 0.95, or (iii) send the offer back (O_SentBack) with probability 0.04. If an offer is sent back, a new offer is created for the customer. If the offer is accepted, the application is finally approved (A_Approved). Declining, cancellation or approval signals the end of the application.

**Resources.** A finite set of resources is provided, each capable of performing a subset of tasks in the process. Different resources can take different time for performing the same task. We may further group resources with the same behaviour and capability into roles.

In the running example, we profiled the top 46 of the busiest resources from the event log. The resource-task matrix with cell values representing the average time taken by the resource on the particular task is shown (truncated) in Table 1 in the appendix. The columns indicate the resources and the rows indicate the tasks. If a cell is empty, it indicates that the task is not assigned to the particular resource. If the value in a cell is 0, the resource performs the task instantly.

**Resource Allocation Strategy.** At any moment, multiple tasks can compete for a number of available resources and, conversely, multiple resources may be available for a single task. The resource allocation strategy is responsible for assigning each task to a single resource. A strategy is said to be deterministic if, given the same snapshot of the system, the strategy always picks the same resource allocation for tasks. While our model can be extended to accommodate randomized schedulers, to keep the formalism simple, we focus on deterministic schedulers. For example, the flexible assignment policy [16] is deterministic: given a specialist-generalist ratio, it assigns the most specialist (available) resource to a given task.

In the running example, we assume that a priority based scheduler is available. This scheduler assumes that there is an ordering among cases—one can think of it as different tiers (platinum/gold/silver) of customers. We also assume that given a case, there is an ordering among its tasks. Hence, there is a total ordering among tasks across all cases. For simplicity, we also assume a total order among the resources. Hence at any configuration of the system, the scheduler goes through the resources in ascending order. For each resource, if a set of assigned tasks are available for that particular resource, it schedules the lowest ranked task among them. We note that such a scheduling policy may not be optimal, but real-life schedulers for business processes are often simple in nature. Also, while our model permits more complex schedulers, we have chosen a simpler one to explain the approach and demonstrate some experimental results.

**The Modeling Formalism and Properties of Interest.** We do not restrict ourselves to any particular modeling formalism and discuss the support of our modeling technique in terms of workflow patterns [19]. We support the core patterns—sequence (WCP-1), parallel split (WCP-2), synchronization (WCP-3), (probabilistic) exclusive choice (WCP-4), structured synchronizing merge (WCP-7). In short, we allow parallel and XOR-splits as well as parallel and XOR-joins as long as each parallel split is matched with a parallel join. With some engineering, our approach can be extended to more unstructured models as long as arbitrarily new thread of controls cannot be spawned. However for this work, we focus on structured business processes.

We are interested in time-bounded temporal properties of rcBPM systems. A fixed number of cases arrive following an arrival process. Each task and resource can have their own time limit. Given a time limit, we are interested to verify linear temporal properties of business processes bounding the probability of error. In the running example, let us assume that we are interested in the following property: when $C$ cases arrive at a rate of $\lambda$ cases per second, with probability at least $p$, at least $x\%$ cases are completed within time $t$. While we can analyze more complex temporal properties, in this work, we focus on reachability properties. Note that we can also find optimal parameter values using binary search. For example, if we are interested in optimizing $x$, we fix the other values $p$, $t$, $\lambda$ and $C$ and use simulations to do a binary search for the optimal value of $x$ in the range $[0, 100]$.

The state-of-the-art approach for analyzing stochastic business process models is to simulate the system an arbitrary number of times. If $p\%$ of the simulations satisfy the desired property, one claims that the property is true with probability at least $p$, perhaps with confidence interval estimates of certain parameters. However, such conclusions are dependent on the sample size of the simulation and correctness is not guaranteed. We apply the theory of hypothesis testing and sequential probability ratio test, which formally connects the sample size to the desired margin for error. The conclusions that we draw from our experiments come with guaranteed bounds on the probability of error due to false positives and false negatives. Our strategy is explained in detail in Sect. 3.

# 3    The Distributed Probabilistic System (DPS) Model

We formulate a model in which a distributed set of agents interact through actions that synchronize subsets of agents. A synchronization results in a probabilistic choice between a set of events, each with its own time duration and cost. We can then use hypothesis testing for statistical model checking of quantitative properties of such systems.

**Definition 1 (Distributed State Space).** *A distributed state space over $n$ agents $[n] = \{1, 2, \ldots, n\}$ is a tuple $\mathcal{S} = (n, \{S_i\}, \{s_i^{in}\})$, such that for each agent $i \in [n]$, $S_i$ denotes its finite set of local states and $s_i^{in}$ denotes its local initial state. We abbreviate $[n]$-indexed sets $\{X_i\}_{i \in [n]}$ as $\{X_i\}$ when the context is clear.*

- *For non-empty $u \subseteq [n]$, $\mathbf{S}_u = \prod_{i \in u} S_i$ denotes the set of joint $u$-states of agents in $u$. We denote $\mathbf{S} = \mathbf{S}_{[n]}$ to be the set of global states.*
- *For a $u$-state $\mathbf{s} \in \mathbf{S}_u$ and $v \subseteq u$, $\mathbf{s}_v$ denotes the projection of $\mathbf{s}$ onto $v$. We do not distinguish between $\mathbf{S}_{\{i\}}$ and $\mathbf{S}_i$, nor between $\mathbf{s}_{\{i\}}$ and $\mathbf{s}_i$.*

**Definition 2 (Events and Actions)**

- *An* event *over a distributed state space $\mathcal{S}$ is a tuple $e = (src_e, tgt_e)$, such that $\emptyset \neq loc(e) \subseteq [n]$ specifies the agents that participate in $e$ and $src_e, tgt_e \in \mathbf{S}_{loc(e)}$ denote the source and target $loc(e)$-states of $e$.*
- *Let $\Sigma$ denote the set of all events over $\mathcal{S}$. Each event $e$ has a duration $\delta(e)$ and a cost $\chi(e)$, given by functions $\delta : \Sigma \to \mathbb{R}_{\geq 0}$ and $\chi : \Sigma \to \mathbb{R}_{\geq 0}$.*
- *An* action *over $(\mathcal{S}, \Sigma)$ is a collection of co-located events with the same source state, equipped with a probability distribution. Formally, an action is a pair $a = (E_a, \pi_a)$, where $E_a \subseteq \Sigma$ is such that for each $e, e' \in E_a$, $src_e = src_{e'}$, and $\pi_a : E_a \to [0, 1]$ is a probability distribution. We write $loc(a)$ for the set of agents participating in action $a$ and $src(a)$ for the common source state of the events in $E_a$. Let $A$ denote the set of all actions over $(\mathcal{S}, \Sigma)$.*
- *We assume that each event belongs to exactly one action. In other words $\Sigma = \bigcup_{a \in A} E_a$ and for each $a, b \in A$ such that $a \neq b$, $E_a \cap E_b = \emptyset$.*
- *For a global state $\mathbf{s} \in \mathbf{S}$, $en(\mathbf{s})$ is the set of actions enabled at $\mathbf{s}$. Formally, $en(\mathbf{s}) = \{a \mid src(a) = \mathbf{s}_{loc(a)}\}$.*
- *At a global state $\mathbf{s} \in \mathbf{S}$, a set of enabled actions $U$ is schedulable if each agent participates in at most one action in $U$. Formally, $U \subseteq en(\mathbf{s})$ is schedulable if for all $a, b \in U$ such that $a \neq b$, $loc(a) \cap loc(b) = \emptyset$. Note that a schedulable set of actions can be executed concurrently since the agents involved do not interfere with each other. Let $sch(\mathbf{s}) \subseteq 2^{en(\mathbf{s})} \setminus \emptyset$ denote the collection of schedulable sets of actions at $\mathbf{s}$.*

**Definition 3 (Distributed Probabilistic Systems).** *A Distributed Probabilistic System (DPS) is a tuple $\mathcal{D} = (\mathcal{S}, \Sigma, \delta, \chi, A)$ such that*

- $\mathcal{S} = (n, \{S_i\}, \{s_i^{in}\})$ *is a distributed state space,*
- $\Sigma$ *is the set of events over $\mathcal{S}$ with duration and cost functions $\delta$ and $\chi$, respectively,*
- $A$ *is the set of actions over $(\mathcal{S}, \Sigma)$.*

A DPS $\mathcal{D}$ evolves as follows. All agents start at their initial state, so the initial global state at time 0 is $\mathbf{s}^{in} = (s_1^{in}, s_2^{in}, \ldots, s_n^{in})$.

Suppose $\mathcal{D}$ is at a global state $\mathbf{s} = (s_1, s_2, \ldots, s_n)$ at time $t$. A set of schedulable actions $U \in sch(\mathbf{s})$ is chosen from the set of enabled actions. We assume the existence of a scheduler that guides this choice.

The actions in $U$ start concurrently and independently. For each action $a \in U$, an event $e_a = (src_e, tgt_e) \in E_a$ is chosen according to the probability distribution $\pi_a$, with an associated duration $\delta(e_a)$ and cost $\chi(e_a)$.

The durations $\{\delta(e_a)\}_{a \in U}$ fix a sequentialization of the events $\{e_a\}_{a \in U}$. In general, there will also be a pending list of partially executed events currently in progress, with their own completion times.

Among the list of pending events, old and new, the events with the earliest time to completion finish first. For each completed event $e$, the local states of agents in $loc(e)$ are changed to $tgt_e$, while the states of the agents $[n] \setminus loc(e)$ are unchanged.

This gives rise to a new global state where potentially a new set of actions are scheduled, and we repeat the process of choosing a set of actions to schedule. We have to ensure that the scheduler respects the decisions made earlier, so that all pending events remain compatible with the new choice.

We denote a global configuration of a DPS as a *snapshot*, consisting of a global state and a list of partially executed events, with their time to completion from the current time point.

**Definition 4 (Snapshots).** *A snapshot of a DPS $\mathcal{D}$ is a tuple $(\mathbf{s}, U, X)$ where $U \subseteq en(\mathbf{s})$ and $X = \{(a, e, t) \mid a \in U, e \in E_a, t \in \mathbb{R}_{\geq 0}\}$ such that:*

- $\mathbf{s}$ *is the current global state and $U \in en(\mathbf{s})$ is the set of actions that are currently being performed, which may not have started together.*
- *For each $a \in U$, there is exactly one entry $(a, e, t) \in X$ denoting that event $e \in E_a$ is in progress with time $t \leq \delta(e)$ till completion.*

Let $\mathcal{Y}$ be the set of snapshots. Though $\mathcal{Y}$ is uncountable, a DPS will give rise to a discrete set of reachable snapshots, determined by the duration function $\delta$.

Nondeterministic choices between actions are resolved by a scheduler. At each snapshot, the scheduler picks a schedulable set of actions that are pairwise independent. For consistency, this choice should include all the actions already in progress, but not necessarily new ones.

**Definition 5 (Schedulers).** *A scheduler $\mathcal{G}$ is defined over snapshots as follows. Let $y = (\mathbf{s}, U, X) \in \mathcal{Y}$ be a snapshot. Then $\mathcal{G}(y) \in sch(\mathbf{s})$ is such that $U \subseteq \mathcal{G}(y) \subseteq en(\mathbf{s})$.*

We also note that, in general, it is hard to define independence-respecting schedulers for distributed systems. This is closely related to defining local winning strategies in distributed games [11]. The main complication is that a sequentially defined scheduler must behave consistently across different linearizations that correspond to the same concurrent execution. However, in a DPS, the durations associated with the events fix a canonical linearization, so there is no need to reconcile decisions of the scheduler across different interleavings.

Once we fix a scheduler, we can associate a transition system associated with a DPS whose states are snapshots.

**Definition 6 (Transition System).** *Given a DPS $\mathcal{D}$ and a scheduler $\mathcal{G}$, we construct the transition system $TS = (\mathcal{S}, y^{in}, \rightarrow)$ where $\mathcal{S} \subseteq \mathcal{Y}$ is a set of snapshots, with the initial snapshot given by $y^{in} = (\mathbf{s}^{in}, \emptyset, \emptyset) \in \mathcal{S}$.*

*Given a snapshot $y = (\mathbf{s}, U, X) \in \mathcal{S}$, where $U = \{a_1, a_2, \ldots, a_m\}$ and $X = \{(a_1, e_1, t_1), \ldots, (a_m, e_m, t_m)\}$, we define the next snapshots from $y$ as follows.*

- *Let $\mathcal{G}(y)$ be the set of actions scheduled. Recall that $U \subseteq \mathcal{G}(y)$. Let $V = \mathcal{G}(y) \setminus U = \{b_1, b_2, \ldots, b_k\}$.*
- *For each action $b = (E_b, \pi_b) \in V$, we pick an event $e_b \in E_b$ according to $\pi_b$, with duration $\delta(e_b)$ and cost $\chi(e_b)$. This generates a list $Y = \{(b, e_b, \delta(e_b)) \mid b \in V, e_b \in E_b\}$. Note that all the events in $X \cup Y$ have pairwise disjoint locations.*
- *From $X \cup Y$, we pick the subset $E_{\min} = \{(a, e, t_{\min}) \mid t_{\min}$ is minimum across all triples in $X \cup Y\}$. We then update the snapshot as follows:*
  - *(i) For each $(a, e, t_{\min})$ in $E_{\min}$, set $\mathbf{s}_{loc(e)}$ to $tgt_e$, and remove $a$ from $U \cup V$ and $(a, e, t_{\min})$ from $X \cup Y$.*
  - *(ii) For each $(a, e, t)$ in $(X \cup Y) \setminus E_{\min}$, update $t$ to $t - t_{\min}$, thus reducing the time to completion of $e$ by $t_{\min}$.*

*This results in a new snapshot $(\mathbf{s}', U', X')$. Note that each probabilistic choice of events for the actions in $V$ deterministically determines the next snapshot.*

We label each transition from snapshot $y$ to $y'$ with a pair of sets $(E_V, E_{\min})$, where $V = \{b_1, b_2, \ldots, b_k\}$ is the new set of actions chosen by the scheduler at snapshot $y$, $E_V = \{e_{b_1}, e_{b_2}, \ldots, e_{b_k}\}$ is the set of events chosen corresponding to $V$ and $E_{\min}$ is the set of triples corresponding to the events that complete their execution at $y'$. We write such a transition in the usual way as $y \xrightarrow{(E_V, E_{\min})} y'$. We associate a probability, time duration and cost with this transition as follows.

The probability associated with the transition is $p = \prod_{e_b \in E_V} \pi_b(e_b)$, where for action $b = (F_b, \pi_b)$ in $V$, the event $e_b$ is probabilistically chosen from the set $E_b$ via $\pi_b$. If $V = \emptyset$, $p = 1$. The time duration associated with the transition is the time $t_{\min}$ attached to each $(a, e, t_{\min}) \in E_{\min}$. The cost associated with the transition is the sum of $\chi(e)$, for all $(a, e, t_{\min}) \in E_{\min}$.

We claim that the probabilities we have attached to transitions transform the system into a Markov chain. Suppose $V = \{b_1, b_2, \ldots, b_k\}$ is the new set of actions chosen by the scheduler at a snapshot $y$. As observed earlier, each combination of events $\{e_1, e_2, \ldots, e_k\}$ generated by applying $\pi_{b_i}$ to $E_{b_i}$, $i \in \{1, 2, \ldots, k\}$, results in a unique next snapshot. Hence the sum of the probabilities across all the successors of $y$ adds up to 1. If $V = \emptyset$, there is only one outgoing transition with probability 1.

## Statistical Analysis of DPS

**Properties.** We are interested in checking linear time properties for distributed probabilistic systems. For traditional transition systems, these are combinations of safety and liveness properties. In a quantitative setting, we would typically like to make assertions about the total duration or the total cost of a run. Since our DPS model is not deterministic, we have to frame these questions in terms probabilities. For instance, we might ask if the probability of reaching a target state within time $t$ is at least $p$.

A natural approach to checking such a property is to estimate the probability by a large number of simulations. For this, we need the additional constraint that the property can be checked within a bounded length run, so that we can effectively terminate each simulation with a yes or no verdict. Bounded duration properties can be formalized in notations such as bounded linear-time temporal logic (BLTL) [24]. In this paper we will not get into the details of BLTL but instead concentrate on reachability properties, defined informally.

The main shortcoming of naively estimating probabilities through random simulations is that we have no guarantee about the accuracy of the estimate. To perform this estimation in a principled manner, we need to frame the property of interest as a hypothesis testing problem.

**Hypothesis Testing.** We briefly overview the preliminaries of hypothesis testing before detailing the simulation procedure. For more details, see [14,24].

Suppose we are given a DPS $(\mathcal{S}, \Sigma, \delta, \chi, A)$, a bounded reachability property described by a formula $\phi$ in a suitable notation, and a threshold $\gamma$. Our goal is to verify if $\phi$ is achieved with probability at least $\gamma$, which we write as $\Phi = Pr_{\geq \gamma}\phi$.

Let $p$ be the probability of satisfying $\phi$. To verify whether $p \geq \gamma$, we test the hypothesis $H : p \geq \gamma$ against its negation $K : p < \gamma$. Since a simulation-based test does not guarantee a correct result, there are two types of errors we encounter: (i) Type-I error: accepting $K$ when $H$ holds, and (ii) Type-II error: accepting $H$ when $K$ holds. We would like to ensure that the probabilities of Type-I and Type-II errors are bounded by pre-defined values (say) $\alpha$ and $\beta$, respectively.

Enforcing exact bounds on Type-I and Type-II errors simultaneously is hard, so we allow uncertainty using an indifference region [24]. We relax the test by providing a range $(\gamma - \delta, \gamma + \delta)$ for a given threshold $\delta$. We now test the hypothesis $H_0 : p \geq \gamma + \delta$ against $H_1 : p \leq \gamma - \delta$. If the value of $p$ is between $\gamma - \delta$ and $\gamma + \delta$, we say that the probability is sufficiently close to $\gamma$, so we are indifferent with respect to which of the hypothesis $K$ or $H$ are accepted.

**Sequential Probability Ratio Test.** Traditional sampling theory fixes the sample size in advance based on the Type-I and Type-II error thresholds $\alpha$ and $\beta$. Computationally, it is often more efficient to estimate the sampling size adaptively, based on the observations made so far. Such an approach was proposed by Wald, called a sequential probability ratio test (SPRT) [23].

Time-bounded SPRT for DPS proceeds as follows. The user provides Type-I and Type-II error bounds $\alpha$ and $\beta$, as well the threshold of indifference $\delta$. We repeatedly simulate the system. We can determine in a bounded amount of time whether a simulation run satisfies the property of interest or not.

After $m$ simulation runs, let $d_m$ be the number of runs with a positive outcome so far. We calculate a ratio $quo = \frac{(\gamma^-)^{d_m}(1-\gamma^-)^{m-d_m}}{(\gamma^+)^{d_m}(1-\gamma^+)^{m-d_m}}$ that takes into account the number of successes and failures seen so far. We accept $H_0$ if $quo \leq \frac{\beta}{1-\alpha}$ and $H_1$ if $quo \geq \frac{1-\beta}{\alpha}$. Otherwise, we continue the simulation. The simulation is guaranteed to halt with probability 1 [24] and will typically converge much before the number of samples required by a traditional static estimate.

Please see Algorithm 1 in the Appendix for a concise algorithm.

## 4    Modeling rcBPM Systems as DPS

We recall that an rcBPM system consists of a business process instantiated as a number of cases and a finite set of resources. Let us assume that in an rcBPM system $\mathcal{B}$, there are $C$ cases numbered $\{1, 2, \ldots, C\}$ each with $k_T$ tasks labelled as follows: $T_{ij}$ denotes the $j^{th}$ task of the $i^{th}$ case, for $1 \leq i \leq C$ and $1 \leq j \leq k_T$. We assume the rate of the arrival process is $\lambda$ cases per second. Let us assume there are $r$ resources denoted $\{R_1, R_2, \ldots, R_r\}$. In the running example of loan/overdraft application, $k_T = 15$ and $r = 46$ (see Fig. 1). For each resource $R_i$, let $tasks(R_i) \in \{T_{ij} \mid 1 \leq i \leq C, 1 \leq j \leq k_T\}$ denote the set of tasks resource $R_i$ is able to perform. Since resources can be shared among tasks, for any $1 \leq i, j \leq r$, $tasks(R_i) \cap tasks(R_j)$ can possibly be non-empty.

Given an rcBPM system, we transform it to a DPS as follows. We model tasks and resources as agents. To facilitate the arrival process and clearly mark the case completion, we also model the start and end states as agents. Hence, the rcBPM system $\mathcal{B}$ can be modeled using $r + C \times (k_T + 2)$ agents.

Each task agent consists of 4 states (i) *ready* to perform, (ii) *waiting* for a resource, (iii) *busy* being executed, and (iv) *finished*. Each resource agent consists of 2 states (i) *available* and (ii) *busy*. An agent modeling the start state is called a starter agent, and has two states *waiting* and *arrived*. Similarly, the agent modeling the end state is named finisher agent with states *pending* and *done*.

We illustrate a part of DPS modeling the running rcBPM example in Fig. 2, We show the agents corresponding to the start state and 4 tasks: A_Submitted, A_Partially Submitted, A_Declined and A_PreAccepted. We also demonstrate a resource that can perform tasks A_Submitted and A_Partially Submitted. The states are depicted in rounded cornered rectangles and the edges between the states are defined as follows: $s \xrightarrow{\;a\;\;\;e\;} s'$ denotes the action at local state $s$, $e \in E_a$ and $s'$ be the next local state after event $e$.

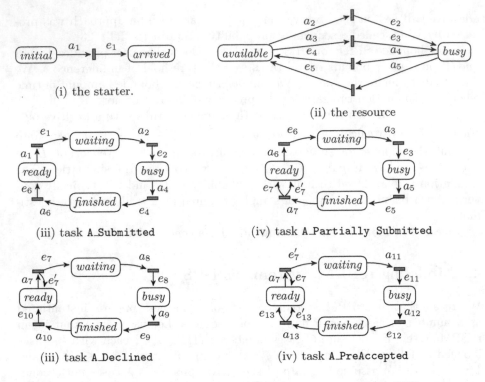

(i) the starter.

(ii) the resource

(iii) task A_Submitted

(iv) task A_Partially Submitted

(iii) task A_Declined

(iv) task A_PreAccepted

**Fig. 2.** Modeling (part of) the running rcBPM system example as DPS agents.

The starter agent mimics the wait for a case. At state *initial*, it synchronizes with the starting task agent at *ready* state followed by an event with probability 1. The duration of the event is equal to the total time before arrival for the particular case. The starter then moves to *arrived* state and the starting task moves to *waiting* state, where it waits for a resource to be scheduled. For example, Fig. 2(i) shows the action $a_1$ between the starter agent and the task A_Submitted followed by event $e_1$ such that $\pi_a(e_a) = 1$ and $\delta(e_1) = 1/\lambda$.

Once a task is in *waiting* state and the scheduler assigns a resource that is in *available* state, the task and the resource synchronize and they both move to *busy* state with probability 1. In Fig. 2, examples of such actions are $a_2$ and $a_3$. There they synchronizes again, performs an event with possibly non-zero time duration, and the task moves to *finished* with probability 1. In Fig. 2, examples of such actions are $a_4$ and $a_5$.

Once a task is finished, it signals the next task(s) in the control flow via synchronization. For example, in Fig. 2, after task A_Partially Submitted is at *finished*, it sychronizes with tasks A_Declined and A_Preaccepted via action $a_7$ such that $E_{a_7} = \{e_7, e_7'\}$ with $\pi_{a_7}(e_7) = 0.84$ and $\pi_{a_7}(e_7') = 0.16$. Hence, with probability 0.84, event $e_7$ is chosen and only task A_Declined moves to *waiting* state. Otherwise, with probability 0.16, event $e_7'$ is chosen and only task

A_Preaccepted moves to *waiting* state. In both cases, A_Partially Submitted moves to *ready* state. A task is ready again after finishing since due to loops in control flow, a task may be executed multiple times in the same case.

When a case finishes, the last task in *finished* state synchronizes with the finisher agent in *pending* state. The task then moves to *ready* as usual and the finisher agent moves to *done*, indicating the completion of the corresponding task. The finisher agent then stays at the state *done* with probability 1.

**Extensibility.** For brevity, we did not illustrate the rest of the DPS corresponding to the loan/draft application, but it can be easily extended. Also, we would like to point out that the described methodology to model rcBPM systems as DPS is only one example among many possibilities. One may easily extend this approach and incorporate an even more complex state space for tasks or resources modeling different scenarios. For example, we can easily model probabilistic error in task execution. Let us assume that when the resource depicted in Fig. 2 performs task A_Submitted, there is a small probability of 0.1 that such an execution fails. This phenomenon can be easily modeled by adding another event $e$ to $E_{a_2}$ such that $\pi_{a_2}(e_2) = 0.9$ and $\pi_{a_2}(e) = 0.1$ where after $e$, they move back to *available* and *waiting* respectively. Also, for simplicity, we assumed that the business processes under consideration are sound, but DPS can be easily used to model unsound rcBPM systems as well.

## 5    Experimental Evaluation

We have tested our SMC procedure on the running example. The property we are interested in is follows:

*With probability at least 0.99, when cases arrive at a fixed rate of 1 case per* 10 s, x *fraction of cases are completed among* C *cases within* t s.

We ensure that the probability of Type-I error and Type-II error of verifying this property is less than 0.01 with indifference region $(0.99 \pm 0.005)$.

We first investigate the relationship between the number of total cases and the number of cases completed within fixed maximum time $t = 500,000$ s. The shaded area in Fig. 3 (left) represents the values $(C, x)$ that satisfies the property when $t = 500,000$ s. The dotted line represents an upper limit for the values $(C, x)$ satisfying the property.

Then, we illustrate the relationship between the minimum total time and the fraction of cases completed within that total time when the total number of cases $C = 100$ is fixed. The shaded area in Fig. 3 (right) represents the values $(x, t)$ that satisfies the property when $C = 100$. The dotted line represents a lower limit for the values $(x, t)$ satisfying the property. We note that the limits are not the tightest, but can be made arbitrary closer with more SMC simulations.

**Extensibility.** Though our experiment is only a proof-of-concept, we would like point out that we can scale the size of the model comfortably to accommodate 500 cases. Since there is no other known approach to verify business processes with provable bounds, we were not able to compare our results with

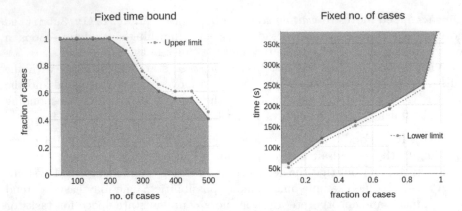

**Fig. 3.** (left) Fraction of cases completed vs total no. of cases when the time bound is fixed, (right) Minimum total time vs fraction of cases completed when the total number of cases is fixed.

existing literature. This methodology also supports verifying a variety of properties, depending on the focus of optimization for the business. For example, one may investigate the performance of resources by verifying the following property: given a resource, in what fraction of cases was it used?

# 6    Conclusion

We have presented a modular approach to modelling resource-constrained BPM systems with multiple cases, using distributed probabilistic systems. We have shown that a real-time distributed probabilistic system under a fixed scheduler behaves like a Markov chain. We have then presented a rigorous technique for time-bounded approximate verification of business processes using statistical model checking, illustrated through a proof-of-concept experiment.

In future, we plan to extend this model to shed light on different types of scheduling policies and their impact on business processes. We would also like to incorporate stochastic durations for events, which will take us to a Continuous Time Markov Chain (CTMC) setting. Finally, we would like to see how approximate verification techniques can also enrich process mining BPM systems [2].

**Acknowledgements.** The authors would like to thank S Akshay for his invaluable comments on the draft and Ansuman Banerjee for the early discussions.

# A     Appendix

---

**Algorithm 1.** Statistical Model Checking for DPS

---

**INPUT:**

1: $\mathcal{D}, \mathcal{G}$                                                                          ▷ a DPS and a scheduler

2: $\Phi = Pr_{\geq \gamma} T_{\leq t} \phi$                                                           ▷ a property

3: $\alpha, \beta, \delta \in (0, 1)$                                             ▷ error bounds and threshold of indifference

**OUTPUT:**

4: YES or NO

5: **procedure** SIMULATE-DPS

6:     $m \leftarrow 0$                                                             ▷ the number of simulations so far

7:     $\gamma^+ \leftarrow \gamma + \delta$ and $\gamma^- \leftarrow \gamma - \delta$

8:     **while** True **do**

9:         $t_{spent} \leftarrow 0$                                                          ▷ time spent so far

10:         $y^{in} = (\mathbf{s}^{in}, \emptyset, \emptyset), y = (\mathbf{s}, U, X) = y^{in}$     ▷ the initial and current snapshot

11:         $\rho \leftarrow y$                                                           ▷ the current execution

12:         $b \leftarrow 0$                                            ▷ the outcome of the Bernoulli random variable

13:         $d_m \leftarrow 0$                                 ▷ accumulator of outcome of the Bernoulli variable

14:         **while** $(t_{spent} \leq t)$ **do**

15:             $\mathcal{G}(y) = U \cup V \leftarrow$ *scheduled actions at y*

16:             $Y \leftarrow$ *set of fresh actions probabilistically chosen from V*

17:             $E_{\min} \subseteq X \cup Y$ *is the set of tuples with minimum time to completion*

18:             **for all** $(a, e, t_{\min}) \in E_{\min}$ **do**

19:                 $\mathbf{s}_{loc(e)} \leftarrow tgt_e$, *remove a from* $U \cup V$, *remove* $(a, e, t)$ *from* $X \cup Y$

20:             **for all** $(a, e, t) \in X \cup Y \setminus E_{min}$ **do**

21:                 $t \leftarrow t - t_{\min}$

22:             $y = (\mathbf{s}', U', X')$ *is the new snapshot*

23:             $t_{spent} = t_{spent} + t_{\min}, \rho \leftarrow \rho y$

24:             **if** $\rho$ satisfies $\phi$ **then**

25:                 $b = 1$

26:                 **break**

27:         $m \leftarrow m + 1$ and $d_m \leftarrow d_m + b$

28:         $quo = \frac{(\gamma^-)^{d_m}(1-\gamma^-)^{m-d_m}}{(\gamma^+)^{d_m}(1-\gamma^+)^{m-d_m}}$

29:         **if** $(quo \geq \frac{(1-\beta)}{\alpha})$ **then**

30:             **return** NO

31:         **else if** $(quo \leq \frac{\beta}{1-\alpha})$ **then**

32:             **return** YES

---

**Table 1.** Resource-task matrix (truncated) with average time taken by the resource.

| | 10228 | 10629 | 10779 | 10859 | 10861 | 10862 | 10863 | 10880 | 10881 | 10889 | 10899 | 10909 | 10910 | 10912 | 10913 | 10929 | 10931 |
|---|---|---|---|---|---|---|---|---|---|---|---|---|---|---|---|---|---|
| A.Accepted | 2521 | | | 30435 | 47184 | 53086 | 94678 | 62187 | 7210 | 18276 | | 24136 | 38871 | 62747 | 71002 | 30092 | 7475 |
| A.Approved | | | | | | | | | | | | | | | | | |
| A.Cancelled | | | 51106 | | | | | | | | | | | | | | |
| A.Declined | | | | | | | | | | | | | | | | | |
| A.Finalized | 0 | | | 35 | 62 | 0 | 59 | 78 | | 86 | | 24 | 26 | 0 | 79 | 94 | 25 |
| A.PartlySubmitted | | | | | | | | | | | | | | | | | |
| A.Preaccepted | 31897 | 2186 | | 5775 | 9823 | 1061 | 3333 | 17990 | 44545 | 21613 | | 16301 | 18480 | 20330 | 17735 | 19720 | 29301 |
| A.Submitted | | | | | | | | | | | | | | | | | |
| O.Cancelled | 47 | | | 173747 | 140840 | 32 | | 37 | 8142 | 53509 | | 19437 | 155656 | | 0 | 0 | 0 |
| O.Created | | | | 1 | 1 | 2 | 1 | 2 | 2 | 2 | | 1 | 2 | 1 | 9 | 1 | 1 |
| O.Selected | 130 | | | 0 | 48 | 38587 | 179 | 83 | 0 | 27032 | | 42117 | 28 | 192 | 18598 | 24747 | 22901 |
| O.Sent | 0 | | | 0 | 0 | 0 | 0 | 0 | 0 | 0 | | 0 | 0 | 0 | 0 | 0 | 0 |
| O.Sent_Back | | | | | | | | | | | 85338 | | | | | | |
| Dummy | 0 | 0 | | 0 | 0 | 0 | 0 | 0 | 0 | 0 | 0 | 0 | 0 | 0 | 0 | 0 | 0 |
| O.Accepted | 0 | 0 | | 0 | 0 | 0 | 0 | 0 | 0 | 0 | 0 | 0 | 0 | 0 | 0 | 0 | 0 |

# References

1. Aalst, W.M.P.: Verification of workflow nets. In: Azéma, P., Balbo, G. (eds.) ICATPN 1997. LNCS, vol. 1248, pp. 407–426. Springer, Heidelberg (1997). doi:10.1007/3-540-63139-9_48
2. Aalst, W.M.P.: Process Mining: Discovery, Conformance and Enhancement of Business Processes. Springer Science & Business Media, New York (2011)
3. Aalst, W.M.P.: Business process management: a comprehensive survey. ISRN Softw. Eng. **2013**, 1–37 (2013)
4. Aalst, W.M.P.: Business process management as the Killer App for Petri nets. Softw. Syst. Model. **14**(2), 685–691 (2014)
5. Aalst, W.M.P.: Business process simulation survival guide. In: vom Brocke, J., Rosemann, M. (eds.) Handbook on Business Process Management 1: Introduction, Methods, and Information Systems, pp. 337–370. Springer, Heidelberg (2015)
6. Aalst, W.M.P., Hee, K.M.V.: Workow Management: Models, Methods, and Systems. MIT Press, Cambridge (2004)
7. Bose, R.P.J.C., Aalst, W.M.P.: Business Process Management Workshops: BPM 2012 International Workshops, Tallinn, Estonia, September 3, 2012. Revised Papers, pp. 221–222. Springer, Heidelberg (2013)
8. Braghetto, K.R., Ferreira, J.E., Vincent, J.-M.: Performance evaluation of resource-aware business processes using stochastic automata networks. Int. J. Innov. Comput. Inf. Control **8**(7B), 5295–5316 (2012)
9. Chuang, L.I.N., Yang, Q.U., Fengyuan, R.E.N., Marinescu, D.C.: Performance equivalent analysis of workflow systems based on stochastic Petri net models. In: Han, Y., Tai, S., Wikarski, D. (eds.) EDCIS 2002. LNCS, vol. 2480, pp. 64–79. Springer, Heidelberg (2002). doi:10.1007/3-540-45785-2_5
10. Dumas, M., Rosa, M.L., Mendling, J., Reijers, H.: Fundamentals of Business Process Management. Springer, Berlin (2013)
11. Gastin, P., Lerman, B., Zeitoun, M.: Distributed games and distributed control for asynchronous systems. In: Farach-Colton, M. (ed.) LATIN 2004. LNCS, vol. 2976, pp. 455–465. Springer, Heidelberg (2004). doi:10.1007/978-3-540-24698-5_49
12. Herbert, L.T.: Specification, verification and optimisation of business processes. a unified framework. Technical University of Denmark (2014)
13. Hinton, A., Kwiatkowska, M., Norman, G., Parker, D.: PRISM: a tool for automatic verification of probabilistic systems. In: Hermanns, H., Palsberg, J. (eds.) TACAS 2006. LNCS, vol. 3920, pp. 441–444. Springer, Heidelberg (2006). doi:10.1007/11691372_29
14. Legay, A., Delahaye, B., Bensalem, S.: Statistical model checking: an overview. In: Barringer, H., Falcone, Y., Finkbeiner, B., Havelund, K., Lee, I., Pace, G., Roşu, G., Sokolsky, O., Tillmann, N. (eds.) RV 2010. LNCS, vol. 6418, pp. 122–135. Springer, Heidelberg (2010). doi:10.1007/978-3-642-16612-9_11
15. Magnani, M., Montesi, D.: BPMN: how much does it cost? An incremental approach. In: Alonso, G., Dadam, P., Rosemann, M. (eds.) BPM 2007. LNCS, vol. 4714, pp. 80–87. Springer, Heidelberg (2007). doi:10.1007/978-3-540-75183-0_6
16. Netjes, M., Aalst, W.M.P., Hajo, A.R.: Analysis of resource-constrained processes with colored petri nets. In: Sixth Workshop and Tutorial on Practical Use of Coloured Petri Nets and the CPN Tools, vol. 576, pp. 251–266 (2005)
17. Oliveira, C.A.L., Lima, R.M.F., Reijers, H.A., Ribeiro, J.T.S.: Quantitative analysis of resource-constrained business processes. IEEE Trans. Syst. Man Cybern. Part A Syst. Hum. **42**(3), 669–684 (2012)

18. Reijers, H.: Design and Control of Workflow Processes: Business Process Management for the Service Industry. Springer, New York (2003)

19. Russell, N., Ter Hofstede, A.H.M., Mulyar, N., Patterns, W.C.: A revised view. Technical report (2006)

20. Saha, R., Esparza, J., Jha, S.K., Mukund, M., Thiagarajan, P.S.: Distributed Markov chains. In: D'Souza, D., Lal, A., Larsen, K.G. (eds.) VMCAI 2015. LNCS, vol. 8931, pp. 117–134. Springer, Heidelberg (2015). doi:10.1007/978-3-662-46081-8_7

21. Sampath, P., Wirsing, M.: Computing the cost of business processes. In: Yang, J., Ginige, A., Mayr, H.C., Kutsche, R.-D. (eds.) UNISCON 2009. LNBIP, vol. 20, pp. 178–183. Springer, Heidelberg (2009). doi:10.1007/978-3-642-01112-2_18

22. van Dongen, B.F.: BPI challenge 2012 (2012)

23. Wald, A.: Sequential tests of statistical hypotheses. Ann. Math. Stat. **16**, 117–186 (1945)

24. Younes, H.L.S.: Verification and planning for stochastic processes with asynchronous events. Ph.D. thesis, Carnegie Mellon University, Pittsburgh, USA (2004)

# Failure Estimation of Behavioral Specifications

Debasmita Lohar$^{(\boxtimes)}$, Anudeep Dunaboyina, Dibyendu Das,
and Soumyajit Dey

Indian Institute of Technology, Kharagpur, India
{debasmita.lohar,danudeep,soumya}@cse.iitkgp.ernet.in,
dibyendud@iitkgp.ac.in

**Abstract.** Behavioral specifications are often employed for modeling complex systems at high levels of abstraction. Failure conditions of such systems can naturally be specified as assertions defined over system variables. In that way, such behavioral descriptions can be transformed to imperative programs with annotated *failure assertions*. In this paper, we present a scalable source code based framework for computing failure probability of such programs under the fail-stop model by applying formal methods. The imprecision in the estimation process resulting from coverage loss due to time, memory bounds and loop invariant synthesis, is also quantified using an upper bound computation. We further discuss the design and implementation of ProPFA (Probabilistic Path-based Failure Analyzer), an automated tool developed for this purpose.

## 1 Introduction

Reactive software systems generally work on data provided by an *input environment* where the environment may be some physical phenomenon, higher layer softwares or a human in the loop. In order to prevent undesired system behaviors, it is common to embed such software with *fail-stop* assertions so that system execution does not progress to unsafe states as defined by some requirement specification and the associated input conditions are handled gracefully. Such *failure runs* of systems, even if handled, lead to degradation in Quality of Service (QoS) where the definition of such QoS measures is dependent on the area of application [1]. For example, in case of Cyber Physical Systems with an underlying networked control infrastructure, environmental noise during transmission or sensor reading leads to unsafe plant state information which needs to be ignored so that an uncalled for control actuation is not passed on to the plant. In that way, the safety of the system is gracefully maintained with an associated degradation in Quality of Control (QoC) [8].

Computing the probability of system level failure provides a direct handle for estimating relevant QoS metrics for the system. Analytical methods for computing such failure probabilities using high level models and parameters are widely established. These approaches focus on reliability models or are based on high level system architecture [7,10]. However, it has the following shortcomings.

© Springer International Publishing AG 2016
M. Fränzle et al. (Eds.): SETTA 2016, LNCS 9984, pp. 315–322, 2016.
DOI: 10.1007/978-3-319-47677-3_20

1. High level models do not capture complex execution semantics of systems as can be specified in the form of a program behavior.
2. In case the system under question is a software program itself, model driven analytical techniques refrain from deriving probability bounds from the source code directly. This implies an added dependence on correctness of model construction given a source implementation.

In case of software systems, existing tools either use failure data during phases of software life cycles to drive one or more of the software reliability growth models [11] or use test coverage measurements [5,12] to estimate reliability. Established techniques like probabilistic risk assessment lack the notion of provability as given by formal techniques which provide sophisticated reasoning mechanisms for working with high level system models as well as behavioral specifications given as imperative programs. A recent approach [6] provides a formal technique for reliability estimation of imperative programs built on the tool Symbolic Path Finder (SPF). However, it does not handle loops in a scalable fashion.

The problem addressed in the present work is formally stated as follows.

**Problem Statement:** *We are given an imperative program $\sigma$ in C-type syntax and a set of failure assertions $\mathcal{A} = \{A_1, \cdots, A_n\}$. Any random execution run of $\sigma$ is considered a failure run iff any one of the failure assertions $A_i \in \mathcal{A}$ actually fail in that run. Given such an instance of $\sigma$ along with the probability distributions of input variables, what is the failure probability of $\sigma$ in any random execution run?*

We summarize the novelty of our framework below.

1. We provide a formal technique for failure probability estimation of behavioral specifications given as imperative programs. The specification can model a complex system as a program or it can be a software system by itself.
2. The technique handles loops in a scalable fashion.
3. The imprecision of the analysis is upper bounded and a relevant confidence measure is computed as a part of the analysis.

While program analysis techniques useful for some of the above steps do exist, there is no end-to-end solution that applies formal techniques and estimates software failure while handling loops gracefully. Our tool suite ProPFA is a scalable solution to this requirement. Further, its implementation using well known, robust formal APIs makes the solution usable and extensible.

## 2    Overview of Failure Analysis Framework

Let an imperative program $\sigma$ be a 2-tuple $\langle S, A \rangle$ where the set of program statements is represented as $S$ and the set of failure assertions is defined as $\mathcal{A}$. The program $\sigma$ is characterized as a Control Flow Graph $G_\sigma = <N, E>$ having edges $\in E$ labeled with program statements and vertices $\in N$ denoting program points at which the program state is captured. $G_\sigma$ has one entry $(n_s)$ and one exit $(n_e)$ node depicting the start and end points of $\sigma$. Generally, in any random run of $\sigma$, the set $V = \{v_1, \cdots v_k\}$ of variables assumes values from the domains $\{D_1, \cdots, D_k\}$ following independent or joint probability distributions. The *branching edges* of $G_\sigma$ represent the conditionals in $\sigma$. Let $\mathcal{I}$ be the set of conditionals. An assertion $(A_i \in \mathcal{A})$ is the label of an edge $(n_i, n_{i+1} \in E)$ in $G_\sigma$. The program point $n_i$ characterizes the program state before $A_i$ is executed and $n_{i+1}$ characterizes the program state after executing $A_i$ successfully. The execution semantics of failure assertions is 'fail stop', i.e., the program fails if any failure assertion in the path do not execute successfully. We first perform the following transformation on $G_\sigma$. For each $A_i$, we introduce a new *failure* node $f_i$ and an edge $e_i = (n_i, f_i)$ labeled with $\overline{A_i}$. This process is continued for all failure assertions $A_i \in \mathcal{A}$. The derived CFG is defined as $G'_\sigma = <N', E'>$ where $N' = N \bigcup \{f_i \mid \exists A_i \in \sigma\}$ and $E' = E \bigcup \{(n_i, f_i) \mid A_i \in \sigma$ and $A_i$ is label of $(n_i, n_{i+1})\}$. An edge $(n_i, f_i) \in E'$ is labeled with $\overline{A_i}$. The conditional set for $G'_\sigma$ is defined as, $\mathcal{I}' = \mathcal{I} \bigcup \mathcal{A} \bigcup \overline{\mathcal{A}}$ where $\overline{\mathcal{A}} = \{\overline{A_i} | A_i \in \mathcal{A}\}$. An example program $\sigma$ with CFGs $G_\sigma$ and $G'_\sigma$ are shown in Fig. 1. A path $\pi$ in the graph $G'_\sigma$ is called a *success path* if the (source, target) pair of $\pi$ is $(n_s, n_e)$. Any path $\pi$ in the graph $G'_\sigma$ is called a *failure path* if the source, target pair of $\pi$ is $(n_s, f_i)$ for $f_i \in N'$.

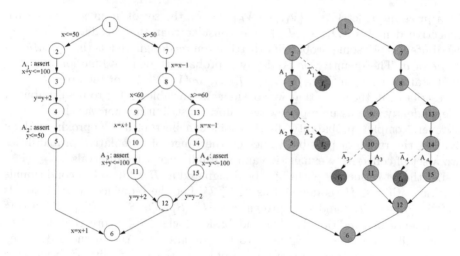

**Fig. 1.** Representations of $G_\sigma$ and $G'_\sigma$

We propose a semi-formal path based approach for program level failure estimation by employing static analysis techniques on $G'_\sigma$. Considering the input profile, there is a specific probability that a program path is taken. We estimate this specific probability associated with each success path. The success probability of the whole program is then enumerated by adding the probabilities of the success paths. Finally, the overall failure probability of the program is estimated.

We develop an associated toolflow ProPFA (Probabilistic Path-based Failure Analyzer) for the proposed framework. The software architecture of ProPFA is shown in Fig. 2. ProPFA computes the failure probability of the program in Fig. 1

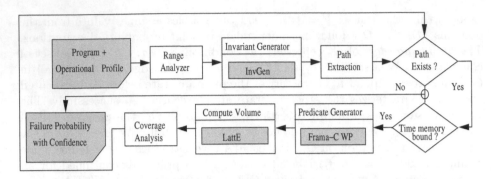

**Fig. 2.** Architecture of ProPFA

as 0.23034979 assuming that the variables $x$ and $y$ are uniformly distributed over the integer ranges [0, 100] and [0, 50] respectively.

## 2.1  Failure Probability Estimation of Program Paths

For a program $\sigma$, let $V_{in} = \{V_1, \cdots, V_k\}$ denote the set of input variables with respective domains $\{D_1, \cdots, D_k\}$. We consider that the variables in $V_{in}$ are distributed as per some probability distribution $op$ designated as the *operational profile* of $\sigma$. The quantity $op$ is ideally a probability density function over the input state space such that, $\int_{V_1 \in D_1} \cdots \int_{V_k \in D_k} op(V_{in}) = 1$. For the present work, we restrict $op$ in the sense that we consider the variables $\in V_{in}$ to be distributed independently. We assume piecewise uniform distributions for variables $\in V_{in}$. However, complex probability distributions can be typically approximated by dividing the range of possible values or the range of cumulative probabilities into a set of collectively exhaustive and mutually exclusive intervals.

Let the set of success paths of $\sigma$ be designated as $\Pi_s$. The set of conditionals in some path $\pi \in \Pi_s$ is denoted as $\mathcal{I}'_\pi \subseteq \mathcal{I}'$. Let the sequence of conditionals $\in \mathcal{I}'_\pi$ be $\{I_1^\pi, \cdots, I_k^\pi\}$ and we represent $\pi$ as $I_1^\pi \cdot S_1^\pi \cdot I_2^\pi \cdot S_2^\pi \cdots S_{k-1}^\pi \cdot I_k^\pi$ writing the sequence of conditionals $(I_j^\pi)$ and basic blocks $(S_j^\pi)$. Note that for $\pi$ to execute, all conditionals $\in \mathcal{I}'_\pi$ need to be evaluated to true in the order they appear in $\pi$. For a given operational profile $op$, the probability $Pr(\pi)$ that a path $\pi \in G'_\sigma$ is taken in any random run of $\sigma$ is defined as, $Pr(\pi) = Pr(n_s \xrightarrow[op]{\pi} n_e) = Pr(I_1^\pi) \times Pr(I_2^\pi/I_1^\pi) \times \cdots Pr(I_k^\pi/(I_1^\pi \wedge I_2^\pi \wedge \cdots \wedge I_{k-1}^\pi))$ This expression essentially captures the probability of the path condition of $\pi$ being satisfied. The path condition of $\pi$ is computed using standard program analysis techniques like Weakest Precondition (WP) analysis. Our tool-flow ProPFA generates the $WP$ $(WP(I_1^\pi \cdot S_1^\pi \cdot I_2^\pi \cdot S_2^\pi \cdots S_{k-1}^\pi, I_k^\pi))$ that characterizes the input subspace of the program driving the execution through path $\pi$. We restrict ourselves to static affine programs so that the input subspace generated by $WP$ computation is a convex polytope.

The path $\pi$ is a *feasible path* if there exists a non-null intersection between the convex polytope $WP(I_1^\pi \cdot S_1^\pi \cdot I_2^\pi \cdot S_2^\pi \cdots S_{k-1}^\pi, I_k^\pi)$ and the input state space.

Let the overall input state space $\mathcal{V}$ be a mutually exclusive disjunction of convex polytopes $\{\mathcal{V}_1, \mathcal{V}_2, \cdots\}$. The volume of this convex polytope intersection is computed by leveraging the concept of exact integration of polynomials over polyhedra regions [3]. Let $Vol(C, op)$ denote the volume of a convex polytope $C$ given a density distribution $op$. The probability of the success path $\pi \in \Pi_s$ is computed as, $Pr(\pi) = \sum_{\mathcal{V}_i \in \mathcal{V}} Vol(WP(I_1^\pi \cdot S_1^\pi \cdot I_2^\pi \cdot S_2^\pi \cdots S_{k-1}^\pi, I_k^\pi) \wedge \mathcal{V}_i, op)/Vol(\mathcal{V}_i, op)$. Hence, the failure probability of $\sigma$ can be estimated as $Pr(\overline{\sigma}) = 1 - \sum_{\pi \in \Pi_s} Pr(\pi)$.

## 2.2   Failure Estimation of Programs with Loops

Till now we have considered loop-free programs. A loop in a program $\sigma$ corresponds to several program paths depending on the number of loop iteration. Loop unrolling replaces the loop by as many instances of its body as the number of iterations and different paths are generated. Hence, for success probability estimation of $\sigma$, all generated paths leading to the success node are considered and success paths and the probabilities of them are computed as discussed in Sect. 2.1. Apart from the additional requirement of loop bound analysis, in general, this method will not scale for large programs with significantly deep loops.

   To accelerate the analysis and minimize the number of success paths to be considered in presence of loops, a technique is proposed which sacrifices some accuracy while improving scalability. This optimization works on simple computational loops which do not contain failure assertions. Consider the loop given as, '$L$ = while $(C)$ do $S$'; The physical significance of such a loop is the program segment $S$ inside $L$ does not fail in any event. We abstract out such *simple* computational loops using invariant relations while computing the WPs. Let $\sigma = \sigma_1 \cdot L \cdot S' \cdot A' \cdot \sigma_2 = \sigma_1 \cdot$ [while$(C)$ do $S$]$\cdot S' \cdot A' \cdot \sigma_2$. Current state-of-the-art invariant synthesis tools require a post-condition for the loop $L$ which is computed as $WP(S', A')$. A disjunction of invariants of the form $\bigvee_{i=1}^n \phi_i \subseteq \phi$ where $\phi$ is the ideal loop invariant of $L$, is synthesized. However, invariant synthesis tools may fail to generate sufficient invariants capturing all possible paths (which satisfy $WP(S', A')$). The exact set of paths for $\sigma$ are captured by the Kleene expression $\sigma_1 \cdot [\overline{C} + (C \cdot S)^+ \cdot \phi] \cdot S' \cdot A' \cdot \sigma_2$ while the subset of paths covered by the analysis are actually $\sigma_1 \cdot [\overline{C} + \sum_i (C \cdot S)^+ \cdot \phi_i] \cdot S' \cdot A' \cdot \sigma_2$. In such situations, certain paths are ignored and their contribution in overall failure probability is not accounted for due to the non-exact nature of the invariant. In that case, we actually underestimate the probability of such success paths (cutting through the loop) and end up overestimating the failure probability of the program thus leading to a safe approximation.

## 2.3   Computing Confidence Measure

Due to loop approximations as discussed above and specified time and memory bounds, all the execution paths of the system may not be explored. We propose a confidence measure to indicate a formal bound of analysis coverage. Let us consider that after approximating all *simple* loops, we were able to explore $k$ success paths with execution probabilities $Pr(\pi_1), \cdots, Pr(\pi_k)$. Let the set of failure assertions covered by these $k$ paths be $\mathcal{A}_k$. For any $A \in \mathcal{A}_k$, let $\overline{A}$ be the label of an edge $(n, f)$ where $f$ is a failure node as per the construction of $G'_\sigma$. We enumerate the set $F_k$ of all such failure nodes and compute the total (success + failure) coverage for $k$ paths as $\sum_{i=1}^{i=k} Pr(\pi_i) + \sum_{f \in F_k} Pr(\pi = (n_s, f))$, where $\pi$ is a failure path with source, target pair $= (n_s, f)$. We report this quantity as the confidence measure.

Let the set of failure assertions present in any path $\pi$ under consideration be $\mathcal{A}_\pi \in \mathcal{I}_\pi$ where $\mathcal{I}_\pi$ is the set of conditionals. It may be noted that confidence can be efficiently estimated by setting all the failure assertions $A_i \in \mathcal{A}_\pi$ as 'TRUE' and computing coverage by considering only success paths. This can be done since the measure thus computed takes into account the success paths labeled with $A_i$ and also all failure paths labeled with $\overline{A_i}$ such that $A_i \in \mathcal{A}_\pi$.

The assertions immediately succeeding simple loops are kept intact so that the approximations due to loop invariants are taken care of on both success and failure paths. This is done because we may miss some paths due to incompleteness of invariant synthesis. Let $A_i$ be the failure assertion just after loop $L_i$ in a path $\pi_i$. For this case, to take care of the under approximation involved, we consider both the success path upto $A_i$ and the failure path ending at the node $f_i$ with the edge labeled as $\overline{A_i}$ individually while computing the confidence measure. We synthesize invariants considering both $A_i$ and $\overline{A_i}$ as postconditions. Because of the incompleteness in invariant generation mechanism, both quantities may turn out to be under-estimates. Hence both cases are considered separately.

## 2.4   Key Features of ProPFA

The ProPFA tool leverages state-of-the-art static analysis tools for failure estimation. It is designed as an integrated framework that takes as input programs annotated with failure assertions and an operational profile of the program. Multiple execution paths are then explored depth-wise and for each path success predicates are generated using third party static analyzer. The probabilities of these predicates under the operating region are computed using lattice point enumeration which in turn is used for program level failure probability estimation.

We mainly use Frama-C WP plug-in [2], a source code analysis platform that generates weakest preconditions of industrial-size C programs, as a third party static analyzer. For loops in the source code, invariants are synthesized by deploying state-of-the-art template based tool InvGen [9]. Finally failure probability computation involves volume computation of convex polytopes which is offloaded to the lattice point counting tool LattE [4].

It may be noted that ProPFA handles static affine programs with both integer and float data types. It transforms the computed WPs to the domain of integers as LattE can only work with integers. The transformation is based on finding the least common multiplier that ensures a lossless `float` to `int` transformation of all WPs. ProPFA is also able to handle piecewise uniform distribution of input variables. In that case it considers all input subregions while computing success probabilities of program execution paths.

**Table 1.** Evaluation of ProPFA (*Restricted to affine input polynomials)

| Program | op | LOC | #Assertion | Failure Prob. |
|---|---|---|---|---|
| Newton-Raphson* | [−2147483648, 2147483647] | ∼50 | 1 | 0.00000000107 |
| Trapezoidal* | [−32768, 32767] | ∼50 | 1 | 0.00003051759 |
| FBW | Recommended op of Airbus A320 [13] in cruising altitude | ∼1000 | 21 | 0.00000000000 |
| FBW | Recommended op [13], but velocity range [520, 555] mph | ∼1000 | 21 | 0.02200000000 |
| FBW | Recommended op [13], but pitch angle between [−90°, −1°] with Prob. 0.001, between [0°, 15°] with Prob. 0.999 | ∼1000 | 21 | 0.00190000000 |

## 3  Results

We have evaluated ProPFA over a set of programs from numerical analysis domain and non-redundant version of Fly-by-wire (FBW) from avionics domain. The results along with the operational profiles (op) are presented in the Table 1. A command line version of ProPFA can be found in the following link: https://github.com/dlohar/ProPFA.git.

## 4  Discussions and Conclusion

In this paper, we discuss a framework for failure probability estimation of imperative programs in C-like syntax. Our path based approach explores program execution paths and computes success probabilities which in turn are used to estimate failure probability of the overall program. As is evident, it may not be feasible to consider all program paths. In that case, a quantitative measure for confidence on the effectiveness of the estimate is provided. We also present a brief description of ProPFA, a toolflow developed for this purpose. It may be noted that, ProPFA is restricted to the limitations of the static analysis tools it integrates with. It works only on linear affine programs. ProPFA avoids unrolling of loops only if the generated templates are sufficient for InvGen to generate a loop invariant. Otherwise, it unrolls the loops. Extending these ideas towards a complete toolset handling concurrent specifications and complex data structures is future work.

# References

1. Cheung, R.C.: A user-oriented software reliability model. IEEE Trans. Softw. Eng. **6**(2), 118–125 (1980)
2. Cuoq, P., Kirchner, F., Kosmatov, N., Prevosto, V., Signoles, J., Yakobowski, B.: Frama-C. In: Eleftherakis, G., Hinchey, M., Holcombe, M. (eds.) SEFM 2012. LNCS, vol. 7504, pp. 233–247. Springer, Heidelberg (2012). doi:10.1007/978-3-642-33826-7_16
3. De Loera, J., Dutra, B., Koeppe, M., Moreinis, S., Pinto, G., Wu, J.: Software for exact integration of polynomials over polyhedra. Commun. Comput. Algebra **45**(3/4), 169–172 (2012)
4. De Loera, J.A., Hemmecke, R., Tauzer, J., Yoshida, R.: Effective lattice point counting in rational convex polytopes. J. Symb. Comput. **38**(4), 1273–1302 (2004)
5. Denton, J.: Robust, an integrated software reliability tool (1997)
6. Filieri, A., Pǎsǎreanu, C.S., Visser, W.: Reliability analysis in symbolic pathfinder. In: International Conference on Software Engineering. IEEE Press (2013)
7. Gokhale, S.S.: Architecture-based software reliability analysis: overview and limitations. Trans. Dependable Secur. Comput. **4**(1), 32–40 (2007)
8. Goswami, D., Müller-Gritschneder, D., Basten, T., Schlichtmann, U., Chakraborty, S.: Fault-tolerant embedded control systems for unreliable hardware. In: International Symposium on Integrated Circuits (ISIC). IEEE (2014)
9. Gupta, A., Rybalchenko, A.: InvGen: an efficient invariant generator. In: Bouajjani, A., Maler, O. (eds.) CAV 2009. LNCS, vol. 5643, pp. 634–640. Springer, Heidelberg (2009). doi:10.1007/978-3-642-02658-4_48
10. Hsu, C.J., Huang, C.Y.: An adaptive reliability analysis using path testing for complex component-based software systems. Trans. Reliab. **60**(1), 158–170 (2011)
11. Lyu, M.R., Nikora, A.P., Farr, W.H.: A systematic and comprehensive tool for software reliability modeling and measurement. In: The 23rd International Symposium on Fault-Tolerant Computing. IEEE (1993)
12. Ramani, S., Gokhale, S.S., Trivedi, K.S.: SREPT: software reliability estimation and prediction tool. Perform. Eval. **39**(1), 37–60 (2000)
13. Sutherland, J.: Fly-by-wire flight control systems. Technical report, DTIC Document (1968)

# Erratum to: Formalization of Fault Trees in Higher-Order Logic: A Deep Embedding Approach

Waqar Ahmad[(⌧)] and Osman Hasan

School of Electrical Engineering and Computer Science,
National University of Sciences and Technology, Islamabad, Pakistan
{waqar.ahmad, osman.hasan}@seecs.nust.edu.pk

Erratum to:
Chapter "Formalization of Fault Trees in Higher-Order Logic:
A Deep Embedding Approach" in:
M. Fränzle et al. (Eds.):
Dependable Software Engineering, LNCS,
DOI: 10.1007/978-3-319-47677-3_17

The original version of this chapter contained an error. The name of the author Waqar Ahmad was spelled incorrectly as Waqar Ahmed in the original publication. The original chapter was corrected.

---

The updated original online version for this chapter can be found at
DOI: 10.1007/978-3-319-47677-3_17

© Springer International Publishing AG 2017
M. Fränzle et al. (Eds.): SETTA 2016, LNCS 9984, p. E1, 2016.
DOI: 10.1007/978-3-319-47677-3_21

# Author Index

Printed in the United States
By Bookmasters

Printed in the United States
By Bookmasters